WORLD AGRICULTURE : TOWARDS 2015/2030

AN FAO PERSPECTIVE

Edited by Jelle Bruinsma

Earthscan Publications Ltd
London

First published in the UK and USA in 2003 by
Earthscan Publications Ltd

Copyright © Food and Agriculture Organization (FAO), 2003

ISBN: 92 5 104835 5 (FAO paperback)
         1 84407 007 7 (Earthscan paperback)
         1 84407 008 5 (Earthscan hardback)

The designations employed and the presentation of the material in this information product do not imply the expression of any opinion whatsoever on the part of the Food and Agriculture Organization of the United Nations concerning the legal status of any country, territory, city or area or of its authorities, or concerning the delimitation of its frontiers or boundaries. The mention of specific companies, their products or brand names does not imply any endorsement by the Food and Agriculture Organization of the United Nations. In the presentation of statistical material, countries are, where appropriate, aggregated in the following main economic groupings: "developed countries" (including the developed market economies or "industrial countries" and the transition countries) and "developing countries". The designations "developed" and "developing" economies are intended for statistical convenience and does not necessarily express a judgement about the stage of development reached by a particular country.

For a full list of publications please contact:

Earthscan Publications Ltd
120 Pentonville Road
London, N1 9JN, UK
Tel: +44 (0)20 7278 0433
Fax: +44 (0)20 7278 1142
E-mail: earthinfo@earthscan.co.uk
Web site: www.earthscan.co.uk

22883 Quicksilver Drive, Sterling, VA 20166-2012, USA

A catalogue record for this book is available from the British Library

Library of Congress Cataloging in Publication Data applied for

Earthscan is an editorially independent subsidiary of Kogan Page Ltd and publishes in association with WWF-UK and the International Institute for Environment and Development

Copies of FAO publications can be requested from:

Sales and Marketing Group
Information Division
FAO
Viale delle Terme di Caracalla
00100 Rome, Italy
E-mail: publications-sales@fao.org
Fax: (+39) 06 57053360
Web site: www.fao.org

This book is printed on elemental chlorine-free paper

Foreword

This report updates and extends the FAO global study *World agriculture: towards 2010*, issued in 1995. It assesses the prospects, worldwide, for food and agriculture, including fisheries and forestry, over the years to 2015 and 2030. It presents the global long-term prospects for trade and sustainable development and discusses the issues at stake in these areas over the next 30 years.

In assessing the prospects for progress towards improved food security and sustainability, it was necessary to analyse many contributory factors. These range from issues pertaining to the overall economic and international trading conditions, and those affecting rural poverty, to issues concerning the status and future of agricultural resources and technology. Of the many issues reviewed, the report concludes that the development of local food production in the low-income countries with high dependence on agriculture for employment and income is the one factor that dominates all others in determining progress or failure in improving the food security of these countries.

The findings of the study aim to describe the future as it is likely to be, not as it ought to be. As such they should not be construed to represent goals of an FAO strategy. But the findings can make a vital contribution to an increased awareness of what needs to be done to cope with the problems likely to persist and to deal with new ones as they emerge. The study can help to guide corrective policies at both national and international levels, and to set priorities for the years ahead.

The world as a whole has been making progress towards improved food security and nutrition. This is clear from the substantial increases in per capita food supplies achieved globally and for a large proportion of the population of the developing world. But, as the 1995 study warned, progress has been slow and uneven. Indeed, many countries and population groups failed to make significant progress and some of them even suffered setbacks in their already fragile food security and nutrition situation. As noted in the 2001 issue of *The State of Food Insecurity in the World*, humanity is still faced with the stark reality of chronic undernourishment affecting over 800 million people: 17 percent of the population of the developing countries, as many as 34 percent in sub-Saharan Africa and still more in some individual countries.

The present study predicts that this uneven path of progress is, unfortunately, likely to extend well into this century. Findings indicate that in spite of some significant enhancements in food security and nutrition by the year 2015, mainly resulting from increased domestic production but also from additional growth in food imports, the 1996 World Food Summit target of halving the number of undernourished persons by no later than 2015 is far from being reached, and may not be accomplished even by 2030.

By the year 2015 per capita food supplies will have increased and the incidence of undernourishment will have been further reduced in most developing regions. However, parts of South Asia may still be in a difficult position and much of sub-Saharan Africa will probably not be significantly better and may possibly be even worse off than at present in the absence of concerted action by all concerned. Therefore, the world must brace itself for continuing interventions to cope with the consequences of local food crises and for action to remove permanently their root causes. Nothing short of a significant upgrading of the overall development performance of the lagging countries, with emphasis on hunger and poverty reduction, will free the world of the most pressing food insecurity problems. Making progress towards this goal depends on many factors, not least among which the political will and additional resource mobilization required. The importance of these factors was reaffirmed in the Declaration of the World Food Summit: five years later, unanimously adopted at the Summit in June 2002 in Rome.

The study also foresees that agricultural trade will play a larger role in securing the food needs of developing countries as well as being a source of foreign exchange. Net cereal imports by developing countries will almost triple over the next 30 years while net meat imports might even increase by a factor of almost five. For other products such as sugar, coffee, fruit and vegetables the study foresees further export potential. How much of this export potential will materialize depends on many factors, not least on how much progress will be made during the ongoing round of multilateral trade negotiations. Developing countries' farmers could gain a lot from lower trade barriers in all areas, not only in agriculture. In many resource-rich but otherwise poor countries, a more export-oriented agriculture could provide an effective means to fight rural poverty and thus become a catalyst for overall growth. But the study also points at potentially large hardships for resource-poor countries, which may face higher prices for large import volumes without much capacity to step up production.

Numerous studies that assessed the impacts of freer trade conclude that lower trade barriers alone may not be sufficient for developing countries to benefit. In many developing countries, agriculture has suffered not only from trade barriers and subsidies abroad but has also been neglected by domestic policies. Developing countries' producers may therefore not benefit greatly from freer trade unless they can operate in an economic environment that enables them to respond to the incentives of higher and more stable international prices. A number of companion policies implemented alongside the measures to lower trade barriers can help. These include a removal of the domestic bias against agriculture; investment to lift product quality to the standards demanded abroad; and efforts to improve productivity and competitiveness in all markets. Investments in transportation and communication facilities, upgraded production infrastructure, improved marketing, storage and processing facilities as well as better food quality and safety schemes could be particularly important, the latter not only for the benefit of better access to export markets, but also for reducing food-borne diseases affecting the local population.

On the issue of sustainability, the study brings together the most recent evaluation of data on the developing countries' agricultural resources, how they are used now and what may be available for meeting future needs. It does the same for the forestry and fisheries sectors. The study provides an assessment of the possible extent and intensity of use of resources over the years to 2030 and concludes that pressure on resources, including those that are associated with degradation, will continue to build up, albeit at a slower rate than in the past.

The main pressures threatening sustainability are likely to be those emanating from rural poverty, as more and more people attempt to extract a living out of dwindling resources. When these processes occur in an environment of fragile and limited resources and when the circumstances for introducing sustainable technologies and practices are not propitious, the risk grows that a vicious circle of poverty and resource degradation will set in. The poverty-related component of environmental degradation is unlikely to be eased before poverty reduction has advanced to the level where people and countries become significantly less dependent on the exploitation of agricultural resources. There is considerable scope for improvements in this direction and the study explores a range of technological and other policy options. Provided such improvements in sustainability are put in place, the prospects point to an easing of pressures on world agricultural resources in the longer term with minimal further buildup of pressures on the environment caused by agricultural practices.

I conclude by reiterating the importance of developing sustainable local food production and of rural development in the low-income countries. Past experience underlines the crucial role of agriculture in the process of overall national development, particularly where a large part of the population depends on the sector for employment and income, as is the case in most low-income countries. Agricultural development is and will be the critical component of any strategy to improve their levels of food security and alleviate poverty. It is for this reason that sustainable agricultural and rural development are given enhanced priority in *The Strategic Framework for FAO: 2000-2015*.

Jacques Diouf
Director-General
Food and Agriculture Organization
of the United Nations

Contents

Boxes

Tables

Figures and maps

Explanatory notes

SYMBOLS AND UNITS

ha	hectare
kg	kilogram
US$	US dollar
tonne	metric ton (1 000 kg)
billion	thousand million
p.a.	per annum
kcal	kilocalories
p.c.	per capita
n.a.	not available
mm	millimetre
km^3	cubic kilometre
mln	million
m^3	cubic metre
mt	metric ton

TIME PERIODS

1998	calendar year
1997/99	average for the three years centred on 1998
1970-90	period from 1970 to 1990
1997/99-2030	period from the three-year average 1997/99 to 2030

GROWTH RATES

Annual percentage growth rates for historical periods are computed from all the annual data of the period using the Ordinary Least Squares (OLS) method to estimate an exponential curve with time as the explanatory variable. The estimated coefficient of time is the annual growth rate. Annual growth rates for projection periods are compound growth rates calculated from values for the begin- and end-point of the period.

COUNTRIES AND COUNTRY GROUPS

The list of countries and the standard country groups used in this report are shown in Appendix 1. In the text, the term "transition countries" is used to denote the countries in Eastern Europe (including the former Yugoslavia SFR) and in the former Soviet Union. The term "industrial countries" is used for the countries referred to formerly as "developed market economies".

LAND DEFINITIONS

Arable area is the physical land area used for growing crops (both annual and perennial). In any given year, part of the arable area may not be cropped (fallow) or may be cropped more than once (double cropping). The area actually cropped and harvested in any given year is the harvested area. The harvested area expressed as a percentage of the arable area is the cropping intensity. Land with (rainfed) crop production potential consists of all land area that is at present arable or is potentially arable, i.e. is suitable for growing crops when developed (see Chapter 4).

DATA SOURCES

All data are derived from FAO sources unless specified otherwise.

Contributors to the book

This book is the product of cooperative work by most technical units of FAO. It was prepared by a team led by Jelle Bruinsma under the general direction of Hartwig de Haen, Assistant Director-General, Economic and Social Department. Members of the team were Nikos Alexandratos (consultant), Josef Schmidhuber, Gerold Bödeker and Maria-Grazia Ottaviani. Paul Harrison edited most of the chapters.

Main contributors to or authors of the individual chapters were as follows: Chapter 1 (Introduction and overview): Nikos Alexandratos and Jelle Bruinsma. Chapter 2 (Prospects for food and nutrition): Nikos Alexandratos with inputs from Jorge Mernies of the Statistics Division (estimates of chronic undernourishment). Chapter 3 (Prospects for aggregate agriculture and major commodity groups): Nikos Alexandratos with inputs from Ali Gürkan, Myles Mielke, Concha Calpe, Nancy Morgan and Peter Thoenes of the Commodities and Trade Division (commodity projections); Samarendu Mohanty of the Food and Agricultural Policy Research Institute (prospects for India); Gregory Scott of the Centro Internacional de la Papa (prospects for roots and tubers); and Klaus Frohberg, Jana Fritzsch and Catrin Schreiber of the Institut für Agrarentwicklung in Mittel- und Osteuropa (prospects for agriculture in Transition countries). Chapter 4 (Crop production and natural resource use): Jelle Bruinsma with inputs from Günther Fischer of the International Institute for Applied Systems Analysis and Freddy Nachtergaele (agricultural land potential), Jean-Marc Faurès and Jippe Hoogeveen (irrigation), Jan Poulisse (fertilizers) of the Land and Water Development Division; Dat Tran, Peter Griffee and Nguu Nguyen (crop land and yield projections) of the Plant Production and Protection Division; Clare Bishop (consultant) and Lawrence Clarke (farm power) of the Agricultural Support Systems Division. Chapter 5 (Livestock production): Henning Steinfeld and Joachim Otte of the Animal Production and Health Division. Chapter 6

(Forestry): Michael Martin and CTS Nair of the Forestry Policy and Planning Division, who coordinated the contribution from the Forestry Department. Chapter 7 (Fisheries): Ulf Wijkstrom and Rebecca Metzner of the Fishery Policy and Planning Division, who coordinated the contributions of specialists in the Fisheries Department. Chapter 8 (Agriculture in poverty alleviation and economic development): Sumiter Broca and Kostas Stamoulis of the Agriculture and Economic Development Analysis Division with inputs from Alberto Zezza (consultant). Chapter 9 (Agricultural trade, trade policies and the global food system): Josef Schmidhuber with inputs from Terri Raney (Commodities and Trade Division), Timothy Josling and Alan Matthews (consultants). Chapter 10 (Globalization in food and agriculture): Josef Schmidhuber with inputs from Bruce Traill (consultant). Chapter 11 (Selected issues in agricultural technology): Jelle Bruinsma with inputs from Nikos Alexandratos (yields), Gerold Bödeker (integrated pest and nutrient management), Nadia Scialabba (organic agriculture), Josef Schmidhuber and Nuria Urquia (biotechnology), and Vivian Timon (research). Chapter 12 (Agriculture and the environment: changing pressures, solutions and trade-offs): David Norse (consultant) with inputs from Jeff Tschirley of the Research, Extension and Training Division. Chapter 13 (Climate change and agriculture: physical and human dimensions): David Norse (consultant) with inputs from René Gommes of the Research, Extension and Training Division.

Members of the Working Group of the FAO Priority Area for Interdisciplinary Action on Global Perspective Studies provided comments on the various drafts. Comments on individual chapters were also provided by Robert Brinkman, Alan Matthews, Vernon Ruttan and Gérard Viatte.

Maria-Grazia Ottaviani was responsible for most of the data preparation, statistical analysis and the appendixes, and Anastasia Saltas for the manuscript preparation.

Introduction and overview

1.1 Introduction

This study is the latest forward assessment by FAO of possible future developments in world food, nutrition and agriculture, including the crops, livestock, forestry and fisheries sectors. It is the product of a multidisciplinary exercise, involving most of the technical units and disciplines present in FAO, as well as specialists from outside FAO. It continues the tradition of FAO's periodical perspective studies for global agriculture, the latest of which was published in 1995 (Alexandratos, 1995). Earlier editions were Alexandratos (1988), FAO (1981a) and FAO (1970). An interim, less complete version of the present study was published in April 2000. Comments received on the interim report helped shape the study in its present form.

The projections were carried out in considerable detail, covering about 140 countries and 32 crop and livestock commodities (see Appendix 1). For nearly all the developing countries, the main factors contributing to the growth of agricultural production were identified and analysed separately. Sources of productivity growth, such as higher crop yields and livestock carcass weights, were distinguished from other growth sources, such as the area of cultivated land and the sizes of livestock herds. Special attention was given to land, which was broken down into five classes for rainfed agriculture and a sixth for irrigated agriculture. This level of detail proved both necessary and advantageous in identifying the main issues likely to emerge for world agriculture over the next 30 years. Specifically, it helped to spot local production and resource constraints, to gauge country-specific requirements for food imports and to assess progress and failure in the fight against hunger and undernourishment. The high degree of detail was also necessary for integrating the expertise of FAO specialists from various disciplines, as the analysis drew heavily on the judgement of in-house experts (see Appendix 2 for a summary account of the methodology). Owing to space constraints however, the results are mainly presented at the regional level and for selective alternative country groups, which of course, masks wide intercountry differences.

Another important feature of this report is that its approach is "positive" rather than "normative". This means that its assumptions and projections reflect the most likely future but not necessarily the most desirable one. For example, the report finds that the goal of the 1996 World Food Summit – to halve the number of chronically undernourished people by no later than 2015 – is unlikely to be accomplished, even though this would be highly desirable. Similarly, the report finds that agriculture will probably continue to expand into wetlands and rainforests, even though this is undoubtedly undesirable. Therefore, the prospective developments presented here are not goals of an FAO strategy but they can provide a basis for action, to cope both with existing problems that are likely to persist and with new ones that may emerge. It should also be stressed that these projections are certainly not trend extrapolations. Instead, they incorporate a multitude of assumptions about the future and often represent significant deviations from past trends (see Appendix 2). To give an impression of how well previous projections compared with actual outcomes, in Chapters 3 and 4, actual outcomes for some major aggregate variables (e.g. developing country cereal production and net imports) for the latest year for which historical data are available are compared with those implied for that same year by the 1988/90-2010 trajectories projected for these variables in 1992-93 for the 1995 study.

A long-term assessment of world food, nutrition and agriculture could deal with a great number of issues, the relevance of which depends on the reader's interest in a particular country, region or topic. As a global study, however, this report had to be selective in the issues it addresses. The main focus is on how the world will feed itself in the future and what the need to produce more food means for the natural resource base. The base year for the study is the three-year average for 1997/99 and projections are made for the years 2015 and 2030. The choice of 2015 allows an assessment of whether or not the goal of the 1996 World Food Summit (WFS) – to halve the (1990/92) number of chronically undernourished people in developing countries no later than 2015 – is likely to be reached. Extending the horizon to 2030 creates a sufficiently long period for the analysis of issues concerning the world's resource base; in other words, the world's ability to cope with further degradation of agricultural land, desertification, deforestation, global warming and water scarcity, as well as increasing demographic pressure. Naturally, the degree of uncertainty increases as the time horizon is extended, so the results envisaged for 2030 should be interpreted more cautiously than those for 2015.

The analysis is, *inter alia*, based on the long-term developments expected by other organizations. The population projections, for instance, reflect the latest assessment (2000 Assessment, Medium Variant) available from the United Nations (UN, 2001a), while those for incomes are largely based on the latest projections of gross domestic product (GDP) from the World Bank. Most of the agricultural data are from FAO's database (FAOSTAT), as of June 2001. Because these assumptions critically shape the projected outcomes, it is important to note that they can change significantly, even over the short term. For example, the historical data and the projections for the growth of population and GDP used in the 1995 study (Alexandratos, 1995), have since been revised in many countries, often to a significant extent. World population in the 1995 study, for instance, was projected at 7.2 billion for 2010, whereas the current UN projections peg the figure at 6.8 billion, i.e. about 400 million less. Similarly, it is now projected that sub-Saharan Africa's population will reach 780 million by 2010, compared with 915 million in the 1995 study. Also, projections for GDP growth (largely based on the latest projections from the World Bank) are different now from what they were in the 1995 study. Finally, FAO's historical data for food production, demand and per capita consumption were often drastically revised for the entire time series as more up-to-date information became available (for an example, see Box 2.2).

In the chapters that follow, particularly Chapters 2 to 4, reference is made on a number of occasions to conceptual and practical problems that arise in the process of making the projections. Many of these problems result from the approach followed in this study, i.e. the fact that the analyses are conducted in considerable country and commodity detail and that all the projections are subjected to a process of inspection, evaluation and iterative adjustment of the numbers by country and discipline specialists. This means that more data problems are encountered than if the analyses were conducted at a more aggregate level. It also means that an alternative scenario cannot be derived without repeating a good part of the whole cycle of inspection-evaluation-adjustment. In Appendix 2 some of these problems and the way they are or are not dealt with are highlighted, so that that the reader can better form an opinion as to how to view and evaluate the statements in this study.

The report is structured as follows. The remainder of this chapter gives an overview of the main findings of the report. The first part of the report (Chapters 2 to 7) presents the main perspective outcomes for food, nutrition and agricultural production and trade, as well as for forestry and fisheries. Chapter 2 presents and discusses expected developments in food demand and consumption and the implications for food security and the incidence of undernourishment. Chapter 3 discusses expected developments in demand, production and trade of total agriculture (crops and livestock) and for the major commodities. Chapter 4 presents the underlying agronomic projections for growth in crop production, including expected developments in crop yields, and resource (land, water) and input (fertilizer, farm power) use in the developing countries. It includes revised and updated estimates of land with rainfed crop production potential, estimates of irrigation potential and possible expansion of irrigated areas and, for the first time, an estimate of the volume of fresh water withdrawals needed to sustain irrigated agriculture in developing countries as projected in this report. Chapter 5 deals with projections of livestock production and related issues. Chapters 6 and 7 present the current state of and plausible developments in world forestry and fisheries.

The second part of the report (Chapters 8 to 10) continues with a discussion of the main issues raised by these developments. Issues of poverty (including international targets for its reduction), nutrition-development interactions, the role of agriculture in the development of the rural and overall economy, and macroeconomic policies and agriculture are the subjects of Chapter 8. Chapter 9 deals with trade policy issues, focusing on lessons from the implementation of the Uruguay Round Agreement on Agriculture (URAA), the possible implications of further liberalization and issues relating to further reforms during the follow-up negotiations in the context of the broader round of multilateral trade negotiations launched by the World Trade Organization (WTO) at its Fourth Ministerial Conference (Doha, November 2001). Chapter 10 then places agricultural trade in the broader context of globalization and discusses the possible effects of globalization on trade and on the concentration and location of the food processing industry.

The final part of the report (Chapters 11 to 13) deals with perspective issues concerning agriculture and natural resources (land, water, air and the genetic base). Chapter 11 examines the potential of existing and incipient agricultural technologies (including modern biotechnology) to underpin further growth in production while conserving resources and minimizing adverse effects on the environment, and the needed directions of agricultural research in the future. Chapter 12 presents an assessment of the environmental implications of agricultural production based, to the extent possible, on the quantitative projections presented in the preceding chapters, and examines options for putting agriculture on a more sustainable path. Finally, Chapter 13 deals with the interactions between climate change and agriculture, the role of agriculture as source of greenhouse gases but also as a carbon sink of potentially growing importance in the context of the Kyoto Protocol, and the broad impact of climate change on agriculture and food security.

1.2 Overview

1.2.1 Prospects for food and nutrition

Historical developments. The 2001 FAO assessment of food insecurity in the world (FAO, 2001a) estimates the incidence of undernourishment in the developing countries at 777 million persons in 1997/99 (17 percent of their population). The estimate was 815 million (20 percent of the population) for the three-year average 1990/92, the base year used by the 1996 WFS in setting the target of halving the number of the undernourished in the developing countries by 2015 at the latest. Obviously, the decline between 1990/92 and 1997/99 has been much less than required for attaining the target of halving undernourishment by 2015 (see further discussion in Box 2.5).

In a longer historical perspective of the last four decades, considerable progress has been made in raising the average world food consumption (measured in kcal/person/day), a variable that is a close correlate of the incidence of undernourishment. The world average kcal/person/day grew by 19 percent since the mid-1960s to 2 800 kcal. What is more important, the gains in the world average reflect predominantly those of the developing countries whose average grew by 31 percent, given that the industrial countries and the transition economies had already reached fairly high levels of per capita consumption in the mid-1960s. This progress in the aggregate of the developing countries has been decisively influenced by the significant gains made by the most populous among them. Of the seven countries with a population of over 100 million (China, Indonesia, Brazil, India, Pakistan, Nigeria and Bangladesh), only Bangladesh remains at very low levels of per capita food consumption. A significant number of countries however failed to participate in this general thrust towards improved national average food consumption levels and, by implication, towards reduced incidence of undernourishment. There are currently still 30 countries with per capita food consumption under 2 200 kcal, most of them in sub-Saharan Africa.

Population, incomes and poverty in the future. The latest assessment of world population prospects by the UN (UN, 2001a) indicates that there is in prospect a rather drastic slowdown in world demographic growth. In the time horizon of this study, the world population of 5.9 billion of our base year (the three-year average 1997/99) will grow to 7.2 billion in 2015, 8.3 billion in 2030 and 9.3 billion in 2050. The growth rate of world population, which had peaked in the second half of the 1960s at 2.04 percent p.a. and had fallen to 1.35 percent p.a. by the second half of the 1990s, is projected to fall further to 1.1 percent by 2015, to 0.8 percent by 2030 and to 0.5 percent by 2050.

This deceleration in demographic growth and the gradual saturation in per capita food consumption for parts of the world population are important factors that will contribute to slow the growth of food demand and, at the world level, also of production. Despite the drastic fall in the growth rates of both population and aggregate demand and production, the absolute annual increments continue to be large. For *population*, 79 million persons are currently added to world population every year. The number will not have decreased much by 2015. Even by 2030, annual additions will still be 67 million. Practically all of the increases in world population will be in the developing countries. Within the developing countries themselves there will be increasing differentiation. East Asia will be reaching a growth of under 0.5 percent p.a. towards the later years of the projection period. At the other extreme, sub-Saharan Africa's population will still be growing at

2.1 percent p.a. in 2025-30. For *demand and production*, the world cereal totals must increase by almost another billion tonnes by 2030, from the current level of 1.9 billion tonnes, an increase almost equal to that of the period since the mid-1960s (see Chapter 3, Table 3.3).

The growth of incomes is the other major determinant of the growth of food demand and of changes in food security and nutrition. The outlook for income growth is mixed. The latest World Bank assessment for the period 2000-15 foresees higher growth rates in per capita GDP than in the 1990s for all regions and country groups (particularly the reversal of declines in the transition economies) with the exception of East Asia. However, for several countries that have low food consumption levels and significant incidence of undernourishment, the economic growth rates that may be achieved are likely to fall short of what would be required for significant poverty reductions. In particular, there is great contrast as regards the prospects of the two regions with high relative concentrations of poverty and food insecurity, South Asia and sub-Saharan Africa. In the former region, a continuation of the relatively high GDP growth holds promise of positive impact on poverty alleviation and increases in per capita food consumption. However, progress in sub-Saharan Africa may be very limited and far from sufficient to make a significant dent on poverty and food insecurity.

The World Bank also projects possible developments in the incidence of poverty – the percentage of the population and numbers of persons below the US$1/day poverty line. This is of particular importance for the evaluation of possible reductions in undernourishment, as poverty is a close correlate of food deprivation and insecurity. The Bank assessment concludes that the *proportion* (not the absolute numbers) of the population living in poverty in the developing countries as a whole may fall from the 32 percent it was in 1990 to 13.2 percent in 2015. This fall, if it materialized, would meet the target of halving the proportion in poverty between 1990 and 2015. It is recalled that the target of halving the proportion of the poor is one of the International Development Goals of the Millennium Declaration adopted by the UN General Assembly (see Chapter 8 for details). However, the absolute numbers in poverty in the developing countries (a measure more directly relevant for the target of reducing undernourishment – see below) are not halved, as they are projected to decline from 1.27 billion in 1990 to 0.75 billion in 2015.

Prospects for food and nutrition. The projections of food demand for the different commodities suggest that the per capita food consumption (kcal/person/day) will grow significantly. The world average will be approaching 3 000 kcal in 2015 and exceeding 3 000 kcal by 2030. These changes in the world averages will reflect above all the rising consumption of the developing countries, whose average will have risen from the 2 680 kcal in 1997/99 to 2 850 kcal in 2015 and close to 3 000 kcal in 2030. These gains notwithstanding, there will still be several countries in which per capita food consumption will not increase to levels that would imply significant reductions in the numbers undernourished from the very high levels they have at present. In 2015 there could still be 6 percent of world population (462 million) living in countries with very low levels of food consumption (under 2 200 kcal). At the regional level, sub-Saharan Africa would still have in 2015 medium-low levels of per capita food consumption, 2 360 kcal/person/day, and even less if Nigeria is excluded from the regional total.

These gains in average consumption mean that the great majority of people will be better fed and the incidence of undernourishment should decline. But will it decline sufficiently to achieve the objectives set by the international community? The 1996 WFS set the target that the numbers undernourished (not just the proportion of the population in that condition) should be reduced by half by 2015 at the latest. Improved nutrition, in addition to being a human right and a final objective of development in its own right, is also an essential precondition for societies to make progress towards overall economic and social development within a reasonable time span. This is because undernourished persons are impeded by their very condition (undernourishment) from fully contributing to, and profiting from, the economic activities that are part and parcel of the development process. There is sufficient empirical evidence (reviewed in Chapter 8) establishing how persons in such condition have smaller earnings and fewer opportunities in life than

others. Removing the causes of undernutrition is a prime area for public policy interventions (e.g. through public health, sanitation and feeding programmes for pregnant women and children) since economic growth, although a necessary condition, is rarely sufficient by itself to achieve this goal within a reasonable time span.

The implication of the projected higher levels of average national food consumption per person is that the proportion of the population undernourished in the developing countries as a whole could decline from the 17 percent in 1997/99 to 11 percent in 2015 and to 6 percent in 2030. All regions would experience declines in these percentages and, by 2030, all of them, except sub-Saharan Africa, should have incidence in the range of 4 to 6 percent of the population, compared with a range of 9 to 24 percent in 1997/99. Sub-Saharan Africa could still have 15 percent of its population undernourished in 2030, down from 34 percent in 1997/99.

Because of population growth, declines in the relative incidence of undernourishment do not necessarily translate into commensurate declines in the absolute numbers (which is of relevance to the WFS target). Notwithstanding the slowdown in their demographic growth, the developing countries' population will still grow from 4 555 million in 1997/99 to 5 804 million in 2015 and to 6 840 million in 2030. Therefore, the numbers undernourished will decline only modestly: from the 776 million in 1997/99 to 610 million in 2015 and to 440 million in 2030.[1] If these projections came true, it would mean that we might have to wait until 2030 before the numbers undernourished are reduced to nearly the target set for 2015 by the WFS, i.e. one half of the 815 million estimated for 1990/92.

Can faster progress than projected here be made? Empirical evidence suggests that in the countries with high dependence on agriculture, assigning priority to the development of food production holds promise of overcoming the constraint to better nutrition represented by unfavourable overall economic growth prospects. Several countries, mainly in sub-Saharan Africa (Nigeria, Ghana, Chad, Burkina Faso, the Gambia, Mali, Benin and Mauritania) have at times achieved in the past quantum jumps in their food consumption per

---

[1] Numbers refer to the population of the countries for which estimates of undernourishment were made.

capita (over 20 percent) over periods comparable in length to the first subperiod of our projections (17 years or less), at a time when national average per capita incomes were not growing or outright falling. The common characteristic of these experiences has been rapid growth in the production of staple foods (cereals, roots and tubers).

Several empirical studies (discussed in Chapter 8) document the mechanism of how agricultural growth can be a potent factor in initiating a process of development that favours poverty reduction and improved food consumption in rural areas in countries with high dependence on agriculture and poverty profiles that are predominantly rural. The key conditions for this to happen are (i) that agricultural growth be somehow initiated, e.g. through policies that develop and diffuse affordable productivity-raising innovations such as improved seeds; and (ii) that the distribution of the ownership of land or other productive assets of agriculture not be too unequal so that benefits from higher agricultural production are widely spread and do not accrue predominantly to large landowners (who would spend most of their additional income on things other than locally produced goods and services).

If these conditions are met, a virtuous cycle of causation can set in: the initial increases of agricultural production, in addition to providing more food for the producers themselves and for others, create incomes that are spent locally and create demand for produce and services from the non-farm rural sector, thus generating incomes there. These, in turn, feed back into increased demand for food and more goods and services from the rural non-farm sector itself. This "pump-priming" role of agriculture is seen as necessary because the rural non-farm sector produces goods and services that are in the general category of non-tradables. This means that in that particular context and stage of development their production can increase only in response to local demand, otherwise it would not increase. The services that can be produced locally are non-tradable almost by definition, even if most goods may be inherently tradable but become non-tradable in the absence of adequate transport and marketing infrastructure.

Returning to the projections, the slow pace of progress in reducing the absolute numbers undernourished notwithstanding, the considerable overall improvement implied by the projected numbers should not be downplayed. It is no mean achievement that in the developing countries the numbers well-fed (i.e. not classified as undernourished according to the criteria used here) could increase from 3.8 billion in 1997/99 (83 percent of their population) to 5.2 billion in 2015 (89 percent of the population) and to 6.4 billion (94 percent) in 2030.

With the reduction of the relative incidence of undernourishment, the problem will tend to become more tractable through policy interventions, both national and international. This is the consequence of the prospect that the number of persons involved (and possibly needing to be targeted by policies) will be smaller, as will be the number of countries with high incidence (see Table 2.3). In addition, many more countries than at present will have relatively low incidence of undernourishment. For example, 16 countries had less than 5 percent undernourishment in 1997/99. Their number will increase to 26 in 2015 and 48 in 2030. By this latter year, such countries (including most of the largest developing countries) will account for three-quarters of the population of the developing countries and for 178 million undernourished out of the total of 440 million (in 1997/99: 8 million out of 776 million). Thus a growing part of the undernourished will be in countries with low relative incidence. For this reason, it will be easier than at present for them to respond to the problem though national policy interventions.

In parallel, the number of countries with high incidence of undernourishment (over 25 percent) and most in need of international policy interventions will be reduced considerably: from 35 in 1997/99 to 22 in 2015 and to only five in 2030. None of them will be in the most populous class (over 100 million population in 1997/99). They will account for an ever-declining proportion of the undernourished, 72 million out of the 440 million in 2030 (1997/99: 250 million out of the 776 million).

1.2.2 Prospects for aggregate agriculture and major commodity sectors

Total agriculture. The bulk of the increases in world consumption of crop and livestock products has been originating in the developing countries. With slower population growth and the gradual attainment of medium to high food consumption levels in several countries, the growth rate of the demand for food (and at the world level, also of production) will be lower than in the past. These are positive factors from the standpoint of human welfare. On the negative side, there is the prospect that in several countries with severe incidence of undernourishment, the growth of the demand for food will be well below what would be required for significant improvement in their food security. How much lower the growth of aggregate global demand will be in relation to the past depends on a number of factors. Foremost among them are the relative share, in world totals and other characteristics, of the countries that have already attained medium to high levels of consumption per capita, say over 2 700 kcal/person/day. Several developing countries (29 of those covered individually in this study) belong to this class, including some of the most populous ones (China, Indonesia, Brazil, Mexico, Nigeria, Egypt, the Islamic Republic of Iran and Turkey). They have one half of the population of the developing countries and account for two-thirds of their aggregate demand.

In the period from the mid-1960s to 1997/99, this group of 29 developing countries made spectacular progress in raising per capita consumption, from an average of 2 075 kcal to 3 030 kcal (see Table 3.2 in Chapter 3). China's performance carried a large weight in these developments. The group's population growth rate was 1.8 percent p.a. and that of its aggregate demand for all uses was 4.2 percent p.a. In the projections, commodity by commodity, the implied per capita consumption for this group in terms of kcal/person/day rises to 3 155 kcal in 2015 and to 3 275 kcal in 2030, i.e. to levels not much below those of the industrial countries today. The implied annual growth rate for the whole period 1997/99-2030 (in terms of kcal per capita) is only 0.25 percent. In parallel, the group's population growth rate is projected to fall to 0.9 percent p.a. The net result is that a drastic deceleration of the growth of aggregate demand is in prospect for this group of countries, from 4.2 percent p.a. in the preceding three decades to 1.7 percent p.a. in the period to 2030.

Given the large weight of this group of countries in the totals of the developing countries and the world, their drastic deceleration is reflected in all the aggregates. Thus, the growth rate of the developing countries as a whole declines from 3.7 percent p.a. in the preceding three decades to 2.0 percent p.a. in the period to 2030. This happens despite the prospect that the demand of the other developing countries (those below 200 kcal in 1997/99 which have the other half of the population) will not decelerate much (less than the decline in their population growth rate) and that their per capita food consumption would rise from 2 315 kcal in 1997/99 to 2 740 kcal in 2030. At the world level, the impact of the deceleration in the developing countries is muted (the deceleration in aggregate demand, from 2.2 percent p.a. to 1.5 percent p.a., is not very different from that of world population, see Table 3.1). This reflects the fact that in the past the world growth rate was depressed because of the collapse of consumption and production in the transition economies. The cessation (and eventual reversal) of this effect in the future offsets in part the deceleration in world totals caused by the slowdown in the developing countries.

At the world level, production equals consumption, so the preceding discussion about global demand growth prospects applies also to that of global production. For the individual countries and country groups, however, the two growth rates differ depending on movements in their net agricultural trade positions. In general, the growth rates of production in the developing regions have been below those of demand, and as a result their imports have been growing faster than their agricultural exports. These trends led to a gradual erosion of their traditional surplus in agricultural trade. In fact, the developing countries have turned in recent years from net agricultural exporters to net importers. This trend continues in the projections. The net imports of the developing countries as a whole of the main commodities in which they are deficit, mainly cereals and livestock products, will continue to rise fairly rapidly. In parallel, their net trade surplus on account of their traditional exports (e.g. tropical beverages, bananas, sugar and vegetable oils) will either rise less rapidly than their net imports of cereals and livestock products or outright decline. This does not mean that the exporting developing countries will not be expanding their net exports. It rather reflects the fact that several large developing countries are turning into growing net importers of products that are exported mainly by other (exporting) developing countries, e.g. vegetable oils, sugar and natural rubber.

Cereals. The preceding discussion about future slow-down in the growth of demand applies in varying degrees to individual commodities.

The deceleration of the world cereal sector has been taking place for some time. It will continue in the future, but the difference between past and future growth rates will not be as pronounced as in other sectors, particularly in livestock and oilcrops (see below). Cereals will continue to be by far the most important source (in terms of calories) of total food consumption. Food use of cereals has kept increasing, albeit at a decelerating rate. In the developing countries, the per capita average food use is now 173 kg, providing 56 percent of total calories, compared with 141 kg and 61 percent in the mid-1960s. The level of around 173 kg has been nearly constant since the mid-1980s. We project that it will remain around that level over the projection period. Cereals will continue to supply some 50 percent of the food consumption (in terms of kcal/person/day) in the developing countries, which is projected to reach nearly 3 000 kcal/person/day in 2030. Within the cereals group, per capita food consumption of rice will tend to stabilize at about present levels and will decline somewhat in the longer term, reflecting developments in, mainly, the East Asia region. In contrast, food consumption of wheat will continue to grow in per capita terms and, in the developing countries, such growth will be associated with growing wheat imports. Increases in the demand and trade of coarse grains will be increasingly driven by their use as animal feed in the developing countries.

As noted earlier, world consumption and production of cereals are projected to increase by almost another billion tonnes by 2030, from the 1.89 billion tonnes of 1997/99. Of this increment, just over one half will be for feed, and about 42 percent for food, with the balance going to other uses (seed, industrial non-food use and waste). Feed use, and within it that of the developing countries, will revert to being the most dynamic element driving the world cereal economy, as it will account for an ever-growing share in aggregate demand for cereals. It had lost this role in the decade to the mid-1990s following events and policies that had depressed feed use of cereals in two major consuming regions, the transition economies and the European Union (EU) (see Chapter 3, Sections 3.2 and 3.3).

The dependence of the developing countries on imports of cereals (wheat and coarse grains) will continue to grow, notwithstanding lower growth of demand compared with the past. This follows from the prospect that in the post-green revolution period, and in the face of growing resource scarcities, particularly of irrigation, developing countries' potential to increase production is also more limited compared with the past. Their net cereal imports are projected to rise from 103 million tonnes in 1997/99 (and the forecast 110 million tonnes for the current trade year July 2001/June 2002) to 190 million tonnes in 2015 and to 265 million tonnes in 2030. These numbers imply a resumption of the growth of the world cereal trade after a period of near stagnation. The latter was mainly the result of the virtual disappearance of net cereal imports of the transition economies in the 1990s as well as the slowdown of the economies and oil export earnings in many countries, particularly in the major importing region Near East/North Africa.

These factors that depressed export markets available to the traditional exporters of cereals will be less limiting in the future, but they will not disappear entirely. Not only are the transition economies unlikely to revert to being the large net importers they were in the pre-reform period, but in the longer term they have the potential of transforming themselves into net exporters of cereals. We have made an allowance for this eventuality by estimating their net exports in 2015 and 2030 at 10 and 25 million tonnes, respectively. The traditional cereal exporters in the industrial world (United States, Canada, the EU and Australia) are expected to increase their net exports from the 144 million tonnes in 1997/99 to 224 million tonnes in 2015 and 286 million tonnes in 2030 (see Table 3.8).

The question is often raised whether these traditional exporting countries have sufficient production potential to continue generating an ever-growing export surplus. Concern with adverse environmental impacts of intensive agriculture is among the reasons why this question is raised. The answer depends, *inter alia*, on how much more they must produce over how many years. Production growth requirements are derived by adding the projected net exports to the projections of their own domestic demand, including demand for cereals to produce livestock products. The result is that these countries are required to increase their collective production from the 629 million tonnes of 1997/99 to 758 million tonnes in 2015 and 871 million tonnes in 2030, an increment of 242 million tonnes over the entire period, of which about 80 million tonnes would be wheat and the balance largely coarse grains. The annual growth rate is 1.1 percent in the period to 2015 and 0.9 percent in the subsequent 15 years, an average of 1.0 percent p.a. for the entire 32-year projection period. This is lower than the average growth rate of 1.6 percent p.a. of the past 32 years (1967-99), although the historical growth rate has fluctuated widely, mostly as a function of the ups and downs of export demand, associated policy changes and occasional weather shocks. The overall lesson of the historical experience seems to be that the production system has so far had the capability of responding flexibly to meet increases in demand within reasonable limits. This is probably also valid for the future.

Livestock. The world food economy is being increasingly driven by the shift of diets towards livestock products. In the developing countries, consumption of meat has been growing at 5-6 percent p.a. and that of milk and dairy products at 3.4-3.8 percent p.a. in the last few decades. However, much of the growth has been taking place in a relatively small number of countries, including some of the most populous ones, especially China and Brazil. Including these two countries, per capita meat consumption in the developing countries went from 11.4 kg in the mid-1970s to 25.5 kg in 1997/99. Excluding them, it went from 11 kg to only 15.5 kg. Including or excluding China in the totals of the developing countries and the world makes a significant difference for the aggregate growth rates of meat, although not of milk and dairy products, given the small weight of these products in China's food consumption. However, many developing countries and whole regions, where the need to increase protein consumption is the greatest, have not been participating in the buoyancy of the world meat sector. In this category are the regions of sub-Saharan Africa (with very low per capita

consumption reflecting quasi permanent economic stagnation) and the Near East/North Africa where the rapid progress of the period to the late 1980s (oil boom) was interrupted and slightly reversed in the subsequent years, helped by the collapse of consumption in Iraq. Similar considerations apply to the developments in the per capita consumption of milk and dairy products.

The world meat economy has been characterized by the rapid growth of the poultry sector. Its share in world meat production increased from 13 percent in the mid-1960s to 28 percent currently, while per capita consumption increased more than threefold over the same period. In more recent years, the world meat trade has been expanding rapidly. This expansion reflected, among other things, significant moves towards meat trade liberalization, including in the context of regional trade agreements. Some drastic changes occurred as to the sources of imports and destination of exports. Japan became the world's largest net importer (it increased net imports fourfold since the mid-1980s), followed by the countries of the former Soviet Union (mainly the Russian Federation). Australia and New Zealand (together) continue to be the world's largest exporters while, in the last decade or so, the United States turned from a large net importer of meat to a large net exporter, mostly on the growing strength of its poultry sector.

Concerning the future, the forces that made for the rapid growth of the meat sector in the past are expected to weaken considerably. The lower population growth is an important factor, as is the natural deceleration of growth following the attainment of fairly high consumption levels in the few major countries that dominated past increases. For example, if China's growth in the 1990s of about 2.6 kg/person/year (leading to the 45 kg of 1997/99)[2] were to continue for much longer, the country would soon surpass the per capita consumption of the industrial countries. Similar considerations apply to other countries, such as Brazil. Therefore, rather drastic deceleration in at least these countries and, given their large weight, also in the global aggregates, is to be expected.

Other countries may do better in the future than in the past, but their weight in the totals is not suffi-

---

[2] It is thought that these very large increases appearing in the FAO food balance sheet data and the discrepancies from independent consumption data result mainly from an apparent overestimation of China's meat production (see Chapter 3).

cient to halt the expected deceleration in the broad aggregates. India, with its very low per capita meat consumption (4.5 kg in 1997/99) and a population rivalling that of China, could be thought of as a potential growth pole of the world meat economy. This is not likely to be, as its per capita consumption would not exceed 10 kg even by 2030, and this under rather optimistic assumptions. At least this is what a number of studies indicate. The result is that the growth rate of world meat demand and production could grow at rates much below those of the past, 1.7 percent p.a. in the period to 2030, down from 2.9 percent in the preceding 30 years. The deceleration in the developing countries would be more drastic, because of the above-mentioned large-country effect, following the projected slower growth of aggregate consumption in China, and to a smaller extent in Brazil. Remove these two countries from the developing countries aggregate and there is only a modest reduction in the growth of their aggregate demand for meat, reflecting essentially the lower population growth rate.

Unlike meat, we project higher growth in the world milk and dairy sector than in the recent past because of the cessation of declines and some recovery in the transition economies. Excluding these latter countries, world demand should grow at rates somewhat below those of the past but, given lower population growth, per capita consumption would grow faster than in the past.

The structural change that characterized the evolution of the meat sector in the past will probably continue, although at attenuated rates. Poultry will continue to increase its share in world meat output and the meat trade will continue to expand. The trend for the developing countries to become growing net importers of meat is set to continue. This is another important component of the broader trend for the developing countries to turn from net exporters to growing net importers of food and agricultural products. Imports of poultry meat will likely dominate the picture of growing dependence on imported meat. Trade in dairy products will also likely recover with the net imports of the developing countries resuming growth after a period of stagnation from the mid-1980s onwards.

As noted earlier, the world feed use of cereals had slowed considerably in the recent past. It had grown less fast than the aggregate production of the livestock sector, although not in the developing countries. The main reasons were the collapse of the livestock sector in the transition economies, high policy prices in the EU up to the early 1990s, the shift of livestock production to poultry, and more general productivity increases creating more output from a given input of cereals. Growth in the world feed use of cereals will resume thanks to the cessation of the downward pressure on world cereals feed use exerted by events in the transition economies and their eventual reversal; the turnaround of the EU to growing feed use of cereals; the increasing weight of the developing countries in world livestock; and the attenuation of structural change towards poultry.

Oilcrops and products. This category of food products with a high calorie content has played an important role in the increases of food consumption in developing countries. Just over one out of every five calories added to their food consumption in the period since the mid-1970s originated in this group of products. In the projections, this trend continues and indeed intensifies: 45 out of every 100 additional calories in the period to 2030 may come from these products. This projection reflects above all the prospect of only modest further growth in the direct food consumption of staples (cereals, roots and tubers, etc.) in the majority of the developing countries in favour of non-staples such as vegetable oils which still have significant scope for further consumption increases.

On the demand side, the major driving force of the world oilcrops economy has been the growth of food demand in the developing countries, with China, India and a few other major countries representing the bulk of such growth. Additional significant demand growth has been in the non-food industrial uses of oils and in the use of oilmeals for the livestock sector. The growth of aggregate world demand and production (in oil equivalent) would continue to be well above that of total agriculture, although it would be much lower than in the past, 2.5 percent p.a. in the next three decades compared with 4.0 percent p.a. in the preceding three. This deceleration will reflect the factors that were discussed earlier in relation to other commodities, i.e. slower population growth, more and more countries achieving medium-high levels of consumption and, of course, persistence of low incomes in many countries limiting their effective demand.

On the production side, the trend has been for four oilcrops (oil palm, soybeans, sunflower seed and rapeseed) and a small number of countries to provide much of the increase in world output (Table 3.16). With the lower demand growth in the future and changes in policies (e.g. limits to subsidized production), the pace of structural change in favour of some of these crops could be less pronounced in the future. The sector accounted for a good part of cultivated land expansion in the past and in the industrial countries it made up for part of the declines of the area under cereals. The projections of land use in the developing countries indicate that oilcrops will continue to account for a good part of future expansion of harvested area in the developing countries. Being predominantly rainfed crops, the expansion of their production depends on area expansion (rather than yield growth) by more than is the case of other crops, such as cereals.

The rapid growth of demand of the developing countries was accompanied by the emergence of several of them as major importers of oils and oilseeds. If we exclude the five major net exporters among the developing countries (Malaysia, Indonesia, the Philippines, Brazil and Argentina), the others increased their net imports of oils and oilseeds (in oil equivalent) from 1 to 17 million tonnes between 1974/76 and 1997/99. In parallel, however, the five major exporters increased their net exports from 4 to 21 million tonnes, so that the net trade balance of all the developing countries increased slightly (see Table 3.20). In the future, these trends are likely to continue and the net trade balance of the developing countries would not change much, despite the foreseen further rapid growth of exports from the main exporter developing countries. The developing countries have so far been net exporters of oil meals, which have enabled them to maintain a positive, although declining, net trade balance in value terms in their combined trade of oilseeds, oils and meals. With the development of their livestock sector, the prospect is that their net exports of oil meals could turn into net imports. This is yet another dimension of the above-mentioned trend for the developing countries to turn into net importers of agricultural products.

Roots, tubers and plantains. These products (mainly cassava, sweet potatoes, potatoes, yams, taro and plantains) represent the mainstay of diets in several countries, many of which are characterized by low overall food consumption levels and food insecurity. The great majority of these countries are in sub-Saharan Africa, with some countries of the region (Ghana, Rwanda and Uganda) deriving 50 percent or more of total food consumption (in terms of calories) from these foods. In general, high dependence on these foods is mostly characteristic of countries that combine suitable agro-ecological conditions for their production and low incomes. With the exception of potatoes, diet diversification away from these products occurs when incomes and overall food consumption levels improve. In a number of countries (e.g. Nigeria, Ghana and Benin), quantum jumps in the production of these products were instrumental in raising food consumption from low or very low levels.

The evolution over time shows declining per capita food consumption of these products for the developing countries and the world as a whole up to about the late 1980s, followed by some recovery in the 1990s. These developments reflected two main factors: (i) the rapid decline in food consumption of sweet potatoes in China (from 94 kg in the mid-1970s to 40 kg at present), only in part counterbalanced by a parallel rise in that of potatoes, in both China and the rest of the developing countries, and (ii) the rapid rise of food consumption of all products in this category in a few countries, such as Nigeria, Ghana and Peru.

Significant quantities of roots are used as feed, including potatoes (mainly in the transition countries and China), sweet potatoes (mainly China) and cassava (mainly Brazil and the EU). The EU's feed consumption of cassava (all imported) amounts to some 10 million tonnes (fresh cassava equivalent). This is less than half the peak it had reached in the early 1990s, when EU policy prices for cereals were high and rendered them uncompetitive in feed use, leading to a process of substitution of cassava and other imported feedstuffs for cereals. The reversal is mostly the result of the policy reforms in the EU which lowered the policy prices for cereals after 1993 and re-established their competitiveness. In Thailand, the main supplier of cassava to the EU, cassava production and exports followed closely the developments in the EU. The rapid expansion of cassava production for export in Thailand is thought to have been a prime cause of land expansion and deforestation, followed by land degradation in certain areas of the country. This link provides a good example of how the effects of policies (or policy distortions such as the high support prices in the EU) in one part of the world can exert significant impacts on production, land use and the environment in distant countries.

The food products in this category will continue to play an important role in sustaining food consumption levels in the many countries that have a high dependence on them and low food consumption levels overall. The main factor responsible for the decline in the average of the developing countries (precipitous decline of sweet potato food consumption in China) will be much weaker, as the scope for further declines is much more limited than in the past. In parallel, the two factors that made for increases in the average, the increase in demand for potatoes when incomes rise and the potential offered by productivity increases in the other roots (cassava and yams), will continue to operate. It will be possible for more countries in sub-Saharan Africa to replicate the experiences of countries such as Nigeria, Ghana, Benin and Malawi, and increase their food consumption. Thus, the recent upturn in per capita consumption of the developing countries is projected to continue.

The main export commodities of the developing countries. The agriculture and often the overall economy, as well as the incidence of poverty and food security of several developing countries, depend to a high degree on the production of one or a few agricultural commodities destined principally for export, e.g. tropical beverages, bananas, sugar, oilseeds and natural rubber. In such cases the overall economies, and often the welfare of the poor are subject to changing conditions in the world markets, i.e. the rate of expansion of such markets and the prices these commodities fetch. For some commodities, the rate of expansion of world consumption has been too slow. For other commodities, such as sugar, protectionism in the main traditional import markets of the industrial countries has been a prime factor in restricting the growth of exports. In the face of such constraints, competition among exporters to capture a market share has resulted in declining and widely fluctuating export prices.

This has been particularly marked for coffee in recent years. The industrial countries account for two-thirds of world coffee consumption and their consumption per capita has been nearly constant for two decades. Given their low population growth, aggregate demand has been growing very slowly while production has kept increasing, including from recent entrants in world markets such as Viet Nam. The result has been that prices have precipitated and this has worsened poverty in the countries where significant parts of the rural population depend on coffee for a living.

The importing developing countries have been increasingly providing market outlets for the exports of other developing countries. As noted, this has been the case for commodities such as sugar, vegetable oils and oilseeds, and natural rubber. It has been much less so for coffee and, to a lesser extent, cocoa. For these latter commodities the growth of export demand over the projection period will be slow, as it will continue to be dominated by consumption trends in the industrial countries. In contrast, higher growth is foreseen in exports of sugar, vegetable oils and oilseeds, and natural rubber, fuelled by demand generated in the importing developing countries where the scope for expansion of consumption is still considerable. The policies of the industrial countries severely restricted imports of commodities such as sugar in the past (including the turning of the EU from net importer to net exporter). These policies will be less restrictive in the future, following policy reforms already agreed (e.g. the Uruguay Round limits on subsidized exports, the North American Free Trade Agreement [NAFTA] eventually leading to tariff-free access of Mexican sugar to the United States, the EU's Everything but Arms Initiative [EBA]) and the new ones that may come in the future.

1.2.3 Issues of trade policy and globalization

As noted in the preceding section, the traditional agricultural trade surplus in the balance of payments of the developing countries has been diminishing over time and turned into a net deficit in recent years. This gradual erosion has reflected the many factors discussed above that influenced demand and supply and associated imports and exports of the individual commodities in the different countries. Among these factors are the agricultural and trade policies of the main players in world markets, foremost among them the Organisation for Economic

Co-operation and Development (OECD) countries. Most OECD countries have traditionally protected their agriculture sectors heavily, partly through policies granting domestic support and partly, and closely related to the former, through trade policies, such as tariffs, quotas and export subsidies.

The impact of these policies on the trade performance and on the welfare of the developing countries has varied widely. There have been clear losers among the countries exporting products competing with those of the OECD (e.g. Argentina for cereals and livestock products, Brazil and Cuba for sugar), and clear gainers among the countries that had preferential access to the protected markets. Possible gainers (from lower prices and plentiful supplies of cereals in world markets) are to be found among the consumers of the food import-dependent developing countries, including those receiving food aid, but often at the expense of farmers in these same countries. More often, the situation is mixed, e.g. countries in North Africa benefiting from low-priced cereal imports but harmed by barriers to their exports of fresh vegetables.

These support and protection policies affected above all the trade performance (changed market shares) as well as consumers (paying higher prices) and the taxpayers (paying the subsidies) of the OECD countries themselves. For this reason, most studies that examined the possible effects of agricultural trade liberalization conclude that the lion's share of gains in welfare would accrue to the high-income countries, and certainly to some developing countries exporters of competing products, e.g. cereals, livestock, sugar and vegetables. However, some developing countries could be harmed, such as those that enjoy preferential access in protected markets or those that have few agricultural exports but import much of their food.

The late 1980s and the 1990s witnessed intensified efforts to discipline policies that distorted trade. The resulting Uruguay Round Agreement on Agriculture (URAA) enshrined a certain measure of discipline. It mandated reductions in border protection, trade-distorting domestic support and export subsidies. These reductions, however, still left the countries that made heavy use of them in the pre-URAA period, i.e. primarily most OECD countries, with considerable scope for continuing them, albeit at lower levels than before. For countries that made little use of domestic support and export subsidies,

overwhelmingly the developing ones, the URAA meant that they were left with very little scope for using such policies in the future, generally within the limits of the *de minimis* clause. In practice, the URAA legitimized and in a sense froze the divide between high-protecting countries and the rest. However, there were some compensations. Developing countries were not required to reduce and could increase, or introduce for the first time, support aimed at their agricultural development, e.g. investment or input subsidies. In addition, the widely diffused practice of binding tariffs at levels well above those effectively applied in the pre-URAA period afforded significant scope for increasing border protection through higher tariffs in the future. These possibilities notwithstanding, some analysts argue that the URAA may have "institutionalized" the production and trade-distorting policies of the OECD countries without addressing the fundamental concerns of developing countries.

Continuing negotiations on agriculture to liberalize trade further are under way. They began in March 2000 and were later subsumed in the broader round of multilateral trade negotiations launched by the WTO at its Fourth Ministerial Conference (Doha, November 2001). In these negotiations the potential exists for the concerns of the developing countries to be addressed more effectively than in the past under the provision for special and differential treatment of the developing countries to reflect their development needs. At the same time, however, further liberalization will tend to erode the gains enjoyed by several developing countries of preferential non-reciprocal access to the protected markets of the major OECD countries (though generally to the benefit of other competing developing countries). Four non-reciprocal preferential arrangements are of particular relevance: the Generalized System of Preferences (GSP) under the WTO; the African, Caribbean and Pacific Group of States (ACP)-EU Cotonou Agreement; the US Trade and Development Act of 2000; and the EBA to provide duty-free and quota-free market access to the EU for the products of the least developed countries (LDCs).

While the importance of "classical" border measures (such as tariffs and quotas) gradually diminishes, the prominence in trade of the safety and quality standards increases. The latter concerns mainly the WTO Agreement on the Application of

Sanitary and Phytosanitary Measures (SPS), but also the quality attributes covered under the Agreement on Technical Barriers to Trade (TBT) and the Agreement on Trade-Related Aspects of Intellectual Property Rights (TRIPS). Recent food safety scares and the advent of genetically modified products have added to the influence of these standards in trade. Their application will be increasingly coming under scrutiny in order to minimize the risk that they will be used inappropriately to protect domestic producers from foreign competition.

The importance of standards in trade has been greatly boosted by the advance of globalization in food and agriculture. What was once a set of national markets linked by raw material trade from land-rich to land-scarce countries is gradually becoming a loosely integrated global market with movements of capital, raw and semi-processed goods, final products and consumer retail services. Intrafirm and intra-industry trade is increasing in importance so that the food trade is assuming certain characteristics of the trade in manufactures. Trade policy with respect to food and agriculture is gradually shifting beyond concentration on issues pertaining to primary farming to encompass issues and interests of the whole food chain, including food processing, marketing and distribution.

The thrust towards a globalized food and agriculture economy is seen as offering opportunities for developing countries to improve the performance of their agricultural and food sectors. This is part of the wider argument that, generally speaking, policies that favour openness of the economy boost economic growth. This has knock-on effects on reducing the numbers below the poverty line, although not necessarily on reducing the income gap between the rich and the poor or, for that matter, between rich and poor countries. These are not arguments in favour of "big bang" liberalization, whether of trade or capital flows. Empirical evidence suggests that openness and outward-oriented policies are not *per se* guarantors of success. More important are the companion policies on the domestic front that facilitate the integration into global markets. These are policies that provide appropriate transition periods towards freer trade; help adapt new, external technologies to the domestic environment; or provide competition policy settings and design contracts that also allow small-scale agriculture to thrive within the operations of transnational corporations (TNCs).

As an important element of globalization in food and agriculture, experience suggests that TNCs can make an important contribution as vehicles of capital, skills, technologies, access to both domestic and export marketing channels, and creation of linkages to the rural economy, for example through contract farming. There are, however, snags such as excessive concentration of market power (and its eventual abuse) in the hands of a few large, sometimes vertically integrated, enterprises operating in many countries. These should be mitigated by appropriate policies, not by closing the economy to the broader benefits and, of course, domestic enterprises are not devoid of monopolistic elements; on the contrary, competition from new entrants, even if foreign, can be a welcome boost to competition.

1.2.4 Factors in the growth of crop production

By 2030, crop production in the developing countries is projected to be 67 percent higher than in the base year (1997/99). In spite of this noticeable increase in the volume of crop production, in terms of annual growth rates this would imply a considerable slowdown in the growth of crop production as compared with the past, for the reasons discussed in Section 1.2.2 concerning the anticipated deceleration in the growth of aggregate demand. Most of this increase (about 80 percent) would continue to come on account of a further intensification of crop production in the form of higher yields and of higher cropping intensities (multiple cropping and reduced fallow periods), with the remainder (about 20 percent) coming on account of further arable land expansion.

The developing countries have some 2.8 billion ha of land with a potential for rainfed agriculture at yields above a "minimum acceptable level". Of this total, some 960 million ha are already under cultivation. Most of the remaining 1.8 billion ha however cannot be considered as land "reserve" since the bulk of the land not used is very unevenly distributed with most of it concentrated in a few countries in South America and sub-Saharan Africa. In contrast, many countries in South Asia and the Near East/North Africa region have virtually no spare land left, and much of the land not in use suffers from one or more constraints making it less suitable for agriculture. In addition, a good part of the land with agricultural potential is under forest or in protected areas, in use for human settlements, or suffers from lack of infrastructure and the incidence of disease. Therefore, it should not be considered as being a reserve, readily available for agricultural expansion.

Taking into account availability of and need for land, arable land in the developing countries is projected to increase by 13 percent (120 million ha) over the period to 2030, most of it in the "land-abundant" regions of South America and sub-Saharan Africa, with an unknown but probably considerable part of it coming from deforestation. In terms of harvested land, the land area would increase by 20 percent (178 million ha) on account of increasing cropping intensity. The latter will reflect, *inter alia*, the growing role of irrigation in total land use and crop production.

Irrigation is expected to play an increasingly important role in the agriculture of the developing countries. At present, irrigated production is estimated to account for 20 percent of the arable land (but about 30 percent of harvested area because of its higher cropping intensities) and to contribute some 40 percent of total crop production (nearly 60 percent of cereal production). This share is expected to increase to 47 percent by 2030. The developing countries are estimated to have some 400 million ha of land which, when combined with available water resources and equipped for irrigation, represent the maximum potential for irrigation extension. Of this total, about one half (some 202 million ha) is currently equipped in varying degrees for irrigation and is so used. The projections conclude that an additional 40 million ha could come under irrigated use, raising the total to 242 million ha in 2030. In principle, by that year the developing countries would be exploiting for agriculture some 60 percent of their total potential for irrigation. Naturally, the harvested area under irrigation will increase by more (33 percent), following fuller exploitation of the potential offered by controlled water use for multiple cropping.

Expansion of irrigation would lead to a 14 percent increase in water withdrawals for agriculture. This latter result depends crucially on the projected increase in irrigation water use efficiency (from 38 to 42 percent on average). Without such efficiency improvements it would be difficult to sustain the above-mentioned rates of expansion of irrigated agriculture. This is most evident in regions such as the Near East/North Africa (where

water withdrawals for irrigated agriculture account for over 50 percent of their total renewable water resources) and South Asia where they account for 36 percent. In contrast they account for only 1 percent in Latin America and 2 percent in sub-Saharan Africa, with East Asia being in the middle (8 percent).

Naturally, these regional averages mask wide intercountry differences in water scarcities. Countries using more than 40 percent are considered to be in a critical situation. There are ten developing countries in that class, including countries such as Saudi Arabia and the Libyan Arab Jamahiriya which use more than 100 percent of their renewable resources (mining of fossil groundwater), and another eight countries are using more than 20 percent, a threshold which could be used to indicate impending water scarcity. By 2030 two more countries will have crossed this threshold and by then 20 developing countries will be suffering actual or impending water scarcity. Within-country differences can be as wide as intercountry ones. Large regions within countries can be in a critical situation even if the national average withdrawals for irrigation are relatively modest. China is in that class with the north facing severe constraints while the south is abundantly endowed with water.

As already mentioned, yield growth will remain the mainstay of crop production growth. For most crops, however, the annual growth rate of yields over the projection period will be well below that of the past. For example, the growth rate of the average cereal yield of the developing countries is projected to be 1.0 percent p.a., as compared with the 2.5 percent p.a. recorded for 1961-99. This deceleration in growth of yields has been under way for some time now. For example, in the last ten years (1991-2001) it was already down to 1.4 percent p.a. Intercountry differences in yields are wide and are projected to remain so. They reflect in part differences in agro-ecological conditions and in part differences in agricultural management practices and the overall socio-economic and policy environment. To the extent that the latter factors change (e.g. if scarcities developed and prices rose), or can be made to change through policy, yields could grow in the countries where the agro-ecological potential exists for this to happen under changed agronomic practices, e.g. better varieties and fertilization.

Chapter 11 provides some illustrative evidence on the existence of such potential in the different countries. It does so by comparing the prevailing yields with those that are attainable under advanced (or high-input) agriculture in those parts of the land that are evaluated as being suitable for the individual crops from the standpoint of agro-ecological conditions. Naturally, the existence of such gaps in yields in no way suggests that countries with yields well below their agro-ecologically attainable ones are less efficient producers economically. Often, the contrary is true. For example, the major exporters of wheat (North America, Argentina and Australia) have low to medium yields well under their attainable ones under high input farming. Yet they are competitive low-cost producers compared with some countries attaining much higher yields, often near their maximum potential, e.g. many countries of the EU.

In conclusion, the yield growth foreseen for the future, although lower than in the past, can still be the mainstay of production increases and need not imply a break from past trends in the balance between demand and supply of food at the world level. This could be so precisely because the demand will also be growing at lower rates. The issue is not really whether the yield growth rates will be slower than in the past. They will. Rather the issue is if such slower growth is sufficient to deliver the required additional production.

Naturally, this slower yield growth may not happen unless we make it happen. In particular, the higher yields of the future cannot come only, or even predominantly, from the unexploited yield potential of existing varieties in the countries and agro-ecologies where such potential exists. It will need also to come from countries and agro-ecologies where such potential is very limited. This requires continued support to agricultural research to develop improved varieties for such environments (including those coming from modern biotechnology, see Section 1.2.5 below).

The preceding discussion may have created the impression that we are saying that there is, or there can be developed, sufficient production potential for meeting the increases in effective demand that may be forthcoming in the course of the next three decades. The impression is correct so long as it refers to the world as a whole. But this is not saying that just because such global potential exists, or can be developed, all people will be food-secure in the future. Far from it, as already discussed. Food security and food production potential are however closely related in poor and agriculturally dependent societies. Many situations exist where production potential is limited (e.g. in the semi-arid areas given existing and accessible technology, infrastructure, etc.) and a good part of the population depends on such poor agricultural resources for food and more general livelihood. Unless local agriculture is developed and/or other income-earning opportunities open up, the food insecurity determined by limited production potential will persist, even in the middle of potential plenty at the world level. The need to develop local agriculture in such situations as the often *sine qua non* condition for improved food security cannot be overemphasized.

In the same vein, the above-mentioned need to continue the agricultural research effort (including in biotechnology) to improve yields, even if yield potentials of existing varieties are not fully exploited in many countries, finds its justification in these same considerations. For example, the existence of significant unexploited yield potential for wheat in Ukraine or Argentina does not obviate the need for research to raise yield ceilings for the agro-ecological and other conditions (e.g. salinity and water shortages) prevailing in the irrigated areas of South Asia. The bulk of the additional demand for wheat in South Asia will not be for the wheat that could be potentially produced in Ukraine or Argentina. It will materialize only as part of the process of increasing production locally for the reasons discussed earlier (increased production stimulates the local rural economy).

Behind all these statements, of course, looms large the issue of whether the continuing intensification of agriculture that proceeds even under decelerating yield growth rates is a sustainable proposition. That is, is the capacity of agriculture to continue to produce as much as is needed for all people to be well-fed now and in the future being put in jeopardy, e.g. because of land degradation, depletion or otherwise deterioration of water resources. Would the externalities generated in the process of agricultural expansion and intensification (e.g. water and air pollution, disturbance of habitat, loss of biodiversity, etc.) not impose unacceptable costs to society? An overview of these issues is given in section 1.2.9 below, and the issues are treated in more detail in Chapter 12. The technological options to minimize risks and transit to a more environmentally benign and resource-conserving agriculture, while still achieving the needed production increases, are explored in Chapter 11. Here suffice to say that we foresee a rather drastic slowdown in the use of mineral fertilizer.

Fertilizer use (nutrients NPK) in the developing countries is projected to increase by 1.1 percent p.a. from 85 million tonnes in 1997/99 to 120 million tonnes in 2030 (at world level from 138 million tonnes to 188 million tonnes). This is a drastic slowdown as compared with the past (e.g. 3.7 percent for 1989-99). The slowdown reflects the expected continuing deceleration in agricultural production growth, the relatively high levels of application that prevail currently, or will be gradually attained, in several countries, and the expected increase in fertilizer use efficiency, partly induced by environmental concerns. Fertilizer use per hectare in the developing countries is projected to grow from 89 kg in 1997/99 to 111 kg in 2030 (near the current level of use in the industrial countries). East Asia would continue to have the highest consumption, reaching 266 kg per ha, while sub-Saharan Africa would still have under 10 kg/ha in 2030, well below what would be required to eliminate nutrient mining and deterioration of soil fertility in many areas.

Significant changes are expected to occur in the mechanization of agriculture which will change the role played by the different sources of power in land tilling and preparation: human labour, draught animals and machines. There is currently wide diversity among countries and regions as to the role of these three power sources. Human labour predominates in sub-Saharan Africa (some two-thirds of the land area is cultivated by hand) and is significant also in Asia, both South (30 percent) and East (40 percent – the estimate for East Asia excludes China). In contrast, it is mostly tractor power in Latin America and the Near East/North Africa. Draught animals account for shares in total power supply comparable to those of human labour in all regions except sub-Saharan Africa, where their use is much less common.

The future is likely to see further shifts towards the use of mechanical power substituting for both human labour and draught animals. The driving forces for such changes are part of the development process (e.g. urbanization and opening of alternative employment opportunities) but also reflect more specific factors pertaining to agriculture and particular socio-economic contexts. These include changes in cultivation methods (spread of no-till/conservation agriculture, change from transplanting to broadcast rice seeding, etc.), in cropping patterns and in some factors affecting the rural workforce such as the impact of HIV/AIDS (an important factor in several countries of sub-Saharan Africa). Only in sub-Saharan Africa will human labour remain the predominant source of power. This is also the only region where draught animals are likely to increase their share, with tractors cultivating no more than about a quarter of the total crop area even in 2030. This compares with shares of 50-75 percent that will have been reached in the other regions.

1.2.5 Agricultural research and biotechnology in the future

The spread of science-based agriculture emanating from the significant past investments in agricultural research underpinned much of the growth of agriculture in the historical period. The need for further increases in production in the future while conserving the resource base of agriculture and minimizing adverse effects on the wider environment, calls for ever greater contributions from agricultural research. The research agenda for the future will be more comprehensive and complex than in the past because the resource base of agriculture and the wider environment are so much more stretched today compared with the past. Research must increasingly integrate current advances in the molecular sciences, in biotechnology and in plant and pest ecology with a more fundamental understanding of plant and animal production in the context of optimizing soil, water and nutrient use efficiencies and synergies. Effective exploitation of advances in information and communication technology will be necessary not only to facilitate interactions across this broad spectrum of scientific disciplines but also to document and integrate traditional wisdom and knowledge in the planning of the research agenda and to disseminate the research results more widely.

Much of the additional production must originate in the developing countries and at least part of it must originate in the agriculture practised by the poor in ways that will also contribute to raising their incomes and food demand as well as those of the wider rural economy. It follows that the research effort must be increasingly oriented in three directions, namely, that it must:

- enhance the capability of world agriculture to underpin further substantial growth in production and also improve nutritional attributes of the produce;
- raise the productivity of the poor in the agro-ecological and socio-economic environments where they practise agriculture and earn a living; and
- maintain the productive capacity of resources while minimizing adverse effects on the wider environment.

These considerations, particularly the last two, suggest a growing role for public sector research. Yet support for public sector research (both national and international) has declined significantly over the past decade at a time when private sector investment in biotechnology research has been growing fast. By now the private sector has built up expertise, technologies and products that are considered essential to the development and growth of tropical agriculture based on rapidly advancing biotechnologies and genetically engineered products. It follows that potential synergies between the private sector and public research offer significant scope for directing more of the research effort in the above-mentioned directions. In particular, such synergies can lead research to underpin a new technology revolution with greater focus on the poor by putting special emphasis on those crop varieties and livestock breeds that were largely ignored throughout the green revolution, but that are specifically adapted to local ecosystems. These include crops such as cassava and the minor root crops, bananas, groundnuts, millets, some oilcrops, sorghum and sweet potato. Indigenous breeds of cattle, sheep, goats, pigs and poultry and locally adapted fish species must also receive much greater priority. A particular focus in the new research agenda should be on plant tolerance to drought, salinity and low soil fertility since nearly half of the world's poor live in dryland regions with fragile soils and irregular rainfall.

The experience to date suggests that biotechnology, if well managed, can be a major contributor to all three objectives. Modern biotechnology is not limited to the much publicized (and often controversial) activity of producing genetically modified organisms (GMOs) by genetic engineering, but encompasses activities such as tissue culture, marker-assisted selection (potentially extremely important for improving the efficiency of traditional breeding) and the more general area of genomics.

The GM crops most diffused commercially at the moment incorporate traits for herbicide tolerance (Ht) or insect resistance (Bt, from *Bacillus thuringiensis*), while maize and cotton combining both Ht and Bt traits (stacked traits) have also been released recently. Ht soybeans dominate the picture (they account for 63 percent of total area under GM crops and for 46 percent of the global area under soybeans), followed by maize, cotton and canola. Diffusion has been very fast although concentrated in a limited number of countries. The United States accounts for over two-thirds of the 53 million ha under GM crops globally. These first-generation GM crops have not been bred for raising yield potential, and any gains in yields and production have come primarily from reduced losses to pests. They have proved attractive to farmers in land-intensive and labour-scarce environments (primarily the developed countries) or those with a high incidence of pests because they save on inputs, including labour for pest control.

The interest for the developing countries even of these first-generation GM technologies lies in the fact that they embody the required expertise for pest control directly into the seeds. This is particularly important for environments where sophisticated production techniques are difficult to implement, are simply uneconomic or where farmers do not command the management skills to apply inputs at the right time, sequence and amount. The spectacular success of Bt cotton in China (higher yields, lower pesticide use and overall production costs, greatly reduced cases of human poisoning) may bear witness to the usefulness of these GM traits for the agriculture of the developing countries.

The potential benefits can be enhanced by further innovations currently in the pipeline in the general area of pest control. These include herbicide tolerance and insect resistance traits for other crops such as sugar beet, rice, potatoes and wheat; virus-resistant varieties for fruit, vegetables and wheat; and fungus resistance for fruit, vegetables, potatoes and wheat. Even more interesting could be the eventual success of current efforts to introduce into crops new traits aimed at enhancing tolerance to abiotic stresses such as drought, salinity, soil acidity or extreme temperatures. Attempting to raise productivity in such situations using GM varieties can be the cheaper, and perhaps the only feasible, option given the difficulties of pursuing the same objective with packages of interventions based on existing technology.

The transition from the first to the second generation of GM crops is expected to shift the focus towards development of traits for higher and better quality output. Many of these new traits have already been developed but have not yet been released to the market. They include a great variety of different crops, notably soybeans with higher and better protein content or crops with modified oils, fats and starches to improve processing and digestibility, such as high stearate canola, low phytate or low phytic acid maize, beta carotene-enriched rice ("golden rice"), or cotton with built-in colours. First efforts have been made to develop crops that allow the production of nutraceuticals or "functional foods", medicines or food supplements directly within the plants. As these applications can provide immunity to disease or improve the health characteristics of traditional foods, they could become of critical importance for an improved nutritional status of the poor.

However, not all is bright with the potential offered by biotechnology for the future of agriculture in its main dimensions (enhancing production, being pro-poor, conserving resources and minimizing adverse effects on the environment). With the present state of knowledge, there persist significant uncertainties about the longer-term impacts and possible risks, primarily for human health (e.g. toxicity, allergenicity) or for the environment, e.g. fears of transmission of pest resistance to weeds, buildup of resistance of pests to the Bt toxins or the toxic effect of the latter on beneficial predators (see Chapter 11 for details). There is, therefore, a well-founded *prima facie* case for being prudent and cautious in the promotion and diffusion of these technologies.

However, the degree of caution any society will have about these products depends on societal preferences about their perceived risks and benefits.

Certain segments of high-income societies with abundant food supplies (and occasional problems of unwanted surpluses) are unwilling to take any risks in order to have more and cheaper food, particularly when it comes to staples such as grains, roots and tubers. In contrast, poor societies with high levels of food insecurity can be expected to attach more weight to potential benefits and less to perceived risks. Obviously, the solution cannot be to let each society make its own choices, because the potential risks affect the global commons of environment, biodiversity and human health. All humanity has a stake in the relevant developments. Moreover, not all stakeholders are equally well informed about the pros and cons of the new technologies. In addition, the absence of a widely shared consensus risks segmenting the world food economy. It may raise obstacles to trade in products that may be acceptable to some countries and not to others.

Finding a solution calls for wide-ranging, well-informed, transparent and fully participatory debate. This is not an easy proposition, because the wealthiest and most risk-adverse societies (or significant segments of such societies) have a disproportionate command over scientific knowledge (including proprietary rights to the technologies) and over the media that can decisively influence the debate. If these distortions remain uncorrected, one cannot feel confident that the debate will lead to optimal results from the standpoint of world welfare. Hence the need for a stronger role of the public sector research system, particularly of the international one, in the generation and diffusion of technologies and related knowledge about the pros and cons.

Besides these possible retarding factors relating to the need for prudence and caution in the face of scientific uncertainties, there are other factors from the socio-economic and institutional spheres, often interacting with the former, that may act in the same direction. The principal among them have to do with the growing control by a small number of large firms of the availability and cost of inputs and technologies farmers will be using and of the use of scientific discoveries for the further development of technology. A whole array of issues pertaining to the establishment and enforcement of intellectual property rights (IPRs) and genetic use restriction technologies (GURTs) are relevant here (see Chapter 11).

1.2.6 Growth in livestock production

Livestock production accounts globally for 40 percent of the gross value of agricultural production (and for more than half in developed countries). Developing countries will continue to increase their part in world production, with their share in meat going up to two-thirds of the world total by 2030 (from 53 percent in 1997/99) and in milk to 55 percent (from 39 percent). The past trend for the livestock sector to grow at a high rate (and faster than the crop sector) will continue, albeit in attenuated form, as some of the forces that made for rapid growth in the past (such as China's rapid growth) will weaken considerably (see discussion above).

The contribution of the increase in the number of animals will remain an important source of growth, but less so than in the past. In the meat sector, higher carcass weights will play a more important role in beef production and higher offtake rates (shorter production cycles) in pig and poultry meat production. The differences in yields (meat or milk output per animal) between developing and industrial countries are, and will likely remain, significant for bovine meat and milk, but much less so for pork and poultry, reflecting the greater ease of transfer and adoption of production techniques. However, factors making for a widening technology gap are also present and may become more important in the future. Among them, the most important is the development and progressive application of biotechnological innovations in the developed countries. The developing countries are not well suited to benefit; in the first place because they often lack the essential human, institutional and other infrastructure, and second because large private companies do not produce such innovations for small farmers in tropical countries.

The broad trends that are shaping the production side of the livestock sector evolution may be summarized as follows:

- an increasing importance of monogastric livestock species compared with ruminants, together with a shift towards increased use of cereal-based concentrate feeds;
- a change, at varying rates according to the region, from multiple production objectives to more specialized intensive meat, milk and egg production within an integrated global food and feed market. The trend for industrial livestock production to grow faster than that from mixed farming systems and, even more, from grazing will continue;
- the spread of large-scale, industrial production with high livestock densities near human population centres brings with it environmental and public health risks, as well as livestock disease hazards. The latter will be enhanced as increasing proportions of world livestock production will be originating in warm, humid and more disease-prone environments;
- animal health and food safety issues associated with these developments are further intensified by the growing role of trade in both live animals and products as well as in feeds;
- increasing pressure on, and competition for, common property resources, such as grazing and water resources, greater stresses on fragile extensive pastoral areas and more pressure on land in areas with very high population densities and close to urban centres.

Policies to address all these issues are required if society is to benefit from the high-growth subsector of agriculture. For the developing countries, besides improved nutrition from higher consumption of animal proteins, the potential exists for livestock sector development to contribute to rural poverty alleviation. It is estimated that livestock ownership currently supports and sustains the livelihoods of 675 million rural poor. Benefits to the rural poor from the sector's growth potential are not, however, assured without policies to support their participation in the growth process. If anything, experience shows that the main beneficiaries of the sector tend to be a relatively small number of large producers in high-potential areas with good access to markets, processors and traders, and middle-class urban consumers. Desirable policy responses are discussed in Chapter 5, as are those relating to the all-important area of health (both animal and human) and food safety.

1.2.7 Probable developments in forestry

During the last two decades, pressures on diminishing forest resources have continued to grow. Traditionally, forestry outcomes were determined by demands for wood and non-wood forest products and the use of forest land for expansion of agriculture and settlements. However, in more recent years there has been a growing recognition of the importance of forestry in providing environmental goods and services such as protection of watersheds, conservation of biodiversity, recreation and its contribution to mitigating climate change. This trend is expected to persist and strengthen during the next 30 years, and increasingly the provision of public goods will gain primacy. In parallel, the expansion of forest plantations and technological progress will greatly help to match demand and supply for wood and wood products and contribute to containing pressures on the natural forest.

FAO's Forest Resources Assessment 2000 estimates the global forest area in 2000 at about 3 870 million ha or about 30 percent of the land area. Tropical and subtropical forests comprise 48 percent of the world's forests, with the balance being in the temperate and boreal categories. Natural forests are estimated to constitute about 95 percent of global forests and plantation forests 5 percent. Developing countries account for 123 million ha (55 percent) of the world's forests, 1 850 million ha of which are in tropical developing countries. During the 1990s, the net annual decline of the forest area worldwide was about 9.4 million ha, the sum of an annual forest clearance estimated at 14.6 million ha (a slight decline from that of the 1980s) and an annual forest area increase of 5.2 million ha. Nearly all forest loss is occurring in the tropics. Population growth coupled with agricultural expansion (especially in Africa and Asia) and agricultural development programmes (in Latin America and Asia) are major causes of forest cover changes.

In most developed countries, deforestation has been arrested and there is a net increase in forest cover. Such a situation is seen also in a number of developing countries, largely because of reduced dependence on land following the diversification of the economies. By 2030 most agriculture-related deforestation in Asia and to some extent also in Latin America could have ceased. In parallel, however, economic growth puts additional pressures on the forest by stimulating the demand for forest products, especially sawnwood, panel products and paper. Things may not move in that direction in Africa, where population growth, combined with limited changes in agricultural technologies and the absence of diversification, could result in continued forest clearance for agricultural expansion.

Current efforts towards wider adoption of sustainable forest management are expected to strengthen, although such efforts may not be uniform and are critically dependent on political and institutional changes. Inadequate investment in capacity building and persistent weaknesses of the political and institutional environment may limit the wider adoption of sustainable forest management in several countries, especially in sub-Saharan Africa.

Industrial wood demand is expected to grow, but this is estimated to be lower than in earlier forecasts. As noted, the area in forest plantations has been growing at a fast rate: it grew from 18 million ha in 1980 to 44 million ha in 1990 and to 187 million ha in 2000. Plantations are now a significant source of roundwood supply and they will become more so in the future. They could double output to some 800 million m^3 by 2030, supplying about a third of all industrial roundwood. There will be substantial involvement of the private sector, including small farmers in the production of wood, also coming from "trees outside forests" (trees planted around farms, on boundaries, roadsides and embankments). The probable further growth of legally protected forest areas should be feasible without too much impact on future wood supplies. On the demand side, technological improvements in processing would help to reduce raw material requirements per unit of final product. Overall, the longer-term outlook for the demand-supply balance looks now much less problematic than in the past.

An estimated 55 percent of global wood production is used as fuelwood. Tropical countries account for more than 80 percent of global fuelwood consumption. Wood will continue to be the most important source of energy in the developing countries, especially for the poor in sub-Saharan Africa and most of South Asia. No significant changes in fuelwood consumption are likely over the period to 2015. Wood will remain a readily accessible source of fuel for millions of poor people throughout the world. Forests close to urban centres may continue to be subjected to heavy exploitation to meet the growing demand for charcoal. Notwithstanding localized shortages, demand will more or less be met, and demand for fuelwood may trigger planting in farmlands and other areas not used for agricultural purposes. The shift towards alternative fuels may accelerate beyond 2015, depending on changes in access to such fuels.

Improved access to manufactured products would reduce the dependence on several of the non-wood forest products traditionally used for subsistence consumption. There will however be an increased demand for certain items, especially medicinal and aromatic plants, ethnic foods and industrial products. While unsustainable exploitation from natural forests could result in a substantial decline in the supply of some of the valuable non-wood forest products, this may eventually lead to domestication and commercial cultivation of some of the more valuable ones. Although there will be significant improvements in the processing technologies, the producers of raw material are unlikely to be the main beneficiaries. This largely stems from persistent weakness in the research and development capacity of the producers.

During the next 30 years, one could witness an increase in the non-consumptive uses of forests, especially for protection of biodiversity, conservation of soil and water, mitigation of global climate change and recreational uses. In fact the cultural and recreational uses of forests will gain prominence, although in many developing countries consumptive use of forest products will remain important. Environmental standards in forest resource management will be widely adopted. Although the expansion of protected areas may not be rapid, protection objectives will be growing in importance as determinants of forest land uses. Increasingly there are efforts to enhance the scope for action by the private sector, communities, non-governmental organizations (NGOs) and civil society at large. These latter changes, coupled with technological improvements, will bring about significant qualitative changes in forestry.

1.2.8 Plausible developments in fisheries

Fish remains a preferred food. Average world per capita consumption continued to increase to 16.3 kg in 1999, up from 13.4 kg in 1990. This development was heavily dominated by events in China. Excluding China, the apparent consumption per person in the rest of the world actually declined from 14.4 kg in 1990 to 13.1 kg in 1999. There are very wide intercountry differences. Reported per capita consumption ranges from less than 1 kg in some countries to over 100 kg in others. The global average per capita consumption could grow to 19-20 kg by 2030, raising total food use of fish to 150-160 million tonnes (97 million tonnes in 1999).

Marine capture fisheries have in the 1990s shown annual catches of between 80 and 85 million tonnes. Inland catches increased slowly to 8.3 million tonnes in 1999. While the gross volume of marine catches has fluctuated and does not show any definite trend, the species composition of the catch has changed, with high-value species (bottom-dwelling, or demersal species and large pelagics) gradually being substituted by shorter-lived surface dwelling (pelagic) and schooling fish.

By the end of the 1990s, an estimated 47 percent of major marine fish stocks were fully exploited, about 18 percent overfished, and another 9 percent depleted. Only a quarter of the fish stocks were moderately exploited or underexploited. The long-term yearly sustainable yield of marine capture fisheries is estimated at approximately 100 million tonnes. This assumes more efficient utilization of stocks, healthier ecosystems and better conservation of critical habitats. Increased landings also depend on improved selectivity of fishing gear, leading to less discarding of unwanted fish, and on sustained higher levels of catch from fisheries on restored stocks. However, this increase will be slow in materializing. Moreover, the estimated potential yield of capture fisheries (100 million tonnes) includes large quantities of currently minimally exploited living aquatic resources in the oceans, of which the most well known are krill, mesopelagic fish and oceanic squids.

The bulk of the increase in supply will therefore have to come from aquaculture. During the 1990s, aquaculture has shown a spectacular development (most of it in Asia, with China accounting for 68 percent of world aquaculture production), and by 1999 aquaculture accounted for 26 percent of world fish production (even 34 percent of fish food supplies). This growth could continue for some time, although constraints (lack of feeding stuffs, diseases, lack of suitable inland sites, environmental problems, etc.) are becoming more binding. Unless these constraints are relaxed, the long-term growth prospects of aquaculture, and hence of fish consumption, could be seriously impeded.

On the feed side, fishmeal producers expect that within a decade or so the aquaculture industry will use up to 75 to 80 percent of all fish oil produced, and about half of the available white fishmeal, while the prospects for growth in supplies are not encouraging. Although considerable research efforts have been undertaken, a satisfactory replacement for fish oil in aquaculture feed has still to be found. The proportion of fish reduced to fishmeal and oil (at present some 30 million tonnes) is likely to fall. In spite of an increasing demand for fishmeal following the intensification of livestock production and further expansion of aquaculture, an increasing share of small pelagic species (normally used for reduction) is likely to be used for food. The fishmeal industry will therefore be forced to find other sources, the most likely of which is zooplankton (initially Antarctic krill).

The single most important influence on the future of wild capture fishery is one of governance. Although in theory renewable, wild fishery resources are in practice finite for production purposes. They can only be exploited so much in a given period; if overexploited, production declines and may even collapse. Hence total fishery production from the wild cannot increase indefinitely. Resources must be harvested at sustainable levels, and access must be equitably shared among producers. So far only very few managers have succeeded in creating sustainable fisheries. As fish resources grow increasingly scarce, conflicts over allocation and sharing are becoming more frequent.

The principal policy challenge is to establish rules for access to fish stocks. Fisheries based on explicit and well-defined rights of access will need to become more common; when rights are well defined, understood and observed, allocation conflicts tend to be minimized. These issues are the subject of an increasing body of international and bilateral agreements dealing with legal access rights.

Another major policy challenge is to bring the global fishing fleet capacity back to a level at which global fish stocks can be harvested sustainably and economically. Past policies have promoted the buildup of excess capacity in the fishing fleet and incited fish farmers to increase the catch beyond sustainable levels. Policies have to react fast to unwind the overcapacity that has been built up over the past to ensure a return to a steady-state fish stock. This process is already under way and has, among other things, led to a contraction of the capture fisheries labour force in developed countries.

A third important policy consideration is the increasing pressure from various quarters in society to reduce environmental damage associated with both capture and culture fisheries. This has also led to a set of national and international rules and laws, some voluntary in nature, regulating fishery methods.

A major factor, affecting both the sustainability of capture fisheries and the expansion of aquaculture, will be the expected improvements in the understanding of the marine ecosystem. Improved knowledge will encompass the working of ecosystems, including the dynamics of fish stocks, and the effects of human intervention on such ecosystems. This knowledge would increase fisheries productivity, facilitate improved fisheries management and enable monitoring of fisheries operations, to ensure compliance with rules and regulations and to assess their impact on the environment.

1.2.9 Environmental aspects of natural resource use in agriculture

The quantum gains in agricultural production and productivity achieved in the past were accompanied by adverse effects on the resource base of agriculture that put in jeopardy its productive potential for the future. Among these effects are, for example, land degradation; salinization of irrigated areas; overextraction of underground water; growing susceptibility to disease and buildup of pest resistance favoured by the spread of monocultures and the use of pesticides; erosion of the genetic resource base when modern varieties displace traditional ones and the knowledge that goes with their use. Agriculture also generated adverse effects on the wider environment, e.g. deforestation, loss or disturbance of habitat and biodiversity, emissions of greenhouse gases (GHGs) and ammonia, leaching of nitrates into waterbodies (pollution, eutrophication), off-site deposition of soil erosion sediment and enhanced risks of flooding following conversions of wetlands to cropping.

The production increases in prospect at the world level for the period to 2030 (in terms of the absolute quantities involved) will be of an order of magnitude similar to those that took place in the comparable historical period (e.g. for cereals and sugar) or even higher (e.g. for meat and vegetable oils). Thus, almost another billion tonnes of cereals must be produced annually by 2030, another 160 million tonnes of meat, and so on. It follows that pressures on the resources and the environment will continue to mount. The challenge facing humanity is how to produce the quantum increases of food in sustainable ways (preserving the productive potential of the resource) while keeping adverse effects on the wider environment within acceptable limits. A *priori*, the task looks more difficult than in the past because:

■ resources are more stretched today than they were 30 years ago (e.g. water shortages are more severe, GHG concentrations are higher, etc.), hence the risk that further deterioration will act as a break to development is much more present;

■ the growing share of the developing countries in world production means that pressures will be increasingly gathering in the agro-ecological environments of the tropics which are more fragile than the temperate ones and contain much of the world's biodiversity;

■ in most of these countries, conventional objectives of agricultural development (food security, employment and export earnings) can easily take precedence over those of sustainability and environmental conservation, no matter that for them the preservation of the productive potential of their agriculture is much more crucial for their survival than it is for the industrial countries where agriculture is a small part of the economy;

■ the research and technology capabilities for finding solutions to respond to the problems reside predominantly in the industrial countries and less so in the developing ones where they are most needed.

The preceding considerations and the magnitudes involved suggest that the increases in production and associated progress in food security cannot be achieved at zero environmental cost. The issue is whether any threats to the resource base of agriculture and the generation of other environmental "bads" associated with more production and consumption can be contained within limits that do not threaten sustainability, that is the ability of future generations to have acceptable food security levels within acceptable more general living standards, including a clean environment. Often the choices present themselves in the form of what are acceptable trade-offs, rather than whether we can have something for nothing. There are trade-offs over time, among the different dimensions of sustainability and over space (e.g. among countries or regions within countries).

An example of trade-offs over time is the eventuality that deforestation (including more general disruption of wildlife habitats) of the past as well as the one now taking place could be reversed in the future when the pressures for further increases in output will be eased and the advancement of yields could lead to less land being needed for any given level of output. Europe's experience seems to point in this direction.

Trade-offs between different dimensions of sustainability include, for example, the higher herbicide applications that could accompany efforts to increase water use efficiency and/or reduce methane emissions by shifting rice cultivation from flooded and transplant to direct seeding. Similarly, the shift to no-till/conservation agriculture in order to reduce soil erosion risks and increase the carbon sequestration capacity of soils (see below), favours the development of weeds that may provoke increased use of herbicides. In like manner, combating the effects of saltwater intrusions into irrigation aquifers in coastal areas (because of overextraction or sea level rise) may require the acceptance of genetically modified salt-tolerant crops together with some as yet unknown risks. In livestock, the shift from ruminant meat to pork and poultry may slow down the growth of methane emissions from ruminant livestock but will aggravate the problem of livestock effluent pollution from large pig and poultry industrial units. Similarly, grazing restraints on extensive rangelands have their counterpart in the increase of feedlot-raised cattle with, as a result, enhanced point source pollution as well as eventual effects on arable land elsewhere, including in other countries, generated in the process of producing more cereals for the feedlots.

Trade-offs among countries relate primarily to the potential offered by international agricultural trade to spread across the globe the environmental pressures from increased production. The individual countries differ as to their capabilities to withstand adverse effects associated with increasing production because they differ in the relative abundance of their resource endowments; they possess agro-ecological attributes that make their resources more or less resilient; or they are at present more or less stretched from past accumulation of environmental damages. Countries also differ as to their technological and policy prowess for finding solutions and responding to emerging problems.

Trade can contribute to minimizing adverse effects on the global system if it spreads pressures in accordance with the capabilities of the different countries to withstand and/or respond to them. Whether it will do so depends largely on how well prices in each country reflect its "environmental comparative advantage". This requires that, in addition to absence of policy distortions that affect trade, the environmental "bads" generated by production be embodied in the costs and prices of the traded products, which is not normally the case when externalities are involved. If all countries meet these conditions, then trade will contribute to minimizing the environmental "bads" globally as they are perceived and valued by the different societies, although not necessarily in terms of some objective physical measure. This latter qualification is important because different countries can and do attach widely differing values to the environmental resources relative to the values of other things, such as export earnings and employment. In the end, the values of environmental resources relative to those of other things are anthropocentric concepts and countries at different levels of development and with differing resource endowments are bound to have differing priorities and relative valuations.

An additional mechanism through which trade can contribute to spread pressures across the globe in ways that minimize adverse effects on sustainability has to do with the enhanced degree of product substitutability afforded by trade, e.g. substituting imported palm oil for locally produced oil from sunflower seed or soybeans. Producing a tonne of palm oil in East Asia requires only a fraction of the agrochemicals (fertilizer and pesticides) used to produce a tonne of oil from sunflower seed or soybeans in Europe, for example.

In considering the prospects for the future, we must be aware that history need not repeat itself as concerns the extent to which the continued growth of production will be associated with adverse effects. The projections to 2030 suggest that some effects will be different from those of the past simply because their determinants will be changing, e.g. the lower growth of rural population will tend to slow down the rate of deforestation. Others will be different because of policy changes; for example, the EU, having paid the environmental cost of transforming itself from a large net importer of sugar to a large net exporter, now has policies that do not favour further expansion of exports. Finally, history need not repeat itself because of the wider adoption of more environmentally benign technologies and of policies that favour them and/or remove incentives for unsustainable practices, e.g. for expansion of ranching into forested areas. Approaches to agricultural production that offer scope for minimizing adverse impacts include those described in the following paragraphs.

Integrated pest management (IPM) promotes biological, cultural, physical and, only when essential, chemical pest management techniques. Naturally occurring biological control is encouraged, for example through the use of alternate plant species or varieties that resist pests, as is the adoption of land management, fertilization and irrigation practices that reduce pest problems. If pesticides are to be used, those with the lowest toxicity to humans and non-target organisms should be the primary option. Precise timing and application of any pesticides used are essential. Broad-spectrum pesticides are only used as a last resort when careful monitoring indicates they are needed according to pre-established guidelines. Given that chemical pesticides will continue to be used, the need for rigorous testing procedures before they are released on to the market (as well as sharing of the relevant information among countries) cannot be overemphasized. The same applies for the need to have comprehensive and precise monitoring systems to give early warning of residue buildup along the food chain, in soils and in water.

Integrated plant nutrient systems (IPNS). The depletion of nutrient stocks in the soil (nutrient mining), which is occurring in many developing countries, is a major but often hidden form of land degradation, making agricultural production unsustainable. In parallel, overuse or inappropriate use of fertilizers creates problems of pollution. IPNS hold promise of mitigating such adverse effects by recycling all plant nutrient sources within the farm and use of nitrogen fixation by legumes to the extent possible, complemented by the use of external plant nutrient input, including manufactured fertilizers.

No-till/conservation agriculture (NT/CA) involves planting and maintaining plants through a permanent cover of live or dead plant material, without ploughing. Different crops are planted over several seasons, to avoid the buildup of pests and diseases and to optimize use of nutrients. NT/CA has positive effects on the physical, chemical and biological status of soils. Benefits include reduced soil erosion, reduced loss of plant nutrients, increase in organic matter levels of soils, higher rainfall infiltration and soil moisture holding capacity and, of course, savings in fossil fuel used in soil preparation.

Organic agriculture. Although the growing demand for organic foods is to a large extent driven by health and food quality concerns, organic agriculture is above all a set of practices intended to make food production and processing respectful of the environment. It is not a product claim that organic food is healthier or safer than that produced by conventional agriculture. Organic agriculture is essentially a production management system aiming at the promotion and enhancement of ecosystem health, including biological cycles and soil biological activity. It is based on minimizing the use of external inputs, representing a deliberate attempt to make the best use of local natural resources. Synthetic pesticides, mineral fertilizers, synthetic preservatives, pharmaceuticals, GMOs, sewage sludge and irradiation are prohibited in all organic standards. Naturally, this by itself does not guarantee the absence of resource and environmental problems characteristic of conventional agriculture. Soil mining and erosion, for example, can also be problems in organic agriculture.

In conclusion, future agro-environment impacts will be shaped primarily by two countervailing forces: mounting pressures because of the continuing increase in demand for food and agricultural products mainly on account of population and income growth; and decreasing pressures via technological change, institutional and policy responses to environmental degradation caused by agriculture, and structural change in the sector. On balance, the potential exists for putting agriculture on a more sustainable pathway than a continuation of past trends would indicate. The main requirement is increasingly to decouple agricultural intensification from environmental degradation through the greater exploitation of biological and ecological approaches to nutrient recycling, pest management and soil erosion control.

1.2.10 Climate change and agriculture

Climate change: agriculture's role in climate change. Agricultural activities contribute to climate change through the emission of GHGs. They also contribute to climate change mitigation through carbon sequestration in cropland and the provision of biofuels that can substitute for fossil fuels. Globally, they generate some 30 percent of total anthropogenic emissions of GHGs. The main ones are carbon dioxide (CO_2 – agriculture accounts for about 15 percent of total anthropogenic emissions); methane (CH_4 – about 50 percent coming from agriculture); and nitrous oxide (N_2O – agriculture accounting for about two-thirds). Both CH_4 and N_2O are gases with warming potentials many times higher than that of CO_2. The main source of CO_2 emissions is tropical forest clearance, related biomass burning and land use change. For methane the chief sources are rice production and livestock through enteric fermentation of ruminants and animal excreta. For nitrous oxide it is mineral fertilizer and animal wastes – manure and deposition by grazing animals.

Future emissions from agriculture will be determined by the evolution of the main variables (land use, fertilizers and numbers of animals) and their qualitative changes, as presented in the preceding sections and discussed in more detail in individual chapters. Thus, carbon dioxide emissions could be stable or even less than in the past because of slower deforestation and land cover change. Likewise, methane emissions from rice cultivation will slow down or possibly even decrease (through changes in paddy field flooding and rice varieties), but annual methane emissions from livestock could increase by 60 percent by 2030. Also, growth of nitrous oxide emissions from fertilizers will be slow (slow growth in consumption combined with a more efficient use of fertilizers) but, as for methane, emissions from livestock (animal waste) could increase by some 50 percent by 2030.

Agricultural land sequesters carbon in the form of vegetation and soil organic matter (SOM) derived from crop residues and manure. The potential exists for it to sequester much more carbon than it actually does under most current cropping practices. This potential is assuming growing economic, political and environmental importance in the context of the Kyoto Protocol. The crop production projections of this study imply an increase in total biomass production and unharvested residues, hence an increase in gross carbon sequestration. Cropland can be made to increase carbon sequestration if managed for this purpose. The above-mentioned no-till/conservation agriculture is a major approach to increasing carbon in the soil, and it also contributes to lower GHG emissions through the reduced use of fuel. Better residue management and changes in cropping patterns can also contribute to carbon sequestration. Even greater gains per hectare could be achieved where marginal cultivated land is taken out of crop production and replaced by grass or legume forages. Permanent set-aside of agricultural land would sequester large amounts of carbon if forested or left to revert to tree scrub. Finally, degraded land that has gone out of production, e.g. saline soils, could be restored to sequester carbon.

In conclusion, managing agricultural land to sequester more carbon could transform it from a net source to a net sink. The rate of sequestration gradually declines before reaching a limit and can be especially high during the first few years. Some of the carbon sequestration will not be permanent and eventually the gains will level off. This notwithstanding, managing agricultural land for carbon sequestration will extend the time available to introduce other measures with longer-term benefits.

Climate change: impacts on agriculture. The main parameters of climate change with potential effects on agriculture and food security are as follows:

- Global average temperatures are projected to rise by about 1°C by 2030. Higher latitudes will warm more quickly than lower ones, land areas will warm more quickly than the oceans, and polar sea ice will decrease more in the Arctic than in the Antarctic. Consequently, average temperatures in the higher latitudes may rise by 2°C.
- Broadly speaking, climate change is projected to increase global mean precipitation and runoff by about 1.5 to 3 percent by 2030. There will be greater increases in the higher latitudes and the equatorial region but potentially serious reductions in the middle latitudes. Parts of Central America, South Asia, northern and southern Africa and Europe may suffer appreciable falls in available water resources.
- Sea levels are rising at about half a centimetre per year, and are likely to continue at this rate for

several decades even if there is rapid implementation of international agreements to limit climate change. Thus sea levels could be about 15 cm higher by 2030.
- There is likely to be an increase in the frequency and intensity of relatively localized extreme events including those associated with El Niño, notably droughts, floods, tropical cyclones and hailstorms.

Recent research has suggested that some impacts of climate change are occurring more rapidly than previously anticipated. For example, average sea temperatures in northern latitudes are rising rapidly (in particular in the North Sea), ocean currents are being disrupted and phytoplankton populations altered.

The main effects of these changes on food and agriculture are foreseen to be the following:
- Temperature and precipitation changes will affect the extents of land that is suitable for growing crops. Suitable areas will increase in higher latitudes because of milder and shorter winters, and will decline in arid and semi-arid areas.
- The effects on yields follow the same pattern as land suitability, with gains in middle to higher latitudes and yield losses in lower latitudes, but with some gains in tropical highlands where at present there are cold temperature constraints.
- Larger changes are predicted in the availability of water from rivers and aquifers because of reductions in runoff and groundwater recharge. Substantial decreases are projected for Australia, India, southern Africa, the Near East/North Africa, much of Latin America and parts of Europe. Thus, irrigation capacity will be negatively affected precisely in the countries where climate change will enhance dependence on irrigation, e.g. because of more frequent and more severe droughts. The main decrease in water availability will be after 2030 but there could be negative effects on irrigation in the shorter term.
- On the positive side, the rise in atmospheric concentrations of carbon dioxide (CO_2), stimulates photosynthesis (the so-called CO_2 fertilization effect) and improves water use efficiency. Up to 2030 this effect could compensate for much or all of the yield reduction coming from temperature and rainfall changes.

■ Effects on livestock will be partly through deterioration of some grasslands in developing countries (mostly after 2030). Of more significance to livestock production is the rise in temperature over the period to 2030, and the CO_2 fertilization effect. These will favour more temperate areas through reduced need for winter housing and for feed concentrates (because of better pasture growth). Many developing countries, in contrast, are likely to suffer production losses through greater heat stress to livestock and lower fodder and forage yields, but this may be compensated by the CO_2 fertilization effect.

■ The substantial rise foreseen for average sea temperatures may have serious effects on fisheries. It could disrupt breeding patterns, reduce surface plankton growth or change its distribution, thereby lowering the food supply for fish, and cause the migration of mid-latitude species to northern waters. The net effect may not be serious at the global level but could severely disrupt national and regional fishing industries and food supplies.

■ Impacts from sea level rise will take the form of progressive inundation of low-lying coastal areas and saltwater intrusion. Increased frequency and intensity of tropical cyclones will also increase the frequency of extreme high-water events as more severe storm surges penetrate further inland.

■ There could be substantial changes in the distribution of major pests even under the small changes in average temperature likely to occur by 2030. Fewer cold waves and frost days could extend the range of some pests and disease vectors and favour the more rapid buildup of their populations to damaging levels. The importance lies both in the larger pest populations *per se* that may arise because of higher temperatures, and in their role as carriers of plant viruses, as in the case of aphids carrying cereal viruses, which are currently held in check by low winter or night temperatures.

■ Finally, greater temperature extremes seem likely to give rise to higher wind speeds, and there may be increases in the occurrence of hurricanes. So, in addition to impacts on plant growth, there may be greater mechanical damage to soil, plants and animals, for example from greater wind erosion, sand blast damage to crops and drowning of livestock.

The main impacts of climate change on global food production capacity are not projected to occur until after 2030, but thereafter they could become increasingly serious. Up to 2030 the impact may be broadly positive or neutral at the global level. However the regional impacts will be very uneven. Food production in higher latitudes will generally benefit from climate change, whereas it may suffer in large areas of the tropics. Over the period to 2030, the most serious and widespread agricultural and food security problems related to climate change are likely to arise from the impact on climate variability, and not from progressive climate change, although the latter will be important where it compounds existing agroclimatic constraints.

Naturally, the prospect that global production potential may not be much affected by climate change, or may even rise, is poor consolation to poor and food-insecure populations with high dependence on local agriculture and whose agricultural resources are negatively affected. Even small declines in the quantity and quality of their agricultural resources can have serious negative impacts on livelihoods and food security. Among the obvious responses to these emerging risks are accelerated economic growth and diversification of economies away from heavy dependence on vulnerable agricultural resources, and promotion of technologies and farming practices to facilitate adaptation to changing agro-ecological conditions.

CHAPTER 2

Prospects for food and nutrition

2.1 The broad picture: Historical developments and present situation

2.1.1 Progress made in raising food consumption per person

Food consumption, in terms of kcal/person/day, is the key variable used for measuring and evaluating the evolution of the world food situation.[1] The world has made significant progress in raising food consumption per person. It increased from an average of 2 360 kcal/person/day in the mid-1960s to 2 800 kcal/person/day currently (Table 2.1). This growth was accompanied by significant structural change. Diets shifted towards more livestock products, vegetable oils, etc. and away from staples such as roots and tubers (Tables 2.7, 2.8). The increase in world average kcal/person/ day would have been even higher but for the declines in the transition economies in the 1990s.

The gains in the world average reflected predominantly those of the developing countries, given that the industrial countries and the transi-

tion economies had fairly high levels of per capita food consumption already in the mid-1960s. This overall progress of the developing countries has been decisively influenced by the significant gains made by the most populous among them. There are currently seven developing countries with a population of over 100 million. Of these, only Bangladesh remains at very low levels of food consumption. China, Indonesia and Brazil have made the transition to fairly high levels (in the range 2 900-3 000 kcal). In more recent years (from the late 1980s), India, Pakistan and Nigeria (but see Box 2.2) also started making progress and have now achieved middling levels of per capita food consumption after decades of near stagnation (Figure 2.1).

An alternative way of looking at changes over the historical period is to observe the distribution of world population living in countries having given levels of kcal/person/day. The relevant data are shown in Table 2.2. In the mid-1960s, 57 percent of the population of the whole world (not only the developing countries), including both China and India, lived in countries with extremely low levels,

---

[1] The more correct term for this variable would be "national average apparent food consumption", since the data come from the national food balance sheets rather than from consumption surveys. The term "food consumption" is used in this sense here and in other chapters.

Table 2.1 Per capita food consumption (kcal/person/day)

	1964/66	1974/76	1984/86	1997/99	2015	2030
World	2 358	2 435	2 655	2 803	2 940	3 050
Developing countries	2 054	2 152	2 450	2 681	2 850	2 980
Sub-Saharan Africa	2 058	2 079	2 057	2 195	2 360	2 540
Near East/North Africa	2 290	2 591	2 953	3 006	3 090	3 170
Latin America and the Caribbean	2 393	2 546	2 689	2 824	2 980	3 140
South Asia	2 017	1 986	2 205	2 403	2 700	2 900
East Asia	1 957	2 105	2 559	2 921	3 060	3 190
Industrial countries	2 947	3 065	3 206	3 380	3 440	3 500
Transition countries	3 222	3 385	3 379	2 906	3 060	3 180
Memo items						
1. World, excl. transition countries	2 261	2 341	2 589	2 795	2 930	3 050
2. Developing countries, excl. China	2 104	2 197	2 381	2 549	2 740	2 900
3. East Asia, excl. China	1 988	2 222	2 431	2 685	2 830	2 980
4. Sub-Saharan Africa, excl. Nigeria	2 037	2 076	2 057	2 052	2 230	2 420

under 2 200 kcal, the great bulk of them being in countries with under 2 000 kcal. At the other extreme, 30 percent of the world population (overwhelmingly in the developed countries) lived in countries with over 2 700 kcal, two-thirds of these in countries with over 3 000 kcal.

It was a world of very pronounced inequality, with at the bottom masses of poor, a very thin middle class and, at the other end, a sizeable group of well-to-do population. By the late 1990s, the situation had changed radically. Only 10 percent of a much larger global population now lives in countries with food consumption below 2 200 kcal, while those in countries with over 2 700 kcal now account for 61 percent of world population. The gains made by some of the very populous developing countries (such as China, Brazil and Indonesia, see Figure 2.1) were largely responsible for this massive upgrading of the world population towards improved levels of per capita food consumption.

Figure 2.1 Per capita food consumption, developing countries with over 100 million population in 1997/99

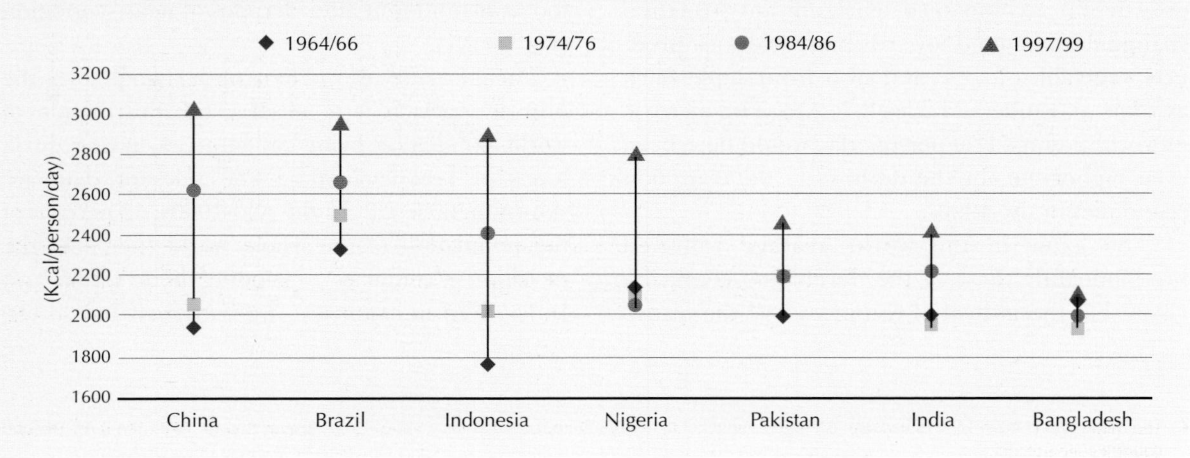

Table 2.2 Population living in countries with given per capita food consumption

Kcal/person/day	1964/66	1974/76	1984/86	1997/99	2015	2030
			Population (million)			
Under 2 200	1 893[a]	2 281[a]	558	571	462	196
2 200-2 500	288	307	1 290[b]	1 487[b]	541	837
2 500-2 700	154	141	1 337[c]	222	351	352
2 700-3 000	302	256	306	1 134	2 397[b]	2 451[b]
Over 3 000	688	1 069	1 318	2 464[c]	3 425[c]	4 392[c]
World total	3 325	4 053	4 810	5 878	7 176	8 229

[a] Includes India and China. [b] Includes India. [c] Includes China.

2.1.2 Failures

A significant number of countries failed to participate in this general thrust towards increasing average food consumption levels. There are currently 30 developing countries where food consumption is under 2 200 kcal/person/day. Figure 2.2 summarizes their historical experience: present (average 1997/99) levels are compared with the highest and lowest ones recorded in any five-year average (in order to smooth out distortions from yearly fluctuations) in the period 1961-1999. The following comments may be made about these 30 countries:

■ Several among them (e.g. the Democratic People's Republic of Korea, the Central African

Republic, Madagascar, Liberia, Malawi and Uganda) had achieved middling levels (over 2 400 kcal) in at least one five-year average in the past. They are now in the under-2 200 kcal class because they suffered declines, some particularly deep ones, in the case of Liberia and the Democratic People's Republic of Korea.

■ For most other countries in Figure 2.2, the highest level ever achieved was totally inadequate to start with, yet they suffered further declines, some very sharp ones, as was the case in Somalia, Burundi, Haiti and Ethiopia/ Eritrea.[2]

■ Finally, a few countries did not suffer declines but have always had very low per capita food

Figure 2.2 Developing countries with under 2 200 kcal in 1997/99. Highest and lowest five-year average kcal recorded during 1961-1999

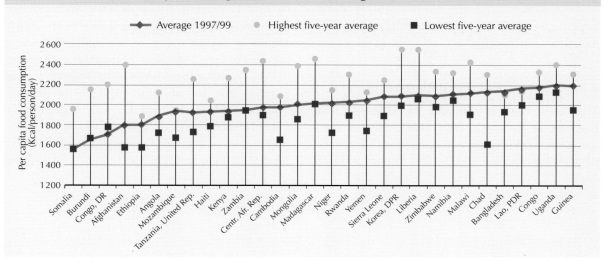

[2] The data used in Figure 2.2 refer to the aggregate Ethiopia and Eritrea, because there are no data for making historical comparisons for the two countries separately.

consumption. That is, they have never had levels that were significantly above the very low ones they have currently. Here belong Bangladesh, Mozambique and the Lao People's Democratic Republic.

The historical evidence from these countries, particularly those that suffered severe declines from better nutritional levels in the past, is a crucial input into the analysis of the evolution of world food insecurity. War or otherwise unsettled political conditions are common characteristics in several of these countries.

Looking at the regional picture, sub-Saharan Africa, excluding Nigeria, stands out as the only region that failed to make any progress in raising per capita food consumption (Table 2.1). Not all countries of the region are in this dire food security situation. Besides Nigeria (but see Box 2.2), a number of other countries made significant progress to over 2 400 kcal/person/ day (Mauritius, Mauritania, the Gambia, Ghana, Gabon, Benin and Togo) but their weight in the regional total is too small to have much effect on the total. The regional aggregate picture is dominated by the failures suffered by the larger countries. Of the 12 countries with a population of over 15 million, most have a per capita food consumption (latest five-year average 1995/99) that is lower than attained in the past – some of them much lower, e.g. the Democratic Republic of the Congo, Madagascar, Côte d'Ivoire, Kenya and the United Republic of Tanzania. Only Nigeria, Ghana and the Sudan among these larger countries have higher levels now than any past five-year average.

2.1.3 The incidence of undernourishment

The 2001 FAO assessment, *The State of Food Insecurity in the World 2001* (FAO, 2001a), estimates the total incidence of undernourishment in the developing countries at 776 million persons in 1997/99 (17 percent of their population, Table 2.3),[3] when average food consumption reached 2 680 kcal/person/day. The number of undernourished in the developing countries is estimated at 815 million (20 percent of the population) for the three-year average 1990/92. This was the base year used by the 1996 WFS in setting the target of halving the numbers undernourished in the developing countries by 2015 at the latest.

Obviously, the decline between 1990/92 and 1997/99 has been much less than required for attaining the target (see further discussion in Box 2.5). In practice, the entire decline has come from East Asia, which is well on its way to halving undernourishment by the year 2015. In contrast, the two regions with the highest incidence in relative terms (percentage of population), sub-Saharan Africa and South Asia, both registered increases in the absolute numbers affected. If these trends continue, the halving target will certainly not be achieved and whatever other reductions take place will further accentuate the differences among regions and countries.

Changes in the incidence of undernourishment are close correlates of changes in food consumption levels (kcal/person/day), as explained in Box 2.1. The historical data in Table 2.1 show that food consumption levels have improved greatly for most regions over the last three decades. It can be deduced that such improvement must have been accompanied by a lowering of the incidence of undernourishment to the current 17 percent. By implication, the incidence of undernourishment must have been much higher in the past, e.g. in the mid-1960s when there were only 2 055 kcal/person/day on average in the developing countries. However, it is unlikely that the absolute numbers of persons undernourished declined by much, given that over the same period (1964/66 to 1997/99) the population of the developing countries doubled from 2.3 billion to 4.6 billion.

---

3 The term "undernourishment" is used to refer to the status of persons whose food intake does not provide enough calories to meet their basic energy requirements. The term "undernutrition" denotes the status of persons whose anthropometric measurements indicate the outcome not only, or not necessarily, of inadequate food intake but also of poor health and sanitation – conditions that may prevent them from deriving full nutritional benefit from what they eat (FAO, 1999a, p.6).

Table 2.3 Incidence of undernourishment, developing countries

	1990/92[1]	1997/99	2015	2030	1990/92[1]	1997/99	2015	2030
	Percentage of population				Million persons			
Developing countries	20	17	11	6	815	776	610	443
Sub-Saharan Africa	35	34	23	15	168	194	205	183
excl. Nigeria	*40*	*40*	*28*	*18*	*156*	*186*	*197*	*178*
Near East/North Africa	8	9	7	5	25	32	37	34
Latin America and Caribbean	13	11	6	4	59	54	40	25
South Asia	26	24	12	6	289	303	195	119
East Asia	16	11	6	4	275	193	135	82

					Undernourishment			
Alternative country groups	**Population (million)**			**Kcal/person/day**			**Percentage of population**			**Million persons**		
	1997/99	2015	2030	1997/99	2015	2030	1997/99	2015	2030	1997/99	2015	2030
I. Countries with kcal in 2015												
Under 2 200 kcal (15 countries)	289	462	671	1 855	2 055	2 260	51	35	23	147	164	152
2 200-2 500 kcal (26 countries)	358	517	676	2 144	2 340	2 525	33	22	13	119	111	89
2 500-2 700 kcal (12 countries)	257	336	395	2 380	2 580	2 780	21	12	6	54	41	25
2 700-3 000 kcal (23 countries)	1 678	2 171	2 561	2 545	2 800	3000	18	9	4	302	190	109
Over 3 000 kcal (21 countries)	1 972	2 317	2 537	3 054	3 200	3 310	8	5	3	154	105	68
Total	4 555	5 804	6 840	2 681	2 850	2 980	17	11	6	776	610	443
II. Countries with percentage undernourishment[2]												
Under 5 percent	349	1158	5129	3 187	3 130	3 150	2	3	3	8	37	178
5-10 percent	1 989	2 162	524	2 999	3 066	2 758	8	6	7	167	134	38
10-25 percent	1 632	1 939	948	2 434	2 644	2 411	21	13	16	349	250	155
Over 25 percent	586	544	239	1 988	2 085	2 149	43	35	30	251	190	72
Total	4 555	5 804	6 840	2 681	2 850	2 980	17	11	6	776	610	443

[1] The estimates for 1990/92 given here differ a little from those used for the same period in the documents for the 1996 WFS (FAO, 1996a). This is due to the revisions after 1996 that take into account new data, mainly for population.

[2] Different countries form each group in the different years.

2.2 The outlook for food and nutrition to 2015 and 2030

2.2.1 Demographics

The latest United Nations assessment of world population prospects (UN, 2001a) indicates that a rather drastic slowdown in world demographic growth is likely. The data and projections are shown in Table 2.4. The world population of 5.9 billion of our base year (the three-year average 1997/99) and the 6.06 billion of 2000 will grow to 7.2 billion in 2015, 8.3 billion in 2030 and 9.3 billion in 2050. The growth rate of world population peaked in the second half of the 1960s at 2.04 percent p.a. and had fallen to 1.35 percent p.a. by the second half of the 1990s. Further deceleration will bring it down to 1.1 percent in 2010-15, to 0.8 percent in 2025-30 and to 0.5 percent by 2045-50.

Despite the drastic fall in the growth rate, the absolute annual increments continue to be large. Seventy-nine million persons were added to the world population every year in the second half of the 1990s and the number will not have decreased much by 2015. Even by 2025-30 annual additions will still be 67 million. It is only by the middle of the century that these increments will have fallen significantly, to 43 million per year in 2045-50. Practically

---

Box 2.1 Measuring the incidence of undernourishment: the key role of the estimates of food available for direct human consumption[1]

The key data used for estimating the incidence of undernourishment are those of food available for direct human consumption. These data are derived in the framework of the national food balance sheets (FBS). The latter are constructed on the basis of countries' reports on their production and trade of food commodities, after estimates and/or allowances are made for non-food uses and for losses. The population data are used to express these food availabilities in per capita terms. The resulting numbers are taken as proxies for actual national average food consumption. For many countries the per capita food consumption thus estimated of the different commodities (expressed in kcal/person/day) are totally inadequate for good nutrition, hence the relatively high estimates of the incidence of undernourishment reported for them, most recently in FAO (2001a).

This conclusion is inferred from a comparison of the estimated kcal/person/day shown in the FBS data with what would be required for good nutrition. The parameters for the latter are well known, although not devoid of controversy. In the first place, there is the amount of food (or dietary) energy that is needed for the human body to function (breathe, pump blood, etc.) even without allowing for movement or activity. This is the basal metabolic rate (BMR). It is in the general range of 1 300-1 700 kcal/day for adults in different conditions (age, sex, height, bodyweight). Taking the age/sex structure and bodyweights of the adult populations of the different developing countries, their national average BMRs for adults are defined. These refer to the amount of energy as a national average per adult person that must be actually absorbed if all were in a state of rest. For children, in addition to the BMR, an allowance is made for growth requirements.

When an allowance for light activity is added – estimated to be about 54 percent of the BMR – this results in a range of between 1 720 kcal and 1 960 kcal person/day for the different developing countries, given their population structures in 1997/99. This will rise to 1 760-1 980 kcal by 2030 when the demographic structure will be different, with a higher proportion of adults. It follows that population groups in which an average individual has an intake below this level (the threshold) are undernourished because they do not eat enough to maintain health, bodyweight and to engage in light activity. The result is physical and mental impairment, characteristics that are evidenced in the anthropometric surveys. Estimating the incidence of undernourishment means estimating the proportion of population with food intakes below these thresholds. It is noted that the notion, measurement and definition of thresholds of requirements are not devoid of controversy. For example, Svedberg (2001, p. 12) considers that the thresholds used in the FAO measurement of undernourishment for the tropical countries are too high, leading to overestimates of the incidence of undernourishment.

In principle, a country having national average kcal/person/day equal to the threshold would have no undernourishment problem provided all persons engage in only light activity and each person had access to food exactly according to his/her respective requirements. However, this is never the case; some people consume (or have access to) more food than their respective "light activity" requirements (e.g. because they engage in more energy-demanding work or simply overeat) and other people less than their requirement (usually because they cannot

all these increases will be in the developing countries. Within the developing countries themselves, there will be increasing differentiation. East Asia will have a growth rate of only 0.4 percent p.a. in the last five years of the projection period. At the other extreme, sub-Saharan Africa's population will still be growing at 2.1 percent p.a. in the same period 2025-30, despite the drastic downward

revision made in recent years in the region's population projections.[4] By that time every third person added annually to the world population will be in that region. By 2050, every second person of the 43 million added annually to the world population will be in sub-Saharan Africa.

The new population projections represent a rather fundamental change in the assumptions

---

[4]   It is tempting to think that a lower population growth rate in the low-income countries where population growth is high would be contributing to improved development. However, in the current projections, the reduced population growth rate is not always a harbinger of good things to come, because in some cases it occurs, at least in part, for the wrong reasons. This is the case of demographic slowdown because of increases in mortality and/or declines in life expectancy, either in relation to present values or to those that would otherwise be in the projections. In the current projections, such cases of increased mortality and reduced life expectancy caused by the AIDS epidemic are a rather significant component of the projected slowdown. Thus, for the 45 most affected countries, the expectation of life at birth by 2015 is projected to stand at 60 years, five years lower than it would have been in the absence of HIV/AIDS (UN, 2001a).

---

afford more). Thus, an allowance must be made for such unequal access. Empirical evidence suggests that the inequality measure used in these estimates – the coefficient of variation (CV) – ranges from 0.2 to 0.36 in the different countries (a CV of 0.2 means, roughly, that the average difference of the food intake of individuals from the national average – the standard deviation – is 20 percent of the average). Even at the lowest level of inequality generally found in the empirical data (CV=0.2), the national average kcal/person/day must be well above the threshold if the proportion of population undernourished is to be very low. For example, a country with threshold 1 800 kcal and CV=0.20, must have a national average of 2 700 kcal/person/day if the proportion undernourished is to be only 2.5 percent, or 2 900 if it is to be 1 percent. Naturally, if inequality were more pronounced, these requirements would be higher (see Fig. 2.4).

These numbers, or norms, are, therefore, a first guide to assessing the adequacy or otherwise of the national average food consumption levels in the FBS data and expressed in kcal/person/day. This latter number is the principal variable used to generate estimates of the incidence of undernourishment as explained elsewhere (FAO, 1996b).[2] Numerous countries fall below the national average energy level (kcal/person/day) required for undernourishment to be very low, in many cases they fall below by considerable margins. Therefore, even if one knew nothing more about the incidence of undernourishment, the inevitable conclusion for these countries is that the incidence must be significant, ranging from moderate to high or very high in the different countries, even when inequality of access to food is moderate. It follows that progress towards reducing or eliminating undernourishment must manifest itself, in the first place, in the form of increased per capita food consumption. Naturally, this is not equivalent to saying that the food consumption shown in the FBS data is itself a variable that can be operated upon directly by policy. For it to rise, somebody must consume more food, and the food must come from somewhere – production or imports. The policies to raise national average consumption are those that enhance the purchasing power and more general access to food of those who would consume more if they had the means, for example, access to resources and technologies to improve their own food production capacities, access to non-farm employment and social policies. The point made here is that changes in the national average kcal/person/day recorded in the FBS data do signal the direction and magnitude of movement towards improved or worsened food security status. This is shown graphically in Fig. 2.4.

How reliable are the FBS data, since in many cases they show very low or very high levels of national average food consumption or sudden spurts or collapses? The answer is: they are as reliable as the primary data on production and trade supplied by the countries, as well as the population data used to express them in per capita terms (see Box 2.2). It is these data that are processed, in the form of the FBS, to derive the indicators of per capita food consumption as national averages used here. Given the primary data, the conclusion that many countries are in a difficult food security situation follows logically and inevitably.

---

[1]   Reproduced with amendments from FAO (1996a).

[2]   These key variables (kcal/person/day and the CV) are used as parameters of the lognormal statistical distribution (with kcal/person/day as the mean) to estimate the percentage of population undernourished.

underlying this and other studies of food and agriculture futures. When our earlier projections study to 2010 (Alexandratos, 1995) was being produced in 1992-93, we were working with a world population projection of 7.2 billion for 2010. The new projections indicate 6.8 billion for the same year, 400 million fewer people. In principle, the lower population projection used now should make for lower growth of demand and production, *ceteris paribus*.

Naturally, other things (incomes, poverty, pressures on resources and the environment) are not expected to be equal; slower demographic growth itself will be a factor for change, for example if it contributes to higher incomes. In this exercise, we assume that whatever effects slower population growth has on the overall. economy have already been taken into account in the derivation of the income (or GDP) growth assumptions (see below). The latter, just like

Table 2.4 Population and GDP data and projections

	Population									
	Million						Annual increments (Million)			
	1964 /66	1974 /76	1984 /86	1997 /99	2015	2030	1995 -2000	2010 -2015	2025 -2030	2045 -2050
World (UN)	3 334	4 065	4 825	5 900	7 207	8 270	79	76	67	43
World (countries with FBS*)	3 325	4 053	4 810	5 878	7 176	8 229	78	76	66	43
Developing countries	2 295	2 925	3 597	4 572	5 827	6 869	74	74	66	45
Sub-Saharan Africa	230	299	400	574	883	1 229	15	20	24	23
Near East/North Africa	160	208	274	377	520	651	8	9	9	7
Latin America and Caribbean	247	318	397	498	624	717	8	7	6	3
South Asia	630	793	989	1 283	1 672	1 969	23	22	19	12
East Asia	1 029	1 307	1 537	1 839	2 128	2 303	20	16	9	-1
Industrial countries	695	761	815	892	951	979	5	2	1	0
Transition countries	335	367	397	413	398	381	0	-1	-1	-2

	Growth rates, percentage p.a.									
	Population					Total GDP		Per capita GDP		
	1969 -99	1979 -99	1989 -99	1997/99 -2015	2015 -2030	1997/99 -2015	2015 -2030	1997/99 -2015	2015 -2030	1997/99 -2030
World	1.7	1.6	1.5	1.2	0.9	3.5	3.8	2.3	2.9	2.6
Developing countries	2.0	1.9	1.7	1.4	1.1	5.1	5.5	3.7	4.4	4.0
Sub-Saharan Africa	2.9	2.9	2.7	2.6	2.2	4.4	4.5	1.8	2.3	2.0
Near East/North Africa	2.7	2.6	2.4	1.9	1.5	3.7	3.9	1.8	2.4	2.1
Latin America and Caribbean	2.1	1.9	1.7	1.3	0.9	4.1	4.4	2.8	3.5	3.1
South Asia	2.2	2.1	1.9	1.6	1.1	5.5	5.4	3.9	4.3	4.1
East Asia	1.6	1.5	1.2	0.9	0.5	6.1	6.3	5.3	5.8	5.5
Industrial countries	0.7	0.7	0.7	0.4	0.2	3.0	3.0	2.6	2.8	2.7
Transition countries	0.6	0.5	0.1	-0.2	-0.3	3.7	4.0	4.0	4.3	4.1

Box 2.2 Data problems and the estimation of undernourishment: the case of Nigeria

In this chapter, Nigeria is singled out as being one of the most populous developing countries, and an exception in sub-Saharan Africa. Along with China, Indonesia, etc., Nigeria has been making progress in raising significantly its per capita food consumption and, by implication, in reducing the incidence of undernourishment. This was not so in earlier projection work (Alexandratos, 1995), nor was it considered that Nigeria could be making significant progress by 2010. At the time of the earlier exercise (1992/93), Nigeria's population was reported in the 1990 UN Assessment (UN, 1991) as being 105 million in the base year of the projections, the three-year average 1988/90. With this population and its food production and trade data, the FBS indicated per capita food consumption of 2 200 kcal in 1988/90. These data implied that Nigeria was in a dire food security situation, just like most other countries of sub-Saharan Africa. Under these initial conditions, and given the very high growth rate of population (projected at 3.15 percent p.a. to reach 201 million by 2010), one could not have been optimistic about the prospects for significant improvements. Even as late as 1996, Nigeria was given as having 43 million undernourished in 1990/92 (38 percent of its population) in the documentation of the 1996 WFS (FAO, 1996c).

The drastic revisions of Nigeria's population estimates came successively after 1996. By the time of the 2000 UN Assessment (UN, 2001a), the population estimate for 1988/90 had been reduced to 83.5 million and the 2010 projection to 147 million (having passed through a projection of 139 million for 2010 in the 1998 Assessment), a growth rate of "only" 2.73 percent p.a. These new data and projections put the assessment of present and future food security prospects of Nigeria in an entirely different light. *Ceteris paribus*, the downward revision of the population by 20 percent for 1988/90 should have raised per capita food consumption for that year, from 2 200 to 2 765 kcal. Yet this was not the case. The reason is that there have also been drastic revisions in the production data for some major food crops of Nigeria. For the 1988/90 average, the production of roots and tubers (which in the unrevised data provided over one-third of the national average calories) was reduced by 38 percent. In parallel, the production of maize was revised upwards no less than 165 percent. The end result is that the revised average kcal consumption for 1988/90 was 2 300, only about 5 percent higher than the previous estimate.

The FBS for the most recent years suggest that Nigeria, after about the mid-1980s, made really spectacular progress and broke out of the long-term pattern of stagnation in per capita food consumption typical of the majority of the countries in sub-Saharan Africa. It moved from 2 050 kcal in 1984/86 (and the 2 300 revised kcal for 1988/90) to 2 815 kcal in 1997/99, implying that undernourishment fell to 7.6 million persons, or 7 percent of the population (FAO, 2001a). Fifty percent of the increase in kcal/person/day came from roots/tubers, 12 percent from maize, 11 percent from rice and 17 percent from oilcrops (rapid production increases of groundnuts, soybeans, cotton seed). Production of all these crops registered three- to sixfold increases in the period from the mid-1980s to 1997/99. If these data are correct, we have a case of a large country registering growth of aggregate food consumption, measured in calories, of 5.4 percent p.a. for over ten years (1984/86-1997/99). At first glance, such rapid growth in food demand/consumption (and associated drastic reduction in undernourishment) would seem to be at variance with what one would expect from movements in other indices of the overall economy, e.g. per capita income. There was no economic miracle of the "Asian tiger" type in Nigeria during this period to explain the phenomenon. The country's per capita income was actually falling in the period 1984-99 (the gross domestic income [GDY] was growing at 2.1 percent p.a. when population was growing at 2.9 percent p.a.; data from the World Bank, 2001b). Nigeria is sometimes given as an example of the wider problem of development failures of sub-Saharan Africa (see *The Economist*, 2000).

One possible explanation for these trends in Nigeria is the rapid growth of food-crop agriculture, which probably has a large subsistence component, particularly in the roots and tubers sector which, as noted, accounted for 50 percent of the improvement[1] (see also Chapter 3, Section on roots and tubers). Before we jump to any conclusions concerning the wider potential for food security improvements based predominantly on agriculture, there is an obvious need to validate the primary data on production as well as to find corroborating evidence (e.g. from surveys) that the improvements in consumption suggested by the FBS are real. If these developments proved to be true, they would imply that rapid progress in food-crop production and demand could be made, at least for some time, even when developments in the overall economy would suggest otherwise. Some

---

[1] The numbers for apparent demand/consumption result largely from the production statistics, hence this explanation, being tautological, crumbles if the production statistics are unreliable.

the demographic projections, are assumptions exogenous to the food and agriculture projections proper. This is not entirely as it should be, but practical reasons preclude any explicit consideration of the interactions between population growth and development (Box 2.3).

2.2.2 Overall economy and poverty

The latest World Bank assessment for the period 2000-15 takes account of the most recent data and views concerning the current (end-2001) slowdown in the world economy. Relatively slow growth in the first five years of the projection period is expected to be followed by faster growth in the subsequent ten years, 2005-15. The current assessment (World Bank, 2001c, Table 1.7) is definitely less optimistic than that of a year earlier (World Bank, 2001a, Table 1.6). Still, it indicates that for the whole period 2000-15 world economic growth is expected to be higher (1.9 percent p.a. in terms of *per capita* GDP) than in the 1990s (1.2 percent p.a.). Higher growth rates in per capita GDP than in the 1990s is foreseen for all regions and country groups (particularly the reversal of declines in the transition economies) with the exception of East Asia. These medium-term projections of the World Bank are shown in Figure 2.3. Earlier versions of these World Bank projections have provided the basis for defining the GDP projections used as exogenous

assumptions in the present study. They are shown in Table 2.4.

There is great contrast in the prospects of the two regions with high relative concentrations of poverty and food insecurity, South Asia and sub-Saharan Africa; in the former, a continuation of the relatively high GDP growth holds promise of positive impact on poverty alleviation and increases in food consumption (see below). However, progress may be very limited in sub-Saharan Africa, with per capita incomes growing at only 1.3 percent p.a. in the period to 2015, according to the latest World Bank study (World Bank, 2001c). This is certainly better than in the past which was characterized by declining incomes. However, it will be far from sufficient to make a significant dent on poverty and food insecurity.

The exogenous economic growth assumptions used here, together with the growth of population, are the major determinants of projected food consumption,[5] hence also of the incidence of undernourishment. One of the important questions we shall be asking below is the extent to which such projected food demand will be associated with reductions in undernourishment. Since undernourishment is more often than not closely correlated with poverty, it is relevant to ask to what extent the economic growth and development outlook used as exogenous assumptions is compatible with poverty reduction.

5   Many other factors besides population and average GDP growth influence the demand for food and have to be taken into account in the process of all phases of analytical and evaluation work concerning nutrition, production and trade. See, for example, Box 2.2 (Nigeria) concerning issues involved in understanding the factors that influence changes in apparent food consumption.

Figure 2.3   Growth rates of per capita GDP, 1990s and 2000-15

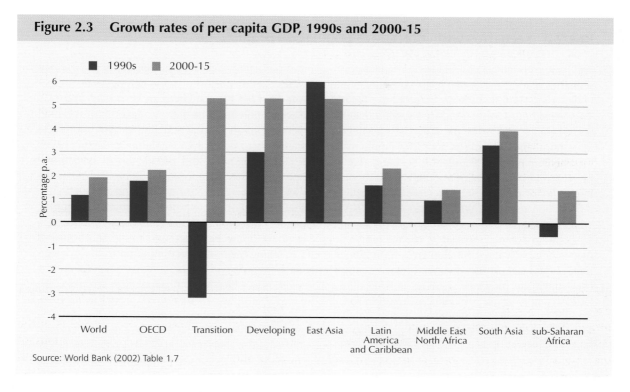

Source: World Bank (2002) Table 1.7

The World Bank has estimated what the baseline economic growth projections may imply for poverty reduction in the year 2015. Their estimates are shown in Table 2.5. They refer to what is commonly known as US$1/day poverty, i.e. the number of persons living in households with per capita expenditure under US$1/day, with dollars defined in units of purchasing power parity (PPP). (For more discussion on concepts and goals relating to poverty, see Chapter 8.) These poverty projections imply that:

■ the goal of halving by 2015 *the proportion* (not the absolute numbers) of the population of the developing countries as a whole living in poverty from that prevailing in 1990 may be achieved (the proportion falls from 32 percent in 1990 to 13.2 percent in 2015);

■ however, the *absolute numbers* in poverty will not be halved. They are expected to decline from 1.27 billion in 1990 to 0.75 billion in 2015;

■ much of the decline is caused by prospective developments in East and South Asia. Indeed, one half of the decline of 400 million foreseen for East Asia from 1990 to 2015 had already occurred by 1999;

■ in contrast, the absolute numbers in poverty in sub-Saharan Africa kept increasing in the 1990s and are projected to continue to do so until 2015.

There is a fairly close parallel between these foreseen developments in the incidence of poverty and those projected here for the incidence of undernourishment, which are the subject of the following section. It is noted, however, that poverty and undernourishment are not identical concepts, in particular as concerns the settings of threshold levels for defining them (for discussion, see FAO, 2001a, p. 10).

2.2.3 Food security outcomes

Higher per capita food consumption in the future, but with significant exceptions. By 2015, and even more by 2030, the key variable used to track developments in food security – per capita food consumption as defined above – will have grown significantly. The world average will be approaching 3 000 kcal/person/day in 2015 and will exceed 3 000 by 2030 (Table 2.1). These changes in world averages will reflect above all the rising consumption of the developing countries, whose average will have risen from 2 680 kcal in 1997/99 to 2 850 kcal in 2015 and close to 3 000 in 2030. More and more people will be living in countries with medium to high levels of per capita food consumption. For example, by 2015 81 percent of the world population will be living in countries

Box 2.3 Scenarios with alternative population projections

The issue of population-development interactions assumes particular importance if one wishes to explore scenarios of food and agriculture futures under alternative population projections. The demographic projections we use in this study are those of the United Nations *Medium Variant*. There are also *High* and *Low Variants* of future population. They suggest that the world population could be in the range of 7.0-7.4 billion by 2015 and by 2030 in the range 7.7-8.9 billion. Projecting the food and agricultural variables for these other population variants is not just a simple matter of scaling up or down the magnitudes projected under the Medium Variant population scenario. For example, world demand and production of cereals are projected to be 2.84 billion tonnes in 2030, or 344 kg per capita. We cannot just assume that under the High Variant population projection it will still be 344 kg, raising aggregate demand and production to 3.06 billion tonnes. If we did, it would be like saying that population growth does not matter for human welfare since per capita consumption (hence, in principle also per capita income) remains the same.

Such an approach would ignore the whole population-development debate concerning the positive or negative impacts of population growth on human welfare. Taking them into account requires that we estimate (or at least express a view) for each country what such impacts will be, i.e. in what direction and by how much the projected incomes will be different from those in the medium population scenario. We cannot simply adopt a blanket assumption for all countries that either: (i) *total GDP growth will be the same*, in which case it would mean that population growth is immiserizing because the higher it is, the lower the per capita income; or (ii) *per capita income growth will be the same,* implying that the population growth rate does not affect income growth. In some countries the effects may be positive, in others negative. Given the great diversity of situations existing in the world, all sorts of permutations between the growth rates of population and other variables are possible.[1] Doing estimates for over 100 countries can be an impossible task, and this would only be the first step in the work required for estimating scenarios with alternative population projections. The great bulk of the additional work would come from revisiting the country-by-country evaluations of such things as nutritional consistency of a new set of consumption projections, or the agronomic considerations (land, water, yields, etc.) underlying the projections of production. These operations are not done mechanically by a model. They involve fairly detailed reviews by country and subject-matter specialists in an interdisciplinary context.

1   For example, in countries which shift to higher population growth rates mainly because of improvements in mortality and life expectancy, such higher rates would probably be indicators of improving economic and social conditions and should be associated with higher, not lower, per capita income. Similar considerations can be relevant for countries facing acute problems of rapidly ageing populations. The opposite case can be made for very poor countries with high population growth rates. A balanced view on the latter seems to be the following: "A slowing of rapid population growth is likely to be advantageous for economic development, health, food availability, housing, poverty, the environment, and possibly education, especially in poor, agrarian societies facing pressure on land and resources" (Ahlburg, 1998). For latest views on this topic see *Population and Development Review* (2001).

with values of this variable exceeding 2 700 kcal/person/day, up from 61 percent at present and 33 percent in the mid-1970s. Those living in countries with over 3 000 kcal will be 48 percent of the world population in 2015 and 53 percent in 2030, up from 42 percent at present (Table 2.2).

These gains notwithstanding, there will still be several countries in which the per capita food consumption will not increase to levels allowing significant reductions in the numbers undernourished from the very high levels currently prevailing (see below). As shown in Table 2.2, in 2015 6 percent of the world population (462 million people) will still be living in countries with very low levels of food consumption (under 2 200

kcal). As discussed earlier (Box 2.1), in these countries a good part of the population is undernourished almost by definition. At the regional level, in 2015 sub-Saharan Africa will still have medium-low levels of per capita consumption, 2 360 kcal/person/day. The disparity between sub-Saharan Africa and the other regions is even more pronounced if Nigeria is excluded from the regional total (but see Box 2.2), in which case the kcal of the rest of the region will only be 2 230 in 2015. Of the 15 countries still remaining in 2015 in the under 2 200 kcal range (Table 2.2), 12 will be in sub-Saharan Africa. Of the 30 countries in the next range of kcal (2 200-2 500), 17 will be in this region.

Table 2.5   Estimates and projections of poverty (US$1/day, World Bank, baseline scenario)

	Million persons			Percentage of population		
	1990	**1999**	**2015**	**1990**	**1999**	**2015**
Developing countries	1 269	1 134	749	32.0	24.6	13.2
Sub-Saharan Africa	242	300	345	47.7	46.7	39.3
Middle East and North Africa	6	7	6	2.4	2.3	1.5
Latin America and Caribbean	74	77	60	16.8	15.1	9.7
South Asia	495	490	279	44	36.9	16.7
East Asia	452	260	59	27.6	14.2	2.8
Memo items						
East Asia, excl China	92	46	6	18.5	7.9	0.9
Developing, excl. China	909	920	696	32.2	27.3	16.4

Source: Adapted from World Bank (2001c), Table 1.8. The definition of regions is not always identical to that used in this study, e.g. Turkey is not included in the developing Middle East/North Africa and South Africa is included in the developing sub-Saharan Africa.

Modest reductions in the numbers undernourished. The relatively high average consumption levels of the developing countries projected for 2015 and 2030 could lead one to expect that the problem of undernourishment will be solved or well on its way to solution, in the sense that the numbers undernourished should show significant declines. This would be the corollary of what was said earlier about the importance of the per capita food consumption as the major variable that is a close correlate of the level of undernourishment. Yet the estimates presented in Table 2.3 show that reductions will be rather modest; the 776 million of 1997/99 (17 percent of the population) may become 610 million in 2015 (11 percent) and 440 million by 2030 (6 percent). For developing countries as a whole, we may have to wait until 2030 before the numbers of undernourished are reduced to nearly the target set for 2015 by the WFS, i.e. one half of the 815 million estimated for the base period of 1990/92.

These findings indicate that achieving significant declines in the incidence of undernourishment may prove to be more arduous than commonly thought. A combination of higher national average food consumption and reduced inequality (see below for assumptions) can have a significant impact on the *proportion* of the population undernourished. However, when population growth is added in, such gains do not necessarily translate into commensurate declines in the absolute numbers, because the population of the developing countries will have grown from 4.55 billion in 1997/99 to 5.8 billion in 2015 and 6.84 billion in 2030.

The numbers of undernourished are expected to remain nearly constant in sub-Saharan Africa, even by 2030. This is no doubt an improvement over the historical trend of nearly stagnant food consumption per capita in the region and, by implication, rising undernourishment. It is, however, far from what is needed to meet the WFS target of reducing the numbers by half by no later than 2015. In contrast, rather significant reductions are expected for both South and East Asia, the two regions that contain the bulk of the world's undernourished population. East Asia is expected to have halved undernourishment by 2015 (it had already reduced it by 30 percent in the period 1990/92-1997/99) and South Asia could achieve this target towards the later part of the period 2015-30.

In order to appreciate why these prospects emerge, let us recall briefly that future estimates are generated by applying the same method used for estimating present undernourishment. The only difference is that we use the future values for those variables for each country that we project, or can assume, to be different from the present ones. As noted (Box 2.1), the variables which, in our method, determine the numbers undernourished are the following:

- The projected population.
- The per capita food consumption in kcal/person/day, taken as a proxy for actual average national consumption. Future values are derived from the projections of per capita food consumption for each commodity discussed in detail elsewhere (Chapter 3) and summarized in Tables 2.7 and 2.8 (major commodities) as well as in Table 2.1 (kcal/person/day).
- The threshold (or cut-off level) of food energy (kcal/day) a person must have in order not to be undernourished. This varies by country depending on age/sex structure of the population. The range of values applicable to different developing countries was given in Box 2.1. It was noted that because of the ageing of the population (growing share of adults in total population) the range will be higher in the projection years than at present. Therefore, this factor would tend to raise the incidence of undernourishment, *ceteris paribus*.
- The coefficient of inequality, as described in Box 2.1. We have no way of knowing how this variable may change in each country in the future. If we applied in the future the same values used for the 1997/99 undernourishment estimates, we would be ignoring the prospect that declining poverty is normally associated with more equal access to food. The World Bank projections of declines in the incidence of poverty (Table 2.5) imply that the share of population below the poverty line (hence also of persons with low food consumption levels) will be smaller in the future compared with the present. Given the nature of food consumption (it increases fast from low levels as incomes rise but then tends to level off as higher levels are attained) it is reasonable to assume that if the reduced poverty projected by the Bank were to materialize, it would be accompanied by reduced inequality in food consumption as measured by the CV. We take this prospect on board by assuming that countries will have lower inequality in the future. How much lower depends on the progress they make in raising their average kcal/person/day (see Box 2.4). The net effect of these assumptions is that the CV of the different developing countries

which are currently in the range of 0.21-0.36 would be in the range of 0.20-0.31 in 2015 and 0.20-0.29 in 2030. The estimates of future undernourishment presented in Table 2.3 are based on such assumptions about changes in inequality.

One factor making for the slow decline in the numbers of undernourished is the gradual rise in the threshold (cut-off level) for classifying a person as undernourished. As noted, this rise is caused by the ageing of the population. The (simple arithmetic) average threshold of the developing countries rises from 1 835 kcal in 1997/99 to 1 882 kcal (2.6 percent) in 2030. This rise has important implications for the future incidence of undernourishment in countries with low average food consumption. It implies that consumption must rise by an equal proportion just to prevent the incidence of undernourishment (in percentage of the population) from increasing. If this ageing of the population and the associated rise in threshold requirements had not intervened, the numbers undernourished estimated for 2030 would be 16 percent lower than shown in Table 2.3, i.e. 370 million rather than 440 million.

A second factor is to be found in the very adverse initial conditions several countries started with in 1997/99. For example, nine developing countries started with estimated base year undernourishment of over 50 percent (FAO, 2001a). They are Somalia, Ethiopia, Eritrea, Mozambique, Angola, the Democratic Republic of the Congo, Afghanistan, Haiti and Burundi. The group's average per capita food consumption is 1 790 kcal and undernourishment is 57 percent of the population or 105 million. The food consumption projections imply (according to the method used here) that the *proportion* of the population affected will fall to 39 percent by 2015. This is a significant decline. However, the *absolute numbers* affected will rise to 115 million in 2015, because of the relatively high growth rate of the group's population, 2.7 percent p.a. in 1997/99-2015. The undernourished may still be 106 million (25 percent of the population) by 2030.

Are we perhaps too pessimistic? Readers may judge for themselves on the basis of the following considerations. The per capita food consumption of this group of countries has moved in the range of 1 735-2 000 kcal in the past three decades. In the projections, it grows from 1 790 kcal in 1997/99 to 2 010 kcal in 2015 and to 2 220 in 2030. Taking into account population growth, aggregate demand for food (expressed in calories) is projected to grow at 3.5 percent p.a. in the 17 years to 2015. This contrasts with the experience of the past three decades when the highest growth rate achieved in any 17-year period (64-81, 65-82, ..., 81-98, 1982-99) was 2.3 percent p.a. In parallel, the production evaluation (Chapter 4) concludes that cereal production in this group of countries could grow at 2.8 p.a., compared with 2.1 percent p.a., the highest rate ever achieved in any 17-year period in the past. Overall, therefore, the projections of food consumption and of production, far from being pessimistic, embody a degree of optimism. This is partly justified by the prospect of recovery of agriculture following eventual cessation of war or warlike activities that are, or were recently, present in most countries in this group. Empirical evidence discussed in the next section suggests that in such situations better performance of agriculture is a key factor in making possible rapid increases in food consumption.

There is also an additional element of optimism embodied in the assumed reductions in distributional inequalities, as already discussed. The average CV of this group is assumed to decline from 0.30 in 1997/99 to 0.26 in 2015 and 0.23 in 2030. But for these reductions, the undernourished would have grown to 123 million in 2015 rather than to 115 million. The difference is admittedly small, indicating the limited impact of reduced inequality on the numbers of undernourished in countries with very low national average kcal. The reasons why this happens and why it does not denote limited welfare value of more equal distribution was explained earlier (Box 2.4).

Similar considerations apply, *mutatis mutandis*, to other countries that start from low or very low per capita food consumption and high undernourishment and also have fairly high population growth rates. By 2015, they will still have low to middle levels of food consumption, and the numbers of undernourished will be either higher or not much below the present ones. The middle part of Table 2.3 provides some more disaggregated information on this aspect of the problem. The first two groups of countries, those that in 2015 will still be below 2 500 kcal (41 countries in all), fall in the above-

Box 2.4 Inequality of access to food and incidence of undernourishment: assumptions about the future

As noted in the text, in deriving future levels of undernourishment from the projected per capita food consumption levels, the assumption is made that countries will have less inequality in the future, if World Bank projections of reduced poverty incidence come about. The measure of inequality (the CV, see Box 2.1) applied in the projections is derived by assuming that the standard deviation (SD, see Box 2.1 for an explanation) will be the same in the future as in 1997/99 in the different countries even as their national average kcal rise, subject to the CV not falling below 0.20. In plain words, this means the following: a country which in 1997/99 had a CV of 0.3 and kcal 2500, had an SD of 750 kcal and 16.5 percent of its population undernourished (assuming its undernourishment threshold is 1 800 kcal). In the future, its average kcal rises to 2 700. If the CV remained at 0.30, undernourishment would fall to 10.9 percent of the projected population. With our assumption (SD constant at 750 kcal), the CV falls to 0.278 and undernourishment falls to 8.8 percent of projected population.

Reductions in inequality have small effects on the incidence of undernourishment when average kcal are very low (e.g. under 2 000). This is so because at that level most of the population is under the undernourishment threshold to start with. The scope of raising many of them above the threshold by redistributing the "surplus" of those above it is limited. Naturally, in no way does this imply that in such countries reduction of inequality has no beneficial effects on the undernourished. It does raise consumption, but not by as much as needed to bring them above the undernourishment threshold. Likewise, the effect is also small when the average kcal is very high, e.g. over 3 000, because at that level the percentage of the population undernourished is already small. The highest beneficial impact from reducing inequality is to be had in the countries in the middle range of per capita food consumption, 2 300-2 600 kcal. These effects are traced in graphic form in Figure 2.4. The vertical distance between the curves shows how far undernourishment (percentage of population) changes when shifting from high to low inequality or vice versa.

Figure 2.4   Paths of change in undernourishment: raising average consumption versus reducing inequality

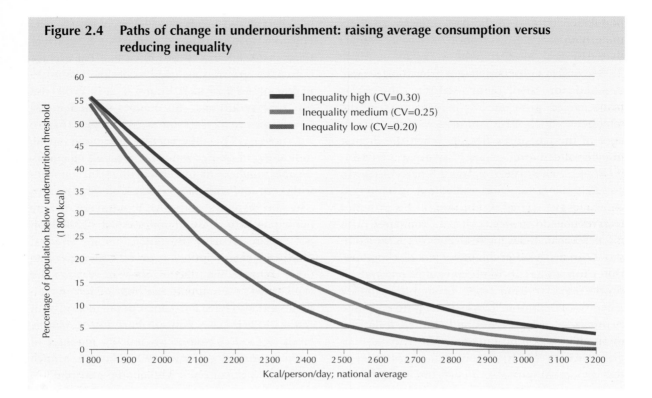

mentioned category, i.e. numbers of undernourished increasing or not declining by much for the reasons just mentioned. The next group (12 countries in the range 2 500-2 700 kcal in 2015) is an intermediate case showing medium reductions in the numbers undernourished.

Almost all the projected reductions in the numbers undernourished by 2015 would occur in the remaining two country groups, those which contain the countries projected to have in 2015 over 2 700 calories. These two groups include some of the most populous developing countries (China, India, Pakistan, Indonesia, Mexico, Nigeria and Brazil) and account for the bulk of the population of the developing countries and some 60 percent of the total numbers of undernourished. The main reason why the gains in these countries are more pronounced than in the earlier groups is that most of them start with kcal in the middle range. As noted, the potential for reducing undernourishment through more equal distribution is highest in this type of countries. It is indeed the projected declines in the numbers in poverty in the most populous countries[6] with already middle to middle-high food consumption levels, and the assumed knock-on effect in reducing inequality in the distribution of food consumption, that generates much of the projected decline in the numbers undernourished in these countries. If inequality were to remain the same, their undernourishment would decline from 456 million in 1997/99 to 400 million in 2015; with the assumed reductions in inequality, undernourishment in 2015 declines by a further 100 million, to 295 million.

In conclusion, rapid reductions in the numbers of undernourished require the creation of conditions that will lead to hefty increases in national average food consumption, particularly in countries starting with low levels, as well as to lower inequality of access to food. Countries with high population growth rates will need stronger doses of policies in that direction than countries with slower growth rates. The projections of population and the overall economic growth used here, and the derived projections of food demand and consumption, indicate that in many countries the decline in the numbers undernourished will be a slow process. Moreover, in several countries with high

---

6   China and the countries of South Asia account for almost all the reductions in poverty projected by the World Bank for 2015 – see Table 2.5.

population growth rates the absolute numbers undernourished are projected to increase rather than decline by 2015.

From a policy perspective, an appropriate way of looking at the problem at hand is to see how many countries, accounting for what part of the total population, will still have significant percentages of their population undernourished. If the number of countries in this category in the future is smaller than at present – particularly if they are among the less populous ones making for a small percentage of aggregate population – then policy interventions to reduce undernourishment will be more feasible. Relevant data and projections are shown in the third section of Table 2.3. The population living in countries with undernourishment over 25 percent will have been reduced from 13 percent of the total of the developing countries at present to only 3.5 percent by 2030. In parallel the number of countries in this category will have declined from 35 at present to 22 in 2015 and to only five in 2030. None of today's most populous countries (over 100 million in 1997/99) will be in this class in the future. At the other extreme, 75 percent of the population of the developing countries could be in countries with undernourishment below 5 percent. At present only 8 percent live in such countries. The shift to the under 5 percent category of the majority of the most populous countries underlies this dramatic change (China, India, Pakistan, Indonesia, Brazil, Mexico and the Islamic Republic of Iran – all with populations of over 100 million in 2030).

Overall, therefore, considerable progress would be made over the longer term, if the projections of food consumption and the assumptions about reduced inequality were to come true. As more and more countries change from medium-high percentages of undernourishment to low ones, the problem will tend to become more tractable and easier to address through policy interventions within the countries themselves. In addition, the greatly reduced number of countries with severe problems holds the promise that policy responses on the part of the international community will tend to become more feasible.

Better outcomes possible with emphasis on agriculture. We have noted that countries that start with very adverse initial conditions (low kcal/person/day, high undernourishment, high population growth) will require very rapid growth in aggregate food consumption if they are to reduce undernourishment significantly, e.g. to halve it by 2015, as per the WFS target – although this target was set for the developing countries as a whole and not for individual countries.

As an example, the Niger starts with very adverse initial conditions. The country's 1997/99 per capita food consumption was only 2 010 kcal/person/day and undernourishment affected 41 percent of its population, or 4.2 million persons (FAO, 2001a). For 1990/92 (the base year of the WFS target) the estimate was 3.3 million, so the situation has worsened since. The 2015 target should be a half of the estimate for 1990/92, i.e. 1.65 million. The Niger is projected to have one of the highest population growth rates in the world, 3.6 percent p.a. to 2015. By then its population will be 18.5 million, up from the current 10.1 million. Reduction of undernourishment to 1.65 million in 2015 would mean reduction from the present 41 percent of the population to only 9 percent in 2015, a really huge change.

What are the characteristics of countries that have around 9 percent of the population undernourished? There are three countries (Gabon, Brazil and China) in this class. They have kcal/person/day of 2 520, 2 970 and 3 040, respectively (FAO, 2001). This implies that even under very low inequality of distribution like Gabon's (CV=0.216), the Niger's per capita food consumption must reach 2 460[7] kcal by 2015 (from the present 2 010 kcal) if it is to halve the numbers suffering undernourishment. In combination with its population growth of 3.6 percent p.a., its aggregate demand for food would need to grow at 4.9 percent p.a. between 1997/99-2015 (1.2 percent p.a. in per capita terms – 22 percent over the entire period of 17 years). The required income growth (normally above 5 percent p.a. given that total demand for food usually grows at rates below those of aggregate income) would be very demanding, if

7 One significant research question looms large in any attempt to project a likely future outcome. Given that the empirical evidence shows that countries go down as well as up, how does one project which countries will be in which category, particularly in the light of the evidence that declines suffered by many countries are often the result of war?

45

at all feasible. Naturally, if inequality of access were to be more pronounced, the national average kcal/person/day would need to be higher, e.g. 2 620 if the CV was to be 0.25.

If overall economic growth were to be the primary force making for growth in food demand/consumption, one would have to be quite pessimistic as to the prospect that the Niger and countries in similar conditions could achieve the quantum jumps in food consumption required for reduction of undernourishment. The Niger has had nearly zero economic growth (and a decline of 2.0 percent p.a. in per capita household final consumption expenditure [HHFCE]/capita[8]) in the last two decades.

Yet, empirical evidence does not support such blanket pessimism. We have referred earlier to the case of Nigeria which achieved quantum jumps in food consumption and associated declines in undernourishment despite falling overall incomes per capita. Other countries have had similar experiences, i.e. achievement of food consumption increases of 22 percent or more in per capita terms in 17 years or even over shorter periods, while per capita incomes were not growing or outright falling. For example, Mali increased kcal/person/day from 1 766 to 2 333 (32 percent) in the nine-year period from 1979/81-1988/90, while its HHFCE/capita was falling at -1.7 percent p.a. over the same period.

---

[8] World Bank (2001b) term for what was previously termed private consumption expenditure in national accounts parlance.

Nine developing countries apparently went through such experiences during some time in their history of the last 30 years (countries 1-9 in Table 2.6) They all achieved rapid growth in their kcal/capita/day (increase of 22 percent or more over periods of 17 years or less), while their HHFCE/ capita was either falling or growing at under 1 percent p.a.. What explains these food consumption gains in the midst of stagnant or deteriorating overall economic conditions? Do these countries share some common characteristics? The following comments can shed some light:

■ These countries all had very low food consumption levels (from 1 600 to 1 950 kcal/person/day, see column 4 in Table 2.6) at the inception of the periods of spurts in their food consumption, hence great potential for such increases in consumption when other conditions were propitious.

■ The countries all have fairly high dependence on agriculture as measured by the percentage of total GDP coming from agriculture and the percentage of agricultural population in the total population (columns 20-21 in Table 2.6).

■ In eight of these nine countries a key common characteristic has been rapid growth in domestic food production. The growth rate of cereal production was in the range of 5.1-8.8 percent p.a. during the periods in question (column 12 in Table 2.6), although in some countries the rapid growth of production of other important staple foods also played a key role, e.g. roots/tubers in Nigeria, Ghana and Benin. In some countries the high growth rates of cereal production reflected recoveries from periods of falling production. For example, in the Gambia cereal production had fallen from 81 thousand tonnes in 1970/72 to 36 thousand tonnes in 1975/77. Recovery led to 96 thousand tonnes in 1985/87 but then there was no further growth for another ten years and production was still 96 thousand tonnes in 1995/97.[9] Other countries had rapid growth following, and continuing beyond, recoveries, e.g. Ghana, Chad and Mauritania.

■ The substantial growth in cereal food consumption per capita was supported in all but two countries by the growth of domestic production, while net imports of cereals (kg per capita) often declined and self-sufficiency improved.

In the light of this evidence it is tempting to conclude that progress in raising food consumption levels is possible in countries facing unfavourable overall economic growth prospects, if domestic food production can be made to grow fairly rapidly for some time. However, before we draw any firm conclusions, we must keep in mind that the data concerning what happened to per capita food consumption come from the food balance sheets, i.e. they are the sum of production plus net imports, minus the non-food uses of food commodities, minus an estimate for waste. It is therefore true by definition that a change in consumption is, in an accounting sense, the counterpart of changed production and/or net imports. Naturally, this is not the same thing as saying that increased production and/or imports "caused" the increases in consumption. What we can be sure of is that if the production and trade data are correct, and if the allowances for non-food uses and waste are of the right order of magnitude, the implied food consumption increases did take place, no matter what the national accounts show concerning national incomes. Otherwise, what has happened to the increased food supplies?

We clearly have a situation where the income changes depicted in the national accounts fail to reflect what actually happens to the capacity of people to have access to food, and indeed of the persons in food insecurity. It may be hypothesized that this is the case in many low-income economies where large parts of the population derive a living from agriculture, including those with significant near-subsistence agriculture and autoconsumption. In such cases increased production can translate into improved incomes and access to food of the persons in agriculture and, through indirect effects, also of the persons in the wider rural economy.

---

9   Latest data to 2001 show a sudden spurt in production in the last three years (1999-2001). The Gambia's 2001 production is given as 179 thousand tonnes (FAOSTAT, update of February 2002).

Table 2.6 Developing countries with increases in food consumption (kcal/person/day) of 22 percent or more over 17 years or less

	Period of growth in food consumption		Kcal/person/day				Income growth during period,[1] % p.a.	
1	Beginning	No. years	Beginning	Final year	% increase	Latest (1997/99)		
	2	3	4	5	6	7	8	
1 Gambia	75/77	11	1 742	2 482	42.4	2 574	-5.4	
2 Nigeria	83/85	14	1 950	2 813	44.3	2 813	-2.9	
3 El Salvador	72/74	17	1 918	2 445	27.5	2 493	-2.0	
4 Mali	79/81	9	1 766	2 333	32.1	2 237	-1.7	
5 Benin	81/83	16	1 947	2 498	28.3	2 498	-0.2	
6 Mauritania	70/72	17	1 878	2 552	35.9	2 690	0.0	
7 Chad	82/84	15	1 596	2 117	32.7	2 117	0.2	
8 Burkina Faso	80/82	12	1 682	2 455	45.9	2 293	0.4	
9 Ghana	81/83	16	1 630	2 546	56.1	2 546	0.8	
10 Jordan	73/75	17	2 210	2 862	29.5	2 812	1.4	
11 Nepal	75/77	14	1 850	2 443	32.0	2 293	1.4	
12 Iran.	69/71	12	2 094	2 829	35.1	2 928	1.5	
13 Syria	69/71	13	2 345	3 246	38.4	3 328	1.7	
14 Myanmar	74/76	12	2 110	2 730	29.4	2 787	1.9	
15 Philippines	69/71	11	1 808	2 244	24.1	2 332	2.0	
16 Peru	90/92	7	1 978	2 551	28.9	2 551	2.4	
17 Morocco	69/71	17	2 474	3 020	22.1	3 030	2.5	
18 Algeria	69/71	17	1 840	2 778	51.0	2 933	3.5	
19 Egypt	69/71	17	2 348	3 105	32.2	3 317	3.8	
20 Tunisia	69/71	17	2 360	3 103	31.5	3 341	4.5	
21 Indonesia	76/78	17	2 056	2 856	38.9	2 903	5.1	
22 China	75/77	17	2 062	2 765	34.1	3 037	7.0	
23 Saudi Arabia	72/74	13	1 774	2 940	65.7	2 957	11.9	
24 Iraq	72/74	14	2 251	3 506	55.7	2 416	no data	
25 Lebanon	74/76	17	2 318	3 211	38.5	3 231	no data	
26 Libya	69/71	7	2 456	3 444	40.2	3 291	no data	
27 Tanzania	70/72	7	1 723	2 253	30.7	1 926	no data	
28 Yemen	70/72	16	1 761	2 166	23.0	2 040	no data	

[1] Household final consumption expenditure per capita, except for Saudi Arabia, Egypt, Myanmar, Syrian Arab Republic and Chad where the growth rates are for per capita gross domestic income.

Cereal food/per capita (kg)			Cereal production growth rates (% p.a.)		Cereal self-sufficiency (%)			Cereal net trade (kg/person)			Agr. GDP % total GDP	Agr. popul. as % of total pop.
Beginning	Final year	Latest (1997/99)	Period of cons. increase	Last ten years (1989-99)	Beginning	Final year	Latest (1997/99)	Beginning	Final year	Latest (1997/99)	1997/99	1997/99
9	**10**	**11**	**12**	**13**	**14**	**15**	**16**	**17**	**18**	**19**	**20**	**21**
119	171	163	8.8	2.7	49	51	44	-49	-105	-121	30	73
122	154	154	5.1	2.6	84	91	91	-26	-19	-19	36	39
120	153	152	2.0	0.2	82	75	69	-22	-55	-76	12	34
146	209	193	6.9	2.3	90	95	94	-14	-11	-12	46	84
96	115	115	5.1	5.4	76	87	87	-30	-25	-25	38	53
104	159	171	5.8	4.9	43	42	22	-65	-114	-231	25	49
99	132	132	6.2	6.7	70	96	96	-34	-7	-7	37	68
156	242	219	7.1	2.8	93	94	91	-14	-16	-20	32	88
58	85	85	6.4	5.5	70	84	84	-17	-17	-17	36	54
153	167	174	1.9	-7.0	36	10	4	-95	-406	-364	3	12
160	208	190	3.4	1.9	112	100	101	8	-1	0	41	91
146	186	191	3.7	2.8	88	71	69	-15	-88	-99	21	30
161	198	221	5.0	7.5	80	60	76	-62	-116	4	no data	28
172	212	216	4.3	2.7	106	101	101	12	9	3	59	70
115	138	138	4.4	0.3	93	88	75	-22	-23	-63	18	40
105	127	127	7.4	4.9	42	47	47	-90	-112	-112	7	31
225	253	250	1.2	-3.0	92	84	54	-20	-76	-118	16	38
151	208	228	-0.4	-1.9	73	27	21	-36	-197	-202	11	24
175	228	251	1.4	4.6	77	49	69	-31	-162	-153	18	38
173	221	222	0.4	1.9	61	36	45	-83	-201	-199	13	26
142	197	202	4.3	2.0	89	89	88	-19	-28	-27	18	45
165	209	210	3.1	2.2	98	99	100	-3	3	4	18	69
110	145	173	22.1	-8.1	20	13	23	-77	-457	-337	7	14
155	246	166	0.2	-1.6	94	34	40	-29	-247	-137	no data	12
128	135	136	1.6	2.0	15	12	10	-181	-253	-224	12	5
148	203	197	10.2	-2.9	27	28	10	-174	-223	-427	no data	9
75	130	115	14.3	0.1	88	99	85	-6	-4	-10	45	72
153	172	165	-2.6	0.0	76	40	24	-36	-109	-143	18	49

Sources: All FAO, except columns 8 and 20 from the World Bank (2001b).

That such links between production and consumption exist and are important for improved food security and development is, of course, nothing new. A body of literature (e.g. Mellor, 1995; de Janvry and Sadoulet, 2000) supports the proposition that, in low-income countries with high dependence on agriculture, facing initial conditions like those of many countries in our sample, strategies promoting in priority agricultural productivity improvements are most appropriate for making progress in poverty reduction and, by implication, in food security. Naturally, one should not just think of production increases in the abstract. The links between increased production and improved food consumption of poor and food-insecure persons are mediated through complex institutional and socio-economic relations. In addition, feedback effects between food production and consumption should be considered, as undernourishment is a handicap to the efforts to improve food production. Better nutrition, in addition to being an end-goal in itself, is also an essential input into the achievement of production increases and overall development (see Chapter 8).

The remaining 19 countries that increased food consumption by 22 percent or more in periods of 17 years or less exhibit a variety of experiences concerning combinations of the different variables underlying the gains in food consumption: the growth of their per capita HHFCE (data not available for the last five countries in Table 2.6), cereal production and net imports. All had positive growth rates in HHFCE/capita. We have here typical cases of the North African countries, where moderate to high growth of incomes fuelled the demand for food, and this was met mainly by quantum jumps in cereal imports rather than production, as in the case of Algeria and Egypt.

In conclusion, if the data used here are anywhere near the reality, the evidence suggests that in the many countries with poor overall economic growth outlook (e.g. most countries of sub-Saharan Africa), priority to raising agricultural productivity holds promise for making progress towards reducing undernourishment. Eventually, sustained agricultural growth will also show up in improved overall national incomes.

2.3 Structural changes in the commodity composition of food consumption

The growth in per capita food consumption was accompanied by significant change in the commodity composition, at least in the countries that experienced such growth. The relevant data and projections are shown in Tables 2.7 and 2.8.

Much of the structural change in the diets of the developing countries concerned the rapid increases of livestock products (meat, milk and eggs), vegetable oils and, to a lesser extent, sugar, as sources of food calories. These three food groups together now provide 28 percent of total food consumption in the developing countries (in terms of calories), up from 20 percent in the mid-1960s. Their share is projected to rise further to 32 percent in 2015 and to 35 percent in 2030. However, structural change was not universal and wide intercountry diversity remains in the share of different commodity groups in total food consumption. The major changes, past and projected, are briefly reviewed below. A more extensive discussion of the forces affecting the main commodity sectors is presented in Chapter 3.

Cereals continue to be by far the most important source (in terms of calories) of total food consumption. Food use of cereals has kept increasing, albeit at a decelerating rate. In the developing countries, the per capita average is now 173 kg, providing 56 percent of total calories. This is up from 141 kg (61 percent of total calories) in the mid-1960s (Table 2.7). Much of the increase in per capita food consumption of cereals in the developing countries took place in the 1970s and 1980s, reflecting *inter alia* the rapid growth of their cereal imports during the period of the oil boom (see Chapter 3). For the developing countries as a whole, average direct food consumption of cereals is projected to stabilize at around present levels (although it would keep rising if feed use of cereals were added), as more and more countries achieve medium-high levels and diet diversification continues. The share of cereals in total calories will continue to decline, but very slowly, falling from 56 percent at present to 53 percent in 2015 and to 50 percent in 2030.

Food consumption of *wheat* grew the fastest of all cereals in the past and will continue to do so in the future. Such growth in consumption will be accompanied by continued growth in wheat imports in many developing countries, particularly those that are non-producers or minor producers for agro-ecological reasons (see Chapter 3). In contrast, per capita food consumption of *rice* should continue its recent trend towards stabilization and gentle decline, reflecting developments in, mainly, the East Asia region.

Food consumption of *coarse grains* has declined on average, but continues to be important mainly in sub-Saharan Africa (where it accounts for 72 percent of food consumption of cereals) and to a lesser extent in Latin America (42 percent). The decline in other regions, particularly in China, has brought down the average for the developing countries. In future, smaller declines in Asia and some recovery in sub-Saharan Africa could halt the trend towards decline of the average of the developing countries. Aggregate demand for coarse grains will be increasingly influenced by the demand for animal feed. As discussed in Chapter 3, the developing countries will be playing a growing role in the world total demand and trade of coarse grains.

Wide intercountry differences in cereal food consumption will continue to persist. Several countries have per capita food consumption of cereals under 100 kg/year and some below 50 kg (the Democratic Republic of the Congo, Burundi and the Central African Republic). These persistently low levels reflect a combination of climatic factors (favouring dependence of diets on roots and tubers, including plantains, in countries mainly in the humid tropics) as well as persistence of poverty and depressed levels of food consumption overall. It is worth noting that Africa includes countries at the two extremes of the cereal consumption spectrum; the countries with the highest food consumption of cereals are also in Africa, namely those in North Africa, with per capita levels in the range of 200 to 250 kg.

The diversification of diet in developing countries has been most visible in the shift towards *live-stock products*. Here again there is very wide diversity among countries as regards both the levels of consumption achieved as well as the speed with which the transformation has been taking place. Several developing countries have traditionally had high meat consumption, comparable to the levels of the industrial countries. They include the traditional meat exporters of Latin America (e.g. Argentina and Uruguay), but also the occasional country with a predominantly pastoral economy, such as Mongolia. However, developments in these countries did not cause the structural change in the diets of the developing countries towards more meat consumption. If anything, they slowed it down as the per capita consumption in many of them either remained flat or actually declined. The real force behind the structural change has been rapid growth in consumption of livestock products in countries such as China[10] (including Taiwan Province of China and Hong Kong SAR), the Republic of Korea, Malaysia, Chile, Brazil and several countries in the Near East/North Africa region. Indeed, as discussed in Chapter 3, the increase in meat consumption of the developing countries from 11 to 26 kg in the period from the mid-1970s to the present was decisively influenced by the rapid growth in China and Brazil. Excluding them from the totals, the average of the other developing countries grew much less over the same period, from 11 kg to only 15 kg (see Chapter 3, Table 3.10).

In the future we may witness a significant slowdown in the growth of demand for meat. This will be the result of slower population growth and of the natural slowdown in consumption accompanying the achievement of high or medium-high levels in the industrial countries but also in some populous developing ones, such as China. The prospects are slim that other large developing countries such as India will emerge as major meat consumers, because of a continuation of low incomes and the influence of dietary preferences favouring meat less than in other societies. Thus, the boost given in the past to world meat consumption by the surge in China (but see footnote 10) is unlikely to be replicated by other coun-

---

[10] See Chapter 3 for doubts concerning the reliability of the meat sector data in China. If the data actually overstate China's meat production by a considerable margin, the country's impact on the world meat economy and particularly the aggregates of the developing countries would have been more modest than suggested here.

tries with the same force in the future. The major structural changes that characterized the historical evolution of the world livestock economy, particularly in the 1990s, are likely to continue, although in somewhat attenuated form. These are the growing role of the poultry sector in total meat production and the growing share of trade in world output and consumption.

The other major commodity group with very high consumption growth in the developing countries has been *vegetable oil*. The rapid growth in consumption and the high calorie content of oilcrop products[11] have been instrumental in bringing about increases in apparent food consumption (kcal/person/day) of the developing countries, which characterized the progress in food security achieved in the past. In the mid-1970s, consumption of oilcrop products (5.3 kg/ person/year, in oil equivalent) supplied only 144 kcal/person/day, or 6.7 percent of the total availability of 2 152 calories of the developing countries. By 1997/99 consumption per capita had grown to 9.9 kg contributing 262 kcal to total food supplies, or 9.8 percent of a total which itself had risen to 2 680 kcal. In practice, just over one out of every five calories added to the consumption of the developing countries over this period originated in this group of products (see further discussion of the oilcrops sector in Chapter 3).

In the future, vegetable oils are likely to retain, and indeed strengthen, their primacy as major contributors to further increases in food consumption of the developing countries: 44 out of every 100 additional calories in the period to 2030 may come from these products. Some important structural changes of the historical period in the world oilcrops economy are likely to continue. These are:

■ the growing share of four oilcrops in the total oilcrops sector (oil palm, soybeans, rape and sunflower);

■ the continued dominance of a few countries as major producers and exporters; and

■ the growing role of imports in meeting the food demand for vegetable oils of many developing countries.

Consumption of *pulses* in the developing countries stagnated overall and registered drastic declines in several countries, mainly in Asia and sub-Saharan Africa. These trends reflected not just changing consumer preferences but also, in several countries, failure to promote production of such crops. Often this was the result of preference for increasing production and self-sufficiency in cereals. It is thought that where these declines in protein-rich pulses were not accompanied by increases in the consumption of livestock products, the result has been a deterioration in the overall quality of diets, even where per capita dietary energy (kcal/person/day) increased (for the case of India, see Hopper, 1999). For the future, no major changes are foreseen in per capita consumption of pulses, with the average of the developing countries remaining at about 7 kg.

Roots, tubers and plantains have traditionally been the mainstay of food consumption in several countries with low-middle levels of overall food consumption, mainly in sub-Saharan Africa and Latin America. Nineteen countries, all in sub-Saharan Africa, depend on these products for over 20 percent of food consumption in terms of calories. These countries account for 60 percent of the region's population. In three of them, the dependence is over 50 percent (the Democratic Republic of the Congo, Rwanda and Ghana). At the same time, the region has countries at the other extreme of the spectrum with only minimal consumption of roots and tubers, such as Mali, Mauritania, the Niger and the Sudan.

The food balance sheet data show that in several of the countries with high dietary dependence on roots and tubers, what happens to the production of these crops is an important determinant of changes in the national average food consumption. As in the case of Nigeria mentioned earlier (Box 2.2), other countries (Ghana, Benin and Peru) also experienced significant increases in per capita food consumption which originated to a large extent in the increases in roots and tubers production. Despite these country examples, the general trend in recent years has been for average per capita food consumption of these products in developing countries to increase only very slowly,

---

[11] The figures given here refer to the consumption of oils as well as that of oilcrops directly (soybeans, groundnuts, etc.) or in the form of derived products other than oil, all measured in oil equivalent. This consumption of oilcrops in forms other than oil is particularly important in some countries.

Table 2.7 Changes in the commodity composition of food consumption, major country groups

Kg/person/year	1964/66	1974/76	1984/86	1997/99	2015	2030
	World					
Cereals, food	147	151	168	171	171	171
Cereals, all uses	283	304	335	317	332	344
Roots and tubers	83	80	68	69	71	74
Sugar (raw sugar equivalent)	21	23	24	24	25	26
Pulses, dry	9	7	6	6	6	6
Vegetable oils, oilseeds and products (oil eq.)	6	7	9	11	14	16
Meat (carcass weight)	24	27	31	36	41	45
Milk and dairy, excl. butter (fresh milk eq.)	74	75	79	78	83	90
Other food (kcal/person/day)	208	217	237	274	280	290
Total food (kcal/person/day)	2 358	2 435	2 655	2 803	2 940	3 050
	Developing countries					
Cereals, food	141	150	172	173	173	172
Cereals, all uses	183	201	234	247	265	279
Roots and tubers	75	77	62	67	71	75
(Developing minus China)	62	61	57	63	69	75
Sugar (raw sugar equivalent)	14	16	19	21	23	25
Pulses, dry	11	8	8	7	7	7
Vegetable oils, oilseeds and products (oil eq.)	5	5	8	10	13	15
Meat (carcass weight)	10	11	16	26	32	37
Milk and dairy, excl. butter (fresh milk eq.)	28	30	37	45	55	66
Other food (kcal/person/day)	122	129	155	224	240	250
Total food (kcal/person/day)	2 054	2 152	2 450	2 681	2 850	2 980
	Industrial countries					
Cereals, food	136	136	147	159	158	159
Cereals, all uses	483	504	569	588	630	667
Roots and tubers	77	68	69	66	63	61
Sugar (raw sugar equivalent)	37	39	33	33	32	32
Pulses, dry	3	3	3	4	4	4
Vegetable oils, oilseeds and products (oil eq.)	11	15	17	20	22	23
Meat (carcass weight)	62	74	81	88	96	100
Milk and dairy, excl. butter (fresh milk eq.)	186	192	212	212	217	221
Other food (kcal/person/day)	461	485	510	516	540	550
Total food (kcal/person/day)	2 947	3 065	3 206	3 380	3 440	3 500
	Transition countries					
Cereals, food	211	191	183	173	176	173
Cereals, all uses	556	719	766	510	596	685
Roots and tubers	148	132	114	104	102	100
Sugar (raw sugar equivalent)	37	45	46	34	35	36
Pulses, dry	5	4	3	1	1	1
Vegetable oils, oilseeds and products (oil eq.)	7	8	10	9	12	14
Meat (carcass weight)	43	60	66	46	54	61
Milk and dairy, excl. butter (fresh milk eq.)	157	192	181	159	169	179
Other food (kcal/person/day)	288	356	384	306	330	350
Total food (kcal/person/day)	3 223	3 386	3 379	2 906	3 060	3 180

Note: Cereal food consumption includes the grain equivalent of beer consumption and of corn sweeteners.

and indeed to stagnate if potatoes are excluded. Increases in some countries were compensated by declines in others. The drastic decline in food consumption of sweet potatoes in China had a decisive influence on these trends. Potatoes were the one commodity with consistent increases in per capita consumption in the developing countries.

These trends are expected to continue, as will the high dependence of several countries on roots and tubers as a major source of food. Per capita food consumption of all roots, tubers and plantains in developing countries should increase slowly, from 67 kg in 1997/99 to 75 kg in 2030 (Table 2.7). This increase partly reflects the fact that the downward pressure exerted in the historical period on the overall average by China's lower food consumption of sweet potatoes will be much weaker in the future. Much of the decline in China's per capita consumption of sweet potatoes (from 94 kg in 1974/76 to 40 kg in 1997/99) has already occurred and any future declines will be much smaller. Potatoes will continue to show relatively high-income elasticity in most developing countries, and average food consumption is projected to increase from 17 kg in 1997/99 to 26 kg in 2030. Another factor that could raise consumption is the potential for productivity increases in the other root crops (cassava and yams). It will be possible for more countries in sub-Saharan Africa to replicate the experiences of countries such as Nigeria, Ghana, Benin and Malawi, and increase their food consumption based on rapid productivity improvements in these crops.

Sugar shares many of the characteristics of vegetable oils as regards food consumption and trade in the developing countries. It is a fast-rising consumption item and a major export commodity of several countries, such as Brazil, Cuba and Thailand. In addition, several developing countries are becoming large and growing net importers (Egypt, the Islamic Republic of Iran and the Republic of Korea), making up for the lack of growth of imports into the industrial countries. The developing countries' average consumption is 21 kg/person/year, but it is higher (26 kg) if China is excluded; China has only 8 kg as a lot of saccharine is used instead of sugar. About a half of the developing countries consume less than 20 kg, and a quarter under 10 kg. The scope for consumption growth is still considerable and we project an increase in the average consumption of developing countries from 21 to 25 kg over the projection period. China's contribution to total growth should be more than in the past since the country could be discouraging the use of saccharine (see also Chapter 3).

2.4 Concluding remarks

Some brief conclusions may be drawn, as follows:

■ There will be significant progress in raising food consumption levels and improving nutrition. There will be significant reductions in the relative incidence of undernourishment (percentage of population affected), but these will not be translated into commensurate declines in the numbers undernourished because of population growth. Reduction in the absolute numbers of undernourished is likely to be a slow process. Numbers will decline from 776 million in 1997/99 to 610 million in 2015 and to 440 million in 2030.

■ The number of undernourished in developing countries stood at 815 million in 1990/92 (the three-year average used as the basis for defining the WFS target). This number is not likely to be halved by 2015, just as the absolute numbers in poverty will not be halved (from the level of 1990) according to the latest World Bank assessment. However, the *proportion* of the population undernourished could be nearly halved by 2015 – from 20 percent in 1990/92 to 11 percent in 2015.

■ The projected slow progress in reducing undernourishment will reflect the failure of many countries to transit to rapid economic development and poverty reduction. However, empirical evidence suggests that in the countries with high dependence on agriculture, assigning priority to the development of food production holds promise of overcoming the constraint to better nutrition represented by the unfavourable overall economic growth prospects.

■ In many countries, including some of the more populous ones, the relative incidence of undernourishment (percentage of the population) will decline significantly. Fewer countries

Table 2.8 Changes in the commodity composition of food consumption, developing regions

Kg/person/year	1964/66	1974/76	1984/86	1997/99	2015	2030
			Sub-Saharan Africa			
Cereals, food	115	115	118	123	131	141
Roots and tubers	186	190	169	194	199	202
Sugar (raw sugar equivalent)	6	8	9	10	11	13
Pulses, dry	10	10	9	9	10	11
Vegetable oils, oilseeds and products (oil eq.)	8	8	8	9	11	12
Meat (carcass weight)	10	10	10	9	11	13
Milk and dairy, excl. butter (fresh milk eq.)	29	28	32	29	31	34
Other food (kcal/person/day)	136	144	135	126	135	145
Total food (kcal/person/day)	2 057	2 079	2 058	2 195	2 360	2 540
			Near East/North Africa			
Cereals, food	172	189	204	209	206	201
Roots and tubers	16	21	31	34	33	33
Sugar (raw sugar equivalent)	19	24	29	28	29	30
Pulses, dry	7	7	7	7	7	7
Vegetable oils, oilseeds and products (oil eq.)	7	9	12	13	14	16
Meat (carcass weight)	12	14	20	21	29	35
Milk and dairy, excl. butter (fresh milk eq.)	69	72	83	72	81	90
Other food (kcal/person/day)	223	247	297	327	335	345
Total food (kcal/person/day)	2 291	2 592	2 953	3 006	3 090	3 170
			Latin America and the Caribbean			
Cereals, food	116	123	132	132	136	139
Roots and tubers	89	79	68	62	61	61
Sugar (raw sugar equivalent)	41	46	46	49	48	48
Pulses, dry	15	12	11	11	11	11
Vegetable oils, oilseeds and products (oil eq.)	6	8	11	13	15	16
Meat (carcass weight)	32	36	40	54	65	77
Milk and dairy, excl. butter (fresh milk eq.)	80	93	94	110	125	140
Other food (kcal/person/day)	228	239	251	262	280	300
Total food (kcal/person/day)	2 393	2 546	2 689	2 824	2 980	3 140
			South Asia			
Cereals, food	146	143	156	163	177	183
Roots and tubers	13	19	19	22	27	30
Sugar (raw sugar equivalent)	20	20	23	27	30	32
Pulses, dry	15	13	12	11	9	8
Vegetable oils, oilseeds and products (oil eq.)	5	5	6	8	12	14
Meat (carcass weight)	4	4	4	5	8	12
Milk and dairy, excl. butter (fresh milk eq.)	37	38	51	68	88	107
Other food (kcal/person/day)	81	85	100	129	150	160
Total food (kcal/person/day)	2 016	1 986	2 204	2 403	2 700	2 900
			East Asia			
Cereals, food	146	162	201	199	190	183
Roots and tubers	94	94	67	66	64	61
Sugar (raw sugar equivalent)	5	6	10	12	15	17
Pulses, dry	8	4	4	2	2	2
Vegetable oils, oilseeds and products (oil eq.)	3	4	6	10	13	16
Meat (carcass weight)	9	10	17	38	50	59
Milk and dairy, excl. butter (fresh milk eq.)	4	4	6	10	14	18
Other food (kcal/person/day)	100	107	149	290	315	340
Total food (kcal/person/day)	1 958	2 105	2 559	2 921	3 060	3 190

Note: Cereal food consumption includes the grain equivalent of beer consumption and of corn sweeteners.

than at present will have high incidence of undernourishment, none of them in the most populous class. The problem of undernourishment will tend to become smaller in terms of both absolute numbers affected and, even more, in relative terms, hence it will become more tractable through policy interventions, both national and international.

■ Despite this slow pace of progress in reducing the incidence of undernourishment, the projections imply a considerable overall improvement. In the developing countries the numbers well fed (i.e. not classified as undernourished according to the criteria used here) could increase from 3.8 billion in 1997/99 (83 percent of their population) to 5.2 billion in 2015 (89 percent of the population) and to 6.4 billion (94 percent) in 2030. That will be no mean achievement.

CHAPTER 3

Prospects for aggregate agriculture and major commodity groups

This chapter deals with the trends and future outlook of world food and agriculture in terms of the main commodity sectors. A brief introduction to the subject is given first, presenting trends and prospects for total agriculture (the aggregates of all crops and livestock products).

3.1 Aggregate agriculture: Historical trends and prospects

The historical evidence suggests that the growth of the productive potential of global agriculture has so far been more than sufficient to meet the growth of effective demand. This is what the long-term term decline in the real price of food suggests (Figure 3.1, see also World Bank, 2000a). In practice, world agriculture has been operating in a demand-constrained environment. This situation has coexisted with hundreds of millions of the world population not having enough food to eat. This situation of un-met demand[1] coexisting with actual or potential plenty is not, of course, specific to food and agriculture. It is found in other sectors as well, such as housing, sanitation and the health services.

Limits on the demand side at the global level reflected three main factors: (i) the slowdown in population growth from the late 1960s onwards (see Chapter 2); (ii) the fact that a growing share of the world population has been attaining fairly high levels of per capita food consumption, beyond which the scope for further increases is rather limited (Table 2.2); and (iii) the fact that those who did not have enough to eat were too poor to afford more food and cause it to be produced, or did not have the resources and other means to produce it themselves.

The first two factors will continue to operate in the future. Their influence will be expressed as lower growth rates than in the past of demand and, at the global level, also of production. The third factor will also continue to play a role, given that the overall economic outlook indicates that poverty will continue to be widespread in the future (see Chapter 2, Section 2.2.2). It follows that for a rather significant part of world population the *potential* demand for food will not be expressed fully as *effective* demand. Thus, the past trends of decelerating growth of demand will likely continue and perhaps intensify.

---

[1] The terms "demand" and "consumption" are used interchangeably. Both terms comprise all forms of use, i.e. food, feed, seed and industrial use as well as losses and waste (other than household waste). Demand for as well as supply from stocks, are disregarded. Given the 30-year time horizon of the study, a separate treatment of stock changes would unnecessarily complicate the analysis.

Figure 3.1 World market prices, 1960-2001 (constant 1990 US$)

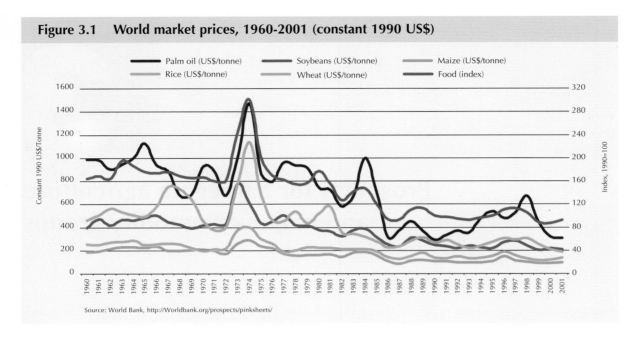

Source: World Bank, http://Worldbank.org/prospects/pinksheets/

However, on the production side, there is no assurance that the past experience, when the world's production potential was more than sufficient to meet the growth of demand, will continue, even when demand growth will be much lower than in the past. The natural resources per head of the growing population (e.g. land or water resources per person) will certainly continue to decline and the yield growth potential is more limited than in the past. It remains to be seen whether advances in technology and related factors (e.g. investment, education, institutions and improved farm management) that underpinned the past growth of production will continue to more than make up for the declining resources per person. The future may be different if we are now nearer critical thresholds, e.g. yield ceilings imposed by plant physiology or availability of water resources for maintaining and/or expanding irrigation. On the positive side, there are those who think that biotechnology has the potential of helping to overcome constraints to further increases in production (Chapters 4 and 11, see also Evenson, 2002).

We present in this section a brief overview of what we can expect in terms of increases of aggregate demand for, and production of, agricultural products. The figures we use refer to the aggregate volume of demand and production of the crop and livestock sectors. They are obtained by multiplying physical quantities of demand or production times price for each commodity and summing up over all commodities (each commodity is valued at the same average international price[2] in all countries in all years). The resulting time series is an index of volume changes over time of aggregate demand and production. The movements in this aggregate indicator are rarely sufficient for us to analyse and understand the forces that shape the evolution of agricultural variables in their different dimensions. The commodities included (see list in Appendix 1) are very diverse from the standpoint of what determines their production, demand and trade. For this reason, the subsequent sections of this chapter analyse and present the historical experience and prospects of world agriculture in terms of the main commodity sectors. Sections 3.2-3.5 deal with the basic food commodities: cereals, livestock products, oilcrops and the roots, tubers and plantains group. Section 3.6 covers more summarily a selection of the main export commodities of the developing countries: sugar, bananas, coffee, cocoa and natural rubber. The commodities dealt with in sections 3.2-3.6 account for 79 percent of the world's aggregate agricultural output.

---

2   International dollar prices, averages for 1989/91, used for constructing the production index numbers in FAOSTAT.

Table 3.1 Growth rates of aggregate demand and production (percentage p.a.)

	1969-99	1979-99	1989-99	1997/99 -2015	2015-30	1997/99 -2030
Demand						
World	2.2	2.1	2.0	1.6	1.4	1.5
Developing countries	3.7	3.7	4.0	2.2	1.7	2.0
idem, excl. China	3.2	3.0	3.0	2.4	2.0	2.2
Sub-Saharan Africa	2.8	3.1	3.2	2.9	2.8	2.9
idem, excl. Nigeria	2.5	2.4	2.5	3.1	2.9	3.0
Near East/North Africa	3.8	3.0	2.7	2.4	2.0	2.2
Latin America and the Caribbean	2.9	2.7	3.0	2.1	1.7	1.9
idem, excl. Brazil	2.4	2.1	2.8	2.2	1.8	2.0
South Asia	3.2	3.3	3.0	2.6	2.0	2.3
East Asia	4.5	4.7	5.2	1.8	1.3	1.6
idem, excl. China	3.5	3.2	2.8	2.0	1.7	1.9
Industrial countries	1.1	1.0	1.0	0.7	0.6	0.7
Transition countries	-0.2	-1.7	-4.4	0.5	0.4	0.5
Production						
World	2.2	2.1	2.0	1.6	1.3	1.5
Developing countries	3.5	3.7	3.9	2.0	1.7	1.9
idem, excl. China	3.0	3.0	2.9	2.3	2.0	2.1
Sub-Saharan Africa	2.3	3.0	3.0	2.8	2.7	2.7
idem, excl. Nigeria	2.0	2.2	2.4	2.9	2.7	2.8
Near East/North Africa	3.1	3.0	2.9	2.1	1.9	2.0
Latin America and the Caribbean	2.8	2.6	3.1	2.1	1.7	1.9
idem, excl. Brazil	2.3	2.1	2.8	2.2	1.8	2.0
South Asia	3.1	3.4	2.9	2.5	2.0	2.2
East Asia	4.4	4.6	5.0	1.7	1.3	1.5
idem, excl. China	3.3	2.9	2.4	2.0	1.8	1.9
Industrial countries	1.3	1.0	1.4	0.8	0.6	0.7
Transition countries	-0.4	-1.7	-4.7	0.6	0.6	0.6
Population						
World	1.7	1.6	1.5	1.2	0.9	1.1
Developing countries	2.0	1.9	1.7	1.4	1.1	1.3
idem, excl. China	2.3	2.2	2.0	1.7	1.3	1.5
Sub-Saharan Africa	2.9	2.9	2.7	2.6	2.2	2.4
idem, excl. Nigeria	2.9	2.9	2.7	2.6	2.3	2.4
Near East/North Africa	2.7	2.6	2.4	1.9	1.5	1.7
Latin America and the Caribbean	2.1	1.9	1.7	1.3	0.9	1.1
idem, excl. Brazil	2.1	1.9	1.8	1.4	1.0	1.2
South Asia	2.2	2.1	1.9	1.6	1.1	1.3
East Asia	1.6	1.5	1.2	0.9	0.5	0.7
idem, excl. China	2.0	1.8	1.6	1.2	0.9	1.0
Industrial countries	0.7	0.7	0.7	0.4	0.2	0.3
Transition countries	0.6	0.5	0.1	-0.2	-0.3	-0.2

The overall picture for total agriculture is presented in Table 3.1. At the world level, the growth of demand for all crop and livestock products is projected to be lower than in the past, 1.6 percent p.a. in the period 1997/99-2015 compared with 2.1 percent p.a. in the preceding 20 years 1979-99. The difference between the past and future growth rates of demand is nearly equal to the difference in the population growth rates. However, the past growth rates of demand had been depressed because of the collapse of production and consumption in the transition economies. We would have expected that the cessation of declines and eventual turnaround of demand in these countries (turning from negative to positive, Table 3.1) would have largely cancelled the effect of lower population growth rate at the global level, at least for the first subperiod of the projections. In practice, it is mainly the slowdown in the growth of demand in the developing countries, and in particular in China, that accounts for a large part of the global deceleration. Why this should be so is shown in the more detailed regional numbers of Table 3.1. They show that deceleration in the developing countries outside China (from 3.0 percent p.a. in 1979-99 to 2.4 percent p.a. in 1997/99-2015) is only a little more than the deceleration in their population growth (from 2.2 percent to 1.7 percent, respectively), an expected outcome given the operation of the factors mentioned earlier. However, a better idea about the roles of the above-mentioned factors making for deceleration (lower population growth and approaching saturation levels) can be had from the data and projections presented in Table 3.2. In it the developing countries are grouped into two sets: those that start in 1997/99 with fairly high per capita food consumption (over 2 700 kcal/person/day) and, therefore, face less scope than before for increasing consumption, and all the rest, that is those with 1997/99 kcal under 2 700.

China carries a large weight in the former group, so its example can be used to illustrate why a drastic deceleration is foreseen for the developing countries. China has already attained a fairly high level of per capita food consumption of the main commodities, a total of 3 040 kcal/person/day in 1997/99. In the projections, it increases further to 3 300 kcal by 2030. This is nearly the level of the industrial countries. Going from 3 040 to 3 300 kcal in 32 years is a growth rate of only 0.3 percent p.a. In contrast, in the preceding three decades the average growth rate of per capita kcal was 1.6 percent p.a. Therefore, the higher level from which China now starts imposes a limit on how fast per capita consumption may grow in the future. In addition, China's population growth in the past was 1.5 percent p.a., but in the projection period it will be only 0.5 percent p.a. These numbers vividly demonstrate the effect of slower population growth and near-saturation levels of per capita food consumption in depressing the aggregate growth of demand for food. This deceleration happened in the historical period in countries transiting from low to high per capita consumption and to low demographic growth. For example, the aggregate food demand growth rate of Japan was 4.7 percent p.a. in the 1960s and fell progressively to 2.2 percent in the 1970s, to 2.0 percent in the 1980s and to 0.8 percent in the 1990s.

When such deceleration occurs in China and in a few other large developing countries, the whole aggregate of the developing countries, and indeed the world, will be affected downwards. There are several other developing countries in situations roughly similar to those of China, i.e. they have fairly high levels of per capita consumption and are experiencing a significant slowdown in their population growth. As noted, the first group shown in Table 3.2 comprises the developing countries starting with over 2 700 kcal/person/day in 1997/99. There are 29 of them (including China) but they account for a half of the population of the developing countries, since the group includes many of the largest developing countries in terms of population.[3] They account for an even larger share of aggregate consumption of the developing countries, 66 percent in 1997/99. As in the case of China, this group of countries has much more limited scope than in the past for increasing per capita consumption, given that the group's average already stands at 3 030 kcal/person/day. We project this average to grow to 3 275 kcal/person/day by 2030. In parallel, the growth rate of their population is projected to be much slower than in the past, 0.9 percent p.a. compared with 1.8 percent p.a. in

---

[3] China, Indonesia, Brazil, Mexico, Nigeria, Egypt, the Islamic Republic of Iran and Turkey.

the preceding three decades. The net effect of these demand-limiting factors is that the group's aggregate demand is projected to decelerate drastically, from 4.2 percent per year between 1969 and 1999, to 1.9 percent p.a. in the period to 2015 and to 1.5 percent p.a. in the following 15 years to 2030.[4]

In contrast, the growth of demand in the other developing countries, those with under 2 700 kcal/person/day in 1997/99, is projected to decelerate less than their population: the growth rate of their demand falls from 2.9 percent p.a. in the preceding three decades to 2.5 percent p.a. in the

period to 2030, while their population growth rate falls from 2.3 percent p.a. to 1.6 percent p.a. This group of countries includes India with its nearly one billion population out of the group's 2.2 billion. The prospect that India will not move much towards meat consumption (see Section 3.3 below) contributes to limit the growth rate of total demand for both food and feed. In the past, the aggregate demand of the developing countries was greatly influenced by the rapid growth of apparent meat consumption in China (see, however, Section 3.3 below for possible overestimation of meat

Table 3.2 Growth rates of demand and production in different country groups

	1969 -99	1979 -99	1989 -99	1997/99 -2015	2015 -2030	1997/99 -2030
Demand (percentage p.a.)						
Developing countries	3.7	3.7	4.0	2.2	1.7	2.0
Countries with over 2 700 kcal/person/day*	4.2	4.2	4.6	1.9	1.5	1.7
Other developing countries	2.9	2.9	2.7	2.7	2.2	2.5
Production (percentage p.a.)						
Developing countries	3.5	3.7	3.9	2.0	1.7	1.9
Countries with over 2 700 kcal/person/day*	4.0	4.2	4.6	1.8	1.4	1.6
Other developing countries	2.7	2.8	2.5	2.5	2.1	2.3
Population (percentage p.a.)						
Developing countries	2.0	1.9	1.7	1.4	1.1	1.3
Countries with over 2 700 kcal/person/day*	1.8	1.6	1.4	1.0	0.7	0.9
Other developing countries	2.3	2.3	2.1	1.8	1.4	1.6

	1964/66	1974/76	1984/86	1997/99	2015	2030
Population (million)						
Developing countries	2 295	2 925	3 597	4 572	5 827	6 869
Countries with over 2 700 kcal/person/day*	1 250	1 594	1 918	2 343	2 798	3 111
Other developing countries	1 045	1 331	1 679	2 229	3 029	3 758
Kcal/person/day						
Developing countries	2 054	2 152	2 450	2 681	2 845	2 980
Countries with over 2 700 kcal/person/day*	2 075	2 243	2 669	3 027	3 155	3 275
Other developing countries	2 029	2 044	2 200	2 316	2 560	2 740

* In 1997/99.

[4] Note that in China's projected 3 300 kcal/person/day in 2030 are included per capita annual meat consumption of 69 kg and 380 kg of cereals (for all uses). To have less deceleration than foreseen here in the growth of aggregate demand would require that these projected levels be even higher (see following commodity sections).

production and consumption in China). The prospect that China's influence will be much weaker in the future and that it will not be replaced by a similar boom in other large countries, is one of the major factors making for the projected deceleration in the aggregate demand of the developing countries.

At the world level, production equals consumption, so the preceding discussion about global demand growth prospects applies also to that of global production. For the individual countries and country groups, however, the two growth rates differ depending on movements in their net agricultural trade positions. In general, the growth rates of production in the developing regions have been below those of demand. As a result their agricultural imports have been growing faster than their exports, leading to gradual erosion of their traditional surplus in agricultural trade (crop and livestock products, primary and processed). The trend has been for this surplus to diminish and to turn into a net deficit in most years in the 1990s (Figure 3.2). In the last 15 years, the net balance reached a peak of US$16 billion surplus in 1986 and a trough of US$6 billion deficit ten years later, before some recovery in the subsequent years. The

net deficit was still US$5.9 billion in 2000, the latest year for which we have data. This prospect had been foreshadowed in our earlier projections to 2010 from base year 1988/90 (Alexandratos, 1995, p. 121, see also Chapter 9).

Behind these trends in the value of the net trade balance of agriculture have been movements in both quantities and prices of the traded commodities and the policies that influenced them. Several factors, often widely differing among commodities, played a role in these developments. For example, the drastic decline in the developing countries' traditional net trade surplus in sugar is in part a result of the fact that several developing countries became major importers of sugar. In parallel, the reduction also reflects the effects of the heavy domestic support and trade protection in major sugar importing countries such as the United States and Japan, or in former net importing countries, such as the EU, which became a significant net exporter as a result of these policies (see Section 3.6 below and Figure 3.16). The emergence of several developing countries as major importers has also helped to cause the rapid declines in the net trade surplus of the developing countries as a whole in the oilcrops complex

Figure 3.2   Developing countries, net agricultural trade balances, 1984-2000

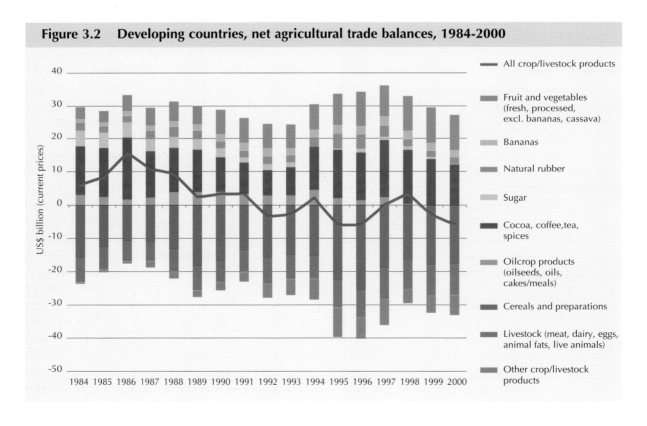

(vegetable oils, oilseeds and cakes/meals, see Section 3.4). In the case of the "non-competing" exportables of the developing countries, such as coffee, it has been the slow growth of demand in the main consuming and importing countries (the industrial ones) – in combination with falling prices – that kept the trade balance from growing.

The evolution of the overall net agricultural trade balance of the developing countries as a whole does not by itself denote overall improvement or deterioration from a developmental standpoint. The aggregate of the developing countries is a composite of very widely differing country and commodity situations. For some countries, a declining agricultural trade balance (or a growing deficit) is an indicator of progress towards improved welfare. This is the case of countries such as the Republic of Korea, in which the growing agricultural deficit has gone hand in hand with high rates of overall development and growing food consumption. The declining overall balance also reflects the rapid growth in such things as China's growing imports of vegetable oils (a positive development overall as they contribute to improve food consumption and are paid for by growing industrial export earnings); or cotton imports into several developing countries, which sustain their growing exports of textiles. However, a declining agricultural balance is a negative developmental outcome in countries that still depend heavily on export earnings from agriculture and/or have to divert scarce foreign exchange resources to pay for growing food imports (eventually building up unsustainable foreign debt). It is an even more negative indicator when such food imports are not associated with rising food consumption per capita and improved food security, but are necessary just to sustain minimum levels of food consumption – which is not an uncommon occurrence.

The projections indicate a continuing deepening of the net trade deficit of the developing countries. Their net imports of the main commodities in which they are deficit, mainly cereals and livestock products, will continue to rise fairly rapidly. In parallel, their net trade surplus in traditional exports (e.g. tropical beverages and bananas) will either rise less rapidly than their net imports of cereals and livestock or outright decline (e.g. vegetable oils and sugar). The particulars relating to future trade outcomes for these

commodities are discussed in the following sections of this chapter. Trade policy issues are discussed in Chapter 9.

Concerning production, evaluation country by country and commodity by commodity suggests that the resource potential and productivity gains required to achieve the aggregate production growth rates shown in Tables 3.1 and 3.2 (and in the commodity sections of this chapter) are by and large feasible. This is based on the assumption that policies will favour rather than discriminate against agriculture. More information on the agronomic dimensions of this proposition is presented in Chapter 4. It is just worth mentioning here that the bulk of the increases in production will come, as they did in the past, from increases in yield and more intensive use of land.

This may sound odd in the light of the widely held view that the potential for further growth of yields is now much more limited than it was in the historical period, which included the heyday of the spread of the green revolution. The two propositions (that yield growth potential is less than in the past and that yield growth will continue to be the mainstay of the production increases) are not necessarily contradictory. What makes them compatible with each other is that in future slower growth in production is needed than in the past. The issue is not really whether the yield growth rates will be slower than in the past. They will. Rather the issue is if such slower growth is sufficient to deliver the required additional production. Naturally, even this slower yield growth may not happen unless we make it happen. This requires continued support to agricultural research and policies and other conditions (education, credit, infrastructure, etc.) to make it profitable and possible for farmers to exploit the yield growth potential.

At the global level sufficient production potential can be developed for meeting the expected increases in effective demand in the course of the next three decades. But this is not to say that all people will be food-secure in the future. Far from it, as Chapter 2 has shown. The interaction between food security and food production potential is very much a local problem in poor and agriculturally dependent societies. Many situations exist where production potential is limited (e.g. in the semi-arid areas, given existing and accessible technology and infrastructure) and a good part of the population depends on such poor agricultural resources for food and more general livelihood. Unless local agriculture is developed and/or other income earning opportunities open up, the food insecurity determined by limited local production potential will persist, even in the middle of potential plenty at the world level. The need to develop local agriculture in such situations as the condition sine qua non for improved food security cannot be overemphasized.

3.2 Cereals

3.2.1 Past and present

Cereals continue to be overwhelmingly the major source of food supplies for direct human consumption. In addition, some 660 million tonnes, or 35 percent of world consumption, are being used as animal feed. Therefore, the growth of aggregate demand for cereals for all uses is a good (though far from perfect) proxy for monitoring trends in world food supplies.

Demand: general historical experience.
Historically, the growth rate of global demand for cereals (for all uses) has been in long-term decline.

This is clearly seen in Tables 3.3 and 3.4 which show the historical developments in all major cereal sector variables for the world and the standard regions used in this report. To gain a better understanding of the main factors that led to the deceleration, we must distinguish between different country groups and historical periods. This is done in Figure 3.3 which plots the growth rates of aggregate demand for successive (moving) 15-year periods of different country groups over the time span 1961-99. The world growth rate declined progressively from 3.1 percent in the first 15-year period (1961-76) to 1.1 percent p.a. in the last 15-year period ending in 1999. We should not be surprised if our projections were to show further declines in the growth rates of world aggregate cereals demand (hence of global production) in the future.

The deceleration in population growth certainly played a role in this slowdown, as has the fact that a growing proportion of the world population has been gradually attaining levels of per capita food consumption that leave less scope than in the past for further increases. However, these two factors explain only part of the decline, given that there are still grossly unsatisfied nutritional needs affecting large parts of the world population. Other

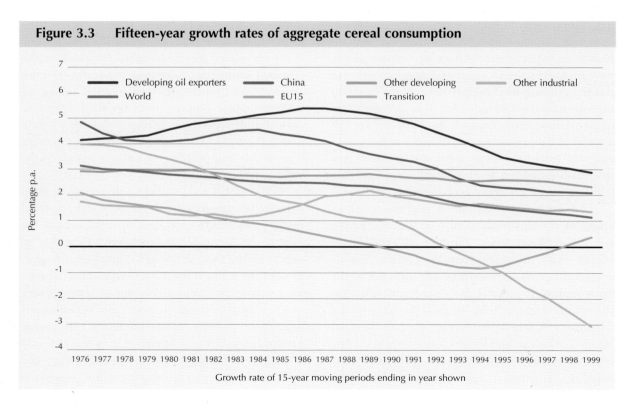

Figure 3.3   Fifteen-year growth rates of aggregate cereal consumption

Growth rate of 15-year moving periods ending in year shown

Table 3.3 Cereal balances, world and major country groups

	Demand				Production	Net trade	SSR[1]		Growth rates, percentage p.a.			
	Per capita (kg)		Total (million tonnes)		(million tonnes)		%			Demand	Production	Population
	Food	All uses	Food	All uses								
World												
1964/66	147	283	489	941	940	4	100		1969-99	1.9	1.8	1.7
1974/76	151	304	612	1 233	1 268	1	103		1979-99	1.4	1.4	1.6
1984/86	168	334	810	1 608	1 659	2	103		1989-99	1.0	1.0	1.5
1997/99	171	317	1 003	1 864	1 889	9	101		1997/99-2015	1.4	1.4	1.2
2015	171	332	1 227	2 380	2 387	8	100		2015-30	1.2	1.2	0.9
2030	171	344	1 406	2 830	2 838	8	100		1997/99-2030	1.3	1.3	1.1
Developing countries												
1964/66	141	183	324	419	399	-17	95		1969-99	3.0	2.8	2.0
1974/76	150	200	438	586	563	-39	96		1979-99	2.6	2.5	1.9
1984/86	172	234	618	840	779	-66	93		1989-99	2.2	2.1	1.7
1997/99	173	247	790	1 129	1 026	-103	91		1997/99-2015	1.9	1.6	1.4
2015	173	265	1 007	1 544	1 354	-190	88		2015-30	1.5	1.3	1.1
2030	172	279	1 185	1 917	1 652	-265	86		1997/99-2030	1.7	1.5	1.3
Industrial countries												
1964/66	136	483	94	335	351	30	105		1969-99	1.0	1.5	0.7
1974/76	136	504	103	384	456	55	119		1979-99	1.0	0.8	0.7
1984/86	147	569	119	464	614	106	132		1989-99	1.7	1.4	0.7
1997/99	159	588	142	525	652	111	124		1997/99-2015	0.8	1.1	0.4
2015	158	630	150	600	785	187	131		2015-30	0.6	0.9	0.2
2030	159	667	155	652	900	247	138		1997/99-2030	0.7	1.0	0.3
Transition countries												
1964/66	211	556	70	186	189	-9	102		1969-99	-0.2	-0.3	0.6
1974/76	191	719	70	263	249	-16	94		1979-99	-1.9	-1.1	0.5
1984/86	183	766	73	304	266	-37	87		1989-99	-4.9	-4.2	0.1
1997/99	173	510	72	211	210	1	100		1997/99-2015	0.7	1.0	-0.2
2015	176	596	70	237	247	10	104		2015-30	0.7	1.0	-0.3
2030	173	685	66	262	287	25	110		1997/99-2030	0.7	1.0	-0.2

Memo item 1. **Growth rates, percentage p.a.**

	Total demand – all cereals						Population				
	1969 -99	1979 -99	1989 -99	1997/99 -2015	2015 -30	1997/99 -2030	1979 -99	1989 -99	1997/99 -2015	2015 -30	1997/99 -2030
Developing oil exporters[2]	4.2	3.4	2.7	2.1	1.8	1.9	2.4	2.1	1.7	1.3	1.5
China	3.2	2.3	2.1	1.3	0.9	1.1	1.3	1.0	0.7	0.3	0.5
Other developing countries	2.6	2.5	2.1	2.1	1.7	1.9	2.1	2.0	1.7	1.3	1.5
EU15	0.3	0.1	2.0	0.4	0.2	0.3	0.3	0.3	0.0	-0.2	-0.1
Other industrial countries	1.5	1.5	1.5	1.0	0.7	0.8	1.0	1.0	0.6	0.4	0.5

Memo item 2. **World demand (all uses) by commodity (million tonnes), food kg/capita and percentage p.a.**

	1964/ 66	1974/ 76	1984/ 86	1997/ 99	2015	2030	1979 -99	1989 -99	1997/99 -2015	2015 -30	1997/99 -2030
Wheat, million tonnes	273	357	504	582	730	851	1.5	0.8	1.3	1.0	1.2
Rice (milled), million tonnes	174	229	308	386	472	533	2.1	1.6	1.2	0.8	1.0
Rice, per capita food only (kg)											
East Asia	84	93	109	106	100	96					
East Asia excl. China	110	125	130	132	129	124					
South Asia	73	69	75	79	84	81					
Coarse grains, million tonnes	493	648	796	896	1 177	1 446	1.0	1.0	1.6	1.4	1.5

1   SSR = self-sufficiency rate = production/demand.
2   Near East/North Africa (excl. Turkey, Morocco, Tunisia, Afghanistan, Jordan, Lebanon and Yemen) plus Mexico, Venezuela, Indonesia, Angola, Bolivia, the Congo, Ecuador, Gabon, Nigeria and Trinidad and Tobago.

Figure 3.4 Per capita consumption (all uses) of individual cereals

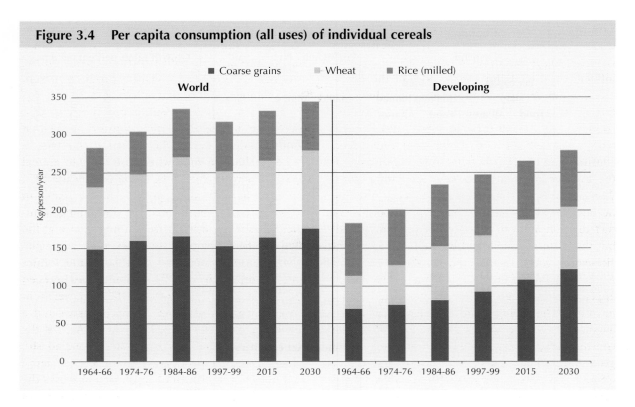

factors must be taken into account if we are to draw valid lessons from the historical experience for exploring the future. The forces that made for this decline may be summarized as follows:

■ In developing countries, the past high growth rate was fuelled up to about the mid-1980s by the rapid growth in production and consumption of China, and of consumption in the developing oil-exporting countries.[5] In this group of countries, the rapid growth of demand was supported by quantum jumps in net cereal imports (see below). Growth of demand became much weaker in subsequent periods. This reflected partly the achievement of mid- to high per capita consumption levels in several countries, and partly changes in economic conditions, mainly the drastic slowdown in the growth of incomes and export earnings from oil. In contrast, the growth of aggregate demand in the remaining developing countries (which, among themselves, account for 30 percent of world consumption) slowed by much less – from a peak of 3.0 percent p.a. in 1966-81 (when population growth was 2.4 percent) to

2.3 percent in the latest 15-year period 1985-99 (population growth 2.1 percent).

■ The collapse of feed use of cereals in the transition economies in the 1990s following the contraction of their livestock sectors (total use declined from 317 million tonnes in 1989/91 to 197 million tonnes in 1997/99) was a major factor in bringing down the growth of world demand in the 1990s.

■ A third major factor was, up to the early 1990s, the decline in cereals used for animal feed (and their replacement by largely imported cereal substitutes – oilmeals, cassava, etc.) in the EU, following the high internal prices of the Common Agricultural Policy (CAP). This trend was reversed from 1993 onwards following the McSharry reform of the CAP.

Evolution of demand: commodities and categories of use. For *rice*, the characteristic feature of the historical evolution is that per capita consumption (for all uses, but overwhelmingly for direct food, although in some countries rice is also used as animal feed) has tended to level off (see Figure 3.4).

5 Countries with over 50 percent of their total merchandise exports coming from export of fuels in the 1980s (data from World Bank, 2001b). In this category belong the countries of the region Near East/North Africa (except Morocco, Tunisia, Afghanistan, Jordan, Lebanon, Turkey and Yemen) plus Angola, the Congo, Gabon, Nigeria, Bolivia, Ecuador, Mexico, Trinidad and Tobago, Venezuela and Indonesia.

This trend has been most evident in several countries of East Asia. Given their large weight in world rice consumption, developments in this region influence the totals for the world and the developing countries in a decisive way. Thus, the levelling off of average consumption in the developing countries since about the mid-1980s reflects essentially small declines in East Asia's per capita food consumption of rice (from 109 to 106 kg), although the average of South Asia kept increasing (from 75 to 79 kg, see Table 3.3, memo item 2). Absolute declines in per capita rice consumption because of diet diversification in Asia are not yet widespread, but the patterns established by the more advanced rice-eating countries (e.g. Japan, the Republic of Korea and Taiwan Province of China) have started appearing, although in much attenuated form, in other developing countries, e.g. China and Thailand. Despite these trends in per capita consumption, the aggregate world demand for rice grew faster than that of both wheat and coarse grains. This reflects essentially the fact that rice is used predominantly for food in the most populous region, Asia, while a good part of coarse grains, but also increasingly of wheat, are used for animal feed. Feed use suffered a severe setback in the last ten

years or so because of the above factors (transition economies, the EU) and, to a lesser extent, other factors having to do with structural change in the livestock economy (shift of meat production from beef to poultry and pork, see Section 3.3).

The per capita consumption of *wheat* kept increasing in the developing countries albeit at a decelerating rate. In the industrial countries, a growing share of total wheat use went to animal feed (37 percent currently, 45 percent in the EU). Growing imports made possible the increases in consumption in the majority of developing countries. This is not evident from a first glance at the totals of the developing countries: their aggregate wheat consumption grew by 185 million tonnes and net imports by only 24 million tonnes between 1974/76 and 1997/99. However, the large weight in these aggregates of India, China, Pakistan and a few other countries (Argentina, Saudi Arabia, the Syrian Arab Republic, Turkey and Bangladesh), which increased production to match, or more than match, the growth of consumption masks the growing dependence of the great majority of the developing countries on wheat imports. In practice, the rest of the developing countries increased consumption by 54 million tonnes and

Figure 3.5   Increases in wheat consumption (all uses) and in net imports, 1974/76 to 1997/99, developing countries with over 500 thousand tonnes increase in consumption

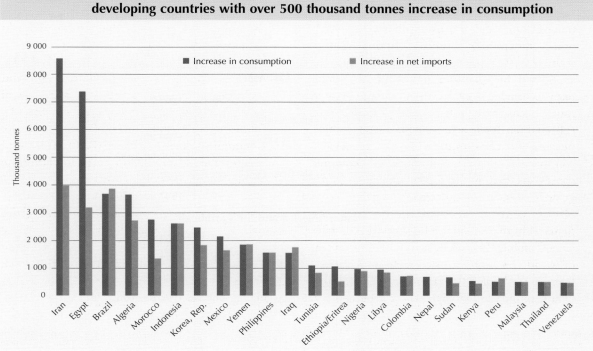

Table 3.4   Cereal balances by developing regions, all cereals (wheat, rice [milled], coarse grains)

	Demand				Production	Net trade	SSR[1]	Growth rates, percentage p.a.			
	Per capita (kg)		Total (million tonnes)		(million tonnes)		%		Demand	Production	Population
	Food	All uses	Food	All uses							
Sub-Saharan Africa											
1964/66	115	143	26	33	32	-2	97	1969-99	3.1	2.6	2.9
1974/76	115	143	34	43	40	-4	94	1979-99	3.4	3.4	2.9
1984/86	118	142	47	57	48	-10	85	1989-99	3.1	2.7	2.7
1997/99	123	150	71	86	71	-14	82	1997/99-2015	2.9	2.8	2.6
2015	131	158	116	139	114	-25	82	2015-30	2.7	2.6	2.2
2030	141	170	173	208	168	-40	81	1997/99-2030	2.8	2.7	2.4
Near East/North Africa											
1964/66	172	292	27	47	40	-5	86	1969-99	3.4	2.4	2.7
1974/76	189	309	39	64	55	-13	85	1979-99	2.7	2.4	2.6
1984/86	204	365	56	100	65	-38	65	1989-99	2.2	1.3	2.4
1997/99	209	352	79	133	83	-49	63	1997/99-2015	2.2	1.5	1.9
2015	206	368	107	192	107	-85	56	2015-30	1.8	1.5	1.5
2030	201	382	131	249	133	-116	54	1997/99-2030	2.0	1.5	1.7
South Asia											
1964/66	146	162	92	102	88	-10	86	1969-99	2.6	2.8	2.2
1974/76	143	162	114	128	125	-10	98	1979-99	2.6	2.7	2.1
1984/86	156	175	154	173	173	-3	100	1989-99	1.8	2.0	1.9
1997/99	163	182	208	234	239	-3	102	1997/99-2015	2.1	1.8	1.6
2015	177	200	295	335	323	-12	97	2015-30	1.5	1.3	1.1
2030	183	211	360	416	393	-22	95	1997/99-2030	1.8	1.6	1.3
East Asia											
1964/66	146	181	150	187	183	-5	98	1969-99	3.2	3.1	1.6
1974/76	162	211	212	275	266	-10	97	1979-99	2.5	2.5	1.5
1984/86	201	263	308	404	391	-12	97	1989-99	2.1	2.1	1.2
1997/99	199	290	366	534	507	-23	95	1997/99-2015	1.4	1.2	0.9
2015	190	317	404	675	622	-53	92	2015-30	1.0	0.9	0.5
2030	183	342	422	787	714	-73	91	1997/99-2030	1.2	1.1	0.7
Latin America and the Caribbean											
1964/66	116	207	29	51	56	5	109	1969-99	2.9	2.2	2.1
1974/76	123	239	39	76	77	-2	101	1979-99	2.3	1.8	1.9
1984/86	132	267	52	106	101	-3	96	1989-99	2.8	3.1	1.7
1997/99	132	285	66	142	125	-14	88	1997/99-2015	2.1	2.4	1.3
2015	136	326	85	203	188	-16	92	2015-30	1.6	1.7	0.9
2030	139	358	99	257	244	-13	95	1997/99-2030	1.9	2.1	1.1

[1] SSR = Self-sufficiency rate = production/demand.

Figure 3.6   Food and non-food use of coarse grains, developing regions and selected countries, average 1997/99

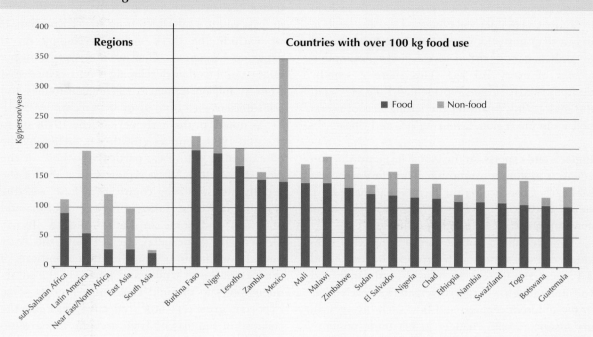

net imports by 39 million tonnes. Figure 3.5 shows some major countries, other than the above-mentioned ones, which increased consumption by over 500 thousand tonnes. They include countries that are major producers themselves (Egypt, the Islamic Republic of Iran, countries of North Africa, Mexico and Brazil) as well as those that are minor or non-producers of wheat – Indonesia, the Philippines, Colombia and the Republic of Korea. In addition, numerous smaller countries in the tropics depend entirely on imports for their consumption of wheat.

Apparent consumption of *coarse grains* grew fairly fast, although the above factors made for significant declines in feed use in certain parts of the world and led to a decline in per capita use in the world as a whole (Figure 3.4). The driving force has been use for animal feed in the developing countries, particularly in the last ten years, with China being a major contributor to such developments. The increase in world consumption between 1984/86 and 1997/99 was 100 million tonnes. This was made up of a 131 million tonnes increase in the developing countries (out of which 77 million tonnes for feed – 43 of which in China); declines of

64 million tonnes in the transition economies; and an increase of 33 million tonnes in the industrial countries. Overall, the pattern of the world coarse grains economy has undergone drastic structural change in the location of consumption. The major export markets shifted increasingly to the developing countries: increases in net coarse grains imports of the developing countries outside China supplied some 30 percent of the increase in their consumption.

Coarse grains as food. About three-fifths of world consumption of coarse grains is used for animal feed, hence the term "feedgrains" often used to refer to them. However, in many countries (mainly in sub-Saharan Africa and Latin America) they play a very important role in the total supplies of food for direct human consumption. Indeed in several countries food consumption of cereals is synonymous with coarse grains. A sample is given in Figure 3.6. The majority of these countries face food security problems, which underlines the importance of coarse grains in food security. At the global level, about a quarter of aggregate consumption of coarse grains is devoted to food,[6] but the share rises to 80 percent in sub-Saharan Africa. Here maize, millet, sorghum

---

[6] Food consumption of coarse grains includes the quantities used to produce beverages (mainly beer) as well as other derived food products, e.g. corn syrup, widely used as a sweetener substitute for sugar.

and other coarse grains (e.g. tef in Ethiopia) account for 3 out of every 4 kg of cereals consumed as food. Coarse grains are also used predominantly as food in South Asia (84 percent of aggregate consumption of coarse grains is for food), but there they account for only a minor part of cereal food consumption (1 out of every 7.5 kg). This share is rapidly declining (coarse grains represented 1 out of 4 kg in the mid-1970s), following the strong bias of cereal policies in the region, favouring rice and wheat.

Imports and exports. At the global level, production equals (roughly) consumption. Therefore what was said earlier concerning the factors that made for a steady decline in the growth rates of world cereal consumption, applies also to production.

However, this is not the case at the level of individual countries and country groups. Tables 3.3 and 3.4 show the extent to which production and consumption growth rates diverged from each other in the different regions, and how such divergences are associated with changes in net trade positions and self-sufficiency rates. In general, in the developing countries demand grew faster than production, so net imports increased from 39 million tonnes in the mid-1970s to 103 million tonnes in 1997/99 (Figure 3.7). Aggregate self-sufficiency (percentage of consumption covered by production) in these countries declined from 96 to 91 percent. If we exclude the three major developing cereal exporters (Argentina, Thailand and Viet Nam) net imports of the other developing countries increased from 51 million tonnes to 134 million tonnes and self-sufficiency fell from 93 to 88 percent.

As noted, in the early period (the 1970s up to the mid-1980s) import growth was fuelled by the oil-exporting countries, particularly those of the Near East/North Africa region, and a few of the rapidly industrializing countries in East Asia (the Republic of Korea, Taiwan Province of China and Malaysia) and some in Latin America (Brazil, Mexico and Venezuela). In addition, the early period saw quantum jumps in the net imports of the former Soviet Union and Japan. In the subsequent years to the mid-1990s, net imports of oil exporters grew very little, reflecting, *inter alia*, the collapse of imports of Iraq, and the turnaround from net importer to net exporter status of the Syrian Arab Republic (all cereals) and Saudi Arabia (wheat). However, growth of net imports of the oil

Figure 3.7   Net cereal imports, developing countries: comparisons of old projections with actual outcomes

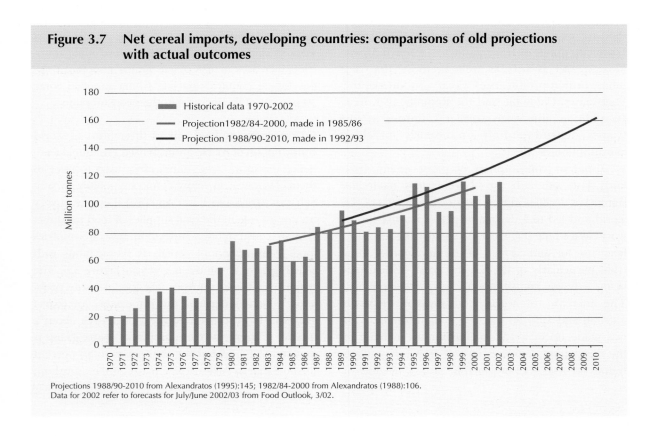

Projections 1988/90-2010 from Alexandratos (1995):145; 1982/84-2000 from Alexandratos (1988):106.
Data for 2002 refer to forecasts for July/June 2002/03 from Food Outlook, 3/02.

Figure 3.8   China's net trade of cereals

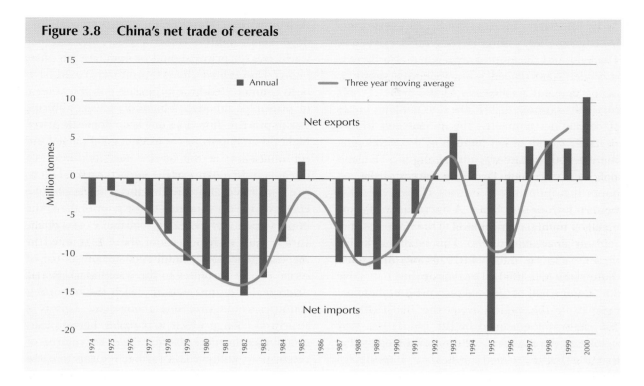

exporters resumed in the second half of the 1990s. The second major factor making for quasi stagnant world cereal trade in much of the 1990s has been the virtual disappearance of the transition economies as major net importers (and their transformation into small net exporters in some years), while the net imports of Japan stagnated, a development clearly associated with Japan's rapid growth of meat imports which substituted for imports of grains for feed (see Section 3.3).

However, other countries continued to expand net imports (the Republic of Korea, Taiwan Province of China, Malaysia and Brazil) and new ones were added to the list of growing net importers (the Philippines, Colombia, Peru and Chile). China must also be added to this list, as it became, albeit temporarily, a big importer in the mid-1990s (net imports of 20 million tonnes in 1995). However, such status of China did not last and the country turned into a small net exporter from 1997 onwards. Its net imports in the last quarter century are plotted in Figure 3.8. Taking smoothed three-year moving averages, its net trade status moved in the range from net imports of 14 million tonnes (early 1980s) to net exports of nearly 5 million tonnes in the three-year average 1997/99. This evidence does not point to China becoming a permanent large net importer in the

future as some studies had indicated (with the knock-on effects on world market prices that would reduce the import capacity of the poor food-deficit countries) (Brown, 1995; for critique see Alexandratos, 1996, 1997). More important for world markets have been the large fluctuations in China's net trade status.

Another remarkable development has been the fact that South Asia has not moved in the direction of becoming the large importer that some commentators thought back in the 1960s it would have needed to become to feed its growing population. In the mid-1960s, the region was a net importer of 10 million tonnes of cereals. This represented a crucial 11 percent of its food consumption of cereals which was very low (146 kg per capita), given that consumption of all food provided only 2 000 kcal/person/day. Thirty-three years on, the region's population had doubled, per capita food consumption had increased to 163 kg and net imports had fallen to only 3.4 million tonnes. This was a result of the miracle of the green revolution but also, on the negative side, a result of the persistence of poverty that prevented demand from growing faster than it did. Finally, sub-Saharan Africa's net imports remained at very low levels, the result of both poverty (lack of effective demand) and import capacity limitations.

By and large, the traditional cereal exporters (North America, Australia, Argentina, Uruguay, Thailand and, in more recent years, also the EU and Viet Nam) coped quite well with spurts in import demand. As shown in Table 3.8, they export currently (average 1997/99) 176 million tonnes of cereals net annually. This is matched by the 135 million tonnes net imports of the developing countries other than Argentina, Uruguay, Thailand and Viet Nam, and 33 million tonnes of the net importing industrial countries (Japan, non-EU western Europe and Israel. A discrepancy of some 9 million tonnes at world level in the trade statistics remains unaccounted for.) This is about double their net exports of the mid-1970s and three times those of the mid-1960s. The data are given in Table 3.8. About a half of the total increment in these net exports in the period from the mid-1970s to 1997/99 was contributed by the EU. It is a very significant development for the world food system that this region turned from a net importer of 21 million tonnes in the mid-1970s to a net exporter. The transformation had been completed by the early 1980s. The EU's net exports reached a peak of 38 million tonnes in 1992 and it was exporting a net 24 million tonnes in 1997/99. In practice, the other traditional exporters have had to increase their net export surplus rather modestly, from 110 million tonnes in the mid-1970s to 151 million tonnes in 1997/99.

We do not have a counterfactual scenario to answer the question how the different variables of the world food system would have actually fared if the EU had not followed a policy of heavy support and protection of its agriculture (in particular prices, production, consumption and trade in the different countries and particularly the per person food availability of the poor countries and those that became heavy importers). This policy led to the region's import substitution and then subsidized exports.[7] The resulting lower world market prices (compared with what they would have been otherwise) are thought to have adversely affected the food security of the developing countries because of the negative effects on the incentives to their producers. However, the structurally import-dependent countries have a clear interest that world market prices should be lower rather than higher. Disincentives to own producers could have been counteracted by appropriate policies, at least in principle, although admittedly a very difficult task in practice (how does one keep domestic prices higher than import prices when a good part of the consumers are in the low-income, food-insecure category?). In looking at the impacts on food security, we should also consider the possible positive effects on the consumption of the poor of the lower import prices and increased availability of food aid. In the end, such policies of the EU resulted in the emergence of an additional major source of cereal export surpluses to the world markets and diversified the sources from which the importing countries could provision themselves. This is a structural change which is probably here to stay, even under the more liberal trade policy reforms of recent years and perhaps further reforms to come (see below).

It is worth noting that (for very different reasons) we are currently witnessing a similar transformation of the group of the transition economies from large net importer to a small one, and, as projected in this study, to a sizeable net exporter in the longer-term future. This group had emerged as a major net importer up to the late-1980s, but the drastic decline in its demand for cereals (no doubt aided by rapidly rising meat imports in the former Soviet Union) has by now led to drastic declines in net imports and occasional net exports, no matter that production also declined drastically.

3.2.2 Prospects for the cereal sector

The preceding discussion has highlighted the main forces that shaped the past. What changes can we expect in the future?

Aggregate demand. As already anticipated, the fundamental forces that made for slowdown in the growth of demand in the past – slower population

---

[7] In retrospect, it seems remarkable that this transformation of western Europe with the aid of heavy subsidies took place without significant trade conflict in its heyday. It is probably explained by the rapid expansion of the demand for cereal imports in that period (oil boom, transition economies becoming large importers), which provided sufficient market outlets for all. Conflict did appear with a vengeance after the mid-1980s when markets stopped expanding. It led to the Agreement on Agriculture under the Uruguay Round which imposed limits on the use of trade-distorting support and protection policies and export subsidies (see Chapter 9).

growth everywhere and the achievement of mid-high levels of per capita consumption in some countries – will continue to operate in the future and contribute to further deceleration in the growth of demand.

Are there any factors that will attenuate or reverse this trend? No major stimulus in this direction is likely to come from the countries and population groups with consumption needs that have still not been met. The overall economic growth outlook and its pattern (see Chapter 2) suggest there will be inadequate growth of incomes, and poverty will persist. However, the downward pressure on world demand exerted by more transient forces (systemic change in the transition economies and the high policy prices of the EU) has already been largely exhausted and will not be there in the future.[8] Indeed, the recovery in demand in these two country groups, already evident in the EU after the early 1990s, as well as eventually in the countries of East Asia, will likely more than compensate for the effects of the more fundamental sources of slowdown. The result will be that, for a time, the growth rate of world demand for cereals could be higher than in the recent past. This shows up in the projections (Table 3.3), where the growth rate of world demand for 1997/99-2015 is 1.4 percent p.a., compared with 1.0 percent p.a. in the preceding ten years.

In the longer term, however, the more fundamental sources of slowdown will predominate and the growth rate of world demand will be lower for the second part of the projection period 2015-30, to 1.2 percent p.a. Tables 3.3 and 3.4 present this information for the standard regions, while memo item 1 in Table 3.3 unfolds these projections in terms of the groups used in the preceding section to analyse deceleration in the historical period. The further slowdown in China's population growth (from 1.3 percent p.a. in the preceding 20 years to 0.7 percent p.a. in the period to 2015 and on to 0.3 percent p.a. in 2015-30), the levelling-off and eventual small decline in its per capita food consumption of rice as well as the growth of production and consumption of meat at rates well below the spectacular ones of the past (see Section 3.3) all contribute to further deceleration in its aggregate

cereal demand. Given China's large weight, the effects are felt in terms of lower growth rates in the aggregates for the world and the developing countries. In conclusion, the role China played in slowing world demand for cereals after about the mid-1980s (Figure 3.3) will probably continue to operate, and so will the deceleration in world demand and trade that in the past was associated with the end of expansion in the oil-exporting countries.

Unless there is another event that will cause effects similar to those of the oil boom (spurt in the demand and imports of a significant number of countries with low levels in per capita food consumption of cereals and livestock products), we cannot expect reversals of the trend towards long-term deceleration in global demand.

Will any such event occur? It would be foolhardy to make predictions. The failure (of ourselves and others) to foresee the collapse of production, demand and trade in the transition economies is instructive. However, we can attempt to see what is implied by some available projections.

Concerning the issue of future commodity booms, or rather significant upward trends in world commodity prices, the following quotation from the World Bank (2000a) is telling: "On balance, we do not see compelling reasons why real commodity prices should rise during the early part of the twenty-first century, while we see reasons why they should continue to decline. Thus, commodity prices are expected to decline relative to manufactures as has been the case for the past century". More recent projections from the Bank to 2015 (World Bank, 2001c, Tables A2.12-14) confirm the view that no significant upwards movements of commodity prices are expected, although this does not exclude the possibility of short-lived cyclical price spikes nor the recovery for some commodities (such as coffee and rubber, see Section 3.6 below) from the very low prices of early 2002.

This leaves the other source of significant spurts in demand and trade that occurred in the past: rapid sustained economic growth (e.g. of the Asian-tiger type) in countries of the above-mentioned typology (low initial levels in per capita food consumption of cereals and livestock products). On this, the latest World Bank view presented in

---

[8] An additional factor that made for slow growth of demand in recent years has been the abrupt reversal of the trend towards growing feed use of cereals in the countries of East Asia hit by the economic crisis of 1998.

Chapter 2 (Figure 2.3) of what the future could hold in terms of economic growth and poverty reductions for the period to 2015 does not permit great optimism. Of the two regions with significant poverty only South Asia may be making progress while little progress is foreseen for sub-Saharan Africa. South Asia could indeed be a source of spurts in cereal demand if it were to behave like typical developing countries in other regions, i.e. undergoing considerable shifts in diets towards meat. However, the prospect that India will not shift in any significant way to meat consumption in the foreseeable future (see Section 3.3) militates against this prospect.

The decline in world per capita production and consumption of cereals that occurred in the decade following the mid-1980s was interpreted by some as foreshadowing an impending world food crisis (e.g. Brown, 1996). However, this trend will probably be reversed, and the reversal has already started. World per capita consumption (all uses) peaked at 334 kg in the mid-1980s (three-year averages) and has since declined to the current 317 kg (three-year average 1997/99).[9] The reasons why this happened were explained above. They certainly do not suggest that the world had run into constraints on the production side and had to live with durable declines in per capita output. In the projections, the declining trend is reversed and world per capita consumption rises again and reaches 332 kg in 2015 and 344 kg in 2030 (Table 3.3, Figure 3.4).[10] This reversal reflects, *inter alia*, the end of declines and some recovery in the per capita consumption of the transition economies.

Demand composition: categories of use. The projected evolution of future demand by commodity (wheat, rice and coarse grains) is given in Table 3.3 (aggregate demand, memo item 2), Figure 3.4 (per capita demand), and by category of use in Figure 3.9. At the world level aggregate consumption of all cereals should increase by 2030 by nearly one billion tonnes from the 1.86 billion tonnes of 1997/99 (Table 3.3). Of this increment, about a half will be for feed, and 42 percent for food, with the balance going to other uses (seed, industrial non-food[11] and waste). Feed use will revert to being the most dynamic element driving the world cereal economy, in the sense that it will account for an ever-growing share in aggregate demand for cereals. It had lost this role in the last decade following the above-mentioned factors that affected feed use in two major consuming regions, the transition economies and the EU. Feed use had contributed only 14 percent of the total increase in world cereal demand between the mid-1980s and 1997/99, down from the 37 percent it had contributed the decade before.

The turnaround of these two regions to growing feed use of cereals (already in full swing in the EU) or, in any case, the cessation of declines in the transition economies, tends to exaggerate the role of feed demand as a driving force in the long-term evolution of the structural relationships of the world cereal economy. With the rather drastic slowdown in the growth of the livestock sector (see Section 3.3), one would have expected that the role of feed demand as a driving force would be less strong than is indicated by the projected world totals in which the increase in feed use accounts for 51 percent of the increment in total cereal demand. As noted, it accounted for only 14 percent in the period from the mid-1980s to 1997/99. For the world without the transition economies and the EU, the jump in the shares is much less pronounced, which is far more representative of the long-term

---

9   Changes in production were much more pronounced (344 and 321 kg, respectively), and the effects on consumption were smoothed by changes in stocks.

10 Average per capita numbers for large aggregates comprising very dissimilar country situations (like the world per capita cereal consumption) have limited value as indicators of progress (or regress) and can be outright misleading. In practice, the world can get poorer on the average even though everyone is getting richer, simply because the share of the poor in the total grows over time. This can be illustrated as follows (example based on approximate relative magnitudes for the developing and the developed countries): in a population of four persons, one is rich, consuming 625 kg of grain, and three are poor, each consuming 225 kg. Total consumption is 1 300 kg and the overall average is 325 kg. Thirty years later, the poor have increased to five persons (high population growth rate of the poor) but they have also increased consumption to 265 kg each. There is still only one rich person (zero population growth rate of the rich), who continues to consume 625 kg. Aggregate consumption is 1 950 kg and the average of all six persons works out to 325 kg, the same as 30 years earlier. Therefore, real progress has been made even though the average did not increase. Obviously, progress could have been made even if the world average had actually declined. Thus, if the consumption of the poor had increased to only 250 kg (rather than to 265), world aggregate consumption would have risen to 1 875 kg but the world average would have fallen to 312.5 kg (footnote reproduced from Alexandratos, 1999).

11 Uses of maize for the production of sweeteners and of barley for beer are included in food, not in industrial use. The latter includes use of maize for the production of fuel ethanol.

Figure 3.9 Aggregate consumption of cereals, by category of use

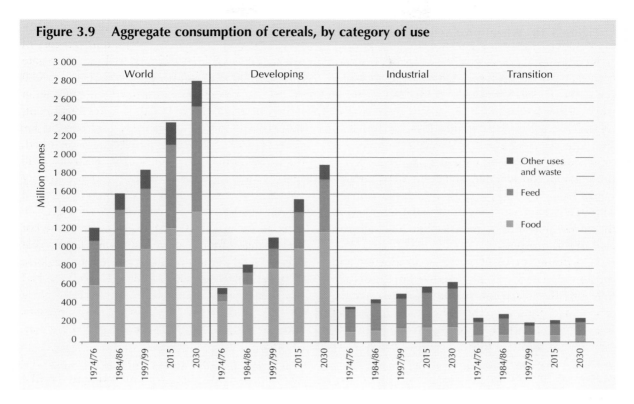

evolution of structural relationships than the global magnitudes would suggest. Such evolution increasingly depends on what happens in the developing countries. In these countries, feed accounted for 21 percent of the increment in total demand in the decade starting in the mid-1970s. It accounted for 29 percent in the period from the mid-1980s to 1997/99. In the projections, the share grows further to 45 percent of the increase in their total projected demand (see Section 3.3 for further discussion).

Commodity composition. The per capita consumption of *rice* will tend to level off in the first part of the projection period and decline somewhat in the second part. In the first subperiod the major factor is the slowdown in China and several other countries in East Asia, while per capita consumption in South Asia continues to increase. Slow declines will probably occur in South Asia after 2015. Here reference is made to the earlier discussion on food demand for all commodities and the related implications for total calories and nutrition: by 2030, East Asia will have moved close to 3 200 calories and South Asia to 2 900 (Chapter 2, Table 2.1). These levels suggest that per capita consumption of rice will not be as high as at present. The end result of these possible developments, together

with the deceleration in population growth, is that the aggregate demand for rice will grow at a much slower rate than in the past, from 1.6 percent p.a. in the 1990s and 2.6 percent in the 1980s to 1.2 percent p.a. in the period to 2015 and on to 0.8 percent p.a. after 2015. Therefore, the pressure to increase production will also ease. With the slowdown in the growth of yields in recent years, maintaining growth even at much lower rates will be no mean task and may require no less effort (research, irrigation, policy, etc.) than in the past (see discussion in Chapter 4).

Wheat consumption per capita (all uses) will continue to increase in the developing countries as well as in the transition economies and the industrial countries, following the cessation of the factors that depressed demand in these two latter groups. In the developing countries, the increases in the consumption of wheat will be partly substituting for rice. The developing countries will continue to increase their dependence on imports. Their net imports would grow from 40 million tonnes in the mid-1970s and 72 million tonnes in 1997/99, to 120 million tonnes in 2015 and 160 million tonnes in 2030 (these numbers exclude Argentina and Uruguay that will be growing net exporters).

Figure 3.10 Wheat production and net imports

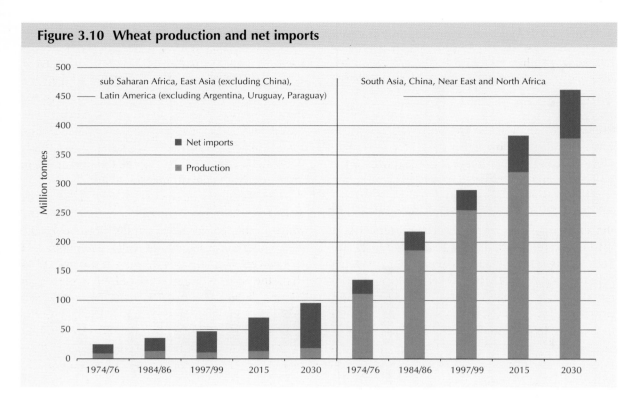

To appreciate why we project these rather significant increases in net imports of wheat, the relevant data and projections are plotted in Figure 3.10 in more disaggregated form. In the first place, in the regions that are not major producers themselves in relation to their consumption (roughly, sub-Saharan Africa, East Asia other than China, Latin America other than Argentina, Uruguay and Paraguay), consumption growth will be accompanied by increases in net imports, as in the past. For example, in these regions a consumption increase of 23 million tonnes between 1997/99 and 2015 will be accompanied by an increase in net imports of 21 million tonnes. In the preceding period (1984/86-1997/99) the comparable figures were 12 million tonnes increase in consumption and 14 million tonnes increase in net imports. Therefore, there is nothing new here. In contrast, what is new is that developments in the rest of the developing countries may diverge from past experience. As noted, production increases and declines in net imports and occasional generation of net exports in some of the major wheat-consuming countries (China, India, some countries in the Near East/North Africa region) masked the growing dependence of consumption growth in the developing countries on imported wheat. This factor will

be much less important in the future. Some of the countries that had this role will probably turn around to be net importers again (e.g. India, Saudi Arabia and the Syrian Arab Republic) or larger net importers (e.g. China, Pakistan and Bangladesh) in the future.

World consumption of *coarse grains* should grow faster that of the other cereals (Table 3.3 and Figure 3.4), following the growth of the livestock sector. The shift of world consumption of coarse grains to the developing countries will continue and their share in world total use will rise from 47 percent at present (and 34 percent in the mid-1980s) to 54 percent in 2015 and 59 percent in 2030. Much of the increase (72 percent) in coarse grains use in developing countries will be for feed, a continuing trend in all regions except sub-Saharan Africa, where food use will continue to predominate. In sub-Saharan Africa, coarse grains will continue to constitute the mainstay of cereal food consumption. If sub-Saharan Africa's production growth rates of the past (3.3 percent p.a. in the past 20 years, 2.8 percent p.a. in the last ten) could be maintained, as is feasible in our evaluation (we project a growth rate of 2.8 percent p.a. to 2015), and given lower population growth, the region could raise per capita food consumption of coarse grains by some 11 kg, to 101 kg by 2030 (Figure 3.11). This may not be impressive and certainly falls short of what is needed for food security, but we must recall that there was no increase in the last 20 years.

Production, imports and exports. World trade in cereals tended to slow down after the mid-1980s. Here we examine net imports and exports of the different country groups.

Our projections anticipate a revival of the net cereal imports of the developing countries and of the exports of the main cereal exporters. FAO's medium-term projections to 2010 (FAO, 2001b) had already anticipated this revival and had net cereals imports of the developing countries growing to 150 million tonnes in 2010. Our own projections to 2010 made in the early 1990s from base year 1988/90 indicated 162 million tonnes in 2010 (Alexandratos, 1995, p. 145). This projection to 2010 remains largely valid in the current work – we now have 190 million tonnes in 2015 and 265 million tonnes in 2030.

The commodity structure (wheat, coarse grains and rice) of the net trade balances is shown in Table 3.5. Net imports of the developing countries are projected to increase by 162 million tonnes between 1997/99 and 2030, roughly 80 million

tonnes each of wheat and coarse grains, while they should increase their net rice exports by some 2 million tonnes (Table 3.5). The latest International Food Policy Research Institute (IFPRI) projections to 2020 paint a somewhat less buoyant outlook for the net imports of the developing countries: they project them at 202 million tonnes for the year 2020 (Rosegrant *et al.*, 2001, Table D.10) compared to our 190 million tonnes in 2015 and 265 million tonnes in 2030.

The quantities of cereals that will need to be traded in the future are certainly large, but the rates of change in the net trade position of the developing countries are not really revolutionary; a 158 percent increase over 32 years is somewhat less than the increase that occurred in a period of 23 years from the mid-1970s to 1997/99. However, these quantities may appear large in the light of the factors discussed above that made for deceleration in the world cereal trade in recent years. How reliable are these projections? We cannot tell in any scientific way, but a comparison of our earlier projections of the cereal deficits of the developing countries to 2000 (made in the mid-1980s from base year 1982/84), and to 2010 (referred to above) with actual outcomes is encouraging. As Figure 3.7 shows, the mid-1980s projection indicated 112 million net imports of the

Figure 3.11 Sub-Saharan Africa, cereal food per capita

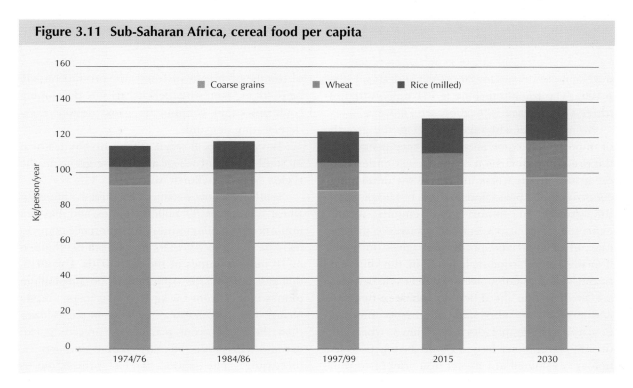

Table 3.5   Net trade balances of wheat, coarse grains and rice

	1974/76	1984/86	1997/99	2015	2030
			Million tonnes		
			Developing countries		
All cereals	-38.8	-66.4	-102.5	-190	-265
Wheat	-37.9	-48.8	-61.8	-104	-141
Coarse grains	-0.2	-17.6	-43.2	-89	-128
Rice (milled)	-0.7	0.0	2.5	3	5
			Industrial countries		
All cereals	55.1	105.9	110.7	187	247
Wheat	41.4	70.8	66.0	104	133
Coarse grains	12.1	33.7	43.4	83	115
Rice (milled)	1.6	1.4	1.4	0	-1
			Transition countries		
All cereals	-15.7	-37.3	0.9	10	25
Wheat	-4.8	-20.2	-0.3	4	12
Coarse grains	-10.4	-16.5	2.1	8	15
Rice (milled)	-0.5	-0.6	-0.9	-1	-1
Memo item.			Developing excl. net exporters[1]		
All cereals	-51.4	-91.5	-134.6	-238	-330
Wheat	-39.5	-55.6	-70.9	-118	-157
Coarse grains	-10.1	-31.3	-54.6	-107	-154
Rice (milled)	-1.8	-4.5	-9.2	-13	-19

[1]   Developing net exporters: those with net cereal exports over 1 million tonnes in 1997/99 (Argentina, Thailand and Viet Nam). India and China, although they met this criterion, are not included in the net exporter category as they are only occasional net exporters.

developing countries for 2000 (Alexandratos, 1988, p.106). The actual outcome is 111 million tonnes (three-year average 1999/2001, see Figure 3.7).

To appreciate why this "revival" in the growth of imports may come about, reference is made to the preceding discussion of the factors underlying the growth in the wheat imports of the developing countries in the projections. In practice, explaining the growth of import requirements means explaining the factors that will cause the growth rates of demand and production (given in Table 3.4) to diverge from each other in the different countries and regions, given that net trade balances are by definition the difference between demand and production. The main arguments affecting demand were amply discussed above, while the (mainly agronomic) factors in projecting production are examined in Chapter 4. A brief discussion

of trends, or breaks in trends, in production will help to explain why deficits of the developing countries could continue to grow despite ever-decelerating demand.

Two examples illustrate what is involved: wheat in South Asia and wheat and coarse grains in the region Near East/North Africa.

South Asia now produces 89 million tonnes of wheat, consumes 89 million tonnes and has net imports of 7 million tonnes (the difference going to increase stocks in 1997/99), down from net imports of 10 million tonnes in the mid-1970s. For 2015, demand (all uses) is projected to be 138 million tonnes (going from 69 kg to 82 kg in per capita terms). The growth rate of 2.6 percent p.a. is lower than the 3.3 percent p.a. of the preceding two decades (1979-99). So why should imports be increasing? The reason is that production is

Table 3.6   South Asia, land-yield combinations of wheat production

	Rainfed land			Irrigated land			Total		
	Area '000 ha	Yield tonnes/ha	Production '000 tonnes	Area '000 ha	Yield tonnes/ha	Production '000 tonnes	Area '000 ha	Yield tonnes/ha	Production '000 tonnes
1974/76				25 459	1.31	33 398			
1984/86				32 129	1.85	59 529			
1997/99	7 450	1.21	9 013	28 889	2.78	80 357	36 339	2.46	89 370
2015	7 437	1.38	10 270	32 763	3.52	115 210	40 200	3.12	125 480
2030	7 416	1.56	11 540	36 434	4.22	153 890	43 850	3.77	165 430

Note: Land refers to harvested area and therefore includes area expansion under wheat from increased double cropping.

unlikely to keep the high growth rates of the past. Much of the wheat is now irrigated and the boost given in the past from expansion of wheat into irrigated areas and the spread of new varieties is becoming much weaker (Mohanty, Alexandratos and Bruinsma, 1998). In addition there are problems in maintaining the productivity of irrigated land, particularly in Pakistan. The growth rate of wheat production in the region has been on the decline: it was 3.2 percent p.a. in the latest ten years (1989-99), down from 4.0 percent in the preceding decade and 5.1 percent in the one before. We project an average production growth rate of 2.0 percent p.a. up to 2015, and 1.9 percent p.a. in 2015-2030. These growth rates are slower than those of demand, hence the growing import requirements even to meet a demand growth much below that of the past. The land-yield combinations underlying these production projections are shown in Table 3.6.

IFPRI projects a turnaround of South Asia to a growing net importer of cereals at higher levels than we project in Table 3.4. In contrast, the latest ten-year projections by the United States Department of Agriculture (USDA) consider that India would continue to be a small net exporter of wheat until the year 2011 (USDA, 2002).[12] The Food and Agricultural Policy Research Institute (FAPRI) 2002 projections for the same year have India as a small net importer (one million tonnes), but the underlying consumption projections are very low (an increase in per capita consumption of only 1 kg in ten years) while production growth is also well below past trends. The small growth in consumption (and hence of import requirements) may be an underestimate, given India's projected relatively high growth of incomes and the prospect that increased food demand will not be diverted to meat in the foreseeable future (see section on livestock below).

Similar considerations apply to wheat and coarse grains in the Near East/North Africa region and, of course, to other regions and crops. Net imports of wheat and coarse grains into Near East/North Africa are projected to grow from 45 million tonnes in 1997/99 to 80 million tonnes in 2015 and to 108 million tonnes in 2030. After a quantum jump in the 1970s up to the mid-1980s, imports stagnated up to the mid-1990s, before resuming rapid growth in the second half of the decade. The projected continuation in the recovery of import growth factors in, among other things, the assumption that the decline in the imports of Iraq will have been reversed by 2015. On the demand side, the region's population growth rate will remain relatively high for some time (1.9 percent p.a. to 2015). Some countries in the region are among the fastest growing in the world. The example of Yemen is instructive: the country had a population of 17 million in the base year 1997/99, but it is projected to be a really large country with 57 million in 2030. Its present consumption of cereals amounts to 180 kg/person (all uses) or some 3 million tonnes p.a. of which only 0.7 million tonnes comes from local produc-

---

[12] "The surpluses of mostly low-quality wheat are generally not exportable without subsidy, but low levels of exports to neighbouring South Asian and Middle Eastern countries are expected to continue" (USDA, 2002, p.103)

tion. No quantum jumps in production are foreseen. Therefore, even without increases in per capita consumption, aggregate demand will be over 10 million tonnes in 2030, more if we factor in some modest increase in per capita consumption. For the whole Near East/North Africa region, the aggregate demand is bound to grow at 2.0 percent p.a. (see Table 3.4), even with a modest increase in per capita consumption of cereals for all uses (under 10 percent over the whole projection period).

The projected growth in Near East/North Africa deficits reflects the prospect that production of wheat and coarse grains may not keep up even with this lower growth of demand. Production, which grew fairly fast in the past (3 percent p.a. in the 1970s and the 1980s) has shown no consistent trend since then; the average growth rate of the latest ten-year period (1991-2001) was -1.7 percent p.a. and production was 67 million tonnes in the latest three-year average 1999/2001, down from the 73 million tonnes at the beginning of the

decade (average 1989/91). Among the major producers, Saudi Arabia's production of wheat and coarse grains declined by 46 percent, and that of North Africa (outside Egypt) by 30 percent. Among the major producers of the region, only Egypt and the Syrian Arab Republic had higher production in 1999/2001 than at the beginning of the decade. The evaluation of the possible land-yield combinations in the future (shown in Table 3.7), as well as the prospect that there will be no return to the heavy production subsidies some countries provided in the past, do not permit optimism concerning the possibility that growth of aggregate wheat and coarse grains production of the region could exceed 1.5 percent p.a. in the projection period. Hence the need for growing net imports to support the modest increase in per capita consumption.

As noted, the production projections that give rise to the import requirements presented here are derived from a fairly detailed analysis of the production prospects of individual developing countries commodity by commodity. The method

Table 3.7   Near East/North Africa: areas and yields of wheat, maize and barley

	Rainfed land			Irrigated land			Total		
	Area '000 ha	Yield tonnes/ha	Production '000 tonnes	Area '000 ha	Yield tonnes/ha	Production '000 tonnes	Area '000 ha	Yield tonnes/ha	Production '000 tonnes
Wheat									
1974/76							26 668	1.20	31 947
1984/86							25 023	1.52	38 090
1997/99	19 201	1.28	25 550	8 008	3.15	24 231	27 210	1.83	49 781
2015	18 965	1.48	28 080	8 980	3.76	33 720	27 945	2.21	61 800
2030	19 065	1.65	31 435	9 935	4.31	42 855	29 000	2.56	74 290
Maize									
1974/76							2 410	2.26	5 450
1984/86							2 205	3.01	6 644
1997/99	655	2.30	1 506	1 563	5.65	8 826	2 218	4.66	10 331
2015	604	2.45	1 482	2 005	6.15	12 333	2 610	5.29	13 815
2030	552	2.69	1 482	2 673	7.15	19 108	3 225	6.39	20 590
Barley									
1974/76							9 585	1.11	10 675
1984/86							12 585	1.20	15 135
1997/99	9 793	1.22	11 906	1 777	1.84	3 266	11 570	1.31	15 172
2015	10 728	1.49	15 942	1 885	2.28	4 304	12 610	1.61	20 245
2030	11 375	1.72	19 577	1 975	2.66	5 243	13 350	1.86	24 820

Figure 3.12   Cereal production, all developing countries: comparison of actual outcomes average 1997/99 with projections to 2010 made in 1993 from base year 1988/90[1]

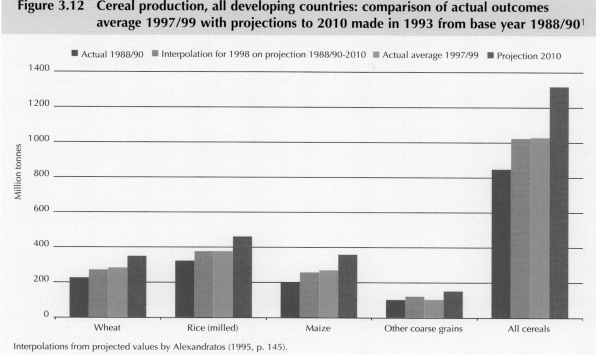

Interpolations from projected values by Alexandratos (1995, p. 145).

is described in Appendix 2. It is the same approach we used in earlier studies. In 1992-93, we had projected total cereal production in the developing countries to grow from 847 million tonnes in 1988/90 to 1318 million tonnes in 2010 (Alexandratos, 1995, p. 145). The interpolation for 1998 on the trajectory 1988/90-2010 (separately for each of the main cereals, see Figure 3.12) is 1 023 million tonnes. The actual outcome for the three-year average 1997/99 is 1 027 million tonnes (data as of February 2002).

Producing the export surplus. To explore how the growing import requirements may be matched by increases on the part of the exporters we need some rearrangement and more detailed setting out of the data and projections. This is attempted in Table 3.8. The following comments refer mainly to the contents of this table.

In the period from the mid-1970s to 1997/99, the net imports of the developing importers (developing countries not including the net exporters Argentina, Uruguay, Thailand and Viet Nam), plus those of the transition economies and the industrial importers (industrial countries minus the EU,

North America and Australia) went from 89 million tonnes to 167 million tonnes, an increment of 78 million tonnes (subtotal 2 in Table 3.8). It was met by increases of the net trade balances of the following country groups which were traditional exporters or became such during that period: the EU 45 million tonnes (from net imports of 21 million tonnes to net exports of 24 million tonnes); North America 10 million tonnes; Australia 12 million tonnes; and combined Argentina, Uruguay, Thailand and Viet Nam 20 million tonnes.[13]

In the projections, the net import requirements of the developing importers and the industrial importers rise from 168 million tonnes in 1997/99 to 275 million tonnes in 2015 and to 368 million tonnes in 2030 (subtotal 1 in Table 3.8), an increase of 107 million tonnes to 2015 and another 93 million tonnes by 2030. These quantities must be generated as additional export surplus by the rest of the world. Where will they come from? The novel element in the projections is that part of the required increase may come from the transition economies, while the rest should come from the traditional exporters, developing and industrial.

---

13 There is a large statistical discrepancy of 9 million tonnes in 1997/99 in the trade statistics.

Table 3.8 World cereal trade: matching net balances of importers and exporters

	Net imports (-) or exports (+)				Increment		
	1974/76	1997/99	2015	2030	1974/76 -1997/99	1997/99 -2015	2015 -2030
	Million tonnes				Million tonnes		
1 Developing importers[1]	-51	-135	-238	-330	-83	-104	-91
2 Industrial importers	-22	-33	-37	-38	-12	-4	-2
3 Subtotal 1 (=1+2)	-73	-168	-275	-368	-95	-107	-93
4 Transition countries	-16	1	10	25	17	10	15
5 Subtotal 2 (=3+4)	-89	-167	-265	-343	-78	-98	-78
6 Argentina +Uruguay + Thailand +Viet Nam	13	32	49	65	20	17	16
7 World imbalance	1	9	8	8	9	-1	0
8 Balance for industrial exporters[2] (=-5-6+7)	77	144	224	286	67	80	62

Memo item. **Production of industrial exporters**

	Million tonnes				Percentage p.a.		
Total	430	629	758	871	1.1	1.1	0.9

[1] Developing countries excl. Argentina, Uruguay, Thailand and Viet Nam.
[2] North America, Australia and EU15.

The transition economies (not included in the net importing regions in the projections) could be net exporters of 10 million tonnes by 2015, a rather modest outcome given their resource potential which could put them in a position to produce even larger surpluses under the right policies. The reasons why this group of countries may eventually turn from the large net importer it was up to the late 1980s into a net exporter in the longer-term future are as follows: per capita consumption or – more correctly – domestic disappearance, will not revert to the very high pre-reform levels following the reduction of the high rates of food losses; the more efficient use of grain in animal feed; and the continued reliance, at least for the medium term, on imports of livestock products to cover part of domestic consumption. On the production side, land resources in several countries in this group are relatively plentiful and yields are well below those achieved in other countries with similar agro-ecological conditions (see Chapter 4 for comparisons with those obtainable under high-input technologies). The eventual integration of some Eastern European countries into the EU has the potential of contributing to this process, mainly in favour of coarse grains (USDA, 1999a).

These considerations suggest that the eventual recovery of production in the transition economies will result in export surpluses. We are rather conservative in our projections, as the production growth rate of cereals required to meet the growth of domestic demand and produce the 10 million tonnes net exports in 2015 is 1.0 percent p.a. This is modest, seeing that they start from the depressed levels to which production had fallen by the late 1990s. Other studies are much more optimistic about this potential. For example the latest IFPRI projections (Rosegrant *et al.*, 2001, Figure 4.9) suggest over 25 million tonnes net exports from Eastern Europe and the countries of the former Soviet Union in 2020. The latest FAPRI (2002) projections foresee 15 million tonnes for the year 2011 (the sum of wheat, barley and maize) as does the most recent USDA study (USDA, 2002, Tables 36-40), while Dyson (1996) has much larger numbers.

The net exports of the developing exporters (Argentina, Uruguay, Thailand and Viet Nam) are projected to rise from 32 million tonnes to some

50 million tonnes by 2015. Thailand and Viet Nam are projected to remain net exporters of cereals because of rice, although they will be growing net importers of the other cereals, mainly wheat. It follows that the more "traditional" industrial country exporters (North America, the EU and Australia) would need to increase net exports by 80 million tonnes by 2015 and another 62 million tonnes by 2030, i.e. by amounts roughly comparable to the increase of 67 million tonnes they recorded in the period 1974/76-1997/99 (Row 8 in Table 3.8).

The question is often raised as to whether these countries have sufficient production potential to continue generating an ever-growing export surplus. Concern with adverse environmental impacts of intensive agriculture is among the reasons for this question. The answer depends, *inter alia*, on how much more these countries must produce over how many years. Production growth requirements are derived by adding the above increments in net exports to the increments in their own domestic demand, including demand for cereals to produce livestock products for export.[14] The resulting projected production is shown in the lower part of Table 3.8. These countries are required to increase their collective production from the 629 million tonnes of 1997/99 to 758 million tonnes in 2015 and 871 million tonnes in 2030, an increment of 242 million tonnes over the entire period, of which about 80 million tonnes are wheat and the balance largely coarse grains. The annual growth rate is 1.1 percent p.a. in the period to 2015 and 0.9 percent p.a. in the subsequent 15 years, an average of 1.0 percent p.a. for the entire 32-year projection period. This is lower than the average growth rate of 1.6 percent p.a. of the past 32 years (1967-99), although the historical growth rate has fluctuated widely, mostly as a function of the ups and downs of export demand,

associated policy changes and occasional weather shocks. The annual growth rates of any ten-year period in the past 35 years moved in the range from 3.4 percent (decade ending in 1982) to minus 0.1 percent (decade to mid-1990s), with the latest being 1.5 percent in 1989-99. The overall lesson of the historical experience seems to be that the production system responds flexibly to meet increases in demand within reasonable limits.

Of the three traditional industrial exporters, the EU faces the additional constraint that it can increase production for export only if it can export without subsidies. A key question is, therefore, whether market conditions will be such as to make possible unsubsidized exports. The relevant variables are the policy prices of the EU, the prices in world markets and the exchange rate €/US$. We have not gone into modelling explicitly these variables, but we project that the EU will be a growing net exporter of wheat and barley without subsidies. There seems to be a fair degree of consensus on this matter. The European Commission's latest projections to 2008 point in the same direction: "total cereal exports would stand substantially above the annual limit for subsidized exports set by the URAA limits (i.e. 25.4 million tonnes for total cereals) as durum wheat, some common wheat and barley/malt would be exported without subsidies" (European Commission, 2001, p. 37).[15]

Other studies agree in their findings that the EU will be a growing net exporter of cereals, at levels exceeding the limits for exports with subsidies[16] as defined under the Uruguay Round Agreement on Agriculture (URAA). The 2002 USDA baseline projection to 2011 concludes that "due to the declines in intervention prices and the weak euro, projected domestic and world prices indicate that EU wheat and barley can be exported without subsidy throughout the baseline period" (USDA,

---

[14] The term "domestic demand" can be misleading if it gives the impression that the inhabitants of the country actually "consume", directly or indirectly, the amounts used domestically. This can be especially misleading when a significant proportion of the cereals consumed goes to produce livestock products for export. For example, Denmark is given in the statistics as having the highest per capita consumption of cereals (all uses) in the world, 1 450 kg. But the country exports net two-thirds of its meat production and over 50 percent of production of milk and dairy products. It is also a net exporter of beer, which uses barley as input.

[15] Last-minute addition (June 2002): the Commission has just published the 2002 edition of its projections to 2009. It has the same projected exports of wheat and coarse grains as in the 2001 edition, i.e. exceeding the URAA limits, but adds: "These projections for cereal exports remain conditional upon an export policy that ensures the full use of the URAA limits" (European Commission, 2002, p.13). Somehow, this could imply the development of dual markets within the EU, if some exports will be with subsidies and others without. However, this need not be so if the non-subsidized cereals (e.g. durum wheat and malting barley) are different from the subsidized ones, e.g. feed barley and soft wheat.

[16] The UR limits for EU exports with subsidies are, roughly, 25 million tonnes. These limits refer to gross exports, while imports will be about 5-7 million tonnes, including those under the UR commitments. It follows that any projection study showing net EU exports over 18-20 million tonnes must assume (implicitly or explicitly) that in the future the right combination of domestic and foreign prices and exchange rates will prevail.

2002, p. 90).[17] It projects net EU exports of wheat and coarse grains of some 35 million tonnes net for the year 2011 (USDA, 2002, Tables 36-40). The FAPRI projections have net exports of 29 million tonnes for the same year (FAPRI, 2002). Longer-term studies point in the same direction: the most recent IFPRI assessment is more conservative with a projection of about 30 million tonnes net EU exports in 2020 (Rosegrant et al., 2001, Table D.10).

When speaking of the need for the traditional industrial food exporters to increase their production for export further, the issue of the environmental effects of more intensification of their agricultures becomes relevant. For example, the EU study's net export outcomes are based on cereal yields rising at an annual rate of 1.3 percent to 2008. This is lower than the historical trend, but still raises the issue of the environmental risks associated with rising yields in the intensively farmed areas that produce much of the EU export surplus, e.g. France. Such risks are mainly related to the excessive use of fertilizer and other chemicals. The risk would be certainly increased if the pursuit of higher yields were to be accompanied by inappropriate use of fertilizer leading to increases in the nitrogen balance in the soil (difference between nitrogen inputs into the soil and uptake by crops). Empirical evidence suggests that this need not be so. OECD work on environmental indicators finds that the nitrogen balance in the EU declined from 69 kg/ha to 58 kg/ha of agricultural land between 1985/87 and 1995/97 (OECD, 2000a, Annex Table 1). Over the same period, the yield of wheat increased from 4.7 tonnes/ha to 5.5 tonnes/ha, and that of total cereals from 4.5 tonnes/ha to 5.2 tonnes/ha. Changes in the structure of incentives (e.g. reduced support prices), advances in technology (precision agriculture, etc.) and imposition of tighter management regimes concerning use of manure, probably explain much of this phenomenon.

Some environmental considerations.[18] It is important that eventual environmental risks associated with the growth of production for export be viewed in a global context, and the associated trade-offs recognized. How do such risks compare with those faced by other countries that would also be raising their production? And how does enhanced production for export contribute to, or detract from, world food security by making world agriculture as a whole more sustainable (or less unsustainable)?

This issue can be addressed schematically with the aid of a simple classification of natural resource/technology combinations used in grain production, on the one hand, and development levels, on the other. The former determines the extent to which the growth of production enhances the risk of adverse environmental impacts (e.g. soil erosion, salinization of irrigated areas, nitrate pollution of water bodies). The latter determines the value people place on resource conservation and on the environment, relative to the more conventional benefits from increased production, e.g. food security, farm incomes, export earnings, etc. This classification is as follows (from FAO, 1996d):

- Intensive high-income systems: for example, the Brie area in the Paris basin of France, characterized by very high grain yields obtained with substantial inputs of fertilizers and pesticides. Main environmental problem: water pollution from fertilizers and other chemicals.

- Extensive high-income systems: for example, Australia and western Canada, characterized by moderate to low grain yields and little or no use of fertilizers. Main environmental problem: soil erosion, mainly caused by wind and occasional intense rainfall.

- Intensive low-income systems: for example, areas of intensive grain cultivation under irrigation in India (Punjab, Haryana and Uttar Pradesh), characterized by medium-high yields and fertilizer applications. Main environmental problem: salinization, but also waterlogging and water scarcities.

- Extensive low-income systems: for example, Côte d'Ivoire, coarse grain production in areas expanding under population pressure (often in slash-and-burn mode), very low yields, virtually no fertilizer. Main environmental problem: deforestation and soil mining, leading to declining yields, abandonment and further expansion into new areas.

---

[17] In the USDA study: "The euro is assumed to strengthen slightly against the dollar in 2002 through 2004, and then to weaken somewhat through the remainder of the projections" (USDA, 2002, p. 89). The FAPRI assumptions are more optimistic about the euro: it reaches parity with the US dollar in 2006 and remains there until 2011.

[18] This section draws heavily on Alexandratos and Bruinsma (1998).

As noted (Table 3.3), in the 32-year period from 1997/99 to 2030, the world will need to increase annual production of cereals (including rice in milled form) by nearly another billion tonnes. This is roughly the amount by which world production increased in the preceding 32 years (1967-1999), a process which led to a better fed world but also brought with it the resource and environmental problems we are facing today.[19] The preceding rough classification shows the very wide diversity of natural and socio-economic conditions and the associated threats to the resource base and the environment, under which humanity will have to extract the additional billion tonnes from the earth. Although sweeping generalizations must be avoided, it would appear that it will be extremely difficult to produce so much more than currently, without putting additional pressure on the environment.

It is conceivable that under the right policies (for incentives, institutions, technology development and adoption) this additional pressure could be minimized or even reversed for some time. However, here we are speaking of very substantial increases in production and, although these are to be achieved over a period of 32 years, it is difficult to visualize how enhanced pressures on the environmental resources can be avoided entirely. In addition, the stark fact has to be faced that, for the world as a whole, adoption of measures to minimize impact will be a slow process, and will perhaps remain for some time beyond the capability of those societies that most need to increase production. These are precisely the countries whose very survival is threatened by the deterioration of their agricultural resources, given the high dependence of their economies on agriculture (Schelling, 1992). It is the high-income countries, in principle those that least need to increase production for their own consumption and food security, that place a high value on minimizing the adverse environmental impacts of agriculture and that also have the means to take action (for the EU, see Brouwer and van Berkum, 1996). These

considerations provide a framework for thinking about the role of traditional exporters as suppliers of growing export surpluses in a world that has to accept the trade-offs between more food and the environment and must seek ways to optimize them (see Chapter 12).

3.3 Livestock commodities

3.3.1 Past and present

Livestock, a major factor in the growth of world agriculture. The world food economy is being increasingly driven by the shift of diets and food consumption patterns towards livestock products. Some use the term "food revolution" to refer to these trends (Delgado *et al.*, 1999). In the developing countries, where almost all world population increases take place, consumption of meat has been growing at 5-6 percent p.a. and that of milk and dairy products at 3.4-3.8 percent p.a. in the last few decades. Aggregate agricultural output is being affected by these trends, not only through the growth of livestock production proper, but also through the linkages of livestock production to the crop sector which supplies the feeding stuffs (mainly cereals and oilseeds), and benefits from the important crop-livestock synergies prevailing in mixed farming systems (de Haan *et al.*, 1998).

On the negative side, and in association with policy distortions or market failures, there are environmental implications associated with the expansion of livestock production. For example, through the expansion of land for livestock development, livestock sector growth has been a prime force in deforestation in some countries such as Brazil, and in overgrazing in other countries. Intensive livestock operations on an industrial scale, mostly in the industrial countries but increasingly in the developing ones, are a major source of environmental problems through the production of point-source pollution (effluents, etc.).[20] In parallel, growth in

---

[19] There are those who hold that the choices made in the past to achieve the increases in production (in essence the pursuit of high yields), although far from perfect, have, on balance, contributed to prevent more serious environmental problems from emerging. The standard example is the amount of additional land that would have been deforested and converted to crops if the additional output had been produced with little growth in yields (Avery, 1997). Naturally, this counterfactual proposition is not always appropriate given the fact that land expansion could not have substituted for intensification in many parts of the world where there was no spare land. Perhaps the trade-off should be conceived between the "bads" of intensification and human suffering (e.g. higher mortality) from reduced food security.

[20] A recent study puts the problem as follows: "In 1964, half of all beef cows in the United States were on lots of fewer than 50 animals. By 1996, nearly 90 percent of direct cattle feeding was occurring on lots of 1 000 head or more, with some 300 lots averaging 16 000-20 000 head and nearly 100 lots in excess of 30 thousand head. These feedlots represent waste management challenges equal to small cities, and most are regulated as point-source pollution sites under the authority of the US Environmental Protection Agency (EPA)" (Commission for Environmental Cooperation-NAFTA, 1999, p. 202).

Table 3.9   Milk and dairy products, production and use: past and projected

	1964/66	1974/76	1984/86	1994/96	1997/99	2015	2030
	Food per capita (kg, whole milk equivalent)						
World	74	75	78	77	78	83	90
Developing	28	30	37	42	45	55	66
Sub-Saharan Africa	28	28	32	29	29	31	34
Near East/North Africa	69	72	83	71	72	81	90
Latin America and the Caribbean	80	93	94	106	110	125	140
South Asia	37	38	51	62	68	88	107
East Asia	4	4	6	10	10	14	18
Industrial countries	186	191	212	212	212	217	221
Transition economies	157	192	181	155	159	169	179
Memo item							
World excl. transition economies	65	64	69	71	72	78	85

	'000 tonnes	Growth rates, % p.a.					
	1997/99	1969-99	1979-99	1989-99	1992-99	1997/99 -2015	2015 -30
		Aggregate consumption (all uses, whole milk equivalent)					
World	559 399	1.3	0.9	0.5	1.1	1.4	1.3
Developing	239 068	3.6	3.4	3.8	4.0	2.7	2.2
Sub-Saharan Africa	18 134	2.7	1.7	2.1	2.8	2.9	2.7
Near East/North Africa	32 979	2.6	1.6	2.0	2.6	2.4	2.2
Latin America and the Caribbean	61 954	2.7	2.6	3.5	3.5	2.0	1.7
South Asia	104 552	4.5	4.8	4.8	5.0	3.1	2.4
East Asia	21 450	5.8	5.6	4.9	4.1	2.7	2.2
Industrial countries	225 797	0.7	0.3	0.3	0.5	0.4	0.3
Transition economies	94 534	-0.4	-1.7	-4.8	-3.6	0.1	0.1
Memo item							
World excl. transition economies	464 865	1.9	1.7	2.0	2.2	1.7	1.5
		Production (whole milk equivalent)					
World	561 729	1.3	0.9	0.6	1.2	1.4	1.3
Developing	219 317	3.6	3.8	4.1	4.3	2.7	2.3
Sub-Saharan Africa	15 752	2.7	2.3	1.9	2.5	3.0	2.8
Near East/North Africa	28 186	2.3	2.2	3.1	3.4	2.2	2.1
Latin America and the Caribbean	56 551	2.6	2.8	3.9	4.1	2.1	1.8
South Asia	103 748	4.5	4.9	4.9	5.0	3.1	2.4
East Asia	15 081	6.9	6.9	4.5	4.4	2.9	2.2
Industrial countries	245 766	0.7	0.3	0.5	0.8	0.5	0.4
Transition economies	96 647	-0.3	-1.6	-4.6	-3.7	0.2	0.2
Memo item							
World excl. transition economies	465 083	1.8	1.7	2.1	2.4	1.7	1.5

the ruminant sector contributes to greenhouse gas concentrations in the atmosphere through methane emissions and nitrous oxide from the waste of grazing animals (see Chapters 12 and 13).

Important exceptions and qualifications. The strength of the livestock sector as the major driving force of global agriculture can be easily exaggerated. Many developing countries and regions, where the need to increase protein consumption is the greatest, are not participating in the process. In 40 developing countries, among those covered individually in this study, per capita consumption of meat was lower in the mid-1990s than ten years earlier. In this category are the regions of sub-Saharan Africa, with very low consumption per capita reflecting quasi-perennial economic stagnation. Also the Near East/North Africa, where the

Table 3.10 Food consumption of meat

	1964/66	1974/76	1984/86	1994/96	1997/99	2015	2030
			kg per capita, carcass weight equivalent				
World	**24.2**	**27.4**	**30.7**	**34.6**	**36.4**	**41.3**	**45.3**
Developing countries	10.2	11.4	15.5	22.7	25.5	31.6	36.7
excl. China	11.0	12.1	14.5	17.5	18.2	22.7	28.0
excl. China and Brazil	10.1	11.0	13.1	14.9	15.5	19.8	25.1
Sub-Saharan Africa	9.9	9.6	10.2	9.3	9.4	10.9	13.4
Near East/North Africa	11.9	13.8	20.4	19.7	21.2	28.6	35.0
Latin America and the Caribbean	31.7	35.6	39.7	50.1	53.8	65.3	76.6
excl. Brazil	34.1	37.5	39.6	42.4	45.4	56.4	67.7
South Asia	3.9	3.9	4.4	5.4	5.3	7.6	11.7
East Asia	8.7	10.0	16.9	31.7	37.7	50.0	58.5
excl. China	9.4	10.9	14.7	21.9	22.7	31.0	40.9
Industrial countries	61.5	73.5	80.7	86.2	88.2	95.7	100.1
Transition countries	42.5	60.0	65.8	50.5	46.2	53.8	60.7
Memo item							
World excl. China	28.5	32.6	34.3	34.1	34.2	36.9	40.3
World excl. China and transition countries	26.5	29.0	30.6	32.4	33.0	35.6	39.1
Meat consumption by type (kg per capita, carcass weight equivalent)							
World							
Bovine meat	10.0	11	10.5	9.8	9.8	10.1	10.6
Ovine and caprine meat	1.8	1.6	1.7	1.8	1.8	2.1	2.4
Pig meat	9.1	10.2	12.1	13.7	14.6	15.3	15.1
excl. China	9.7	10.8	11.3	10.4	10.3	9.9	9.7
Poultry meat	3.2	4.6	6.4	9.3	10.2	13.8	17.2
Developing countries							
Bovine meat	4.2	4.3	4.8	5.7	6.1	7.1	8.1
Ovine and caprine meat	1.2	1.1	1.3	1.6	1.7	2.0	2.4
Pig meat	3.6	4.1	6.4	9.6	10.8	12	12.2
excl. China	2.1	2.4	2.8	3.3	3.4	4.0	4.7
Poultry meat	1.2	1.8	2.9	5.8	6.9	10.5	14.0
excl. China and Brazil	1.2	1.9	3.2	4.8	5.2	8.1	11.6

rapid progress of the period to the late 1980s (oil boom) was interrupted and slightly reversed in the subsequent years, helped by the collapse of consumption in Iraq. Similar considerations apply to developments in per capita consumption of milk and dairy products (Table 3.9). In the great majority of countries failing to participate in the upsurge of the livestock products consumption, the reason has simply been lack of development and income growth (including failures to develop agriculture and production of these products) that would translate their considerable latent demand for what are still luxury items into effective demand. Cultural and religious factors have also stood in the way of wider diffusion of consumption of meat in general in some countries (such as India) or of particular meats (such as beef in India and pork in Muslim countries).

The second major factor limiting the growth of world meat consumption is the fact that such consumption is heavily and disproportionately concentrated in the industrial countries. They account for 15 percent of world population but for 37 percent of world meat consumption and 40 percent of that of milk. Their average per capita consumption is fairly high – that of meat is 88 kg compared with 25 kg in developing countries. This leaves rather limited scope for further increases in their per capita consumption, while their population grows very slowly at 0.6 percent p.a. currently and 0.4 percent p.a. in the coming two decades. These characteristics of the industrial countries have meant that a good part of world demand has been growing only slowly. The aggregate meat consumption of the industrial countries grew at 1.3 percent p.a. in the last ten years (0.3 percent p.a. for milk), compared with 6.1 percent (3.8 percent for milk) in developing countries. This slow growth in the industrial countries has partly offset the accelerating growth in several developing countries that have been rapidly emerging as major meat consumers, such as China, Brazil and the Republic of Korea. The net effect of these contrasting trends has been a

deceleration in the growth of world average per capita consumption of meat, going from 24 kg in the mid-1960s to 36 kg at present (Table 3.10). This deceleration is clearly seen in the growth rates of world aggregate consumption of meat (Table 3.11). The deceleration has been even more pronounced in the case of milk (Table 3.9), mostly because of developments in the transition economies (see below).

World averages conceal as much as they reveal. In the case of meat the strong growth in production and implied apparent consumption of pig meat in China in the 1980s and 1990s (which many observers believe to be grossly overstated in the country's statistics),[21] has shifted world meat consumption averages upwards rather significantly, from 30.7 kg in the mid-1980s to 36.4 kg at present. Without China, the average for the rest of the world would have actually stagnated in the same period (see memo item in Table 3.10). Again, this stagnation reflects the other extraordinary event of the 1990s, the collapse of consumption in the transition economies which went from 73 kg in the pre-reform period (late 1980s, when it had been boosted by heavy subsidies) to an estimated 46 kg in 1997/99. Excluding also the transition economies and the downward bias they impart to world totals, the per capita meat consumption in the rest of the world has been growing at a much slower, but always decelerating, pace: by 2.5 kg in the first decade (mid-1960s to mid-1970s), and by 1.6 kg in the second and third decades (see memo item in Table 3.10). Meat sector trends in the developing countries as a whole have been decisively influenced not only by China's rapid growth in the last two decades, but also by a similar performance in Brazil (from 32 kg in the mid-1970s to 71 kg at present). Including these two countries, the per capita meat consumption in the developing countries went over the same period from 11.4 to 25.5 kg. Excluding them, it went from 11 kg to only 15.5 kg (Table 3.10).

[21] According to its production and trade statistics, China's per capita meat consumption, resulting from the food balance sheets, was 45 kg in 1997/99. Independent consumption statistics show per capita consumption of "pork, beef, mutton" for 1997 of 19.04 kg for urban residents and 12.72 kg for the rural ones (UNDP, 2000a). The food balance sheet data we use here show for 1997/99 36.5 kg for the same meats plus another 8.5 kg for poultry meat. For a discussion of discrepancies see Feng Lu (1998); Fuller, Hayes and Smith (2000); and Colby, Zhong and Giordano (1998). It is indicative of the reservations with which the official production (and implied consumption) statistics are received by those concerned with world trade in livestock products, feedgrains and soybeans, that in FAPRI's latest projection study the data for per capita consumption of meat in China have been revised downwards drastically, to 31.5 kg in 1997/99 (FAPRI, 2002, livestock tables).

Table 3.11 Meat, aggregate production and demand: past and projected

	Production 1997/99 '000 tonnes	Production 1969-1999	1979-1999	1989-1999	1997/99-2015	2015-2030	Consumption 1997/99 '000 tonnes	Consumption 1969-1999	1979-1999	1989-1999	1997/99-2015	2015-2030
		Growth rates, % p.a.						Growth rates, % p.a.				
World												
Bovine	58 682	1.4	1.2	0.8	1.4	1.2	57 888	1.4	1.2	0.7	1.4	1.2
Ovine	10 825	1.9	2.2	1.4	2.1	1.8	10 706	1.9	2.2	1.4	2.1	1.8
Pig meat	86 541	3.2	2.9	2.7	1.4	0.8	86 392	3.2	2.9	2.7	1.4	0.8
Poultry meat	61 849	5.2	5.1	5.4	2.9	2.4	60 809	5.2	5.0	5.2	2.9	2.4
Total meat	217 898	2.9	2.8	2.7	1.9	1.5	215 795	2.9	2.8	2.7	1.9	1.5
Developing countries												
Bovine	27 981	3.0	3.4	3.8	2.3	2.0	28 074	3.4	3.5	4.1	2.3	2.0
Ovine	7 360	3.4	3.9	3.7	2.5	2.1	7 625	3.5	3.8	3.7	2.7	2.2
Pig meat	49 348	6.1	6	5.7	2.0	1.2	49 522	6.1	6.0	5.8	2.1	1.2
excl. China	10 892	3.7	3.3	3.4	2.7	2.4	11 393	3.6	3.2	3.7	2.7	2.4
Poultry meat	31 250	7.9	8.3	9.4	3.8	3.1	31 920	7.8	8.0	9.4	3.9	3.1
Total meat	115 938	5.2	5.5	5.9	2.7	2.1	117 141	5.3	5.6	6.1	2.7	2.1
excl. China	59 896	3.8	3.8	3.9	3.0	2.7	61 591	4.0	3.8	4.1	3.0	2.7
excl. China and Brazil	47 122	3.5	3.4	3.3	3.1	2.9	49 845	3.8	3.4	3.6	3.2	2.9
Total meat by region												
Sub-Saharan Africa	5 320	2.3	2.0	2.2	3.3	3.5	5 408	2.6	2.1	2.1	3.4	3.7
Near East/North Africa	6 956	4.4	4.4	3.8	3.6	2.9	8 164	4.7	3.3	3.3	3.6	2.9
Latin America and the Caribbean	27 954	3.5	3.4	4.5	2.6	2.1	27 296	3.8	3.7	4.8	2.4	2.0
excl. Brazil	15 180	2.5	2.2	3.1	2.7	2.3	15 551	3.0	2.6	4.0	2.6	2.2
South Asia	6 974	3.7	3.9	2.8	3.6	3.9	6 801	3.6	3.8	2.7	3.8	4.0
East Asia	68 734	7.1	7.6	7.6	2.4	1.6	69 472	7.1	7.7	7.8	2.5	1.6
excl. China	12 692	5.1	5.1	4.1	3.0	2.8	13 923	5.1	5.1	4.6	3.0	2.7
Memo items												
World livestock production (meat, milk, eggs)[1]		2.2	2.1	2.0	1.7	1.5						
World cereal feed demand (million tonnes)	657	1.3	0.6	0.6	1.9	1.5						
Population (million)												
World							5 878	1.7	1.6	1.5	1.2	0.9
Developing countries							4 572	2.0	1.9	1.7	1.4	1.1
excl. China							3 340	2.3	2.2	2.0	1.7	1.3
excl. China and Brazil							3 174	2.3	2.2	2.1	1.7	1.3

[1] Growth rates from aggregate production derived by valuing all products at 1989/91 international prices.

For milk and dairy products, there has been no "China effect" on world totals (given the small weight of these products in China's food consumption), but a very strong negative one on account of the transition economies, leading to a sharp slow-down in the growth rate of world production and consumption. Without them, there has been no deceleration in world production and consumption (Table 3.9, memo items).

In conclusion, the modest and decelerating growth in world per capita consumption of meat has been taking place for a wide variety of reasons. For the high-income countries, the reasons include the near saturation of consumption (e.g. in the EU and Australia), policies of high domestic meat prices and/or preference for fish (Japan and Norway), and health and food safety reasons every-where. However, by far the most important reasons have been the above-mentioned failure of many low-income countries to raise incomes and create effective demand, as well as the cultural and religious factors affecting the growth of meat consumption in some major countries.

Rapid growth of the poultry sector. Perhaps the perception of revolutionary change in the meat sector reflects the extraordinary performance of world production and consumption of poultry meat. Its share in world meat production increased from 13 percent in the mid-1960s to 28 percent currently. Per capita consumption increased more than threefold over the same period. That of pork also increased from 9.1 kg to 14.6 kg (China's statistics helping, but from 9.7 kg to only 10.3 kg for the world without China, Table 3.10). In contrast, per capita consumption of ruminant meat (from cattle, sheep and goats) actually declined a little. The most radical shifts in consumption in favour of poultry meat took place in countries that were the traditional producers, and often major exporters, of bovine meat: Latin America, North America and Oceania (accompanied in the latter two by deep cuts in the consumption of beef), as well as in the mutton-eating region of the Near East/North Africa. Significant increases in beef consumption were rare. They occurred in the Republic of Korea, Japan, Malaysia, Kuwait, Saudi

Table 3.12   World exports of livestock products and percentage of world consumption

	1964/66	1974/76	1984/86	1997/99
Total meat				
Exports ('000 tonnes)	5 996	8 869	14 011	27 440
% of consumption	7.4	7.9	9.4	12.7
Bovine				
Exports ('000 tonnes)	3 134	4 626	6 225	9 505
% of consumption	9.4	10.3	12.2	16.4
Pig meat				
Exports ('000 tonnes)	1 734	2 522	4 665	8 270
% of consumption	5.7	6.0	7.9	9.6
Poultry meat				
Exports ('000 tonnes)	436	887	1 973	8 465
% of consumption	4.0	4.7	6.3	13.9
Ovine				
Exports ('000 tonnes)	691	835	1 148	1 200
% of consumption	11.1	12.6	14.1	11.2
Milk and dairy (liquid milk equivalent)				
Exports ('000 tonnes)	21 606	31 769	57 004	71 364
% of consumption	6.0	7.6	11.1	12.8

Note: Meat exports include meat equivalent of live animal exports.

Table 3.13 Net trade positions of the major importers and exporters of livestock products ('000 tonnes)

	Major meat importers					Major meat exporters			
	1964/66	1974/76	1984/86	1997/99		1964/66	1974/76	1984/86	1997/99
Japan					**United States**				
Beef	-12	-85	-221	-867	Beef	-563	-887	-854	-475
Mutton	-69	-119	-78	-34	Pig meat	-116	-125	-493	-159
Pig meat	-1	-118	-214	-862	Poultry meat	101	100	247	2 548
Poultry meat	-9	-28	-130	-666	Total meat	-602	-916	-1 109	1 895
Total meat	-91	-350	-643	-2 430	Milk/dairy products[1]	2 547	-2 040	-719	-2 909
Milk/dairy products[1]	-847	-1 351	-2 129	-2 137	**EU-15**				
Former Soviet Union					Beef	-879	-220	552	504
Beef	-101	-407	-470	-704	Mutton	-377	-290	-233	-219
Pig meat	-4	-4	-333	-694	Pig meat	-46	-17	476	1 243
Poultry meat	-15	-61	-143	-956	Poultry meat	-79	59	253	915
Total meat	-95	-489	-1 036	-2 366	Total meat	-1 381	-468	1 048	2 444
Milk/dairy products[1]	-79	70	-502	529	Milk/dairy products[1]	-396	5 846	11 821	10 408
Mexico					**Australia and New Zealand**				
Beef	106	44	32	-141	Beef	558	989	918	1 959
Pig meat	0	1	-1	-112	Mutton	469	524	713	716
Poultry meat	0	-1	-17	-297	Total meat	1 033	1 523	1 637	2 681
Total meat	105	41	7	-586	Milk/dairy products[1]	4 729	5 584	7 764	13 302
Milk/dairy products[1]	-254	-684	-1 951	-2 231	**Brazil**				
Republic of Korea					Beef	57	83	263	302
Beef	0	0	-13	-181	Pig meat	1	6	-6	113
Pig meat	1	6	0	-5	Poultry meat	0	7	266	621
Poultry meat	0	0	0	-40	Total meat	58	97	518	1 028
Total meat	0	-1	-17	-231	Milk/dairy products[1]	-230	-224	-1 044	-1 913
Milk/dairy products[1]	-68	-30	-67	-205	**Argentina**				
Saudi Arabia					Beef	583	348	280	376
Beef	-3	-10	-70	-52	Total meat	620	380	282	267
Mutton	-6	-6	-47	-89	Milk/dairy products[1]	59	444	36	1 217
Poultry meat	-3	-47	-169	-292	**Eastern Europe**				
Total meat	-12	-62	-286	-433	Beef	217	327	310	78
Milk/dairy products[1]	-62	-213	-1 169	-877	Mutton	17	51	74	24
					Pig meat	211	287	398	215
					Poultry meat	59	169	287	38
					Total meat	504	833	1 069	355
					Milk/dairy products[1]	214	828	2 388	1 683

Note: Data include the meat equivalent of trade in live animals.
[1] In liquid milk equivalent (excludes butter).

Arabia, Mexico and Taiwan Province of China (all of them somehow linked to increased beef imports, often the result of more liberal trade policies), while Brazil is an example of fast growth in both production and consumption of beef.

Buoyancy of meat trade in recent years. The rapid growth in consumption of several countries was supported by even faster growth in trade. Some drastic changes occurred in the sources of exports and destination of imports, particularly in the last ten years or so. For example, Japan increased per capita meat consumption from 32.6 kg in 1984/86 to 41.5 kg in 1997/99, while its net imports quadrupled and self-sufficiency fell from 84 to 56 percent. At the global level, trade (world exports, including the meat equivalent of live animal exports) increased from 9.4 percent of world consumption in the mid-1980s to 12.7 percent in 1997/99, with poultry increasing from 12.2 to 16.4 percent and beef from 6.3 to 13.9 percent (Table 3.12). The major actors in this expansion of the meat trade are shown in Table 3.13. Japan tops the list of importers. Recent surges in the poultry meat (and to a lesser extent pig meat) imports of the countries of the former Soviet Union (overwhelmingly the Russian Federation), put this group of countries second in the league of importers, with its net imports rivalling those of Japan. On the export side, the combined exports of beef and mutton of Australia and New Zealand put them at the top of world meat exporters. However, the really extraordinary development of the 1990s has been the turnaround of the United States from a sizeable net importer of meat to a sizeable net exporter, a result reflecting its declining net imports of beef and pig meat and skyrocketing exports of poultry meat. In a sense, although the policies are different, the United States is replicating the earlier experience of the EU, which turned from a big net importer of meat up to the late 1970s to a large and growing net exporter.

The developing countries did not participate as much as the developed countries in this buoyancy of the world meat trade, although there have been some notable exceptions on both the import and the export side. In *poultry meat*, Brazil and Thailand became significant exporters, while Mexico became a large importer together with the more traditional importers of the Near East region (Saudi Arabia, Kuwait and the United Arab Emirates) and Hong Kong SAR. In *pig meat*, the largest developing net exporter continued to be China (mainland, including trade in live animals), although this has declined in recent years. China was rivalled in recent years by growing net exports from Brazil. Taiwan Province of China became a major exporter (mostly to Japan) in the decade to the mid-1990s before turning into a net importer after 1997[22] following the outbreak of foot-and-mouth disease (Fuller, Fabiosa and Premakumar, 1997).

On the import side, Hong Kong SAR has continued to be the predominant developing importer, while Mexico and Argentina became fast-growing net importers of pig meat in recent years. Overall, the pig meat trade has not been buoyant in the developing countries, an outcome that has partly reflected the lack of consumption in the major meat importers of the Near East/North Africa region. In *bovine meat*, India joined the more traditional developing exporters of South America as a significant exporter (mostly buffalo meat). The Republic of Korea became the largest developing net importer, surpassing Egypt. Several other developing countries became significant importers of bovine meat in recent years, including some countries of Southeast Asia (the Philippines, Malaysia and Indonesia) as well as Chile. Most recently, Mexico turned from net exporter to net importer of beef (including the meat equivalent of trade in live animals – on this latter point, see USDA, 2001a). Finally, only a few of the traditional importers of the Near East/North Africa region (Saudi Arabia, the United Arab Emirates and Kuwait) continued to be significant net importers of *mutton* (including live animals), but the imports of other countries collapsed (the Islamic Republic of Iran, Iraq and the Libyan Arab Jamahiriya) so that net imports of the region as a whole declined.

Slow growth in the dairy trade. In contrast to the buoyancy of the meat trade in recent years, trade in dairy products virtually stagnated. There was no growth in net imports of the developing countries. Increases in East Asia and modest ones in Latin America just compensated for declines in the other

---

[22] In the latest FAPRI projections, the net, and growing, net importer status of Taiwan Province of China continues to 2011 (FAPRI, 2002).

regions. There was no boost from increased imports on the part of the transition economies as was the case with meat. On the contrary, the former Soviet Union turned from net importer to net exporter. The decline of production of subsidized surpluses and the associated decline in food aid shipments on the side of the major exporters were an integral part of these trade outcomes.

Growth of livestock output achieved with modest increases in the feed use of cereals. We referred earlier to the importance of the livestock sector in creating demand for grains and oilseeds. Feed demand for cereals is often considered as the dynamic element that conditions the growth of the cereal sector. Occasionally, such use of cereals is viewed as a threat to food security, allegedly because it "subtracts" cereal supplies (or the resources going into their production) that would otherwise be available to food-insecure countries and population groups. We have argued elsewhere that this way of viewing things is not entirely appropriate, although, where economies are closed to trade, negative effects on food supplies available to food-insecure population groups can be produced (Alexandratos, 1995, p. 91-92).

Estimates put the total feed use of cereals at 657 million tonnes, or 35 percent of world total cereal use. Demand for feed in recent years has been a much less dynamic component of aggregate demand for cereals than commonly believed. The main reasons for these developments in the 1990s were discussed in the preceding section on cereals: the collapse of the livestock sector in the transition economies and high policy prices for cereals in the EU that favoured use of non-cereal feedstuffs (see also the discussion on cassava in Section 3.5 below). An additional factor that slowed down the growth of cereal use as feed has been the shift of meat production away from beef and towards poultry meat and pork, particularly in the industrial countries, the major users of cereals for feed. Pigs and poultry are much more efficient converters of feed to meat than cattle (see

Smil, 2000, Chapter 5). World totals have been decisively influenced by developments in the United States where the shift was most pronounced (poultry now accounts for 44 percent of total meat output, up from 30 percent in the mid-1980s, with the share of beef declining from 43 to 33 percent). Given the predominance of the feedlot system in the United States for producing bovine meat with high feedgrain conversion rates (5-7 kg of grain per kg of beef are the numbers usually given in the literature), it is easy to see why the shift to poultry has had such a pronounced impact on the average meat/grain ratios. Finally, productivity increases (reduction of the amount of feed required to produce 1 kg of meat), resulting from animal genetic improvements and better management, also played a role, at least in the major industrial countries.

3.3.2 Prospects for the livestock sector

Slower growth in world meat consumption. The forces that shaped the rapid growth of meat demand in the past are expected to weaken considerably in the future. Slower population growth compared with the past is an important factor. Perhaps more important is the natural deceleration of growth because fairly high consumption levels have already been attained in the few major countries that dominated past increases. As noted, China went from 10 kg in the mid-1970s to 45 kg currently, according to its statistics. If it were to continue at the same rate, it would soon surpass the industrial countries in per capita consumption of meat, an unreasonable prospect given that China will still be a middle-income country with significant parts of its population rural and in the low-income category for some time to come.[23] These characteristics suggest that further growth leading to about 60 kg in 2015 and 69 kg in 2030 as a national average for China is a more reasonable prospect than the much higher levels that would result from a quasi continuation of past trends (see also Alexandratos, 1997).[24] As another example and for similar reasons, Brazil's current average

---

[23] The poverty projections of the World Bank (see Chapter 2), suggest that despite the expected rapid decline in poverty in China, the country may still have some 200 million persons in the "under US$2 a day" poverty line (source as given in Table 2.5).

[24] The latest global food projections study of IFPRI projects China's per capita meat consumption to reach 64.4 kg in 2020 (Rosegrant *et al.*, 2001, p. 131). A higher estimate of 71 kg in 2020 is used in another IFPRI study (Delgado, Rosegrant and Meijer, 2001, Table 7). FAPRI, starting from the radical downward revisions of the historical meat production/consumption data (which bring the consumption per capita to 31 kg in 1997/99), projects 41 kg in 2011.

meat consumption of 71 kg suggests that the scope for the rapid increases of the past to continue unabated through the coming decades is rather limited.

The next question is whether any new major developing countries with present low meat consumption will emerge as major growth poles in the world meat economy.

The countries of South Asia come readily to mind. India has the potential to dominate developments in this region and indeed the world as a whole. It should be recalled that India is expected to rival China in population size by 2030 (1.41 billion versus 1.46 billion) and indeed surpass it ten years later, reaching 1.5 billion by 2040. It is also recalled that South Asia's projected growth rate of GDP per capita (overwhelmingly reflecting that for India) is, in the latest World Bank assessment, a respectable 3.5 percent p.a. for 2000-2004 and 4.0 percent p.a. after that until 2015. These rates are higher than those achieved in the 1990s (World Bank, 2001c, Table 1.7). India's meat consumption is very low – currently 4.5 kg per capita – and it has grown by only 1 kg in the last 20 years.

Can India play the role China has had so far in raising world meat demand? On this point, there are widely differing views. The first viewpoint, downplaying this prospect (Mohanty, Alexandratos and Bruinsma, 1998), is essentially based on the analysis of the differences in meat consumption among different income groups of Indian society. They show that high-income Indians, whether urban or rural, do not consume significantly more meat than low-income ones, although the differences in milk consumption are wide. Tomorrow's middle- and high-income population groups are likely to behave in a similar fashion. This would seem to preclude significant increases in national meat consumption because of income growth. Some support for this view is provided by recent marketing studies indicating that traditional consumption habits in Indian society are more resistant to change than one would expect from macroeconomic indicators (Luce, 2002). Other studies disagree, pointing to changing tastes and the prospect that the rapidly emerging middle classes will tend to adopt diets with higher meat content. For example, Bhalla, Hazell and Kerr (1999, Table 3) use "best guess" higher demand elasticities and project per capita food consumption

of meat and eggs to increase from 5.8 kg in 1993 to 15 kg in 2020, in a scenario with 3.7 percent p.a. growth of per capita income. The latest IFPRI study of global food projections expects per capita meat consumption of 7.4 kg in 2020 (baseline scenario) or, in a rather improbable "India high-meat scenario", 18.0 kg (Rosegrant *et al.*, 2001, Table 6.23). Yet another IFPRI study has 7.0 kg for India in 2020 (Delgado, Rosegrant and Meijer, 2001, Table 7)

The only generalization we can make with some confidence is that the recent high-growth rates of per capita consumption of poultry meat in India (admittedly from the very low base of 0.2 kg in the mid-1980s to 0.6 kg in 1997/99) is bound to continue unabated in the coming decades. That is, India's participation in the global upsurge of the poultry sector, being at its incipient stage, still has still a long way to go. Consumption of other meats will probably grow by much less, with beef and pork subject to cultural constraints for significant parts of the population of India and indeed the whole of South Asia. In parallel, consumption of the preferred mutton/goat meat faces production constraints, implying rising real relative prices compared with poultry meat. Overall, the force of the growth of poultry meat consumption has the potential of raising India's average consumption of all meat by 2 kg in the period to 2015 (compared with 1 kg in the preceding two decades) and by another 4 kg in the subsequent 15 years to 10 kg in 2030 (meat plus eggs: 15 kg). This kind of growth will perhaps be viewed as revolutionary in a national context, since it would raise the very low intake of animal protein in the structure of the country's diet. However, it will be far from having an impact on world averages and those of the developing countries anywhere near that exerted in the historical period by developments in China.

Other countries or regional groups that played a role in raising global consumption of meat in the past are those of *East Asia other than China* (mainland). Some countries in this group have attained mid- to high consumption per capita, e.g. Hong Kong SAR, Taiwan Province of China, Malaysia and the Republic of Korea. However, in the most populous country of the region, Indonesia, as well as in several others (Thailand, Malaysia, Korea, Rep., Taiwan Province of China) the process of rapid growth in meat consumption came to an abrupt end, often followed by reversals, in the late 1990s because of the economic crisis. The recent food crisis in the Democratic People's Republic of Korea accentuated the regional slowdown. A period of slow growth in the per capita consumption of this region (East Asia excluding China) may ensue before rapid growth resumes to reach the levels indicated in Table 3.10. This is an additional factor militating against the continuation of the rapid growth in world meat consumption at the high rates of the past.

Of the other regions, *Latin America and the Caribbean*, excluding Brazil, is still in a middling position as regards per capita consumption of meat (45 kg per capita), with only the traditional meat producers and exporters (Argentina and Uruguay) having levels comparable to those of the industrial countries. The swing to poultry consumption has been fairly strong, often substituting other meats. Per capita consumption of poultry meat is still at a middling level (17.2 kg, up from 9 kg in the mid-1980s), so the process still has some way to go. This will raise the group's overall meat average, as the decline in the consumption of beef, mutton and pork that characterized past developments is not likely to continue at past rates.

Average per capita consumption in the *Near East and North Africa* region grew little since the mid-1980s, in contrast with the sharp increases experienced in the preceding decade of the oil boom. The recent slowdown of the regional average reflected the sharp declines in Iraq, and the near stagnation of per capita consumption in several other countries. Most countries of the region are in a middling position as regards per capita consumption (in the range of 17-47 kg in 1997/99), although some of the smaller oil-rich countries (such as Kuwait and the United Arab Emirates) have fairly high levels. The three most populous countries of the region, Egypt, Turkey and the Islamic Republic of Iran (which between them have 53 percent of the region's population) are in the range of 20-22 kg. In this region there has also been a consumption trend towards poultry meat – in fact all growth in average meat

consumption came from poultry – although the shift has not been as strong as in Latin America. The income growth prospects of the region for the next ten years are somewhat better than in the preceding ten years (see Figure 2.3). Therefore, some resumption of the growth in meat consumption, accompanied with further shifts towards poultry meat, may be expected.

Finally, *sub-Saharan Africa's* economic prospects suggest that little growth in its per capita consumption of meat is likely. There have been no improvements in the past 30 years, with per capita consumption stagnant at around 10 kg. Although some countries did increase consumption of poultry meat significantly (e.g. Gabon, Mauritius, Senegal and Swaziland), the region has hardly benefited from the options offered by the poultry sector – a situation that will probably persist for some time. For the longer term, the projections suggest only very modest gains.

Per capita meat consumption in the *transition economies* will eventually reverse its downward trend of the 1990s (having fallen from a peak of 73 kg in 1990 to 45 kg in 1999). However, at the projected level of 61 kg, it will not have reverted to the pre-reform levels even by 2030. In the *industrial countries*, average per capita consumption of meat at 88 kg is fairly high, although, as noted, countries with high fish consumption (Japan and Norway) have much lower levels. In principle, the achievement of near-saturation levels of overall food consumption, as well as concerns about health, suggest that there is very little scope for further increases. Yet the data indicate that such increases do take place even in countries that have passed the 100-kg mark, probably reflecting a mix of overconsumption and growing post-retail waste or feeding of pets. For example, the United States went from 112 to 123 kg in the last ten years (1989-99) and the latest FAPRI projections foresee an increase of 5 percent by 2011 (FAPRI, 2002). The USDA projections, however, foresee a slight decline (about 1 kg in retail weight) by 2011 (USDA, 2002, Table 22).[25] Even in the more food quality/safety conscious EU, per capita consumption of meat is projected to continue growing, from 89 kg to 94 kg

[25] Long-term historical series (1909-2000) of meat consumption in the United States can be found in www.ers.usda.gov/Data/FoodConsumption/Spreadsheets/mtpcc

Table 3.14   Net trade in meat[1] and milk/dairy ('000 tonnes)

Type of meat	1964/66	1974/76	1984/86	1997/99	2015	2030
			Developing countries			
Bovine	859	706	13	-114	-310	-650
Ovine	52	-29	-229	-259	-700	-1 170
Pig meat	90	55	516	-173	-550	-830
Poultry meat	-33	-200	-430	-693	-2 340	-3 250
Total meat	969	532		-129-1 238	-3 900	-5 900
			Industrial countries			
Bovine	-907	-375	370	1 491	1 860	1 840
Ovine	-20	86	372	375	820	1 270
Pig meat	-167	-273	25	891	930	1 010
Poultry meat	-9	110	314	2 624	4 320	4 800
Total meat	-1 103	-452	1 080	5 380	7 930	8 920
			Transition countries			
Bovine	116	-80	-160	-626	-810	-590
Ovine	43	34	-16	12	10	20
Pig meat	207	282	65	-479	-100	100
Poultry meat	44	108	143	-918	-1 000	-620
Total meat	409	344	33	-2 012	-1 900	-1 090
Total meat by developing region						
Sub-Saharan Africa	111	180	-60	-92	-280	-740
Near East/North Africa	-97	-337	-1 437	-1 246	-2 360	-3 520
Latin America and the Caribbean	829	672	867	658	1770	2 770
South Asia	-6	0	47	173	-80	-410
East Asia	132	18	454	-732	-2 950	-4 000
Milk and dairy products in whole milk equivalent (excluding butter)						
Developing countries	-5 300	-8 735	-20 040	-19 848	-29 600	-38 900
Sub-Saharan Africa	-522	-1 206	-2 785	-2 321	-3 600	-5 200
Near East/North Africa	-753	-2 031	- 757	-4 980	-8 900	-12 500
Latin America and the Caribbean	-1 879	-2 571	-5 500	-5 374	-6 350	-6 700
South Asia	-662	-553	-1 247	-804	-1 200	-1 500
East Asia	-1 486	-2 374	-3 751	-6 370	-9 550	-13 000
Industrial countries	**6 920**	**8 973**	**18 420**	**19 665**	**28 000**	**35 700**
Transition countries	**135**	**898**	**1 886**	**2 212**	**3 500**	**5 200**

[1] Includes the meat equivalent of trade in live animals.

in the ten years 1998-2008 (European Commission, 2001, Table 1.19). These trends have to be taken into account, even if they make little sense from the standpoint of nutrition and health. In the projections for the industrial countries we have factored in rather more modest increases than the EU projections indicate: a 12 kg increase in per capita consumption for the entire 34-year period, raising it from 88 kg in 1997/99 to 100 kg in 2030. The whole of the increase will be in the poultry sector.

These prospects for changes in the per capita consumption of meat, in combination with slower population growth, suggest that the strength of the meat sector as a driving force of the world food economy will be much weaker than in the past. Thus, world aggregate demand for meat is projected to grow at 1.7 percent p.a. in the period to 2030, down from 2.9 percent in the preceding 30 years. The reduction is even more drastic for the developing countries, in which the growth rate of aggregate demand is reduced by half, from 5.3 to 2.4 percent. Much of this reduction is a result of the projected slower growth of aggregate consumption in China and, to a lesser extent, in Brazil. If these two countries are removed from the developing countries aggregate, there is very little reduction in the growth of aggregate demand for meat, from 3.8 percent p.a. in the preceding three decades to 3.1 percent p.a. in the next three. All this reduction reflects the slowdown in population growth from 2.0 p.a. to 1.3 percent p.a. (Table 3.11).

In conclusion, the projected slowdown in the world meat economy is based on the following assumptions: (i) relatively modest further increases in per capita consumption in the industrial countries; (ii) growth rates in per capita consumption in China and Brazil well below those of the past; (iii) persistence of relatively low levels of per capita consumption in India; and (iv) persistence of low incomes and poverty in many developing countries. If these assumptions are accepted, the projected slowdown follows inevitably. Naturally, a slower growth rate applied to a large base year world production (218 million tonnes in 1997/99, of which 116 million in the developing countries) will still produce large absolute increases (some 160 million tonnes by 2030, of which some 130 in the developing countries). These quantitative increases will accentuate environmental and other

problems associated with such large livestock sectors (see Chapter 5).

No slowdown in the consumption of dairy products.
The average dairy consumption of the developing countries is still very low (45 kg of all dairy products in liquid milk equivalent), compared with the average of 220 kg in the industrial countries. Few developing countries have per capita consumption exceeding 150 kg (Argentina, Uruguay and some pastoral countries in the Sudano-Sahelian zone of Africa). Among the most populous countries, only Pakistan has such a level. In South Asia, where milk and dairy products are preferred foods, India has only 64 kg and Bangladesh 14 kg. East Asia has only 10 kg. In this region, however, food consumption preferences do not favour milk and dairy products, but the potential for growth is still there with growing urbanization. Overall, therefore, there is considerable scope for further growth in consumption of milk and dairy products. The projections show higher world growth than in the recent past (Table 3.9) because of the cessation of declines and some recovery in the transition economies. Excluding these latter countries, world demand should grow at rates somewhat below those of the past but, given slower population growth, per capita consumption would grow more quickly than in the past.

Meat trade expansion will continue, some recovery in dairy trade. Despite the projected slowdown in meat demand growth, some of the forces that made for the buoyancy in the world meat trade in the recent past discussed above are likely to continue to operate – in particular the changes in trade policy regimes. The projected net trade positions are shown in Table 3.14. They reflect the above-mentioned factors making for growth in demand and the analysis of production possibilities, as well as the prospect that the more liberal trade policies of recent years will continue to prevail. Overall, the trend for the developing countries to become growing net importers of meat is set to continue. This is another important component of the broader trend for developing countries to turn from net exporters to growing net importers of food and agricultural products (see Section 3.1). Imports of poultry meat are likely to dominate the picture of growing dependence on imported meat.

Trade in dairy products will also likely recover, with net imports of the developing countries resuming growth after a period of stagnation from the mid-1980s onwards (Table 3.14). This would reflect continuation of the growth of imports of East Asia, as well as the resumption of import growth into the major deficit region, the Near East/North Africa, following recovery in the growth of demand.

Growth of feed use of cereals will resume. Demand for cereals for feed is projected to grow at 1.9 percent a year between 1997/99 and 2015, and at 1.5 percent p.a. thereafter to 2030. The growth rate of the first subperiod is a little higher than that of livestock production (1.7 and 1.5 percent in the two subperiods, respectively). The projected growth rates of feed demand for cereals are higher than the depressed ones of the historical period, despite the projected slowdown in livestock production (Table 3.11, memo items). The main reasons are as follows:

- As noted in the section on cereals, the turnaround of the transition economies and the EU from declines in the preceding decade to renewed growth in feed use of cereals (or at least, the cessation of declines) will lead to resumption of growth of demand for cereals in this sector. This process has been under way in the EU since the early 1990s (INRA, 2001, p. 46). These two regions have a large weight (33 percent) in world cereal feed consumption. Therefore, the continued turnaround will be reflected in higher growth rates of the global demand for feed compared with the past (European Commission, 2001, Table 1.6).
- In parallel, the shift in meat production away from bovine meat towards poultry meat will exert a much weaker downward pressure on the demand for feedstuffs than in the past. As shown in Table 3.11, in the past 20 years poultry production increased at 5.1 percent p.a. and bovine meat at 1.2 percent p.a. In the period to 2015, the difference in the growth rates will be much smaller, 2.9 and 1.4 percent, respectively, hence the downward pressure

exerted on the aggregate grain-meat conversion rates will also be weaker.

In the developing countries, the growth of cereal use for feed and the growth of livestock production have moved much more in unison than in the industrial countries. If these trends were to continue, the growing weight of the developing countries in world livestock production would imply that the global totals and relationships will increasingly reflect developments in these countries. Will the trends continue? It is likely that, for some time, there will be a tendency for the use of concentrates in total animal feed to increase more quickly than aggregate livestock output in the developing countries. This will reflect the gradual shift of their production from grazing and "backyard" systems to stall-fed systems using concentrate feedstuffs (see Chapter 5). Such structural change in the production systems will tend to raise the average grain-meat ratios of the developing countries and perhaps compensate for opposite trends resulting from improvements in productivity. A strong case for this prospect is made in a recent analysis by the Centre for World Food Studies in the Netherlands (Keyzer, Merbis and Pavel, 2001).

3.4 Oilcrops, vegetable oils and products

3.4.1 Past and present

Fastest growth of all subsectors of global agriculture. The oilcrops sector has been one of the most dynamic parts of world agriculture in recent decades. In the 20 years to 1999 it grew at 4.1 percent p.a. (Table 3.15), compared with an average of 2.1 percent p.a. for all agriculture.[26] Its growth rate exceeded even that of livestock products. The major driving force on the demand side has been the growth of food demand in developing countries, mostly in the form of oil but also direct consumption of soybeans, groundnuts, etc. as well as in the form of derived products other than oil.[27] Food demand in the developing countries accounted for half the

---

[26] For the derivation of the growth rates of the entire oilcrops sector, the different crops are added together with weights equal to their oil content. This is what the expression "oil equivalent", used in this study, means.

[27] For example, in China it is estimated that out of total domestic consumption of soybeans (production plus net imports) of about 15 million tonnes, only about nine million tonnes are crushed for oil and meal and the balance is consumed as food directly or in other forms (Crook, 1998, Table 8.1).

increases in world output of the last two decades, with output measured in oil content equivalent (Table 3.16). China, India and a few other countries represented the bulk of this increase. No doubt, the strong growth of demand for protein products for animal feed was also a major supporting factor in the buoyancy of the oilcrops sector. The rapid growth of this sector reflects the synergy of the two fastest rising components of the demand for food (Table 3.16, lower part): food demand for oils favouring the oil palm and for livestock products favouring soybeans.

Growing contribution to food supplies and food security. World production, consumption and trade in this sector have been increasingly dominated by a small number of crops (soybeans, oil palm, sunflower and rapeseed) and countries. However, the more traditional and less glamorous

oilcrops continue to be very important as major elements in the food supply and food security situation in many countries, such as groundnuts and sesame seed in the Sudan and Myanmar, coconuts in the Philippines and Sri Lanka and olive oil in the Mediterranean countries.

Rapid growth of food demand in the developing countries, in conjunction with the high calorie content of oil products, has contributed to the increases achieved in food consumption in developing countries (measured in the national average kcal/person/day, see Chapter 2). In the mid-1970s, consumption of these products (5.3 kg/person/year, in oil equivalent, Table 3.17) supplied 144 kcal/person/day, or 6.7 percent of the total availability of 2 152 calories of the developing countries. By 1997/99 consumption per capita had grown to 9.9 kg contributing 262 kcal to total food supplies, or 9.8 percent of the total, which itself had

Table 3.15 Oilcrops, vegetable oils and products, production and demand

	Million tonnes	Growth rates, percentage p.a.				
	1997/99	1969-99	1979-99	1989-99	1997/99-2015	2015-2030
Aggregate consumption (all uses, oil equivalent)						
World	98.3	4.0	3.9	3.7	2.7	2.2
Developing countries	61.8	5.0	4.8	4.6	3.2	2.5
Sub-Saharan Africa	6.7	3.2	3.4	4.3	3.3	3.2
Near East/North Africa	6.2	5.1	4.3	3.2	2.5	2.2
Latin America and the Caribbean	9.0	4.7	3.7	3.2	3.2	2.4
South Asia	13.6	4.5	4.5	4.2	3.5	2.5
East Asia	26.2	6.2	6.1	5.8	3.2	2.3
Industrial countries	30.6	3.2	3.4	3.1	1.7	1.8
Transition countries	6.0	1.1	-0.4	-1.4	1.3	1.4
Production (oilcrops, oil equivalent)						
World	103.7	4.1	4.1	4.3	2.5	2.2
Developing countries	67.7	4.8	5.0	4.7	2.8	2.4
Sub-Saharan Africa	6.0	1.5	3.0	3.5	3.2	3.0
Near East/North Africa	1.8	2.0	2.4	2.4	2.3	2.1
Latin America and the Caribbean	14.6	5.7	4.8	5.3	2.9	2.6
South Asia	9.7	3.6	4.6	2.4	3.2	2.4
East Asia	35.5	6.2	5.8	5.5	2.7	2.2
Industrial countries	30.2	3.6	3.1	4.6	1.7	1.7
Transition countries	5.8	0.7	0.9	-0.5	1.3	1.6

risen to 2 680 kcal. In practice, just over one out of every five calories added to the consumption of the developing countries originated in this group of products. This trend is set to continue and intensify: 44 out of every 100 additional calories in the period to 2 030 may come from these products. This reflects the prospect of only modest growth in the direct food consumption of staples (cereals, roots and tubers, etc.) in most developing countries, while non-staples such as vegetable oils still have significant scope for consumption increases.

Non-food uses. The second major driving force on the demand side has been the non-food industrial use of vegetable oils, with China and the EU being major contributors to this growth (Table 3.16). The existing data do not permit us to draw even a partial balance sheet of the non-food industrial products for which significant quantities of vegetable oil products are used as inputs.[28] The main industrial products involved (paints, detergents, lubricants, oleochemicals in general) are commodities for which demand can be expected to grow as fast, if not faster, than the demand for food uses of vegetable oil products, particularly in the developing countries. In addition, the rapid demand growth for industrial uses in the EU probably reflects the incentives given in recent years to farmers to grow crops (oilseeds, rape, linseed) for non-food uses on land set aside under the CAP rules for limiting excess production of food crops. There have been serious efforts in several European countries to expand the market for biofuel from rapeseed oil as a substitute for diesel fuel (Koerbitz, 1999, Raneses *et al.*, 1999). This option has been viewed not only as a way out of the impasse of surplus food production but also as a desirable land use for environmental reasons, although the energy efficiency and the net effect on the environment remain to be established (Giampietro, Ulgiati and Pimentel, 1997). In terms of actual oil produced and used (rather than of oil equivalent of oilcrops) the world is apparently using some 24 million tonnes for non-food indus-

trial uses out of a total use of 86 million tonnes. In the mid-1970s the comparable figures were 6 and 33 million tonnes, respectively.

Concentration of growth in a small number of crops and countries. The demand for protein meals for animal feed also contributed to a change in the geographic distribution of oilseeds production. This shifted towards countries that could produce and export oilseeds of high protein content, in which oilmeals are not by-products but rather joint products with oil, e.g. soybeans in South America. In addition, support policies of the EU helped to shift world production of oilseeds in favour of rapeseed and sunflower seed. Overall, four oilcrops, oil palm, soybeans, rapeseed and sunflower seed account for 72 percent of world production. In the mid-1970s they accounted for only 55 percent (Table 3.18). These four crops contributed 82 percent of the aggregate increase in oilcrops production since the mid-1970s (Table 3.16). Moreover, a good part of these increases came from a small number of countries, as shown in the lower part of Table 3.16: palm oil mainly from Malaysia and Indonesia; soybeans from the United States, Brazil, Argentina, China and India; rapeseed from China, the EU, Canada, India and Australia; and sunflower seed from Argentina, the EU, the United States, Eastern Europe, China and India. For many countries, including some major producers, these fast expanding oilseeds are new crops that were hardly cultivated at all, or in only insignificant amounts, 20 or even ten years ago.[29]

Growing role of trade. The rapid growth of demand in the developing countries was accompanied by the emergence of several countries as major importers, with net imports rising by leaps and bounds. Thus, by the mid-1990s there were ten developing countries, each with net imports of over 0.7 million tonnes (India, China,[30] Mexico, Pakistan, etc., see Table 3.20). These ten together now have net imports of 13 million tonnes, a 12-fold increase in the period since the mid-1970s. Numerous other

---

[28] One should be careful with these numbers as this category of demand is often used by statisticians as the dumping ground for unexplained residuals of domestic disappearance and statistical discrepancies. There is no doubt, however, that non-food industrial uses are a dynamic element of demand.

[29] For example, soybeans in India and even Argentina, sunflower seed in China, Pakistan, Brazil, Myanmar, oil palm in Thailand, rapeseed in the United States and Australia.

[30] It is believed that China's net imports of vegetable oils are much larger than reported in the trade statistics because of considerable smuggling (1.5-2.0 million tonnes, OECD, 1999, p. 18).

Table 3.16 Sources of increases in world production and consumption of oilcrops (in oil equivalent)

		Major countries/regions
Increase in world consumption, 1974/76-1997/99		**% contribution to increment of each item, 1974/76 to 1997/99**
Total world increase (=100), of which:	100	
Developing countries, food	50	China 29, India 18, N. East/N. Africa 10, Indonesia 8, Brazil 5, Nigeria 5, Pakistan 5
Developed countries, food	13	United States 34, EU 34, Japan 9, E. Europe 7
Non-food industrial uses, world	29	EU 17, China 10, Indonesia 9, United States 12, Brazil 5, Malaysia 4, India 4
Other uses (feed, seed, waste), world	8	
Increase in world production, 1974/76-1997/99		**% contribution to increment of each item, 1974/76 to 1997/99**
Total world increase (=100), of which:	100	
Oil palm (palm oil and palm kernel oil)	28	Malaysia 51, Indonesia 34, Thailand 3.2, Nigeria 2.6, Colombia 2.5
Soybeans	27	United States 38, Brazil 21, Argentina 17, China 8, India 7
Rapeseed	18	EU 27, China 26, Canada 21, India 12, Australia 5
Sunflower seed	9	Argentina 33, EU 20, United States 10, E. Europe 8, China 9, India 5
All other oilcrops	18	

developing countries are smaller net importers, but still account for another 4 million tonnes of net imports, up from small net exports 20 years ago. This group includes a number of countries that turned from net exporters to net importers over this period, e.g. Senegal, Nigeria and Sri Lanka. With these rates of increase of imports, the traditional net trade surplus of the vegetable oils/oilseeds complex (oils, oilmeals and oilseeds) of the developing countries was reduced to zero in both 1999 and 2000, compared with a range of US$1.0-US$4.4 billion in the period 1970-98. This happened despite the spectacular growth of exports of a few developing countries that came to dominate the world export scene, i.e. Malaysia and Indonesia for palm oil and Brazil and Argentina for soybeans. As happened with the livestock sector, the overall evolution of trade of oilseeds and products has contributed to the agricultural trade balance of the developing countries diminishing rapidly and becoming negative (Figure 3.2).

Oilcrops responsible for a good part of agricultural land expansion. On the production side, these four oilcrops expanded mainly, although not exclu-sively, in land-abundant countries (Brazil, Argentina, Indonesia, Malaysia, the United States and Canada). Particularly notable is the rapid expansion of the share of oil palm products (in terms of oil palm fruit) in Southeast Asia (from 40 percent of world production in 1974/76 to 79 percent in 1997/99) and the dramatically shrinking share of Africa (from 53 to 14 percent). Africa's share in terms of actual production of palm oil (9 percent of the world total – down from 37 percent in the mid-1970s) remained well below that of its share in oil palm fruit production. This denotes the failure to upgrade the processing industry – but also the potential offered by more efficient processing technology to increase oil output from existing oil palm areas. The contrast of these production shares with the shares of land area under oil palm is even starker: Africa still accounts for 44 percent of the world total, three-quarters of it in Nigeria.

The oil palm and the other three fast growing crops (soybeans, rapeseed and sunflower) have been responsible for a good part of the expansion of cultivated land under all crops in the developing countries and the world as a whole. In terms of

Table 3.17   Vegetable oils, oilseeds and products, food use: past and projected

	1964/66	1974/76	1984/86	1997/99	2015	2030
	Food use (kg/capita, oil equivalent)					
World	**6.3**	**7.3**	**9.4**	**11.4**	**13.7**	**15.8**
Developing countries	4.7	5.3	7.5	9.9	12.6	14.9
Sub-Saharan Africa	7.7	8.0	7.9	9.2	10.7	12.3
Near East/North Africa	6.7	9.4	12.1	12.8	14.4	15.7
Latin America and the Caribbean	6.2	8.0	11.1	12.5	14.5	16.3
South Asia	4.6	5.0	6.2	8.4	11.6	14.0
East Asia	3.4	3.5	6.4	9.7	13.1	16.3
East Asia, excl. China	4.9	5.4	8.4	11.2	13.6	16.3
Industrial countries	11.3	14.5	17.3	20.2	21.6	22.9
Transition countries	6.9	8.2	10.1	9.3	11.5	14.2

	Million tonnes	Growth rates, percentage p.a.				
	1997/99	1969-99	1979-99	1989-99	1997/99 -2015	2015 -2030
		Total food use (oil equivalent)				
World	**66.9**	**3.7**	**3.3**	**2.8**	**2.3**	**1.9**
Developing countries	45.1	4.8	4.3	3.6	2.9	2.2
Sub-Saharan Africa	5.3	3.3	3.5	4.2	3.5	3.2
Near East/North Africa	4.8	4.5	3.3	2.5	2.6	2.1
Latin America and the Caribbean	6.2	4.4	3.0	2.0	2.2	1.7
South Asia	10.8	4.4	4.4	3.9	3.5	2.4
East Asia	17.9	6.0	5.5	4.3	2.7	2.0
East Asia, excl. China	6.8	5.5	4.9	2.8	2.4	2.1
Industrial countries	18.0	2.3	2.1	1.8	0.8	0.6
Transition countries	3.8	1.3	-0.1	-0.7	1.0	1.2

harvested area,[31] land devoted to the main crops (cereals, roots and tubers, pulses, fibres, sugar crops and oilcrops) in the world as a whole expanded by 59 million ha (or 6 percent) since the mid-1970s. A 105-million ha increase in the developing countries was accompanied by a 46-million ha decline in the industrial countries and the transition economies. The expansion of land under the four major oilcrops (soybeans, sunflower, rape and oil palm) was 63 million ha, that is, these four crops accounted for all the increase in world harvested area and more than compensated for the drastic declines in the area under cereals in the industrial countries and the transition economies (Table 3.19). In these countries, the expansion of oilseed area (25 million ha) substituted and compensated for part of the deep decline in the area sown to cereals. But in the developing countries, it seems likely that it was predominantly new land that came under cultivation, as land under the other crops also increased.

These numbers clearly demonstrate the revolutionary changes in cropping patterns that occurred, particularly in the developed countries, as a result

[31] The increase of harvested area implies not only expansion of the cultivated land in a physical sense (elsewhere in this report referred to as arable area) but also expansion of the land under multiple cropping (in the harvested or sown area definition, a hectare of arable land is counted as two if it is cropped twice in a year). Therefore, the harvested area expansion under the different crops discussed here could overstate the extent to which physical area in cultivation has increased. This overstatement is likely to be more pronounced for cereals (where the arable area has probably declined even in the developing countries) than for oilcrops, as the latter include also tree crops (oil and coconut palms and olive trees).

Table 3.18   Major oilcrops, world production

	Production of oilcrops in oil equivalent (Million tonnes)						Actual oil production	Growth rates, % p.a.				
	1964/66	1974/76	1984/86	1997/99	2015	2030	1997/99	1969-99	1979-99	1989-99	1997/99 -2015	2015 -2030
Soybeans	5.8	10.5	17.2	27.7	42	58	22.5	4.1	3.2	4.5	2.5	2.2
Oil palm	2.1	3.7	8.7	21.6	35	49	21.6	8.2	7.7	6.5	2.8	2.3
Rapeseed	1.7	3.0	7.1	14.5	22	32	11.9	6.9	6.6	5.6	2.4	1.8
Sunflower seed	3.4	4.2	7.5	10.3	15	21	9.3	3.7	3.1	2.3	2.2	2.4
Groundnuts	4.8	5.4	6.1	9.4	15	20	4.8	2.3	3.2	4.1	2.6	2.2
Coconuts	3.1	3.7	4.3	6.0	9	12	3.2	2.3	2.7	2.5	2.4	1.8
Cottonseed	3.4	3.7	5.0	5.3	7	9	3.8	1.6	1.2	0.1	2.0	1.4
Sesame seed	0.7	0.8	1.0	1.2	2	3	0.7	1.5	1.9	2.4	3.0	2.5
Other oilcrops	3.7	4.3	4.8	7.6	10	13	5.1	1.3	1.6	2.7	1.6	1.8
Total	29.0	39.0	62.0	104.0	157	217	83.0	4.1	4.1	4.3	2.5	2.2
Oils from non-oilcrops (maize, rice bran)							2.6					

of policies (e.g. the EU support to oilseeds) and of changing demand patterns towards oils for food in the developing countries and oilcakes/meals for livestock feeding everywhere. They also demonstrate that land expansion still can play an important role in the growth of crop production. The 200 percent increase in oilcrop output between 1974/76 and 1997/99 in developing countries was brought about by a 70 percent (50 million ha) expansion of land under these crops, at the same time as land under their other crops also increased by an almost equal amount (Table 3.19).

3.4.2 Prospects for the oilcrops sector

Food demand. The growth of food demand in the developing countries was the major driving force behind the rapid growth of the oilcrops sector in the historical period. The most populous countries played a major role in these developments (Table 3.16).

Will these trends continue in the future? In the first place, slower population growth, particularly in the developing countries, will be reflected in slower growth rates of their aggregate demand for food, *ceteris paribus*. Naturally, others things will not be equal: in particular, the per capita consumption of the developing countries was only 5.3 kg in the mid-1970s. This afforded great scope for the increases in consumption that took place.

However, in the process, per capita consumption grew to 9.9 kg in 1997/99. There is reduced scope for further increases and this will lead to slower growth in the future. A deceleration in growth has already been evident for some time: the annual growth rate of per capita demand declined to 1.9 percent p.a. in the decade to 1999, down from 2.9 percent p.a. in the preceding ten-year period. Even this reduced growth rate was well above that for other food products. The corresponding growth rate for cereals (direct food demand per capita in the developing countries) has been near zero in recent years. Only demand for meat has had higher growth rates, heavily influenced by developments in China where meat production data may grossly overstate actual consumption (see Section 3.3).

Despite the increases already achieved, vegetable oils and oil products still have relatively high income elasticities in the developing countries, particularly those in which consumption of livestock products and animal fats is and will likely remain at low levels. In our earlier projections to 2010, the per capita food consumption of all vegetable oils, oilseeds and products (expressed in oil equivalent) was projected to rise from 8.2 kg in 1988/90 to 11 kg in 2010 (Alexandratos, 1995, Table 3.5). By 1997/99 it had grown to 9.9 kg. In the current projections, the per capita food demand for the developing countries as a whole

Table 3.19 Harvested area increases: main oilseeds versus other main crops

	Developing countries			Rest of world		
	1974/76	1997/99	Change	1974/76	1997/99	Change
Main crops			Million ha			
Cereals	406	441	35	306	242	-63
Roots and tubers	33	40	7	15	11	-4
Pulses	53	60	7	10	9	-1
Sugar beet and cane	12	20	7	9	7	-2
Fibre crops	30	28	-2	9	9	0
Oilcrops	70	120	50	38	63	25
Total above	603	708	105	387	342	-46
of which:						
Soybeans	15.6	38.9	23.3	22.2	31.0	8.8
United States				20.8	28.6	7.8
Brazil	5.8	12.6	6.8			
China	7.0	8.3	1.3			
Argentina	0.4	7.2	6.8			
India	0.1	6.1	6.0			
Sunflower	2.4	7.8	5.4	7.0	13.3	6.3
EU (15)				0.7	2.2	1.5
Former Soviet Union				4.4	6.8	2.4
Eastern Europe				1.1	2.2	
India	0.3	1.8	1.5			
Argentina	1.2	3.5	2.3			
Rapeseed	6.5	14.3	7.7	2.6	11.5	8.9
China	2.2	6.6	4.4			
India	3.5	6.7	3.2			
Canada				1.3	5.3	4.0
EU (15)				0.8	3.1	2.3
Oil palm	3.5	9.0	5.4			
Malaysia	0.4	2.6	2.2			
Indonesia	0.1	1.8	1.6			
Nigeria	2.1	3.0	0.9			
Groundnuts	18.6	22.5	3.9	1.0	0.7	-0.2
India	7.1	7.2	0.1			
China	1.8	4.0	2.2			
Nigeria	1.3	2.5	1.2			
Sudan	0.8	1.5	0.6			
Senegal	1.3	0.7	-0.5			
Indonesia	0.4	0.6	0.2			
United States				0.6	0.6	0.0

Note: In this table, cotton is included in fibre crops and not in oilcrops.

rises further to 12.6 kg in 2015 and to nearly 15 kg by 2030 (Table 3.17). We have already noted earlier that average per capita food consumption (all food products) in developing countries may rise from the 2 680 kcal of 1997/99 to 2 850 kcal in 2015 and to 2 980 kcal by 2030 (see Chapter 2). Vegetable oils and products would contribute some 45 percent of this increase. This is an acceleration of the historical trend for these commodities to account for an ever-increasing part of the growth in food consumption in the developing countries. They had contributed 18 percent of the total increment in the decade from the mid-1970s to the mid-1980s and 27 percent in the subsequent decade.

It follows that this group of products will play an important role in increasing dietary energy availability. However, it is clear that, given lower growth rates of both population and per capita demand, the growth of aggregate food demand for these products will probably be well below that of the past. The relevant data and projections are shown in Table 3.17. The growth rate of food demand in developing countries is projected to be 2.9 percent p.a. in the period to 2015, down from 3.6 percent p.a. in the 1990s. Obviously, the explosive growth in the demand for food of the past two decades in countries such as China and India (which contributed 47 percent of the increase in total food demand of the developing countries, see Table 3.16) cannot be sustained at the same rates in the decades to come, even though further substantial growth is in prospect. What was said earlier in relation to the slowdown of aggregate demand for all agricultural products (Section 3.1) applies *a fortiori* to the food demand for vegetable oils. That is, much higher growth rates than projected here of these calorie-rich products would drive the average food consumption of many countries to excessive levels.

Non-food industrial uses. We noted earlier the inadequacy of the statistics on vegetable oil products used for non-food industrial purposes. However, we also noted that some of the industrial products resulting from such use have high income elasticities of demand. There is, therefore, a prima facie case to believe that the share of total

vegetable oil production going to non-food industrial uses will continue to grow fairly rapidly. In the projections, we make an allowance for this category of demand to grow at rates above those of the demand for food (3.5 percent p.a. versus 2.1 percent p.a.). Even so, this growth rate is lower than the historical one. This is an additional factor contributing to the slowdown of the growth of aggregate demand for all uses (Table 3.15).

Trade. The projected fairly buoyant growth in demand, and the still considerable potential for expansion of production in some of the major exporters, suggest that past trade patterns will continue for some time, i.e. rapidly growing imports in most developing countries, matched by continued export growth of the main exporters. The projections are shown in Table 3.20. For the developing countries, they suggest a continuation of past patterns. Further rapid growth of exports from the major developing exporters will be more than compensated by the equally rapid growth of imports by other developing countries. The positive net trade balance of the developing countries, as measured here,[32] will stagnate or even decline somewhat. With the development of the livestock sector in developing countries and growing use of feed concentrates, their demand for oilmeals will increase and their net trade surplus in this subsector of the oilcrops complex will tend to disappear and eventually become negative. It is considered that China will become a major contributor to these developments. According to a recent USDA report, "Expansion of its feed manufacturing sector will require China to import more oilseed meals and more oilseeds for crushing. China's meal production from domestically grown soybeans is currently about 6 million tonnes, far short of the country's estimated demand for 20 to 30 million tonnes of oilseed meals annually over the next decade" (USDA, 1999b).

Production. Production issues are discussed in Chapter 4. Production analysis of oilcrops is conducted separately in terms of the individual crops listed in Table 3.18. Cotton is included among these crops because it contributes some 4 percent of

---

[32] The trade numbers in Table 3.20 comprise the oils traded as well as the oil equivalent of the trade in oilseeds and products made from vegetable oils. They do not include trade in oilmeals in order to avoid double counting when all numbers are expressed in oil equivalent.

world oil production, although projected production is determined in the context of world demand-supply balance of cotton fibre rather than oil. The production projections for these major oilcrops are shown in Table 3.18. It was noted in Table 3.19 that oilcrop production has been responsible for a good part of the area expansion under crops in, mainly, the developing countries.

Will these crops continue to play this role? The analysis presented in Chapter 4 suggests that they will indeed continue to be a major force in the expansion of harvested area in developing countries. As shown in Table 3.19, in the preceding two decades oilcrops accounted for 47 percent of the total increase in the harvested area under the major crops of the developing countries. In the projection period, they will account for 50 percent. The relatively land-intensive nature of oilcrop production growth reflects in large part the fact that they are predominantly rainfed crops (less than 10 percent irrigated, compared with about 40 percent for cereals).

Table 3.20   Net trade balances for oilseeds, oils and products

	1974/76	1997/99	2015	2030
		Million tonnes (oil equivalent)[1]		
Developing countries	3.0	4.0	3.4	3.5
Malaysia	1.3	9.0		
Argentina	0.2	4.9		
Indonesia	0.4	3.9		
Brazil	1.0	2.6		
Philippines	1.1	0.9		
Subtotal, five major exporters	*4.0*	*21.3*	*31.6*	*44.0*
India	0.0	-2.9		
China	0.0	-2.8		
Mexico	-0.1	-1.5		
Pakistan	-0.2	-1.3		
Egypt	-0.2	-0.9		
Bangladesh	-0.1	-0.8		
Korea, Rep.	0.0	-0.7		
Iran, Islamic Rep.	-0.2	-0.7		
Taiwan Province of China	-0.2	-0.7		
Hong Kong SAR	-0.1	-0.7		
Subtotal, ten major importers	*-1.1*	*-13.1*	*-21.3*	*-30.0*
Other developing countries	*0.0*	*-4.1*	*-6.9*	*-10.5*
Industrial countries	-2.9	-0.9	-0.5	-1.3
United States	2.8	4.9		
Canada	0.2	2.3		
Japan	-1.3	-2.6		
EU-15	-4.3	-5.0		
Other industrial countries	-0.3	-0.6		
Transition countries	0.0	-0.2	-0.2	0.0
World balance (stat. discrepancy)	0.2	2.9	2.7	2.3

[1]   Trade in oils and products derived from oils plus the oil equivalent of trade in oilseeds; trade in oilmeals not included in order to avoid double counting in the equation: production + net trade = consumption expressed in oil equivalent.

3.5 Roots, tubers and plantains

3.5.1 Past and present

Food consumption of roots, tubers and plantains.
This category of basic foods comprises a variety of products; the main ones are cassava, sweet potatoes, potatoes, yams, taro and plantains.[33] Average world food consumption is 69 kg per capita, providing 6 percent of total food calories. However, these products represent the mainstay of diets in several countries, many of which are characterized by low overall food consumption levels and food insecurity. The great majority of these countries are in sub-Saharan Africa. The region's per capita consumption is 194 kg, providing 23 percent of total calories, but some countries depend overwhelmingly on these products for food. Thus, Ghana, Rwanda and the Democratic Republic of the Congo derive 50 percent or more from these foods. Another 16 countries, all in sub-Saharan Africa, derive between 20 and 50 percent.

Figure 3.13 shows the relevant data for the 19 countries that derive more than 20 percent of total food consumption from these products. Collectively, they have a population of 345 million (60 percent of sub-Saharan Africa's total population). Most have low overall per capita food consumption (under 2 200 calories, several of them under 2 000) and, consequently, high incidence of undernourishment. Cereals, which in the developing countries as a whole provide 56 percent of total food calories, account in these countries for much smaller proportions, typically 20-45 percent, rising to just over 50 percent only in the United Republic of Tanzania (mostly maize), Togo (maize, sorghum and rice) and Madagascar (rice).

The high dependence on roots, tubers and plantains reflects the agro-ecological conditions of these countries, which make such products suitable subsistence crops, and to a large extent also the persistence of poverty and lack of progress towards diet diversification. There are significant differences as to which of these starchy products provide the mainstay of diets in the 19 countries dependent on this family of products. Cassava predominates in most of them (the Congo, the Democratic Republic of the Congo, Angola, Mozambique, the United Republic of Tanzania, the Central African Republic, Liberia and Madagascar). In contrast it is mostly plantains in Rwanda and Uganda, and cassava and sweet potatoes in Benin, Togo, Nigeria and Burundi, while there is more balance among the different products (cassava, plantains, sweet potatoes and other roots and tubers – mostly yams) in the other countries (Ghana, Cameroon, Côte d'Ivoire, Guinea and Gabon).

It should be noted that although high dependence on these foods is typical of many countries with low overall food consumption, the phenomenon is far from universal. Many countries with low food intakes consume only minuscule quantities of starchy roots. For example, there are 13 countries[34] with low calories (under 2 200), which derive less than 10 percent from roots, and some less than 2 percent. The explanation of these wide disparities is straightforward for some of these countries: their agro-ecologies are not suitable for rainfed cultivation of these products. In several countries (e.g. Yemen, Somalia, Afghanistan, Eritrea, the Niger and Mongolia) the potential land suitable for any rainfed crops (at the intermediate level of technology as defined in the agro-ecological zones study discussed in Chapter 4) is extremely scarce (in per capita terms) and very little of it is suitable for any of the starchy crops considered here. Irrigated agriculture in these countries is devoted to other more valuable crops.

However, there are also several low food-intake countries that consume few starchy products, even though high proportions of their potential agricultural land are suitable for their rainfed cultivation. The explanation is probably to be found, at least in part, in the fact that even higher proportions are suitable for other more preferred crops, e.g. rice (Bangladesh, Cambodia and the Lao People's Democratic Republic) or maize (Zimbabwe and Kenya). Beyond that, the factors that contributed to the formation of traditional agricultural production patterns and present prevailing food consumption habits, probably explain why fairly similar agro-ecological conditions can result in widely differing diet preferences for starchy foods.

---

[33] Plantains are included with the roots and tuber crops because "... Plantains and cooking bananas are grown and utilized as a starchy staple mainly in Africa ..." (FAO, 1990).

[34] Yemen, Afghanistan, Cambodia, Bangladesh, Somalia, Zimbabwe, Mongolia, the Niger, the Lao People's Democratic Republic, Eritrea, the Democratic People's Republic of Korea, Chad and Kenya.

This is most evident in Thailand which produces some 17 million tonnes of cassava but consumes only 4 percent of it as food (11 kg per capita), with the rest going to export as animal feed (see below).

In conclusion, a positive correlation between dependence of food consumption on starchy foods and agro-ecology certainly exists in countries with low incomes and food consumption, but it is rather weak. It is more significant if we control for the effects of the level of total food consumption (kcal/person/day from all foods.) The higher the level, the lower the percentage coming from starchy foods, ceteris paribus – this is a proxy for the negative income elasticity of demand for these products. Another factor causing a difference in "food habits" is the share of calories coming from rice and maize.

The preceding discussion reflects the current situation (average 1997/99) across different countries. The evolution over time shows declining per capita food consumption of starchy foods for the developing countries and the world as a whole up to about the late 1980s, followed by some recovery in the 1990s. These developments were a result of two main factors:

■ the rapid decline in food consumption of sweet potatoes in China (from 94 kg in the mid-1970s to 40 kg at present), the parallel rise in that of potatoes, in both China (from 14 kg to 34 kg) and the rest of the developing countries (from 9 kg to 15 kg); and

■ the rapid rise of food consumption of all products in a few countries, e.g. Nigeria, Ghana and Peru, with Nigeria having a major weight in shaping the aggregate for sub-Saharan Africa.

These developments are plotted in Figure 3.14. Some studies highlight the high income elasticity of demand for potatoes in the developing countries, the majority of which have very low levels of per capita consumption.[35] This contrasts with the position of the other starchy foods (particularly sweet potatoes but also cassava), whose per capita food consumption in the developing countries has

Figure 3.13 Countries with over 20 percent of calories from roots, tubers and plantains in 1997/99

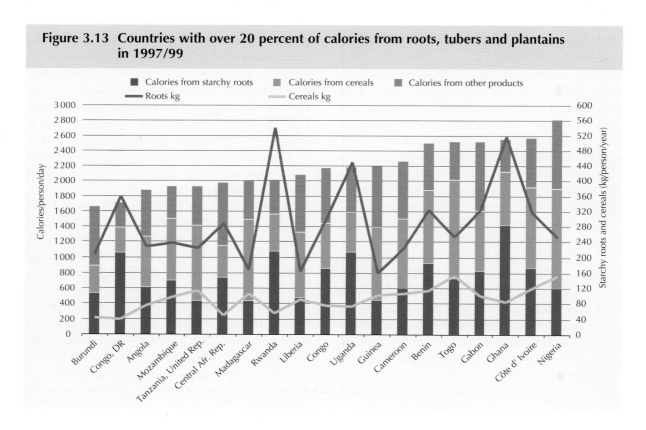

---

[35] "Whereas potatoes are typically considered a cheap, starchy staple in industrialized countries, they tend to be high-priced and sometimes are luxury vegetables in the developing world … Consumption of potatoes increases as income increases. The relationships for cassava and sweet potato are different. As per capita incomes increase, per capita consumption declines" (Scott, Rosegrant and Ringler, 2000).

apparently stagnated or declined.[36] However, caution is required in drawing firm inferences from these numbers because of the particularly poor quality of data as regards the production and consumption of several of these crops.

Recent efforts to improve the cassava data in Africa in the context of the collaborative study of cassava in Africa (COSCA), initiated in 1989, suggest that cassava is far from being the inferior good put forward in traditional thinking. "The COSCA study found that the income elasticities of demand for cassava products were positive at all income levels" (Nweke, Spencer and Lyman, 2000). Indeed, cassava played an important role in the nutrition gains made by a number of countries that faced severe food insecurity problems. For example, gains in per capita food consumption in Ghana (from 1925 calories in 1984/86 to 2545 calories in 1997/99) and in Nigeria (from 2060 to 2815) came largely from increases in the production and consumption of cassava – 80 and 50 percent of the total calorie increases, respectively. Indeed, these two countries are presented in *The State of Food Insecurity in the World 2000* (FAO, 2000a) as success cases in improving food security based on the diffusion of improved high-yielding cassava cultivars, largely developed by the International Institute of Tropical Agriculture (IITA) (see also Nweke, Spencer and Lyman, 2000).

However, such gains were the exception rather than the rule in the many countries with food insecurity problems and dependence on starchy foods. Of the 19 countries in Figure 3.13 only three (Ghana, Nigeria and Benin[37]) registered significant increases in the per capita food consumption of these products. The others had no gains, indeed most of them suffered outright declines according to the reported production statistics. In conclusion, the experiences of the "success" countries indicate that these crops have a promising potential to contribute to improved food security.

Analysing why most countries with high dependence on these crops (over 20 percent of calories in 1997/99) failed to benefit from such potential can throw some light on the more general issue of conditions that must be met if progress in food security is to be made. The fact that some of these countries have been experiencing unsettled political conditions and war is certainly part of the problem.

Feed uses of root crops. Significant quantities of roots are used as feed, mostly potatoes (14 percent of world production goes to feed), sweet potatoes (36 percent) and cassava (19 percent). A small number of countries or country groups account for the bulk of this use. As regards potatoes, this is mostly represented by the countries of the former Soviet Union, Eastern Europe and China. Potato feed use has declined in recent years in absolute tonnage as well as percentage terms, mainly as a result of the decline of the livestock sector in the transition economies. For sweet potatoes, China accounts for almost the totality of world feed use and for about 70 percent of world production. Feed use in China has been expanding rapidly following the fast growth of its livestock sector and the shift of human consumption to potatoes and other preferred foods.

For cassava, it is mostly Brazil (50 percent of its production goes to feed) and the EU (all imported), each accounting for one-third of world feed use. The EU's feed consumption currently amounts to some 10 million tonnes (fresh cassava equivalent). This is less than half the peak it reached in the early-1990s. The rapid growth of cassava as feed in the EU provides an interesting story of the power of policies to change radically feeding patterns and exert significant impacts on trade as well as on production and land use in distant countries. High domestic cereal prices of the CAP, in combination with low import barriers for oilseeds/meals and cassava, reduced the EU's feed use of cereals (see previous discussion) and increased the use of imported cereal substitutes (such as cassava, oilmeals and corn gluten feed).

The cassava trade (mostly of dried cassava) skyrocketed from less than 5 million tonnes (in fresh

---

[36] "Outside of Kerala (India) and isolated mountain areas of Viet Nam and China, most cassava in Asia for direct food purposes is first processed. As incomes increase over time, also these areas will reduce their non-processed cassava intake in favour of the preferred rice. On-farm cassava flour consumption seems to behave in a similar way to non-processed cassava in Asia, as it is also substituted for rice as economic conditions improve. Nonetheless, on-farm, in the poorer Asian rural areas (Indonesia, Viet Nam and China) cassava may remain as an emergency or buffer crop in times of rice scarcity. However, this is not the primary nor the preferred use" (Henry, Westby and Collinson, 1998). Also, "the general tendency is that cereals are preferred to root crops" (FAO, 1990, p. 24) and "In general, cassava is not well regarded as a food, and in fact there is often a considerable stigma against it" (Plucknett, Phillips and Kagbo, 1998).

[37] If we extend the sample of countries to include those with dependence between 10-20 percent (another 11 countries) then Malawi, Zambia and Sierra Leone also registered significant increases in production and per capita food consumption of these products.

Figure 3.14 Roots, tubers and plantains, food consumption (kg/person/year), 1970-99

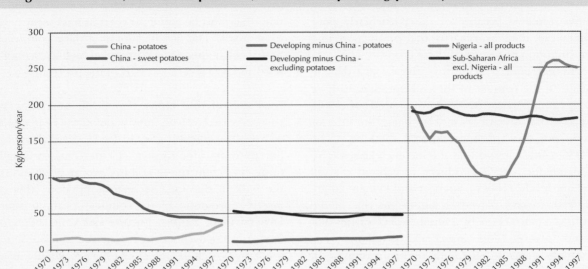

root equivalent) in the early 1970s to a peak of 30 million tonnes at the beginning of the 1990s, before falling to half that level in recent years. Much of the increase came from the EU on the import side and from Thailand (and, to a much lesser extent, Indonesia) on the export side. The same holds for the contraction of world trade in recent years following the CAP policy reforms, which reduced cereal prices and improved the competitiveness of cereals vis-à-vis cassava (Figure 3.15). Some compensation for the contraction of the European markets has been provided in recent years by increased imports of cassava for feed use in a number of Asian countries (China, Taiwan Province of China, the Republic of Korea and Japan). However, the problem of low cereal prices in world markets continues to be an obstacle to further expansion (Plucknett, Phillips and Kagbo, 2000).

In Thailand, cassava production has been one of the major factors in the expansion of agricultural land. Between the early 1970s and the peak of the late 1980s land under cassava increased sevenfold to 1.5 million ha, approximately the same rate as production, as yields remained virtually stagnant at around 15 tonnes/ha. In the same period, deforestation in Thailand (a heavily forested country up to the middle of the last century) proceeded very rapidly. The country's forest cover is estimated to

have fallen from 53 percent in 1961 to 25 percent in 1998 (Asian Development Bank, 2000, Chapter 2). The two processes are certainly not unrelated, particularly as cassava can be grown on soils with limited alternative agricultural uses, low fertility, slopes denuded of tree cover and subject to erosion and degradation.[38] A recent study on the environmental impact of cassava production concludes that in Thailand "the area expansion [took place] through deforestation of frontier areas, mainly in the northeast ... The massive deforestation continued until the late 1980s, when most land within Thailand's borders, up to the borders with Laos and Cambodia, had already been cleared" (Howeler, Oates and Costa Allem, 2000).

The preceding paragraph provides an illustration of how the environmental impacts of agricultural policies and policy distortions spread far and wide across the world and are not confined locally. The environmental effects of high agricultural support and protection in Europe have been mostly debated in terms of the adverse effects these policies have on the local environment (as a consequence of the more intensive agriculture they promote), such as nitrates in waterbodies, eutrophication, landscape deterioration and the effects of pesticides. However, environmental impacts of policies, particularly those that affect trade, must

---

[38] "Cassava often winds up in hill-lands, lands with low soil fertility, or lands susceptible to periodic or seasonal drought or flooding" (Plucknett, Phillips and Kagbo, 2000).

Figure 3.15 Cassava in Thailand and the EU

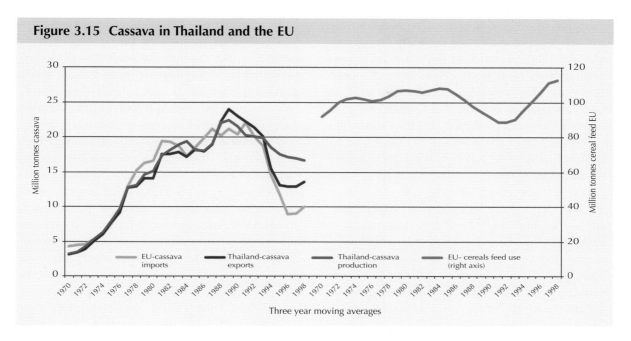

Three year moving averages

be evaluated in a global context, with particular reference as to how those impacts manifest themselves in countries that receive the economic stimuli but have no environmental policies in place to internalize costs.

3.5.2 Roots, tubers and plantains in the future

These products will continue to play an important role in sustaining food consumption levels in the many countries that have a high dependence on them and low food consumption levels overall. The possible evolution in food consumption per capita is shown in Tables 2.7-2.8. The main factor causing the decline in the average of the developing countries (precipitous decline of sweet potato food consumption in China) will be much weaker. The scope for further declines is much more limited than in the past. In parallel, the two factors creating increases on the average – the positive income elasticities of the demand for potatoes and the potential offered by productivity increases in the other roots (cassava and yams) – will continue to operate. It will be possible for more countries in sub-Saharan Africa to replicate the experiences of countries such as Nigeria, Ghana, Benin and Malawi, and increase their food consumption. Thus, the recent upturn in

per capita consumption of the developing countries is projected to continue (Table 2.7), while the declining trend in sub-Saharan Africa (other than Nigeria and Ghana) may be reversed (Table 2.8).

3.6 Main export commodities of the developing countries

3.6.1 General

Agriculture, and often the overall economy, poverty incidence and food security of several developing countries depend heavily on the production of one or a few agricultural commodities destined principally for export. For example, in Côte d'Ivoire a small number of commodities (cocoa, coffee, cotton, rubber, bananas and palm oil) account for 45 percent of the country's aggregate agricultural output, cocoa predominating with 27 percent. In the past two decades, the dependence of agriculture on these commodities actually increased (from 38 percent in the early 1980s). Agriculture carries a large weight in the country's overall economy (26 percent of GDP), while exports of these commodities provide 50 percent of the country's aggregate export earnings (aggregate, not only agricultural, earnings).[39] There has

---

[39] Aggregate export earnings are the sum of goods (merchandise) exports, exports of (non-factor) services and income (factor) receipts (definition and data from World Bank, 2001b).

been little movement towards diversification; the levels of dependence of the economy on agriculture, and of the balance of payments on export earnings from these commodities have remained static since the early 1980s. The importance of these commodities as earners of foreign exchange is further enhanced by the fact that the country has a heavy external debt to service; it is classified by the World Bank in the "severely indebted" category (World Bank, 2001b).

Côte d'Ivoire exemplifies the case of countries with high dependence on exports of a few agricultural commodities (both for the incomes generated in agriculture and for foreign exchange earnings) and showing no signs of diversification from such dependence. The country is the world's largest producer and exporter of cocoa, so its own production and exports impact world market prices directly. An additional problem is that some of these export commodities (e.g. coffee and cocoa) are consumed mainly in the industrial countries. The latter have reached fairly high consumption levels, and have low income and price elasticities of demand and low population growth rates. There is, therefore, limited potential for market expansion. In such conditions, attempts to expand production and exports by the main producers/ exporters and/or new entrants can lead to falling prices rather than increased incomes and export earnings.

Another example is Malaysia, which shares Côte d'Ivoire's high dependence of agriculture on a few export commodities, of which the country is a major world supplier. Malaysia is the world's largest producer and exporter of palm oil and the third largest exporter of natural rubber (it was the first until the late 1980s). These two commodities account for 36 percent of the country's agriculture. But here the similarities with Côte d'Ivoire end. Malaysia's per capita income is five times that of Côte d'Ivoire, and the country has been diversifying its economy away from agriculture (from 21 percent of GDP in the early 1980s to 11 percent currently) and reducing its dependence on palm oil and rubber as sources of export earnings. These two commodities now account for 5.6 percent of its aggregate export earnings, down from 20 percent in the early 1980s, even though palm oil exports have tripled in the same period.

In a smaller number of countries there is high dependence of agriculture on a single export commodity. For example, 56 percent of the aggregate agricultural output in Mauritius is sugar, almost all of it exported. Sugar is 46 percent of Swaziland's agriculture (85 percent exported), while in Cuba the shares are 33 and 74 percent, respectively.

The most common case is of countries with high dependence of the balance of payments on the earnings of one or a few agricultural commodities which, however, account for modest percentages of their total agriculture, since domestic crops predominate. For example, Burundi's export earnings are dominated by coffee and, to a much lesser extent, tea (over 80 percent together) but these two crops represent a very small part of the country's agriculture (9 percent) in which bananas, roots/tubers, beans and maize predominate. Likewise, Malawi derives 60 percent of its aggregate export earnings from tobacco and, to a much lesser extent, tea. However, its agriculture is mostly potatoes and maize; tobacco and tea account for only 18 percent of total agricultural production.

It is obvious that in the countries with high dependence on exports of one or a few commodities, the overall economy, and often the welfare of the poor,[40] are subject to changing conditions in world markets, i.e. commodity prices and the rate of expansion of markets. For some commodities, the rate of expansion of world consumption has been too slow, while producers/exporters have been competing with one another to capture a market share. The result has been declining and widely fluctuating export prices. This has been particularly marked for coffee in recent years: per capita consumption in the industrial countries, accounting for two-thirds of world consumption, has been nearly constant for two decades at around 4.5 kg (green beans equivalent), while production increased, including from recent entrants in world markets such as Viet Nam (see below). The result has been that the robusta coffee price had precipitated to US$0.50/kg by January 2002 (monthly average), down from a range of US$1.3-2.0/kg in the monthly averages of the three years 1997-99, which itself was not a period of peak prices.

Coffee and cocoa belong to the class of commodities produced exclusively in the devel-

---

[40] Several export commodities, e.g. coffee and natural rubber, are more often than not smallholder crops and significant parts of the agricultural population in many countries depend on them for a living. Even when they are not smallholder crops, the rural poor depend on them as labourers.

oping countries and consumed overwhelmingly in the industrial countries, where there is no close substitute in the agricultural commodities produced locally (although vegetable oils are increasingly making inroads in the chocolate industry as substitutes for cocoa butter). This means that trade policy issues for the countries dependent on the production and exports of these commodities transcend the field of agricultural protectionism and export subsidization often practised by the industrial countries, e.g. for sugar. The main handicap of exporting countries dependent on these commodities is that world demand has been, and will probably continue to be for some time, tantamount to the demand of the industrial countries, and this is not growing at anywhere near the rates required to provide profitable export expansion opportunities for all supplying countries.

Other export commodities are much less subject to these constraints because import demand for them originates increasingly in markets with rapidly growing consumption, notably the importing developing countries. We mentioned earlier the role of these latter as rapidly growing outlets for vegetable oils originating in other developing countries. A similar case is represented by sugar, where imports of the importing developing countries are now equal to almost three-quarters of the net exports of the developing exporters, up from only one tenth in the mid-1970s. The policy issues relevant to sugar and similar commodities (produced in both developed and developing countries) are very much at the heart of the agricultural trade policy debate, because of high support and protection granted to them in the main industrial countries that are, or used to be in the recent past, large importers. For example, under such policies the EU turned from a large net importer of sugar up to the second half of the 1970s to a large net exporter in subsequent years. In the United States aggregate consumption of sweeteners continued to grow, but little of it went to sugar, as the strong protection of the sector from external competition favoured corn-based sweeteners (high fructose corn syrup [HFCS], glucose syrup and dextrose).

It is a common theme in the development literature that countries that have not managed to diversify their economies away from heavy dependence on primary commodity exports have been handicapped in their development. However, this need not be so for all countries. Those that achieve cost-reducing productivity gains can increase export earnings (often, in the case of slowly growing global demand, by gaining a market share at the expense of other exporting countries), and thus pave the way for diversification and an eventual sharp reduction in their dependence on these commodities. Thus, diversification and reduced dependence more often than not pass through a stage of increased and more efficient production of the very commodities from which the country seeks to diversify (World Bank, 1994). The above-mentioned examples of Malaysia and the Côte d'Ivoire are telling. Naturally, for commodities such as coffee, this avenue is not open to all exporting countries simultaneously (adding-up problem or fallacy of composition). The poorer countries in this category face formidable problems in their development because more often than not they are at a disadvantage vis-à-vis other and more advanced competitors.

There is an ongoing need to increase productivity and eventually diversify out of export crops (in relative, not absolute, terms). The long-term trend towards falling real prices of primary farm commodities is not likely to be reversed.[41] Limited demand growth together with virtually incessant growth in agricultural productivity has been fuelling this trend. The pipeline of technological innovations, including genetic technologies, is apparently not drying up, indicating further potential for cost reductions (Evenson, 2002). The problem is that countries that do not or cannot benefit from genetic technology will have to accept ever-decreasing returns to the labour producing these commodities. This will further aggravate poverty problems, as there are few alternative employment opportunities in an undiversified economy. If these countries remain in the business of producing "unprofitable" (for them) commodities for export (e.g. tapping rubber trees even more

---

[41] "On balance, we do not see compelling reasons why real commodity prices should rise during the early part of the twenty-first century, while we see reasons why they should continue to decline. Thus, commodity prices are expected to decline relative to manufactures as has been the case for the past century" (World Bank, 2000a, p. 8-29).

intensively when prices are low), they will face continued difficulties. For more discussion on policies relating to these matters, see Chapter 9.

In what follows, we present possible future outcomes for a few of the main commodities. This is not a comprehensive list but endeavours to cover commodities that span a wide range of policy concerns, from strictly "non-competing"[42] commodities (coffee, cocoa) to the "semi-competing" ones (bananas, competing with temperate zone fruit, and natural rubber, competing with synthetic rubber) to fully competing ones (sugar). In recognition of the great diversity of individual developing countries as concerns trade positions and interests in these commodities, a distinction is made between net exporters and net importers. Thus, the sum of the net exports of all net exporter developing countries is taken here as the meaningful measure of performance. It is a much more appropriate indicator than the often used net trade status of all the developing countries together. For example, sugar exports of the net exporting developing countries have been booming in recent years, at least in quantities, mainly because of fast growth of imports of other developing countries (see Table 3.23 below).

3.6.2 Coffee

Coffee is the commodity *par excellence* produced in one zone (the exporting developing countries in the tropics) and consumed in another (the developed industrialized countries). The growth rate of aggregate demand for coffee has been slow and declining. In this situation, competition to capture a market share among low-income producers (including relatively new entrants such as Viet Nam[43]) increases production more quickly than consumption, leading to stock accumulation and accentuating the long-term trend towards price declines.

Such price declines are hitting hard the economies of the countries with major dependence on coffee production and exports. The following quotation from *The Economist* (21 September 2001) is telling: "A slump in the price of coffee is adding to economic misery in Latin America – and no

relief is at hand. In Nicaragua, coffee pickers with malnourished children beg for food at the roadside. In Peru, some families have abandoned their land, while others have switched to growing drug crops in search of cash, just as they have in Colombia. From Mexico to Brazil, tens of thousands of rural labourers have been laid off, swelling the peripheries of the cities in a desperate search for work". See also Oxfam (2001).

At the same time, consumers are not getting much benefit from such price declines since retail prices in the main consuming countries have fallen by only a small fraction of the corresponding declines in the world price of coffee beans (Oxfam, 2001, Figure 5). At the same time, coffee producers are getting prices below the world price and only a minuscule part of the retail price of coffee. These large gaps and difference in the behaviour between world trade prices and those received by producers or paid by the consumers are said to reflect, *inter alia*, the dominance of the world coffee trade by a few giant multinational companies (see Chapter 10, Box 10.2). The relative position of producers has worsened following the collapse of the International Coffee Agreement in 1989. In the pre-1989 period producers were getting around 20 percent of the total income generated by the coffee industry and importing country operators around 50 percent. It is estimated that after 1989 the shares shifted dramatically in favour of the latter (Ponte, 2001, Table 3).

Coffee has benefited much less than other commodities (such as sugar, bananas and natural rubber) from the factors that contributed to market expansion, i.e. the rapid growth of demand and imports in importing developing countries and, in more recent years, in the transition economies. The importing developing countries were taking 4 percent of the net coffee exports of the exporting developing countries in the mid-1970s. This share has now risen to only 7 percent. In contrast, for sugar the net imports of the importing developing countries amount to three-quarters of the net exports of the exporting developing countries, up from only 10 percent in the mid-1970s.

In the projections shown in Table 3.21 there are

---

[42] That is, non-competing with locally produced close substitutes in the main importing/consuming countries, although there is no commodity that does not face some degree of competition in consumption, such as soft drinks for coffee and non-cocoa confectionery for chocolate. As noted, cocoa now faces the competition of vegetable fats in the production of chocolate.

[43] Viet Nam had become the world's second largest producer after Brazil by 2000. Its exports now account for 10 percent of aggregate exports of the developing countries (sum of net exports of all the net exporting developing countries, see Table 3.21), up from only 1 percent ten years earlier.

Table 3.21 Coffee and products, production, consumption and trade: past and projected

	'000 tonnes 1997/99	Growth rates, percentage p.a.				
		1969-99	1979-99	1989-99	1997/99 -2015	2015 -2030
		Production*				
Developing countries	6 452	1.6	1.1	1.0	1.2	1.2
Industrial countries	3	3.1	9.1	16.1	2.3	3.2
World	6 455	1.6	1.1	1.0	1.2	1.2

	Food consumption (kg/person/year)*					
	1964/66	1974/76	1984/86	1997/99	2015	2030
Developing countries	0.5	0.4	0.4	0.4	0.4	0.5
Industrial countries	3.8	4.1	4.4	4.4	4.7	5.0
Transition countries	0.3	0.5	0.6	1.2	1.6	2.2
World	1.2	1.1	1.1	1.0	1.1	1.1

	Net trade ('000 tonnes)*					
Industrial countries	-2 620	-3 106	-3 550	-3 969	-4 475	-4 875
Transition countries	-90	-193	-220	-502	-640	-830
Developing countries	2 681	3 247	3 902	4 555	5 200	5 800
Exporters in 1997/99	2 790	3 395	4 088	4 947	5 825	6 660
Brazil	910	867	1 003	1 205		
Colombia	352	426	664	649		
Viet Nam	2	5	12	420		
Indonesia	88	125	294	351		
Guatemala	95	127	152	243		
Mexico	95	145	206	238		
Côte d'Ivoire	192	287	234	228		
Uganda	155	186	142	202		
India	27	53	82	191		
Costa Rica	51	79	110	126		
El Salvador	103	149	148	121		
Honduras	22	41	73	119		
Peru	37	39	63	119		
Ethiopia	75	60	77	114		
Other exporters	585	805	828	620		
Importers in 1997/99	-109	-148	-186	-392	-625	-860
Algeria	-26	-34	-62	-81		
Morocco	-9	-12	-14	-23		
Other Near East/North Africa	-23	-44	-62	-101		
Korea, Rep.	0	-2	-19	-76		
Argentina	-33	-38	-34	-39		
Hong Kong SAR, Taiwan Province of China	-3	-1	-5	-23		
Other importers	-14	-19	11	-50		

* Coffee and products in green bean equivalent

no radical changes from past trends. World coffee consumption per capita may increase a little and aggregate demand may grow at 1.2 percent p.a., a rate slightly above that of the last two decades. The weight of the industrial countries in world consumption and imports will continue to be dominant, although the importing developing countries and the transition economies will continue to increase their role as consumers and importers, but at a very slow pace. The present very low price levels are not sustainable, and some recovery is expected as producing countries take measures to control the growth of production. The latest World Bank price projections foresee such an upturn in prices. By 2010 they may recover (in real terms) the ground lost in 2000 and 2001, years of sharp declines (World Bank, 2001c, Table A2.13).

3.6.3 Cocoa

Cocoa shares some of the characteristics of coffee in the sense that it is produced exclusively in tropical developing countries and consumed mainly (two-thirds of world consumption) in the industrial countries. However, consumption growth, at 2.4 percent p.a. in the last ten years, has been faster than that of coffee (0.7 percent p.a.). In parallel, the importing developing countries are increasingly providing export outlets: they now account for 12 percent of the exporting countries' net exports, up from only 3 percent in the mid-1970s.

On the exporter side, there have been some radical changes in the relative positions of the different countries. Côte d'Ivoire continues to be the world's largest exporter and has increased its share considerably, to almost 50 percent of the aggregate net exports of the exporting developing countries, up from 33 percent in the mid-1980s and only 17 percent in the mid-1970s (Table 3.22). In contrast, Brazil, the world's second largest exporter up to the late 1980s, had almost disappeared as a net exporter by the late 1990s, mainly as a result of disease that hit production but also because a growing part of production went to increase domestic consumption. Meanwhile, first Malaysia (from the early 1980s) and then Indonesia (ten years later) emerged as major and growing exporters. Between them (but with Indonesia growing and Malaysia declining recently), they have been

providing 15-20 percent of the aggregate exports of the net exporting countries in recent years, up from less than 4 percent in the early 1980s.

Consumption per capita is likely to continue to grow in all country groups, although at slower rates than in the past (Table 3.22). The growth of world production, which decelerated sharply in the second half of the 1990s (to under 1 percent p.a. in the five years to 2001), will resume higher rates, but still below those of the longer-term historical period, given the slower growth of per capita consumption and of population. Cocoa prices have been characterized by short booms and long periods of oversupply with depressed prices. The latest trough in prices was in the second half of 2000, with prices around US$900/tonne, down from the previous peak of around US$1 700 in mid-1998. Prices recovered to US$1 380/tonne in January 2002. World Bank projections foresee little further recovery up to 2015 in real terms.

3.6.4 Sugar

Consumption has been growing fast in the developing countries, which now account for 72 percent of world consumption (up from 52 percent in the mid-1970s), including the sugar equivalent of some 60-65 percent of Brazil's sugar-cane production used in ethanol production (USDA, 1998a). In contrast, consumption has grown very little in the industrial countries, and has declined in the transition economies in the 1990s. An important factor in the stagnation of sugar consumption in the industrial countries has been the rapid expansion of corn-based sweeteners in the United States, where they now exceed consumption of sugar (11.8 million short tonnes dry weight, versus 9.4 million short tonnes of refined sugar). Production of the major sweetener, HFCS, shot up from negligible quantities in the mid-1970s to 5 million short tonnes in the mid-1980s and to 9.7 million short tonnes currently.[44]

Sugar is produced under heavy protection in the industrial countries, with the exception of the traditional exporters among them (Australia and South Africa) (OECD, 2002). Under this shelter, production grew at 1.5 percent p.a. in the last three decades, at a time when total consumption in indus-

[44] Data from www.ers.usda.gov/briefing/sugar/Data/data.htm/

Table 3.22 Cocoa and products, production, consumption and trade: past and projected

	'000 tonnes	Growth rates, percentage p.a.				
	1997/99	1969-99	1979-99	1989-99	1997/99 -2015	2015 -2030
		Production*				
Developing countries	3 000	3.1	3.8	2.1	1.8	1.4
World	3 000	3.1	3.8	2.1	1.8	1.4

	Food consumption (kg/person/year)*					
	1964/66	1974/76	1984/86	1997/99	2015	2030
Developing countries	0.08	0.06	0.08	0.14	0.17	0.21
Industrial countries	1.37	1.42	1.65	2.08	2.38	2.62
Transition countries	0.40	0.70	0.73	0.87	1.15	1.48
World	0.38	0.37	0.40	0.49	0.52	0.55

	Net trade ('000 tonnes)*					
Industrial countries	-1 011	-1 070	-1 352	-1 802	-2 310	-2 625
Transition countries	-133	-266	-288	-375	-475	-580
Developing countries	1 114	1 310	1 614	2 126	2 725	3 160
Exporters in 1997/99	1 159	1 354	1 688	2 413	3 145	3 715
Côte d'Ivoire	133	233	557	1 170		
Ghana	458	369	194	329		
Indonesia	0	0	28	322		
Nigeria	232	228	157	163		
Cameroon	86	99	108	122		
Malaysia	-1	9	102	95		
Ecuador	34	64	86	63		
Dominican Republic	26	26	35	43		
Brazil	107	229	325	31		
Other	83	98	97	75		
Importers in 1997/99	-45	-43	-75	-287	-420	-560
Mexico	6	7	4	-40		
China (incl. Taiwan Prov. of China and Hong Kong SAR)	-10	-10	-17	-39		
Argentina	-11	-10	-14	-26		
Philippines	-9	-3	0	-25		
Turkey	-1	-2	-4	-20		
Korea, Rep.	0	-1	-7	-14		
Chile	-2	-1	-4	-13		
Other	-20	-23	-33	-110		

* Cocoa and products in bean equivalent

trial countries was not growing. The result has been that these countries turned from net importers of 7.4 million tonnes in the mid-1970s to net exporters of 3.8 million tonnes in 1997/99 (Figure 13.16). This reflected partly the growing exports of Australia, declining imports of the United States and nearly stagnant imports in Japan, but above all it reflected the shifting of the EU from a net importer of 1.9 million tonnes to a net exporter of 4 million tonnes. As a result, the net exports of the developing exporters showed little growth from the late 1970s to the mid-1990s and shot up only in the second half of the 1990s as several developing countries became major importers.

A major characteristic of these developments is that the low prices prevailing in world markets acted as a disincentive to production in countries that failed to improve productivity and, together with the rapid growth of their own consumption, contributed to turning several traditional exporting developing countries into net importers. These include countries such as the Philippines, Peru and Taiwan Province of China. Collectively, they were net exporters of 2.2 million tonnes in the mid-1970s. They now have net imports of 0.7 million tonnes.

Consumption in the developing countries is projected to continue to grow, from 21 kg person/year currently to 25 kg in 2030 (Table 3.23). Growth could be higher if China's policy to limit saccharin consumption succeeds (Baron, 2001, p. 4). Much of the growth would occur in Asia, as Latin America and the Near East/North Africa have already attained fairly high levels of consumption (Table 2.8). Per capita consumption will probably remain constant in the industrial countries, compared with declines in part of the historical period during which corn sweeteners were substituting for sugar in the United States. This process, very pronounced up to the mid-1980s, has by now been largely exhausted. It could be reversed if sugar prices were not to be supported at the high levels set by policy. Some increases are expected in the transition countries, making up for some of the declines suffered during the 1990s.

A significant unknown that may influence the sugar aggregates in the future is the fate of the use of sugar cane and its main by-product, molasses, as feedstocks for the production of fuel ethanol. There has been renewed interest in this option during the recent (year 2000) peak of world petroleum prices, which coincided with low sugar prices in world markets.[45] Several countries, including "Australia, Thailand and India began to consider the feasibility of large-scale ethanol production while Mexico had proceeded as far as to undertake pilot ethanol programmes to combat urban air pollution" (Jolly, 2001). Enthusiasm with this option weakened as price relationships returned to more "normal" levels. Obviously, the prospect that interest will be rekindled in the future will depend on developments in the oil/sugar price ratios. The latest World Bank price projections to 2015 indicate that developments will probably not favour this option. If anything, price prospects point in the opposite direction: one barrel of oil may be worth about 80 kg of sugar in 2015 (World Bank, 2001c, Table A2.12).

The trade implications of these trends in consumption are shown in Table 3.23. By and large, recent trends in the pattern and rates of expansion of the world sugar trade would continue. This means that the fairly rapid expansion of imports into the deficit developing countries will provide scope for the main developing sugar-exporting countries to continue to expand production for export. The main divergence from past patterns is the likely arrest and some reversal of the trend for the industrial countries to be growing net exporters. The traditional exporters in this group (Australia and South Africa) should continue to expand exports, but the EU is unlikely to continue on the past path of rising net exports and may see some reduction in net exports under the WTO rules (Poonyth et al., 2000).

Further pressure leading to increased EU imports (hence lower net exports) will come from the implementation of the Everything but Arms Initiative (EBA) of free access to exports from the least developed countries, with sugar liberalization being phased in during the period 2006-09 (Baron, 2001, p. 6). However, the potential effect of EBA on the EU's sugar imports from this source is unlikely

[45] In the first half of 1999 a barrel of oil (US$14, average monthly price January-June 1999) was worth about 100 kg of sugar at the then prevailing world free market price of US$140/tonne. A year later (first half of 2000) it was worth almost twice as much (195 kg of sugar). The latest six-month average price data (August 2001-January 2002) indicate a return to more normal levels (127 kg). Price data from the World Bank: http://Worldbank.org/prospects/pinksheets/

Table 3.23   Sugar, production, consumption and trade: past and projected

	'000 tonnes 1997/99	Growth rates, percentage p.a.				
		1969-99	1979-99	1989-99	1997/99 -2015	2015 -2030
Production*						
Developing countries	128 814	3.2	2.8	2.6	1.8	1.4
Industrial countries	36 049	1.5	1.1	1.7	0.1	0.3
Transition countries	8 553	-1.2	-2.4	-5.6	0.8	0.7
World	173 415	2.4	2.0	1.8	1.4	1.2

	Food consumption (kg/person/year)*					
	1964/66	1974/76	1984/86	1997/99	2015	2030
Developing countries	14	16	19	21	23	25
Industrial countries	37	39	33	33	33	33
Transition countries	37	45	46	34	35	36
World	21	23	23	24	25	2

	Net trade ('000 tonnes)*					
Industrial countries	-7 425	-7 519	12	3 755	2 850	3 570
Australia	1 228	1 919	2 539	4 304		
EU15	-2 675	-1 857	2 595	4 125		
South Africa	426	780	861	1 142		
Japan	-1 451	-2 535	-1 880	-1 608		
United States	-3 536	-4 196	-2 246	-2 309		
Other industrial	-1 417	-1 630	-1 857	-1 899		
Transition countries	-748	-3 281	-5 281	-5 863	-5 350	-4 270
Developing countries	7 908	11 107	6 993	5 833	6 000	4 000
Exporters in 1997/99	8 803	12 391	16 143	22 696	28 850	34 400
Brazil	672	1 783	3 535	9 304		
Thailand	64	725	1 702	3 348		
Cuba	4 570	5 667	6 706	2 924		
Guatemala	46	215	259	1 179		
Colombia	85	142	239	975		
Mexico	509	196	11	877		
Pakistan	-28	-16	-77	592		
Mauritius	575	573	566	531		
Other	2 310	3 106	3 203	2 969		
Importers in 1997/99	-895	-1 283	-9 150	-16 864	-22 850	-30 350
Korea, Rep.	-41	-215	-591	-1 021		
Iran, Islamic Rep.	-395	-377	-496	-1 119		
Egypt	-103	-114	-790	-1 236		
Indonesia	62	-137	-23	-1 535		
Other	-417	-440	-7 249	-11 953		

* Sugar in raw sugar equivalent

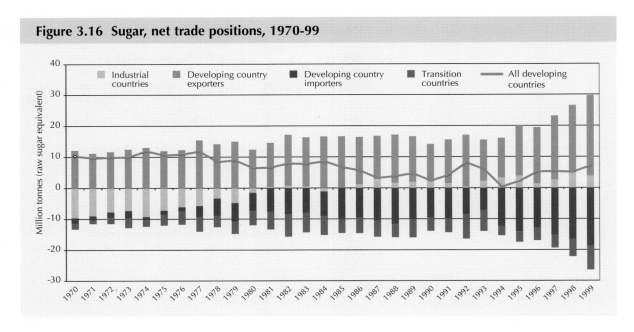

Figure 3.16 Sugar, net trade positions, 1970-99

to be dramatic (Stevens and Kennan, 2001). In addition, the United States may revert to being a growing net importer, mainly as a result of the gradual reduction of tariffs on sugar imports from Mexico under the NAFTA rules leading to tariff-free market access from 2008 onwards (USDA, 2002, p. 45). Finally, the world's largest importer, Russia (1997/99 net imports 4.5 million tonnes, 73 percent of consumption) is likely to continue to hold this role for some time and only lose it in the second half of the projection period, as other countries become larger importers and Russia itself moves towards 50 percent self-sufficiency (Gudosnikov, 2001).

3.6.5 Bananas

Bananas are an important food crop in several tropical developing countries (see earlier discussion of roots and tubers). There are 12 countries with a production of over 1 million tonnes annually, ranging from India (12 million tonnes) to Venezuela (1 million tonnes). However, in only four of them are bananas produced primarily for export, with the part of total production going to exports ranging from over 90 percent in Colombia and Costa Rica, to 70 percent in Ecuador and 33 percent in the Philippines.[46] Among the major

producers, China is a significant net importer.

Traditionally, markets for banana-exporting countries were overwhelmingly in the industrial countries (western Europe, North America and Japan). Only in recent years have a number of developing countries and transition economies become significant importers. These two groups now take 26 percent of exports from exporting countries, up from only 9 percent ten years ago (see Table 3.24). These trends are likely to continue and their share could rise further to 33 percent by 2030, as both the transition economies and the importing (largely non-producing) developing countries will be increasing further their per capita consumption and imports. However, the industrial countries will continue to hold a dominant position as major importers, with their consumption per capita increasing further, although at a slower pace than in the past.

The trade policy reform in prospect in the EU (shift from the preferential tariff-rate quotas granted to the African, Caribbean and Pacific Group of States [ACP] to a tariffs-only regime from 2006 onwards) will remove one of the major sources of friction and may encourage consumption in the EU by increasing supplies at lower prices from the more efficient Latin American producers. Naturally, this shift in policy has the

46 Other countries that produce predominantly for export include Guatemala, Panama, Côte d'Ivoire, Jamaica and several of the smaller Caribbean countries whose banana production, and often their overall economies, depend on exports to the EU under the preferential access regime (tariff-rate quotas) of the ACP, which however is to terminate in 2006 when the EU will shift to a tariffs-only regime (WTO, 2001f).

Table 3.24   Bananas, production, consumption and trade: past and projected

	'000 tonnes 1997/99	Growth rates, percentage p.a.				
		1969-99	1979-99	1989-99	1997/99 -2015	2015 -2030
			Production			
Developing countries	57 933	2.5	3.0	3.0	2.0	1.5
Industrial countries	996	1.3	0.8	0.9	0.7	0.7
World	58 929	2.5	2.9	3.0	2.0	1.5

	Food consumption (kg/person/year)					
	1964/66	1974/76	1984/86	1997/99	2015	2030
Developing countries	7.1	6.9	7.7	8.5	9.6	10.4
Industrial countries	6.1	7.2	7.8	9.3	11.8	13.3
Transition countries	0.2	0.6	0.3	3.2	4.5	5.9
World	6.2	6.4	7.1	8.2	9.6	10.6

	Net trade ('000 tonnes)					
Industrial countries	-3 954	-5 310	-5 964	-4 012	-10 900	-12 600
Transition countries	-52	-219	-133	-1 421	-1 900	-2 350
Developing countries	3 922	5 361	5 951	10 447	13 600	16 000
Exporters in 1997/99	4 113	5 765	6 417	12 240	16 800	20 400
Ecuador	1 287	1 236	1 129	4 250		
Costa Rica	323	1 070	917	2 272		
Colombia	245	388	932	1 522		
Philippines	0	761	815	1 204		
Guatemala	65	282	351	663		
Panama	328	490	648	555		
Honduras	560	572	788	353		
Mexico	14	5	52	220		
Côte d' Ivoire	130	136	98	204		
Others	1 161	825	685	997		
Importers in 1997/99	-192	-403	-465	-1 793	-3 200	-4 400
Korea, Rep.	0	-6	-4	-129		
Saudi Arabia	-12	-35	-88	-144		
Chile	-33	-43	-37	-148		
Iran, Islamic Rep.	-3	-103	0	-190		
Argentina	-177	-123	-102	-263		
China	23	2	-28	-491		
Others	11	-96	-206	-429		

potential of damaging the ACP countries that will lose preferential access to the EU market, particularly those that are not in the LDC category. As noted, the latter will have tariff-free access under the EU's EBA.

3.6.6 Natural rubber

Developments in natural rubber have some similarities with those in the sugar sector. In the first place, a good part of the growth in world consumption originated in developing countries, which now account for 50 percent of the world total, up from only 28 percent in the mid-1970s. This growth reflected, among other things, large increases in consumption in some of the main producing countries such as Malaysia and India including, in the case of Malaysia, domestic consumption used in the production of rubber manufactures for export. Second, the growth of exports of the developing exporting countries has become increasingly dependent on the growth of imports of other developing countries (Table 3.25). And third, natural rubber faces strong competition from synthetic rubber in the industrial countries and transition economies. However, unlike sugar, such competition does not come from substitute agricultural products produced under protection in the industrial countries. Therefore, in contrast to the case of sugar, policy issues relating to the rubber trade and export earnings of the developing countries transcend the agriculture-specific aspects of the trade policy debate, and extend into the more general issues of, for example, price stabilization schemes, commodity development and tariff escalation with degree of processing.

The demand for rubber (both natural and synthetic) is a close correlate of overall economic growth and industrialization, in particular the growth of the automotive sector (65-70 percent of total rubber use is for tyres). The share of natural rubber in total rubber consumption is influenced by prices, technology and the mix of final rubber products. This latter factor can be particularly important; the increase in the share of radial tyres and those for heavy trucks (which require a higher component of natural rubber) in total tyre production will favour natural rubber. This now accounts for about 40 percent of aggregate rubber consumption,[47] up from some 30 percent in the 1980s. This increase in the share reflected some radical changes in the geographic patterns of consumption of both natural rubber (above-mentioned increases in consumption in the major producing countries) and synthetic rubber, following the decline of total rubber consumption (mostly synthetic) in the transition economies. These factors, particularly the second, will be less influential in the future and the rise in the share of natural rubber is not likely to continue.

In the longer term, the competitive position of natural rubber vis-à-vis synthetic will depend on its price relative to that of the main feedstock for the production of synthetic rubber – petroleum. The latest World Bank projections foresee a recovery in the natural rubber price by 2015 from its very low current levels, and a decline in the price of petroleum.[48] Therefore, the share of natural rubber may not continue growing and indeed may fall,[49] leading to a slowdown in total demand compared with the past. This is reflected in the projections of Table 3.25: world demand is projected to grow at around 2.0 percent p.a. compared with 2.6 percent p.a. in the last ten years. The historical trends of faster growth in the consumption of natural rubber in the developing countries compared with the industrial ones are likely to continue, leading to further increases in their share in world consumption, to 60 percent by 2030.

The predominance of the industrial countries as major importers will continue although it will be somewhat less pronounced than at present. These countries now take 65 percent of the exports of the producing/exporting developing countries. They may still account for 56 percent in 2030. The importing developing countries undergoing rapid industrialization (e.g. China, the Republic of Korea and Brazil) will be gradually increasing their share in world imports. However, prospective changes in the role of the importing developing countries as growing outlets for the exports of the producing/exporting countries of natural rubber will be nowhere near as revolutionary as those that have characterized, and are still in prospect for, the sugar sector.

47 Average 1999/2000, from *Rubber Statistical Bulletin,* 56 (6). March 2002.
48 Changes to 2015 from the three-year average of 1999/2001 (World Bank, 2001c, Table A2.12).
49 A recent projections study suggests that the share of natural rubber may fall back to 35 percent by 2020 (Burger and Smit, 2000, Figure 5.4).

Table 3.25 Natural rubber, production, consumption and trade: past and projected

	'000 tonnes	Growth rates, percentage p.a.				
	1997/99	1969-99	1979-99	1989-99	1997/99 -2015	2015 -2030
		Production				
Developing countries	6 601	3.0	3.4	2.9	2.1	1.9
World	6 601	3.0	3.4	2.9	2.1	1.9

	Total demand ('000 tonnes)					
	1964/66	1974/76	1984/86	1997/99	2015	2030
Developing countries	3 252	6.0	6.1	3.5	2.9	2.4
Industrial countries	3 091	1.7	2.0	2.2	1.1	1.2
Transition countries	170	-4.9	-6.8	-3.5	2.0	2.3
World	6 512	2.8	3.2	2.6	2.1	1.9

	Net trade ('000 tonnes)					
Industrial countries	-1 500	-2 044	-2 349	-3 090	-3 750	-4 500
Transition countries	-383	-493	-379	-170	-240	-350
Developing countries	1 902	2 452	2 728	3 424	4 150	5 000
Exporters in 1997/99	2 220	3 033	3 555	4 767	6 250	7 950
Thailand	210	356	680	1 982		
Indonesia	684	798	990	1 498		
Malaysia	928	1 507	1 493	589		
Viet Nam	56	17	36	193		
Côte d'Ivoire	3	16	41	96		
Other exporters	339	339	316	409		
Importers in 1997/99	-318	-581	-827	-1 343	-2 100	-2 950
China*	-171	-275	-294	-516		
Korea, Rep	-13	-72	-163	-304		
Brazil	-6	-45	-65	-104		
Other importers	-129	-189	-305	-419		

* Includes net trade of Hong Kong SAR and Taiwan Province of China.

Crop production and natural resource use

4.1 Introduction

This chapter discusses the main agronomic factors underlying the projections of crop production presented in Chapter 3. The focus is on crop production in developing countries, for which the projections were unfolded into land and yield projections under rainfed (five land classes) and irrigated conditions. Although the underlying analysis was carried out at the level of individual countries, the discussion here is limited to presenting the results at the level of major regions, which unavoidably masks wide intercountry differences. The parameters underlying the livestock production projections will be discussed in Chapter 5. Selected technology issues such as the scope for further yield increases, technologies in support of sustainable agriculture and the role of biotechnology are discussed in Chapter 11. Issues of environment and the possible impact of climate change on crop production are the subjects of Chapters 12 and 13.

4.2 Sources of growth in crop production

Aggregate crop production at the world level is projected to grow over the period to 2030 at 1.4 percent p.a., down from the annual growth of 2.1 percent of the past 30 years (Table 4.1). For the developing countries as a group, the corresponding growth rates are 1.6 and 3.1 percent p.a., respectively (or 1.8 and 2.7 percent p.a., excluding China). The reasons for this continuing deceleration in crop production growth have been explained in Chapter 3.

The projected increase in world crop production over the period from 1997/99 to 2030 is 55 percent, against 126 percent over the past period of similar length. Similar increases for the developing countries as a group are 67 and 191 percent, respectively. The only region where the projected increase would be about the same as the historical one would be sub-Saharan Africa, namely 123 and 115 percent, respectively. The faster growth in the developing countries, as compared to the world average, means that by 2030 this group of countries will account for almost three-quarters (72 percent) of world crop production, up from two-thirds (67 percent) in 1997/99 and just over half (53 percent) 30 years earlier.

Table 4.1   Annual crop production growth

	1969-99	1979-99	1989-99	1997/99 -2015	2015-30	1997/99 -2030
				Percentage		
All developing countries	3.1	3.1	3.2	1.7	1.4	1.6
excl. China	2.7	2.7	2.5	2.0	1.6	1.8
excl. China and India	2.7	2.6	2.5	2.0	1.7	1.9
Sub-Saharan Africa	2.3	3.3	3.3	2.6	2.5	2.5
Near East/North Africa	2.9	2.9	2.6	1.8	1.5	1.6
Latin America and the Caribbean	2.6	2.3	2.6	1.8	1.6	1.7
South Asia	2.8	3.0	2.4	2.1	1.5	1.8
East Asia	3.6	3.5	3.7	1.3	1.1	1.2
Industrial countries	1.4	1.1	1.6	0.9	0.9	0.9
Transition countries	-0.6	-1.6	-3.7	0.7	0.7	0.7
World	2.1	2.0	2.1	1.5	1.3	1.4

There are three sources of growth in crop production: arable land expansion which, together with increases in cropping intensities (i.e. increasing multiple cropping and shorter fallow periods), leads to an expansion in harvested area; and yield growth. About 80 percent of the projected growth in crop production in developing countries will come from intensification in the form of yield increases (67 percent) and higher cropping intensities (12 percent, Table 4.2). The share due to intensification will go up to 90 percent and higher in the land-scarce regions of the Near East/North Africa and South Asia. The results for East Asia are heavily influenced by China. Excluding the latter, intensification will account for just over 70 percent of crop production growth in East Asia. Arable land expansion will remain an important factor in crop production growth in many countries of sub-Saharan Africa, Latin America and some countries in East Asia, although much less so than in the past. The estimated contribution of yield increases is partly a result of the increasing share of irrigated agriculture in total crop production (see Section 4.4.1), and irrigated agriculture is normally more "intensive" than rainfed agriculture.

The results shown in Table 4.2 should be taken as rough indications only. For example, yields here are weighted yields (1989/91 price weights) for 34 crops and historical data for arable land for many countries are particularly unreliable.[1] Data on cropping intensities for most countries are non-existent and for this study were derived by comparing data on harvested land, aggregated over all crops, with data on arable land. The projections are the end result of a detailed investigation of present and future land/yield combinations for 34 crops under rainfed and irrigated cultivation conditions, for 93 developing countries.[2] In the developed countries, the area of arable land in crop production has been stagnant since the early 1970s and recently declining. Hence growth in yields and more intensive use of land accounted for all of their growth in crop production and also compensated for losses in their arable land area.

Growth in wheat and rice production in the developing countries increasingly will have to come from gains in yield (more than four-fifths), while expansion of harvested land will continue to be a major contributor to production growth of maize,

---

[1]   See Alexandratos (1995, p. 161, 168) for a discussion on problems with land use data.

[2]   Unfortunately, revised data for harvested land and yields by crop for China (mainland) are not available until the results of the 1997 Chinese Agricultural Census have been processed and published. Therefore, ad hoc adjustments had to be made to base year data based on fragmentary non-official information on harvested land and yield by crop.

Table 4.2   Sources of growth in crop production (percentage)

	Arable land expansion (1)		Increases in cropping intensity (2)		Harvested land expansion (1+2)		Yield increases	
	1961-1999	1997/99-2030	1961-1999	1997/99-2030	1961-1999	1997/99-2030	1961-1999	1997/99-2030
All developing countries	23	21	6	12	29	33	71	67
excl. China	23	24	13	13	36	37	64	63
excl. China and India	29	28	16	16	45	44	55	56
Sub-Saharan Africa	35	27	31	12	66	39	34	61
Near East/North Africa	14	13	14	19	28	32	72	68
Latin America and the Caribbean	46	33	-1	21	45	54	55	46
South Asia	6	6	14	13	20	19	80	81
East Asia	26	5	-5	14	21	19	79	81
World	15		7		22		78	
All developing countries								
Crop production – rainfed		25		11		36		64
Crop production – irrigated		28		15		43		57

possibly even more so than in the past (Table 4.3). These differences are partly because the bulk of wheat and rice is produced in the land-scarce regions of Asia and the Near East/North Africa while maize is the major cereal crop in sub-Saharan Africa and Latin America, regions where many countries still have room for area expansion. As discussed in Chapter 3, an increasing share of the increment in the production of cereals, mainly coarse grains, will be used in livestock feed. As a result, maize production in the developing countries is projected to grow at 2.2 percent p.a. against "only" 1.3 percent for wheat and 1.0 percent for rice. Such contrasts are particularly marked in China where wheat and rice production is expected to grow only marginally over the projection period, while maize production is expected to nearly double. Hence there will be a corresponding decline in the wheat and rice areas but an increase of 36 percent in the maize area.

The actual combination of the factors used in crop production (land, labour and capital) in the different countries will be determined by their relative prices. For example, taking the physical availability of land as a proxy for its relative scarcity and hence price, one would expect land to play a greater role in crop production the less scarce and cheaper it is. For the 60 countries out of the 93 developing countries studied in detail, which at present use less than 60 percent of their land estimated to have some rainfed crop production potential (see Section 4.3.1), arable land expansion is projected to account for one-third of their crop production growth. In the group of 33 land-scarce countries – defined here as countries with more

Table 4.3   Sources of growth for major cereals in developing countries (percentage)

	Harvested land expansion		Yield increases	
	1961 - 1999	1997/99 - 2030	1961 - 1999	1997/99 - 2030
Wheat	22	17	78	83
Rice	23	14	77	88
Maize	30	49	70	51

Table 4.4   Shares of irrigated production in total crop production of developing countries

Shares (percentage)	Arable land	All crops Harvested land	Production	Cereals Harvested land	Production
Share in 1997/99	21	29	40	39	59
Share in 2030	22	32	47	44	64
Share in increment 1997/99–2030	33	47	57	75	73

than 60 percent of their suitable land already in use – the contribution of land expansion is estimated to be less than 10 percent.

For the developing countries, this study made an attempt to break down crop production by rainfed and irrigated land in order to analyse the contribution of irrigated crop production to total crop production. It is estimated that in the developing countries at present, irrigated agriculture, with about a fifth of all arable land, accounts for 40 percent of all crop production and almost 60 percent of cereal production (Table 4.4). It should be emphasized that, apart from some major crops in some countries, there are only very limited data on irrigated land by crops and the results presented in Table 4.4 are almost entirely based on expert judgement (see Appendix 2 for the approach followed in this study). Nevertheless, the results suggest an increasing importance of irrigated agriculture, which accounts for a third of the total increase in arable land and for over 70 percent of the projected increase in cereal production.

4.3   Agricultural land

At present some 11 percent (1.5 billion ha) of the globe's land surface (13.4 billion ha) is used in crop production (arable land and land under permanent crops). This area represents slightly over a third (36 percent) of the land estimated to be to some degree suitable for crop production. The fact that there remain some 2.7 billion ha with crop production potential suggests that there is still scope for further expansion of agricultural land. However, there is also a perception, at least in some quarters, that there is no more, or very little, land to bring under cultivation. In what follows, an attempt is made to shed some light on these contrasting views by first discussing the most recent estimates of

land with crop production potential and some constraints to exploiting such suitable areas (Section 4.3.1). Then the projected expansion of the agricultural area during the next three decades (to 2030) is presented in Section 4.3.2, while Section 4.3.3 speculates about whether or not there will be an increasing scarcity of land for agriculture.

4.3.1 Land with crop production potential for rainfed agriculture

Notwithstanding the predominance of yield increases in the growth of agricultural production, land expansion will continue to be a significant factor in those developing countries and regions where the potential for expansion exists and the prevailing farming systems and more general demographic and socio-economic conditions favour it. One of the frequently asked questions in the debate on world food futures and sustainability is: how much land is there that could be used to produce food to meet the needs of the growing population? Since the late 1970s, FAO has conducted a series of studies to determine the suitability of land for growing various crops. Recently, a new study was undertaken together with the International Institute for Applied Systems Analysis (IIASA) to refine the methods, update databases and extend the coverage to all countries in the world by including also countries in temperate and boreal climates, which were previously not covered. A summary description of the method is given in Box 4.1 and a full description and presentation of results can be found in Fischer, van Velthuizen and Nachtergaele (2000).

Table 4.5 gives some results for selected crops and input levels. At a high input level (commercial farm operations, see Box 4.1), over 1.1 billion ha would be suitable for growing wheat at an average maximum attainable yield level of 6.3 tonnes/ha, i.e. taking into account all climate, soil and terrain constraints. At the low technology level (subsistence farming), 1.5 billion ha would be suitable, but at an average maximum yield level of only 2.3 tonnes/ha. The suitable area at this lower input level is greater because, for example, tractors (high input level) cannot be used on steep slopes. For developing countries alone, the estimates are 314 million ha and 5.3 tonnes/ha under the high technology level because most of the suitable area for wheat at this input level is in the developed countries. For the other crops shown, the bulk of the suitable area is in the developing countries.

Summing over all crops and technology levels considered (see Box 4.1), it is estimated that about 30 percent of the world's land surface, or 4.2 billion ha, is suitable for rainfed agriculture (Table 4.6). Of this area, the developing countries have some 2.8 billion ha of land of varying qualities that have

potential for growing rainfed crops at yields above an "acceptable" minimum level. Of this land, nearly 960 million ha are already in cultivation. The remaining 1.8 billion ha would therefore seem to provide significant scope for further expansion of agriculture in developing countries. However, this favourable impression must be much qualified if a number of considerations and constraints are taken into account.

First, the method of deriving the land suitability estimates: it is enough for a piece of land to support a single crop at a minimum yield level for it to be deemed suitable. For example, large tracts of land in North Africa permit cultivation of only olive trees. These lands therefore are counted as "suitable" although one might have little use for them in practice (see also Box 4.2 for further similar qualifications).

Table 4.5   Land with rainfed crop production potential for selected crop and input levels

	Actual 1997/99		Total suitable			Very suitable		Suitable		Moderately suitable		Marginally suitable	
	A	Y	% of land	A	Y	A	Y	A	Y	A	Y	A	Y
Wheat – high input													
World	226	2.6	8.5	1139	6.3	160	9.4	397	7.8	361	5.2	221	3.0
All developing	111	2.5	4.1	314	5.3	38	8.2	97	6.7	105	4.7	74	2.7
Transition countries	51	2.0	15.6	359	6.3	44	9.6	107	8.2	130	5.3	78	3.4
Industrial countries	65	3.3	14.3	466	6.9	78	9.9	193	8.1	125	5.4	69	2.9
Wheat – low input													
World	226	2.6	11.3	1510	2.3	175	4.1	403	2.9	487	2.0	445	1.2
All developing	111	2.5	6.1	467	1.7	31	3.1	101	2.4	152	1.7	183	1.0
Transition countries	51	2.0	18.5	425	2.6	47	4.4	127	3.2	151	2.3	100	1.3
Industrial countries	65	3.3	19.0	617	2.5	97	4.2	175	3.1	183	2.1	161	1.2
Rice – high input													
World	161	3.6	12.5	1678	4.3	348	6.2	555	4.9	439	3.6	337	2.2
All developing	157	3.6	21.4	1634	4.3	347	6.2	549	4.9	423	3.6	315	2.2
Transition countries	0.5	2.5	0.0	1	3.2	0	0.0	0	7.6	0	4.3	1	2.6
Industrial countries	4	6.5	1.3	44	4.1	1	9.0	6	6.4	16	4.6	21	2.6
Maize – high input													
World	144	4.2	11.6	1557	8.2	246	13.2	439	10.3	393	7.4	479	4.3
All developing	99	2.8	18.2	1382	8.0	221	13.2	359	10.3	339	7.3	463	4.3
Transition countries	9	3.9	0.5	11	6.9	1	13.4	8	7.3	2	5.3	1	3.4
Industrial countries	38	7.7	5.0	163	9.6	24	13.9	73	10.7	52	7.7	15	4.6
Soybean– high input													
World	72	2.1	10.3	1385	2.4	183	4.0	353	3.1	415	2.2	434	1.3
All developing	41	1.8	16.8	1277	2.4	173	4.0	324	3.1	372	2.2	407	1.3
Transition countries	0.7	1.3	0.1	3	3.0	1	4.2	1	3.2	1	2.3	0	1.5
Industrial countries	30	2.6	3.2	105	2.6	10	4.1	27	3.3	42	2.4	26	1.5

Notes: A=area in million ha; Y=average attainable yield in tonnes/ha. The 1997/99 data are not distinguished by input level as information does not exist. The area data for 1997/99 refer to harvested area and elsewhere in the table to arable area.

Box 4.1 Summary methodology of estimating land potential for rainfed agriculture

For each country an evaluation was made of the suitability of land for growing 30 crops[1] under rainfed conditions and various levels of technology. The basic data for the evaluation consist of several georeferenced data sets: the inventory of soil characteristics from the digital FAO-UNESCO Soil Map of the World (SMW; FAO, 1995a), an inventory of terrain characteristics contained in a digital elevation model (DEM; EROS Data Center, 1998), and an inventory of climate regimes (New, Hulme and Jones, 1999). The data on temperature, rainfall, relative humidity, wind speed and radiation are used, together with information on evapotranspiration, to define the length of growing periods (LGPs), i.e. the number of days in a year when moisture availability in the soil and temperature permit crop growth.

The suitability estimates were carried out for grid cells at the 5 arc minute level (9.3 by 9.3 km at the equator), by interfacing the soil, terrain and LGP characteristics for each grid cell with specific growth requirements (temperature profile, moisture, nutrients, etc.) for each of the 30 crops under three levels of technology. These levels of technology are: low, using no fertilizers, pesticides or improved seeds, equivalent to subsistence farming; intermediate, with some use of fertilizers, pesticides, improved seeds and mechanical tools; and high, with full use of all required inputs and management practices as in advanced commercial farming. The resulting average attainable yields for each cell, crop and technology alternative were then compared with those obtainable under the same climate and technology on land without soil and terrain constraints, termed here the maximum constraint-free yield (MCFY). The land in each grid cell is for each crop (and technology level) subdivided into five suitability classes on the basis of the average attainable yield as a percentage of the MCFY, as follows – very suitable (VS): at least 80 percent; suitable (S): 60 to 80 percent; moderately suitable (MS): 40 to 60 percent and marginally suitable (mS): 20 to 40 percent. Not suitable (NS) land is that for which attainable yields are below 20 percent of the MCFY.

The result of this procedure is an inventory of land suitability by grid cell for each crop and technology level. To make statements of the overall suitability for rainfed agriculture, one has to aggregate the suitability estimates for all crops and technology levels.[2] There are various ways of doing this (e.g. one could add up over crops by applying value (prices) or energy (calories) weights to arrive at an "average" crop). Here the method applied in Fischer, van Velthuizen and Nachtergaele (2000) was followed: For each grid cell, first the largest (i.e. out of all the crops considered) extent of very suitable and suitable area under the high technology level was taken. Then the part of the largest very suitable, suitable and moderately suitable area under the intermediate technology, exceeding this first area, was added. Finally the part of the largest very suitable, suitable, moderately suitable and marginally suitable area under the low technology, exceeding this second area, was added. The rationale for this methodology is that it is unlikely to make economic sense to cultivate moderately and marginally suitable areas under the high technology level, or to cultivate marginally suitable areas under the intermediate technology level. The result of this is the maximum suitable area in each grid cell under what was dubbed the "mixed" input level. Table 4.6 shows the results of aggregating over all grid cells in each country.

It is noted, however, that some of the land classified as not suitable on the basis of this evaluation is used for rainfed agriculture in some countries, e.g. where steep land has been terraced or where yields less than the MCFY are acceptable under the local economic and social conditions (see also Box 4.2). For these reasons, land reported as being in agricultural use in some countries exceeds the areas deemed here as having rainfed crop production potential.

[1] These crops are: wheat (2 types), rice (3 types), maize, barley (2 types), sorghum, millet (2 types), rye (2 types), potato, cassava, sweet potato, phaseolus bean, chickpea, cowpea, soybean, rapeseed (2 types), groundnut, sunflower, oil palm, olive, cotton, sugar cane, sugar beet and banana.

[2] For a full explanation of the methodology, see Fischer, van Velthuizen and Nachtergaele (2000). The estimate of total potential land of 2 603 million ha in developing countries, excluding China, is about 3 percent higher than the 2 537 million ha estimated for the 1995 edition of this study (Alexandratos, 1995). This is due to the more refined methodology followed in the present study, new climate and terrain data sets, the increase in the number of crops for which suitability was tested (from 21 in 1995 to 30 in the present study), and a different method of aggregation.

Second, the land balance (land with crop production potential not in agricultural use) is very unevenly distributed among regions and countries. Some 90 percent of the remaining 1.8 billion ha is in Latin America and sub-Saharan Africa, and more than half of the total is concentrated in just seven countries (Brazil, the Democratic Republic of the Congo, the Sudan, Angola, Argentina, Colombia and Bolivia). At the other extreme, there is virtually no spare land available for agricultural expansion in South Asia and the Near East/North Africa. In fact, in a few countries in these two latter regions, the land balance is negative, i.e. land classified as not suitable is made productive through human intervention such as terracing of sloping land, irrigation of arid and hyperarid land, etc. and is in agricultural use. Even within the relatively land-abundant regions, there is great diversity of land availability, in terms of both quantity and quality, among countries and subregions.

Third, much of the land also suffers from constraints such as ecological fragility, low fertility, toxicity, high incidence of disease or lack of infrastructure. These reduce its productivity, require high input use and management skills to permit its sustainable use, or require prohibitively high investments to be made accessible or disease-free. Alexandratos (1995, Table 4.2) shows that over 70 percent of the land with rainfed crop production

potential in sub-Saharan Africa and Latin America suffers from one or more soil and terrain constraints. Natural causes as well as human intervention can also lead to deterioration of the productive potential of the resource, for example through soil erosion or salinization of irrigated areas. Hence this evaluation of suitability may contain elements of overestimation (see also Bot, Nachtergaele and Young, 2000) and much of the land balance cannot be considered to be a resource that is readily usable for food production on demand.

There is another cause for the land balance to be overestimated: it ignores land uses other than for growing the crops for which it was evaluated. Thus, forest cover, protected areas and land used for human settlements and economic infrastructure are not taken into account. Alexandratos (1995) estimated that forests cover at least 45 percent, protected areas some 12 percent and human settlements some 3 percent of the land balance, with wide regional differences. For example, in the land-scarce region of South Asia, some 45 percent of the land with crop production potential but not yet in agricultural use is estimated to be occupied by human settlements. This leaves little doubt that population growth and further urbanization will be a significant factor in reducing land availability for agricultural use in this region.

Table 4.6 Land with rainfed crop production potential

	Total land surface	Share of land suitable (%)	Total land suitable	Very suitable	Suitable	Moderately suitable	Marginally suitable	Not suitable
				Million ha				
Developing countries	7 302	38	2 782	1 109	1 001	400	273	4 520
Sub-Saharan Africa	2 287	45	1 031	421	352	156	103	1 256
Near East/North Africa	1 158	9	99	4	22	41	32	1 059
Latin America and the Caribbean	2 035	52	1 066	421	431	133	80	969
South Asia	421	52	220	116	77	17	10	202
East Asia	1 401	26	366	146	119	53	48	1 035
Industrial countries	3 248	27	874	155	313	232	174	2 374
Transition countries	2 305	22	497	67	182	159	88	1 808
World*	13 400	31	4 188	1 348	1 509	794	537	9 211

* Including some countries not covered in this study.

Box 4.2 Estimating the land potential for rainfed agriculture: some observations[1]

The evaluation of land potential undertaken in the global agro-ecological zones (GAEZ) study starts by taking stock of (i) the biophysical characteristics of the resource (soil, terrain, climate); and (ii) the growing requirements of crops (solar radiation, temperature, humidity, etc.). The data in the former set are interfaced with those in the second set and conclusions are drawn on the amount of land that may be classified as suitable for producing each one of the crops tested (see Fischer, van Velthuizen and Nachtergaele, 2000).

The two data sets mentioned above can change over time. Climate change, land degradation or, conversely, land improvements, together with the permanent conversion of land to non-agricultural uses, all contribute to change the extent and characteristics of the resource. This fact is of particular importance if the purpose of the study is to draw inferences about the adequacy of land resources in the longer term.

In parallel, the growth of scientific knowledge and the development of technology modify the growing requirements of the different crops for achieving any given yield level. For example, in the present round of GAEZ work the maximum attainable yield for rainfed wheat in subtropical and temperate environments is put at about 12 tonnes/ha in high input farming and about 4.8 tonnes/ha in low input farming. Some 25 years ago, when the first FAO agro-ecological zone study was carried out (FAO, 1981b), these yields were put at only 4.9 and 1.2 tonnes/ha, respectively. Likewise, land suitable for growing wheat at, say, 5 tonnes/ha in 30 years time may be quite different from that prevailing today, if scientific advances make it possible to obtain such yields where only 2 tonnes/ha can be achieved today. A likely possibility would be through the development of varieties better able to withstand stresses such as drought, soil toxicity and pest attack. Scientific knowledge and its application will obviously have an impact on whether or not any given piece of land will be classified as suitable for producing a given crop.

Land suitability is crop-specific. To take an extreme example, more than 50 percent of the land area in the Democratic Republic of the Congo is suitable for growing cassava but less than 3 percent is suitable for growing wheat. Therefore, before statements can be made about the adequacy or otherwise of land resources to grow food for an increasing population, the information about land suitability needs to be interfaced with information about expected demand patterns – volume and commodity composition of both domestic and foreign demand. For example, the Democratic Republic of the Congo's ample land resources suitable for growing cassava will be of little value unless there is sufficient domestic or foreign demand for the country's cassava, now or in the future.

Declaring a piece of land as suitable for producing a certain crop implicitly assumes that people find it worthwhile to exploit the land for this purpose. In other words, land must not only possess minimum biophysical attributes in relation to the requirements of the crops for which there is, or will be, demand, but it must also be in a socio-economic environment in which people consider it an economic asset. For example, in low-income countries, people will exploit land even if the yields or, more precisely, the returns to their work, are low relative to the urgency to secure their access to food. This means that the price of food is high relative to their income and that the opportunities of earning higher returns from other activities are limited as well. Thus, what qualifies as land with an acceptable production potential in a poor country may not be so in a high-income one. An exception would be if poor quality of land were compensated by a larger area per person with access to mechanization[2] so that returns to work in farming would generate income not far below earnings from other work. Obviously, the socio-economic context within which a piece of land exists and assumes a given value or utility, changes over time: what qualifies today as land suitable for farming may not be so tomorrow.

It is no easy task to account fully for all these factors in arriving at conclusions concerning how much land with crop production potential there is. For example, if food became scarce and its real price rose, more land would be worth exploiting and hence be classified as agricultural than would otherwise be the case. Therefore, depending on how such information is to be used, one may want to adopt different criteria and hence generate alternative estimates.

---

[1] Adapted from Alexandratos and Bruinsma (1999).

[2] Relatively low-yield rainfed but internationally competitive agriculture (wheat yields of 2.0-2.5 tonnes/ha compared with double that in western Europe) is practised in such high-income countries as the United States, Canada and Australia. But this is in large and fully mechanized farms permitting the exploitation of extensive areas that generate sufficient income per holding even if earnings per ha are low.

These considerations underline the need to interpret estimates of land balances with caution when assessing land availability for agricultural use. Cohen (1995) summarizes and evaluates all estimates made of available cultivable land, together with their underlying methods, and shows their extremely wide range. Young (1999) offers a critique of the more recent estimates of available cultivable land, including those given in Alexandratos (1995), and states that "an order-of-magnitude estimate reaches the conclusion that in a representative area with an estimated land balance of 50 percent, the realistic area is some 3 to 25 percent of the cultivable land".

4.3.2 Expansion of land in crop production

There is a widespread perception that there is no more, or very little, new land to bring under cultivation. Some of this perception may be well grounded in the specific situations of land-scarce countries and regions such as Japan, South Asia and the Near East/North Africa. Yet this perception may not apply, or may apply with much less force, to other parts of the world. As discussed above, there are large tracts of land with varying degrees of agricultural potential in several countries, most of them in sub-Saharan Africa and Latin America, with some in East Asia. However, this land may lack infrastructure, be partly under forest cover or be in wetlands that have to be protected for environmental reasons, or the people who would exploit it for agriculture lack access to appropriate technological packages or the economic incentives to adopt them.

In reality, expansion of land in agricultural use takes place all the time. It does so mainly in countries that combine growing needs for food and employment with limited access to technology packages that could increase intensification of cultivation on land already in agricultural use. The data show that expansion of arable land continues to be an important source of agricultural growth in sub-Saharan Africa, South America and East Asia, excluding China (Table 4.7).[3]

The projected expansion of arable land in crop production shown in Tables 4.7, 4.8 and 4.9, has been derived for the rainfed and irrigated land classes. In each country the following factors have been taken into account: (i) actual data or, in many cases, estimates for the base year 1997/99 on harvested land and yield by crop in each of the two classes; (ii) total arable land and cropping intensity in each class; (iii) production projections for each crop; (iv) likely increases in yield by crop and land class; (v) increases in the irrigated area; (vi) likely increases in cropping intensities; and (vii) the land balances for rainfed agriculture described in the preceding section, and for irrigated land discussed in the following section. This method was used only for the 93 developing countries covered in this study (see Appendix 1). For the developed countries or country groups, only projections of crop production have been made, which were then translated into projections for total harvested land and yield by crop.

The overall result for developing countries is a projected net increase in the arable area of 120 million ha (from 956 in the base year to 1 076 in 2030), an increase of 12.6 percent (see Table 4.7).[4] The increase for the period 1961/63 to 1997/99 was 172 million ha, an increase of 25 percent. Not surprisingly, the bulk of this projected expansion is expected to take place in sub-Saharan Africa (60 million), Latin America (41 million) and East Asia, excluding China (14 million), with almost no land expansion in the Near East/North Africa and South Asia regions and even a decline in the arable land area in China. The slowdown in the expansion of arable land is mainly a consequence of the projected slowdown in the growth of crop production and is common to all regions.

The projected increase of arable land in agricultural use is a small proportion (6.6 percent) of the total unused land with rainfed crop production potential. What does the empirical evidence show concerning the rate and process of land expansion for agricultural use in the developing countries? Microlevel analyses have generally established that under the socio-economic and institutional condi-

---

[3] Historical data for China have been drastically revised upwards from 1985 onwards, which distorts the historical growth rates in Table 4.7 for East Asia (and for the total of developing countries).

[4] As mentioned in Section 4.2, data on arable land are unreliable for many countries. Therefore base year data were adjusted and are shown in column (4) as "1997/99 adjusted".

Table 4.7   Total arable land: past and projected

	Arable land in use					Annual growth			Land in use as % of potential		Balance	
	1961 /63	1979 /81	1997 /99	1997 /99 adj.	2015	2030	1961 -1999	1997/99 -2030	1997 /99	2030	1997 /99	2030
	(million ha)						(% p.a.)		(%)		(million ha)	
	(1)	(2)	(3)	(4)	(5)	(6)	(7)	(8)	(9)	(10)	(11)	(12)
Sub-Saharan Africa	119	138	156	228	262	288	0.77	0.72	22	28	803	743
Near East/ North Africa	86	91	100	86	89	93	0.42	0.23	87	94	13	6
Latin America and the Caribbean	104	138	159	203	223	244	1.22	0.57	19	23	863	822
South Asia	191	202	205	207	210	216	0.17	0.13	94	98	13	4
excl. India	29	34	35	37	38	39	0.37	0.12	162	168	-14	-16
East Asia	176	182	227	232	233	237	0.89	0.06	63	65	134	129
excl. China	72	82	93	98	105	112	0.82	0.43	52	60	89	75
Developing countries	676	751	848	956	1 017	1 076	0.68	0.37	34	39	1 826	1 706
excl. China	572	652	713	822	889	951	0.63	0.46	32	37	1 781	1 652
excl. China and India	410	483	543	652	717	774	0.81	0.54	27	32	1 755	1 633
Industrial countries	379	395	387				0.07		44		487	
Transition countries	291	280	265				-0.19		53		232	
World	1 351	1 432	1 506				0.34		36		2 682	

Source: Column (1)-(3): FAOSTAT, November 2001.
Note: "World" includes a few countries not included in the other country groups shown.

tions (land tenure, etc.) prevailing in many developing countries, increases in output are obtained mainly through land expansion, where the physical potential for doing so exists. For example, in a careful analysis of the experience of Côte d'Ivoire, Lopez (1998) concludes that "the main response of annual crops to price incentives is to increase the area cultivated". Similar findings, such as the rate of deforestation being positively related to the price of maize, are reported for Mexico by Deininger and Minten (1999). Some of this land expansion is taking place at the expense of long rotation periods and fallows, a practice still common to many countries in sub-Saharan Africa, with the result that the natural fertility of the soil is reduced. Since fertilizer use is often uneconomic, the end result is soil mining and stagnation or outright reduction of yields.

The projected average annual increase in the developing countries' arable area of 3.75 million ha (120/32), compared with 4.8 million (172/36) in the historical period, is a net increase. It is the total of gross land expansion minus land taken out of production for various reasons, for example because of degradation or loss of economic viability. An unknown part of the new land to be brought into agriculture will come from land currently under forests. If all the additional land came from forested areas, this would imply an annual deforestation rate of 0.2 percent, compared with the 0.8 percent (or 15.4 million ha p.a.) for the 1980s and 0.6 percent (or 12.0 million ha p.a.) for the 1990s (FAO, 2001c). The latter estimates, of course, include deforestation from all causes, such as informal non-recorded agriculture, grazing, logging, gathering of fuelwood, etc.

The arable area in the world as a whole expanded between 1961/63 and 1997/99 by 155 million ha (or 11 percent), the result of two opposite trends: an increase of 172 million ha in the developing countries and a decline of 18 million ha in the developed ones. This decline in the arable area in the latter group has been accelerating over time (-0.3 percent p.a. in the industrial countries and -0.6 percent p.a. in transition countries during 1989-99). The longer-term forces determining such declines are sustained yield growth combined with a continuing slowdown in the growth of demand for their agricultural products. In addition, there are more temporary phenomena such as policy changes in the industrial countries and political and economic transition problems in the former centrally planned countries. No projections were made for arable land in the developed countries but, assuming a continuation of these trends, one would expect a further decline in the developed countries' arable area. However, this decline in arable area could in part be offset by the emerging trend towards a de-intensification of agriculture in these countries through increasing demand for organic products and for environmentally benign cultivation practices, and a possible minor shift of agriculture to temperate zones towards the end of the projection period because of climate change. The net effect of these countervailing forces could be a roughly constant or only marginally declining arable area in the developed countries. Arable area expansion for the world as a whole therefore would more or less equal that of the developing countries.

Although the developing countries' arable area is projected to expand by 120 million ha over the projection period, the harvested area will expand by 178 million ha or 20 percent, because of increases in cropping intensities (Table 4.8). The increase of harvested land over the historical period (1961/63 to 1997/99) was 221 million ha or 38 percent. Sub-Saharan Africa alone accounts for 63 million ha, or 35 percent, of the projected increase in harvested land, the highest among all regions. This is a consequence of the high and sustained growth in crop production projected for this region (see Table 4.1) combined with the region's scope for further land expansion. The other region for which a considerable expansion of the harvested area is foreseen, albeit at a slower pace than in the past, is Latin America with an increase of 45 million ha.

As mentioned before, the quality of the data for arable land use leaves much to be desired (see also Young, 1998). The data of harvested or sown areas for the major crops are more reliable. They show that expansion of harvested area continues to be an important source of agricultural growth, mainly in sub-Saharan Africa, but also in Southeast Asia and, to a lesser extent, in Latin America. Overall, for the developing countries, excluding China, the harvested area under the major crops (cereals, oilseeds, pulses, roots/tubers, cotton, sugar cane/beet, rubber and tobacco) grew by 10 percent during the ten years from 1987/89 to 1997/99, or about 1 percent p.a. This is only slightly higher than the growth rate of 0.9 percent p.a. projected for all crops for the 21-year period 1988/90-2010 in Alexandratos, 1995 (p. 165).

The overall cropping intensity for developing countries will rise by about 6 percentage points over the projection period (from 93 to 99 percent). Cropping intensities continue to rise through shorter fallow periods and more multiple cropping. An increasing share of irrigated land in total agricultural land contributes to more multiple cropping. About one-third of the arable land in South and East Asia is irrigated, a share which is projected to rise to 40 percent in 2030. The high irrigation share is one of the reasons why the average cropping intensities in these regions are considerably higher than in the other regions. Average cropping intensities in developing countries, excluding China and India which together account for more than half of the irrigated area in the developing countries, are and will continue to be much lower.

The rise in cropping intensities has been one of the factors responsible for increasing the risk of land degradation and threatening sustainability, when it is not accompanied by technological change to conserve the land, including adequate and balanced use of fertilizers to compensate for soil nutrient removal by crops. It is expected that this risk will continue to exist because in many cases the socio-economic conditions will not favour the promotion of the technological changes required to ensure the sustainable intensification of land use (see Chapter 12 for a further discussion of this issue).

Table 4.8   Arable land in use, cropping intensities and harvested land

		Total land in use			Rainfed use			Irrigated use		
		A	CI	H	A	CI	H	A	CI	H
Sub-Saharan Africa	1997/99	228	68	154	223	67	150	5	86	4.5
	2030	288	76	217	281	75	210	7	102	7
Near East/North Africa	1997/99	86	81	70	60	72	43	26	102	27
	2030	93	90	83	60	78	46	33	112	37
Latin America and the Caribbean	1997/99	203	63	127	185	60	111	18	86	16
	2030	244	71	172	222	68	150	22	100	22
South Asia	1997/99	207	111	230	126	103	131	81	124	100
	2030	216	121	262	121	109	131	95	137	131
East Asia	1997/99	232	130	303	161	120	193	71	154	110
	2030	237	139	328	151	122	184	85	169	144
All above	1997/99	956	93	885	754	83	628	202	127	257
	2030	1 076	99	1 063	834	87	722	242	141	341
excl. China	1997/99	822	83	679	672	76	508	150	114	171
	2030	951	90	853	769	81	622	182	127	230
excl. China/India	1997/99	652	75	489	559	70	392	93	105	97
	2030	774	83	641	662	77	507	112	119	134

Note: A=arable land in million ha; CI=cropping intensity in percentage; H=harvested land in million ha.

4.3.3 Is land for agriculture becoming scarcer?[5]

As noted in the preceding section, land in agricultural use (arable land and land under permanent crops) in the world as a whole has increased by only 155 million ha or 11 percent to about 1.5 billion ha between the early 1960s and the late 1990s. Nevertheless there were very significant changes in some regions. For example, the increase was over 50 percent in Latin America, which accounted for over one-third of the global increase. During the same period, the world population nearly doubled from 3.1 billion to over 5.9 billion. By implication, arable land per person declined by 40 percent, from 0.43 ha in 1961/63 to 0.26 ha in 1997/99. In parallel, there is growing preoccupation that agricultural land is being lost to non-agricultural uses. In addition, the ever more intensive use of land in production through multiple cropping, reduced fallow periods, excessive use of agrochemicals, spread of monocultures, etc. is perceived as leading to land degradation (soil erosion, etc.) and the undermining of its long-term productive potential.

These developments are seen by many as having put humanity on a path of growing scarcity of land as a factor in food production, with the implication that it is, or it will be in the near future, becoming increasingly difficult to produce the food required to feed the ever-growing human population. Are these concerns well founded? Any discourse about the future should be as precise as possible concerning the magnitudes involved: how much land there is (quantity, quality, location) and how much more food, what type of food and where it is required, now or at any given point of time in the future. The brief discussion of historical developments and, in particular, of future prospects in world food and agriculture presented in Chapters 2 and 3, provides a rough quantitative framework for assessing such concerns.

The evidence presented above about historical developments does not support the notion that it has been getting increasingly difficult for the world to extract from the land an additional unit of food. Rather the contrary has been happening, as shown by the secular decline in the real price of food. This

5   Adapted from Alexandratos and Bruinsma (1999).

secular decline indicates that it has been getting easier for humanity to produce an additional unit of food relative to the effort required to produce an additional unit of an "average" non-food product. This statement applies to the world as a whole, not necessarily to particular locations, and is valid only under particular conditions which are, essentially, the absence of market failures and ethical acceptability of the resulting distribution of access to food by different population groups.

The notion that resources for producing food, in which land is an important constituent, have been getting more abundant rather than scarcer in relative terms, i.e. in relation to the aggregate stocks of resources of the global economy, appears counter-intuitive. How can it be reconciled with the stark fact that the world population nearly doubled while land in agricultural use increased by only 11 percent, meaning that land per capita declined by some 40 percent? The answer is to be found in the fact that over the same period yields per ha of cropped area increased, as did the cropping intensity in the areas where a combination of irrigation and agro-ecological conditions permitted it and the growth of the demand for food justified it economically. For example, during the 36-year period when world average grain yields more than doubled from 1.4 tonnes/ha in 1961/63 to 3.05 tonnes/ha in 1997/99 and the overall cropping intensity probably increased by some 5 percentage points, the amount of arable land required to produce any given amount of grain declined by some 56 percent. This decline exceeded the above-mentioned 40 percent fall in the arable land per person which occurred during the same period.

In this comparison of physical quantities, land for food production is seen to have become less scarce, not scarcer. The economic evidence, a declining real price of food, corroborates in a general sense the conclusion that it has also become less scarce relative to the evolution of the demand for food and relative to what has been happening in the other sectors of the economy. However, as noted, such economic evidence properly refers to the decreasing relative scarcity of the aggregate resource base for food production in which land is only one component together with capital, labour, technology, etc. rather than to land alone.[6] In practice, what we call land today is a composite of land in its natural form and capital investments embodied in it such as irrigation infrastructure, levelling, fencing and soil amendments. It follows that any further discussion of the prospective role of land in meeting future food needs has to view it as just one component, indeed one of changing and probably declining relative weight, in the total package of factors that constitute the resource base of agriculture which, as the historical record shows, is flexible and adaptable.

Concerning the future, a number of projection studies have addressed and largely answered in the positive the issue as to whether the resource base of world agriculture, including its land component, can continue to evolve in a flexible and adaptable manner as it did in the past, and also whether it can continue to exert downward pressure on the real price of food (see, for example, Pinstrup-Andersen, Pandya-Lorch and Rosegrant, 1999). The largely positive answers mean essentially that for the world as a whole there is enough, or more than enough, food production potential to meet the growth of effective demand, i.e. the demand for food of those who can afford to pay farmers to produce it.

The preceding discussion refers to the evidence about land scarcities that can be deduced from the evolution of global magnitudes, whether aggregates such as world population, averages such as world per capita values of key variables, or food price trends observable in world markets. However, observing, interpreting and projecting the evolution of global aggregates can go only part of the way towards addressing the issues often raised in connection with the role of land in food production, essentially those issues pertaining to the broader nexus of food security and the environment. A more complete consideration of the issue, which goes beyond the scope of this report, will require an analysis at a more disaggregated level and going beyond the use of conventional economic indicators of scarcity or abundance. It should also address the following issues. First, whether land availability for food production is likely to become, or has been already, a significant

---

[6] The role of agricultural land as a resource contributing to human welfare, as the latter is conventionally measured by GDP, has been on the decline. Johnson (1997) says that "agricultural land now accounts for no more than 1.5 percent of the resources of the industrial nations".

Table 4.9   Irrigated (arable) land: past and projected

	Irrigated land in use					Annual growth		Land in use as % of potential		Balance	
	1961 /63	1979 /81	1997 /99	2015	2030	1961 -1999	1997/99 -2030	1997 /99	2030	1997 /99	2030
	(million ha)					(% p.a.)		(%)		(million ha)	
	(1)	(2)	(3)	(5)	(6)	(7)	(8)	(9)	(10)	(11)	(12)
Sub-Saharan Africa	3	4	5	6	7	2.0	0.9	14	19	32	30
Near East/ North Africa	15	18	26	29	33	2.3	0.6	62	75	17	11
Latin America and the Caribbean	8	14	18	20	22	1.9	0.5	27	32	50	46
South Asia	37	56	81	87	95	2.2	0.5	57	67	61	47
excl. India	12	17	23	24	25	1.9	0.2	84	89	4	3
East Asia	40	59	71	78	85	1.5	0.6	64	76	41	27
excl. China	10	14	19	22	25	2.1	0.9	40	53	29	23
All above	103	151	202	221	242	1.9	0.6	50	60	200	161
excl. China	73	106	150	165	182	2.1	0.6	44	54	188	157
excl. China/India	48	67	93	102	112	2.0	0.6	41	50	132	114
Industrial countries	27	37	42			1.3					
Transition countries	11	22	25			2.6					
World	142	210	271			1.8					

Source: Columns (1)- (3): FAOSTAT, November 2001.

constraint to solving problems of food insecurity at the local level. Second, whether the market signals which tell us that the resources for producing food, land among them, have been getting relatively less scarce, are seriously flawed because they fail to account for the environmental costs and eventual future risks associated with the expansion and intensification of agriculture.

4.4   Irrigation and water use

4.4.1 Expansion of irrigated land

The projections of irrigation presented below reflect a composite of information on existing irri-gation expansion plans in the different countries, potentials for expansion and need to increase crop production. The projections include some expan-sion in informal (community-managed) irrigation, which is important in sub-Saharan Africa. Estimates of "land with irrigation potential" are notoriously difficult to make for various reasons (see Alexandratos, 1995, p. 160-61) and should be taken as only rough orders of magnitude.[7]

The aggregate result for the group of devel-oping countries shows that the area equipped for irrigation in this group of countries will expand by 40 million ha (20 percent) over the projection period (Table 4.9). This means that some 20 percent of the land with irrigation potential not yet

[7] FAO (1997a) states concerning such estimates: "Irrigation potential: area of land suitable for irrigation development (it includes land already under irrigation). Methodologies used in assessing irrigation potential vary from one country to another. In most cases, it is computed on the basis of available land and water resources, but economic and environmental considerations are often taken into account to a certain degree. Except in a few cases, no consideration is given to the possible double counting of water resources shared by several countries, and this may lead to an over-estimate of irrigation potential at the regional level. Wetlands and floodplains are usually, but not always, included in irrigation potential".

equipped at present will be brought under irrigation, and that 60 percent of all land with irrigation potential (403 million ha) would be in use by 2030.

The expansion of irrigation will be strongest (in absolute terms) in the more land-scarce regions hard-pressed to raise crop production through more intensive cultivation practices, such as South Asia (+14 million ha), East Asia (+14 million ha) and the Near East/North Africa. Only small additions will be made in the more land-abundant regions of sub-Saharan Africa and Latin America, although they may represent an important increase in relative terms. The importance of irrigated agriculture has already been discussed in Section 4.2. Because of a continuing increase in cropping intensity on both existing and newly irrigated areas, the harvested irrigated area will expand by 84 million ha and will account for almost half of the increase in all harvested land (Table 4.8).

The projected expansion of irrigated land by 40 million ha is an increase in net terms. It assumes that losses of existing irrigated land resulting from, for example, water shortages or degradation because of salinization, will be compensated through rehabilitation or substitution by new areas for those lost. The few existing historical data on such losses are too uncertain and anecdotal to provide a reliable basis for drawing inferences about the future. However, if it is assumed that 2.5 percent of existing irrigation must be rehabilitated or substituted by new irrigation each year, that is, if the average life of irrigation schemes were 40 years, then the total irrigation investment activity over the projection period in the developing countries must encompass some 200 million ha, of which four-fifths would be for rehabilitation or substitution and the balance for net expansion.

The projected net increase in arable irrigated land of 40 million ha is less than half of the increase over the preceding 36 years (100 million ha). In terms of annual growth it would be "only" 0.6 percent, well below the 1.9 percent for the historical period. The projected slowdown reflects the projected lower growth rate of crop production combined with the increasing scarcity of suitable areas for irrigation and of water resources in some countries, as well as the rising costs of irrigation investment.

Most of the expansion of irrigated land is achieved by converting land in use in rainfed agri-culture or land with rainfed production potential but not yet in use, into irrigated land. Part of the irrigation, however, takes place on arid and hyper-arid land which is not suitable for rainfed agriculture. It is estimated that of the 202 million ha irrigated at present, 42 million ha are on arid and hyperarid land and of the projected increase of 40 million ha, about 2 million ha will be on such land. In some regions and countries, irrigated arid and hyperarid land form an important part of the total irrigated land at present in use: 18 out of 26 million ha in the Near East/North Africa, and 17 out of 81 million ha in South Asia.

The developed countries account for a quarter of the world's irrigated area, 67 out of 271 million ha (Table 4.9). Their annual growth of irrigated area reached a peak of 3.0 percent in the 1970s, dropping to 1.1 percent in the 1980s and to only 0.3 percent in 1990-99. This evolution pulled down the annual growth rate for global irrigation from 2.4 percent in the 1970s to 1.3 percent in the 1980s and 1990-99. Perhaps it is this sharp deceleration in growth which led some analysts to believe that there is only limited scope for further irrigation expansion. As already said, no projections by land class (rainfed, irrigated) were made for the developed countries. However, given the share of developing countries in world irrigation and the much higher crop production growth projected for this group of countries, it is reasonable to assume that the world irrigation scene will remain dominated by events in the developing countries.

4.4.2 Irrigation water use and pressure on water resources

One of the major questions concerning the future of irrigation is whether there will be sufficient freshwater to satisfy the growing needs of agricultural and non-agricultural users. Agriculture already accounts for about 70 percent of the freshwater withdrawals in the world and is usually seen as the main factor behind the increasing global scarcity of freshwater.

The estimates of the expansion of land under irrigation presented in the preceding section in part provide an answer to this question. The assessment of irrigation potential already takes into account water limitations and the projections to 2030 assume that agricultural water demand will not exceed available water resources. Yet, as discussed above, the concept of irrigation potential has severe limitations and estimates of irrigation potential can vary over time, in relation to the country's economic situation or as a result of competition for water for domestic and industrial use. Estimates of irrigation potential are also based on renewable water resources, i.e. the resources replenished annually through the hydrological cycle. In those arid countries where mining of fossil groundwater represents an important part of water withdrawal, the area under irrigation is usually larger than the irrigation potential.

Renewable water resources available to irrigation and other uses are commonly defined as that part of precipitation which is not evaporated or transpired by plants, including grass and trees, which flows into rivers and lakes or infiltrates into aquifers. The annual water balance for a given area in natural conditions, i.e. without irrigation, can be defined as the sum of the annual precipitation and net incoming flows (transfers through rivers from one area to another) minus evapotranspiration.

Table 4.10 shows the renewable water resources for 93 developing countries. Average annual precipitation is around 1 040 mm. In developing regions, renewable water resources vary from 18 percent of precipitation and incoming flows in the most arid areas (Near East/North Africa) where precipitation is a mere 180 mm per year, to about 50 percent in humid East Asia, which has a high precipitation of about 1 250 mm per year. Renewable water resources are most abundant in Latin America. These figures give an impression of the extreme variability of climatic conditions facing the 93 developing countries, and the ensuing differences observed in terms of water scarcity: those countries suffering from low precipitation and therefore most in need of irrigation are also those where water resources are naturally scarce. In addition, the water balance presented is expressed in yearly averages and cannot adequately reflect seasonal and interannual variations. Unfortunately, such variations tend to be more pronounced in arid than in humid climates.

The first step in estimating the pressure of irrigation on water resources is to assess irrigation water requirements and withdrawals. Precipitation provides part of the water crops need to satisfy their transpiration requirements. The soil, acting as a buffer, stores part of the precipitation water and returns it to the crops in times of deficit. In humid climates, this mechanism is usually sufficient to ensure satisfactory growth in rainfed agriculture. In arid climates or during the dry season, irrigation is required to compensate for the deficit resulting from insufficient or erratic precipitation. *Consumptive water use in irrigation* therefore is defined as the volume of water needed to compensate for the deficit between potential evapotranspiration and effective precipitation over the growing period of the crop. It varies considerably with climatic conditions, seasons, crops and soil types. In this study, consumptive water use in irrigation has been computed for each country on the basis of the irrigated and harvested areas by crop as estimated for the base year (1997/99) and projected for 2030 (see Box 4.3 for a brief explanation of the methodology applied). As mentioned before, in this study the breakdown by crop over rainfed and irrigated land was performed only for the 93 developing countries.

However, it is *water withdrawal for irrigation*, i.e. the volume of water extracted from rivers, lakes and aquifers for irrigation purposes, which should be used to measure the impact of irrigation on water resources. Irrigation water withdrawal normally far exceeds the consumptive water use in irrigation because of water lost during transport and distribution from its source to the crops. In addition, in the case of rice irrigation, additional water is used for paddy field flooding to facilitate land preparation and for plant protection.

For the purpose of this study, *irrigation efficiency* has been defined as the ratio between the estimated consumptive water use in irrigation and irrigation water withdrawal. Data on country water withdrawal for irrigation has been collected in the framework of the AQUASTAT programme (see FAO, 1995b, 1997a, 1997b and 1999b). Comparison of these data with the consumptive use of irrigation was used to estimate irrigation efficiency at the regional level. On average, for the 93 developing countries, it is estimated that irrigation efficiency was around 38 percent in 1997/99, varying from 25 percent in areas of abundant water resources (Latin America) to 40 percent in the Near East/North Africa region and 44 percent in South Asia where water scarcity calls for higher efficiencies (Table 4.10).

To estimate irrigation water withdrawal in 2030, an assumption had to be made about possible

Table 4.10 Annual renewable water resources (RWR) and irrigation water requirements

		Sub-Saharan Africa	Latin America and the Caribbean	Near East/ North Africa	South Asia	East Asia	All developing countries
Precipitation	mm	880	1 534	181	1 093	1 252	1 043
Internal RWR	km³	3 450	13 409	484	1 862	8 609	28 477
Net incoming flows	km³	0	0	57	607	0	0
Total RWR	km³	3 450	13 409	541	2 469	8 609	28 477
Irrigation water withdrawal							
Irrigation efficiency 1997/99	%	33	25	40	44	33	38
Irrigation water withdrawal 1997/99	km³	80	182	287	895	684	2 128
idem as percentage of RWR	%	2	1	53	36	8	7
Irrigation efficiency 2030	%	37	25	53	49	34	42
Irrigation water withdrawal 2030	km³	115	241	315	1 021	728	2 420
idem as percentage of RWR	%	3	2	58	41	8	8

Note: RWR for all developing countries exclude the regional net incoming flows to avoid double counting.

developments in the irrigation efficiency of each country. Unfortunately, there is little empirical evidence on which to base such an assumption. Two factors, however, will have an impact on the development of irrigation efficiency: the estimated levels of irrigation efficiency in 1997/99 and water scarcity. A function was designed to capture the influence of these two parameters, bearing in mind that improving irrigation efficiency is a very slow and difficult process. The overall result is that efficiency will increase by 4 percentage points, from 38 to 42 percent (Table 4.10). Such an increase in efficiency would be more pronounced in water-scarce regions (e.g. a 13 percentage point increase in the Near East/North Africa region) than in regions with abundant water resources (between 0 and 4 percentage points in Latin America, East Asia and sub-Saharan Africa). Indeed, it is expected that, under pressure from limited water resources and competition from other uses, demand management will play an important role in improving irrigation efficiency in water-scarce regions. In contrast, in humid areas the issue of irrigation efficiency is much less relevant and is likely to receive little attention.

For the 93 countries, irrigation water withdrawal is expected to grow by about 14 percent, from the current 2 128 km³/yr to 2 420 km³/yr in

2030 (Table 4.10). This increase is low compared to the 33 percent increase projected in the harvested irrigated area, from 257 million ha in 1997/99 to 341 million ha in 2030 (see Table 4.8). Most of this difference is explained by the expected improvement in irrigation efficiency, leading to a reduction in irrigation water withdrawal per irrigated hectare. A small part of this reduction is also a result of changes in cropping patterns for some countries such as China, where a substantial shift in the irrigated area from rice to maize production is expected: irrigation water requirements for rice production are usually twice those for maize.

Irrigation water withdrawal in 1997/99 was estimated to account for only 7 percent of total water resources for the 93 countries (Table 4.10). However, there are wide variations between regions, with the Near East/North Africa region using 53 percent of its water resources in irrigation while Latin America barely uses 1 percent of its resources. At the country level, variations are even higher. Of the 93 countries, ten already used more than 40 percent of their water resources for irrigation in the base year (1997/99), a situation which can be considered critical. An additional eight countries used more than 20 percent of their water resources, a threshold sometimes used to indicate impending water scarcity. Yet the situation should not change

Box 4.3 Summary methodology of estimating water balances

The estimation of water balances for any year is based on five sets of data, namely four digital georeferenced data sets for precipitation (Leemans and Cramer, 1991), reference evapotranspiration (Fischer, van Velthuizen and Nachtergaele, 2000), soil moisture storage properties (FAO, 1998b), extents of areas under irrigation (Siebert and Döll, 2001) and irrigated areas for all major crops for 1997/99 and 2030 from this study. The computation of water balances is carried out by grid cells (each of 5 arc minutes, 9.3 km at the equator) and in monthly time steps. The results can be presented in statistical tables or digital maps at any level of spatial aggregation (country, river basin, etc.). They consist of annual values by grid cell for the actual evapotranspiration, water runoff and consumptive water use in irrigation.

For each grid cell, the actual evapotranspiration is assumed to be equal to the reference evapotranspiration (ET_0, in mm; location-specific and calculated with the Penman-Monteith method; Allen *et al.*, 1998, New, Hulme and Jones, 1999) in those periods of the year when precipitation exceeds reference evapotranspiration or when there is enough water stored in the soil to allow maximum evapotranspiration. In drier periods of the year, lack of water reduces actual evapotranspiration to an extent depending on the available soil moisture. Evapotranspiration in open water areas and wetlands is considered to be equal to reference evapotranspiration.

For each grid cell, runoff is calculated as that part of the precipitation that does not evaporate and cannot be stored in the soil. In other words, runoff is equal to the difference between precipitation and actual evaporation. Runoff is always positive, except for areas identified as open water or wetland, where actual evapotranspiration can exceed precipitation.

Consumptive use of water in irrigated agriculture is defined as the water required in addition to water from precipitation (soil moisture) for optimal plant growth during the growing season. Optimal plant growth occurs when actual evapotranspiration of a crop is equal to its potential evapotranspiration.

Potential evapotranspiration of irrigated agriculture is calculated by converting data or projections of irrigated (sown) area by crop (at the national level) into a cropping calendar with monthly occupation rates of the land equipped for irrigation.[1] The table below gives, as an example, the cropping calendar of Morocco for the base year 1997/99:[2]

Crop under irrigation	Irrigated area ('000 ha)	J	F	M	A	M	J	J	A	S	O	N	D
Wheat	592	47	47	47	47						47	47	47
Maize	156			12	12	12	12	12					
Potatoes	62					5	5	5	5	5			
Beet	34				3	3	3	3	3	3			
Cane	15	1	1	1	1	1	1	1	1	1	1	1	1
Vegetables	156					12	12	12	12	12			
Citrus	79	6	6	6	6	6	6	6	6	6	6	6	6
Fruit	88	7	7	7	7	7	7	7	7	7	7	7	7
Groundnuts	10						1	1	1	1	1		
Fodder	100	8	8							8	8	8	8
Sum over all crops[3]	1 305	70	69	74	77	49	49	49	36	44	70	70	70
Equipped for irrigation	1 258												
Total cropping intensity	104%												

[1] India and China have been subdivided into respectively four and three units for which different cropping calendars have been made to distinguish different climate zones in these countries.
[2] For example, wheat is grown from October to April and occupies 47 percent (592 thousand ha) of the 1 258 thousand ha equipped for irrigation.
[3] Including crops not shown above.

The (potential) evapotranspiration (ET_c in mm) of a crop under irrigation is obtained by multiplying the reference evapotranspiration with a crop-specific coefficient ($ET_c = K_c * ET_0$). This coefficient has been derived (according to FAO, 1998b) for four different growing stages: the initial phase (just after sowing), the development phase, the mid-phase and the late phase (when the crop is ripening before harvesting). In general, these coefficients are low during the initial phase, high during the mid-phase and again lower in the late phase. It is assumed that the initial, the development and the late phase all take one month for each crop, while the mid-phase lasts a number of months. For example, the growing season for wheat in Morocco starts in October and ends in April, as follows: initial phase: October (K_c = 0.4); development phase: November (K_c = 0.8); mid-phase: December – March (K_c = 1.15); and late phase: April (K_c = 0.3).

Multiplying for each grid cell its surface equipped for irrigation with the sum over all crops of their evapotranspiration and with the cropping intensity per month results in the potential evapotranspiration of the irrigated area in that grid cell. The difference between the calculated evapotranspiration of the irrigated area and actual evapotranspiration under non-irrigated conditions is equal to the consumptive use of water in irrigated agriculture in the grid cell.

The method has been calibrated by comparing calculated values for water resources per country (i.e. the difference between precipitation and actual evapotranspiration under non-irrigated conditions) with data on water resources for each country (as given in FAO 1995b, 1997b and 1999b). In addition, the discharge of major rivers as given in the literature was compared with the calculated runoff for the drainage basin of these rivers. If the calculated runoff values did not match the values as stated in the literature, correction factors were applied to one or more of the basic input data on precipitation, reference evapotranspiration, soil moisture storage and open waters.

Finally, the water balance for each country and year is defined as the difference between the sum of precipitation and incoming runoff on the one hand and the sum of actual evapotranspiration and consumptive use of water in irrigated agriculture in that year on the other. This is therefore the balance of water without accounting for water withdrawals for other needs (industry, household and environmental purposes).

drastically over the period of the study, with only two more countries crossing the threshold of 20 percent. If one adds the expected additional water withdrawals needed for non-agricultural use, the picture will not be much different since agriculture represents the bulk of water withdrawal.

Nevertheless, for several countries, relatively low national figures may give an overly optimistic impression of the level of water stress: China, for instance, is facing severe water shortages in the north while the south still has abundant water resources. Already by 1997/99, two countries (the Libyan Arab Jamahariya and Saudi Arabia) used volumes of water for irrigation larger than their annual renewable water resources. Groundwater mining also occurs in parts of several other countries of the Near East, South and East Asia, Central America and in the Caribbean, even if at the national level the water balance may still be positive. In a survey of irrigation and water resources in the Near East region (FAO, 1997c), it was estimated that the amount of water required to produce the net amount of food imported in the region in 1994 would be comparable to the total annual flow of the Nile river at Aswan.

In concluding this discussion on irrigation, for the 93 developing countries as a whole, irrigation currently represents a relatively small part of their total water resources and there remains a significant potential for further irrigation development. With the relatively small increase in irrigation water withdrawal expected between 1997/99 and 2030, this situation will not change much at the aggregate level. Locally and in some countries, however, there are already very severe water shortages, in particular in the Near East/North Africa region.

4.5 Land-yield combinations for major crops

As discussed in Section 4.2, it is expected that growth in crop yields will continue to be the mainstay of crop production growth, accounting for nearly 70 percent of the latter in developing countries. Although the marked deceleration of crop production growth foreseen for the future (Table 4.1) points to a similar deceleration in growth of yields, such growth will continue to be needed. Questions often asked are: will yield increases continue to be possible? and what is the potential for a continuation of such growth?

There is a realization that the chances of a new green revolution or of one-off quantum jumps in yields are now rather limited. There is even a belief that for some major crops, yield ceilings have been, or are rapidly being reached. At the same time, empirical evidence has shown that the cumulative gains in yields over time resulting from slower, evolutionary annual increments in yields have been far more important than quantum jumps in yields, for all major crops (see Byerlee, 1996).

In the following sections, the land-yield combinations underlying the production projections for major crops will first be discussed. Subsequently some educated guesses will be made about the potential for raising yields and for narrowing existing yield gaps.

4.5.1 Harvested land and yields for major crops

As explained in Section 4.3.2, for the developing countries the production projections for the 34 crops of this study[8] are unfolded into and tested against what FAO experts think are "feasible" land-yield combinations by agro-ecological rainfed and irrigated environment, taking into account whatever knowledge is available. Major inputs into this evaluation are the estimates regarding the availability of land suitable for growing crops in each country and each agro-ecological environment, which come from the FAO agro-ecological zones work (see Section 4.3.1). In practice they are introduced as constraints to land expansion but they also act as a guide to what can be grown where. It is emphasized that the resulting land and yield projections, although they take into account past performance, are not mere extrapolations of historical trends since they take into account all present knowledge about changes expected in the future. Box 4.4 shows an example of the results, tracked against actual outcomes.

The findings of the present study indicate that in developing countries, as in the past but even more so in the future, the mainstay of production

Table 4.11 Area and yields for the ten major crops in developing countries

	Production (million tonnes)			Harvested area (million ha)				Yield (tonnes/ha)			
	1961 /63	1997 /99	2030	1961 /63	1997 /99	1997/99 adj.*	2030	1961 /63	1997 /99	1997/99 adj.*	2030
Rice (paddy)	206	560	775	113	148	157	164	1.82	3.77	3.57	4.73
Wheat	64	280	418	74	104	111	118	0.87	2.70	2.53	3.53
Maize	69	268	539	59	92	96	136	1.16	2.92	2.78	3.96
Pulses	32	40	62	52	60	60	57	0.61	0.66	0.67	1.09
Soybeans	8	75	188	12	39	41	72	0.68	1.93	1.84	2.63
Sorghum	30	44	74	41	39	40	45	0.72	1.13	1.11	1.66
Millet	22	26	42	39	35	36	38	0.57	0.76	0.73	1.12
Seed cotton	15	35	66	23	25	26	31	0.67	1.44	1.35	2.17
Groundnuts	14	30	65	16	22	23	39	0.83	1.34	1.28	1.69
Sugar cane	374	1 157	1 936	8	18	19	22	46.14	63.87	61.84	88.08
Cereals	419	1 210	1 901	358	440	464	528	1.17	2.75	2.61	3.60
All 34 crops				580	801	848	1 021				

Notes: * 1997/99 adj. For a number of countries for which the data were unreliable, base year data for harvested land and yields were adjusted. Ten crops selected and ordered according to harvested land use in 1997/99, excluding fruit (31 million ha) and vegetables (29 million ha). "Cereals" includes other cereals not shown here.

[8] For the analysis of production, the commodities sugar and vegetable oil are unfolded into their constituent crops (sugar cane, beet, soybeans, sunflower, groundnuts, rapeseed, oil palm, coconuts, sesame seed, etc.), so that land-yield combinations are generated for 34 crops.

increases will be the intensification of agriculture in the form of higher yields and more multiple cropping and reduced fallow periods. This situation will apply particularly in the countries with appropriate agro-ecological environments and with little or no potential of bringing new land into cultivation. The overall result for yields of all the crops covered in this study (aggregated with standard price weights) is roughly a halving of the average annual rate of growth over the projection period as compared to the historical period: 1.0 percent p.a. during 1997/99 to 2030 against 2.1 percent p.a. during 1961-99. This slowdown in the yield growth is a gradual process which has been under way for some time and is expected to continue in the future. It reflects the deceleration in crop production growth explained earlier.

Discussing yield growth at this level of aggregation however is not very helpful, but the overall slowdown is a pattern common to most crops covered in this study with only a few exceptions such as pulses, citrus and sesame. These are crops for which a strong demand is foreseen in the future or which are grown in land-scarce environments. The growth in soybean area and produc-tion in developing countries has been remarkable, mainly as a result of explosive growth in Brazil and, more recently, in India (Table 4.11). Soybean is expected to continue to be one of the most dynamic crops, albeit with its production increasing at a more moderate rate than in the past, bringing by 2030 the developing countries' share in world soybean production to 58 percent, with Brazil, China and India accounting for three-quarters of their total.

For cereals, which occupy 58 percent of the world's harvested area and 55 percent in developing countries (Table 4.11), the slowdown in yield growth would be particularly pronounced: down from 2.1 to 0.9 percent p.a. at the world level and from 2.5 to 1.0 percent p.a. in developing countries (Table 4.12). Again this slowdown has been under way for quite some time. The differences of sources of growth and some regional aspects of the various cereal crops have been discussed in Section 4.2. Suffice it here to note that irrigated land is expected to play a much more important role in increasing maize production, almost entirely because of China which accounts for 45 percent of the developing countries' maize production and

Table 4.12 Cereal yields in developing countries, rainfed and irrigated

		Share in production		Average (weighted) yield				Annual growth			Annual growth excluding China		
		%		tonnes/ha				% p.a.			% p.a.		
		1997/99	2030	1961/63	1997/99	1997/99 adj.	2030	1961-99	1989-99	1997/99-2030	1961-99	1989-99	1997/99-2030
Wheat	total			0.87	2.70	2.53	3.55	3.3	2.0	1.1	2.6	1.7	1.2
	rainfed	35	25			1.86	2.26			0.6			0.8
	irrigated	65	75			3.11	4.44			1.1			1.2
Rice (paddy)	total			1.82	3.77	3.57	4.73	2.1	1.1	0.9	2.0	1.2	1.1
	rainfed	24	21			2.20	2.82			0.8			0.8
	irrigated	76	79			4.45	5.78			0.8			1.0
Maize	total			1.16	2.92	2.78	3.96	2.6	2.6	1.1	1.8	2.5	1.2
	rainfed	68	51			2.34	2.99			0.8			1.2
	irrigated	32	49			4.52	5.96			0.9			0.8
All cereals	total			1.17	2.75	2.61	3.60	2.5	1.7	1.0	2.0	1.7	1.1
	rainfed	41	36			1.76	2.29			0.8			1.0
	irrigated	59	64			3.93	5.30			0.9			1.1

Note: Historical data are from FAOSTAT; base year data for China have been adjusted.

Box 4.4 Cereal yields and production: actual and as projected in the 1995 study

Since, contrary to the practice in most other projection studies, the projections presented here are not based on formal analytical methods, it may be of interest to see how well the projections of the preceding study (Alexandratos, 1995), which were based on a similar approach, tracked actual outcomes to date. The base year of the preceding study was the three-year average 1988/90 and the final projection year 2010. The detailed projections for the land-yield combinations for cereals in the 90 developing country sample, excluding China,[1] which was not covered in detail in the 1995 study, was as follows. The average yield of cereals was projected to grow by 1.5 percent p.a., from 1.9 tonnes/ha in 1988/90 to 2.6 tonnes/ha in 2010 (see table below), compared with 2.2 percent p.a. in the preceding 20 years. Ten years into the projection period, both the actual average cereal yield and cereal production in 1997/99 were close to the projected values.

	Base year: average 88/90	Projected 2010	Projected Interpolated average 97/99	Actual outcome: average 97/99
Yields (excl. China)	kg/ha	kg/ha	kg/ha	kg/ha
Wheat	1 900	2 700	2 209	2 220
Rice (paddy)	2 800	3 800	3 192	3 080
Maize	1 800	2 500	2 072	2 190
All cereals	1 900	2 600	2 173	2 184
Production (incl. China*)	million tonnes	million tonnes	million tonnes	million tonnes
Wheat	225	348	271	280
Rice (milled)	321	461	375	375
Maize	199	358	256	269
Other cereals	102	151	121	103
All cereals	847	1 318	1 023	1 027

Source: Base year data and 2010 projections from Alexandratos (1995, p. 145,169).

* China's production was projected directly, not in terms of areas and yields.

[1] Problems with the land and yield data of China (Alexandratos, 1996) made it necessary to project the country's production directly, not in terms of land-yield combinations as was done for the other developing countries. The resulting projection of China's production of cereals implied a growth rate of 2.0 percent p.a. from 1988/90 to 2010. The actual outcome to 1999 has been 2.2 percent p.a.

where irrigated land allocated to maize could more than double. Part of the continued, if slowing, growth in yields is a result of a rising share of irrigated production, with normally much higher cereal yields, in total production. This fact alone would lead to yield increases even if rainfed and irrigated cereal yields did not grow at all.[9]

It is often asserted (see, for example, Borlaug, 1999) that thanks to increases in yield, land has been saved with diminished pressure on the environment as a result, such as less deforestation than otherwise would have taken place. To take cereals as an example, the reasoning is as follows. If the average global cereal yield had not grown since 1961/63 when it was 1 405 kg/ha, 1 483 million ha would have been needed to grow the 2 084 million tonnes of cereals produced in the world in 1997/99. This amount was actually obtained on an area of only 683 million ha at an average yield of 3 050 kg/ha. Therefore, 800 million ha (1 483 minus 683) have been saved because of yield increases for cereals alone. This conclusion should be qualified, however; had there been no yield growth, the most probable outcome would have been much lower production because of lower demand resulting from higher prices of cereals, and

[9] This is seen most clearly for rice in developing countries, excluding China (Table 4.12) where the growth in the overall average yield of rice exceeds that of rainfed and irrigated rice. This is because the rainfed rice area is projected to remain about the same but the irrigated area is projected to increase by about one-third.

somewhat more land under cereals. Furthermore, in many countries the alternative of land expansion instead of yield increases does not exist in practice.

4.5.2 Yield gaps

Despite the increases in land under cultivation in the land-abundant countries, much of agricultural production growth has been based on the growth of yields, and will increasingly need to do so. What is the potential for a continuation of yield growth? In countries and localities where the potential of existing technology is being exploited fully, subject to the agro-ecological constraints specific to each locality, further growth, or even maintenance, of current yield levels will depend crucially on further progress in agricultural research. In places where yields are already near the ceilings obtained on research stations, the scope for raising yields is widely believed to be much more limited than

in the past (see, for example, Sinclair, 1998). However, this has been true for some time now, but average yields have continued to increase, albeit at a decelerating rate. For example, wheat yields in South Asia, which accounts for about a third of the developing countries' area under wheat, increased by 45 kg p.a. in the 1960s, 35 kg in the 1970s, 55 kg in the 1980s and 45 kg in 1990-99. Yields are projected to grow by 41 kg per year over 1997/99 to 2030.

Intercountry differences in yields remain very wide, however. This can be illustrated for wheat and rice in the developing countries. Current yields in the 10 percent of countries with the lowest yields (excluding countries with less than 50 000 ha under the crop), is less than one-fifth of the yields of the best performers comprising the top decile (Table 4.13). If subnational data were available, a similar pattern would probably be seen for intranational differences as well. For wheat this gap

Table 4.13   Average wheat and rice yields for selected country groups

	1961/63		1997/99		2030	
	tonnes/ ha	as % of top decile	tonnes/ ha	as %of top decile	tonnes/ ha	as % of top decile
Wheat						
No. of developing countries included	32		32		33	
Top decile	2.15	100	5.31	100	7.44	100
Bottom decile	0.40	18	0.80	15	1.25	17
Decile of largest producers (by area)	0.87	40	2.60	49	3.89	52
All countries included	0.97	45	2.15	41	3.11	42
Major developed country exporters	1.59		3.19		4.13	
World	1.23		2.55		3.47	
Rice (paddy)						
No. of developing countries included	44		52		55	
Top decile	4.51	100	6.57	100	7.93	100
Bottom decile	0.72	16	1.14	17	2.12	27
Decile of largest producers (by area)	1.82	40	3.51	53	4.84	61
All countries included	1.88	42	3.17	48	4.30	54
World	2.07		3.43		4.52	

Notes: Only countries with over 50 000 harvested ha are included. Countries included in the deciles are not necessarily the same for all years. Average yields are simple averages, not weighted by area.

between worst and best performers is projected to persist until 2030, while for rice the gap between the top and bottom deciles may be somewhat narrowed by 2030, with yields in the bottom decile reaching 27 percent of yields in the top decile. This may reflect the fact that the scope for raising yields of top rice performers is more limited than in the past. However, countries included in the bottom and top deciles account for only a minor share of the total production of wheat and rice. Therefore it is more important to examine what will happen to the yield levels obtained by the countries which account for the bulk of wheat and rice production. Current unweighted average yields of the largest producers,[10] are about half the yields achieved by the top performers (Table 4.13). In spite of continuing yield growth in these largest producing countries, this situation will remain essentially unchanged by 2030 for wheat, with rice yields reaching about 60 percent of the top performers' yields.

Based on this analysis, a *prima facie* case could be made that there has been and still is, considerable slack in the agricultural sectors of the different countries. This slack could be exploited if economic incentives so dictated. However, the fact that yield differences among the major cereal producing countries are very wide does not necessarily imply that the lagging countries have scope for yield increases equal to intercountry yield gaps. Part of these differences may simply reflect differing agro-ecological conditions. For example, the low average yields in Mexico of its basic food crop, maize (currently 2.4 tonnes/ha), are largely attributable to agro-ecological constraints that render it unsuited for widespread use of the major yield-increasing technology, hybrid seeds, a technology which underlies the average 8.3 tonnes/ha of the United States. Hybrids are at present used in Mexico on about 1.2 million ha, out of a total harvested area under maize of 7 million ha, while the area suitable for hybrid seed use is estimated to be about 3 million ha (see Commission for Environmental Cooperation, 1999, p.137-138).

However, not all, or perhaps not even the major part, of yield differences can be ascribed to such conditions. Wide yield differences are present even among countries with fairly similar agro-ecological

environments. In such cases, differences in the socio-economic and policy environments probably play a major role. The literature on yield gaps (see, for example, Duwayri, Tran and Nguyen, 1999) distinguishes two components of yield gaps: one due to agro-environmental and other non-transferable factors (these gaps cannot be narrowed); and another component due to differences in crop management practices such as suboptimal use of inputs and other cultural practices. This second component can be narrowed provided that it is economic to do so and therefore is termed the "exploitable yield gap". Duwayri, Tran and Nguyen (1999) state that the theoretical maximum yields for both wheat and rice are probably in the order of 20 tonnes/ha. On experimental stations, yields of 17 tonnes/ha have been reached in subtropical climates and of 10 tonnes/ha in the tropics. FAO (1999c) reports that concerted efforts in Australia to reduce the exploitable yield gap increased rice yields from 6.8 tonnes/ha in 1985/89 to 8.4 tonnes/ha in 1995/99, with many individual farmers obtaining 10 to 12 tonnes/ha.

In order to draw conclusions on the scope for narrowing the yield gap, one needs to separate its "non-transferable" part from the "exploitable" part. One way to do so is to compare yields obtained from the same crop varieties grown on different locations of land that are fairly homogeneous with respect to their physical characteristics (climate, soil and terrain), which would eliminate the "non-transferable" part in the comparison. One can go some way in that direction by examining the data on the suitability of land in the different countries for producing any given crop under specified technology packages. The required data comes from the GAEZ analysis discussed in Section 4.3.1. These data make it possible to derive a "national maximum obtainable yield" by weighting the yield obtainable in each of the suitability classes with the estimated land area in each suitability class. The derived national obtainable yield can then be compared with data on the actual national average yields. This comparison is somewhat distorted since the GAEZ analysis deals only with rainfed agriculture, while the national statistics include irrigated agriculture as

---

[10] Top 10 percent of countries ranked according to area allocated to the crop examined: China, India and Turkey for wheat; and India, China, Indonesia, Bangladesh and Thailand for rice.

well. However, the findings seem to confirm the hypothesis that a good part of the yield gap is of the second, exploitable type. For a further discussion on this topic, see Section 11.1 in Chapter 11.

4.6 Input use

4.6.1 Fertilizer consumption

As discussed in Section 4.2, the bulk of the projected increases in crop production will have to come from higher yields, with the remaining part coming from an expansion in harvested area. Both higher yields, which normally demand higher fertilizer application rates, and land expansion will lead to an increase in fertilizer use. Increases in biomass require additional uptake of nutrients which may come from both organic and mineral

sources. Unfortunately, for most crops there are not enough data to estimate the relation between mineral fertilizer consumption and biomass increases. The historical relationship between cereal production and mineral fertilizer consumption is better known. One-third of the increase in cereal production worldwide and half of the increase in India's grain production during the 1970s and 1980s have been attributed to increased fertilizer consumption. The application of mineral fertilizers needed to obtain higher yields should complement nutrients available from other sources and match the needs of individual crop varieties.

Increased use of fertilizer is becoming even more crucial in view of other factors, such as the impact on soil fertility of more intensive cultivation practices and the shortening of fallow periods. There is empirical evidence that nutrient budgets[11]

Table 4.14   Fertilizer consumption by major crops

	1997/99 Share (%) in total	1997/99	2015	2030	1997/99-2030 % p.a.
		Nutrients, million tonnes			
Wheat	18.4	25.3	30.4	34.9	1.0
Rice	17.3	23.8	26.5	28.1	0.5
Maize	16.3	22.5	29.0	34.5	1.3
Fodder	6.2	8.5	9.3	10.0	0.5
Seed cotton	3.5	4.9	6.2	7.1	1.2
Soybeans	3.4	4.6	7.6	11.5	2.9
Vegetables	3.3	4.6	5.3	6.1	0.9
Sugar cane	3.2	4.4	5.5	6.6	1.3
Fruit	2.9	4.1	4.3	7.5	1.9
Barley	2.9	4.0	4.4	4.8	0.6
Other cereals	2.9	3.9	9.2	8.3	2.3
Potato	2.0	2.7	3.3	3.8	1.1
Rapeseed	1.5	2.1	3.5	5.1	2.8
Sweet potato	1.3	1.8	2.0	2.1	0.5
Sugar beet	1.0	1.4	1.6	1.7	0.6
All cereals		79.5	99.5	110.6	1.0
% of total	57.7	57.7	64.8	58.8	
All crops above		118.5	148.2	172.1	1.2
% of total	86.0	86.0	89.8	91.5	
World total		137.7	165.1	188.0	1.0

Notes: Crops with a 1997/99 share of at least 1 percent, ordered according to their 1997/99 share in fertilizer use.

[11] A nutrient budget is defined as the balance of nutrient inputs such as mineral fertilizers, manure, deposition, biological nitrogen fixation and sedimentation, and nutrient outputs (crops harvested, crop residues, leaching, gaseous losses and erosion).

Table 4.15 Fertilizer consumption: past and projected

	1961 /63	1979 /81	1997 /99	2015	2030	1961 -1999	1989- 1999	1997/99 -2030
Total			Nutrients, million tonnes				% p.a.	
Sub-Saharan Africa	0.2	0.9	1.1	1.8	2.6	5.3	-1.8	2.7
Latin America and the Caribbean	1.1	6.8	11.3	13.1	16.3	6.1	4.4	1.2
Near East/ North Africa	0.5	3.5	6.1	7.5	9.1	7.3	0.8	1.3
South Asia	0.6	7.3	21.3	24.1	28.9	9.6	4.5	1.0
excl. India	0.2	1.6	4.2	5.4	6.9	9.2	4.6	1.5
East Asia	1.7	18.2	45.0	56.9	63.0	9.3	3.8	1.1
excl. China	0.9	4.1	9.4	13.8	10.3	7.0	3.2	0.3
All above	4.1	36.7	84.8	103.5	119.9	8.5	3.7	1.1
excl. China	3.3	22.6	49.2	60.4	67.3	7.6	3.5	1.0
excl. China and India	2.9	16.9	32.1	41.6	45.3	6.9	3.1	1.1
Industrial countries	24.3	49.1	45.2	52.3	58.0	1.4	0.1	0.8
Transition countries	5.6	28.4	7.6	9.3	10.1	0.7	-14.9	0.9
World	34.1	114.2	137.7	165.1	188.0	3.6	0.2	1.0
Per hectare			kg/ha (arable land)				% p.a.	
Sub-Saharan Africa	1	7	5	7	9	4.5	-2.4	1.9
Latin America and the Caribbean	11	50	56	59	67	6.0	0.0	0.6
Near East/North Africa	6	38	71	84	99	5.7	3.9	1.0
South Asia	6	36	103	115	134	9.5	4.5	0.8
excl. India	6	48	113	142	178	8.8	4.3	1.4
East Asia	10	100	194	244	266	8.3	3.6	1.0
excl. China	12	50	96	131	92	6.1	3.3	-0.1
All above	6	49	89	102	111	7.7	3.3	0.7
excl. China	6	35	60	68	71	6.9	3.2	0.5
excl. China and India	7	35	49	58	58	6.0	2.6	0.5
Industrial countries	64	124	117			1.3	0.3	
Transition countries	19	101	29			0.9	-14.4	
World	25	80	92			3.3	0.1	

Note: Kg/ha for 1997/99 are for developing countries calculated on the basis of "adjusted" arable land data. For industrial and transition countries no projections of arable land were made.

change over time and that higher yields can be achieved through reduction of nutrient losses within cropping systems. That is, increases in food production can be obtained with a less than proportional increase in fertilizer nutrient use. Frink, Waggoner and Ausubel (1998) showed this situation for maize in North America. Farmers achieve such increased nutrient use efficiency by adopting improved and more precise management practices. Socolow (1998) suggests that management techniques such as precision agriculture offer abundant opportunities to substitute information for fertilizer. It is expected that this trend of increasing efficiency of nutrient use through better nutrient management, by improving the efficiency of nutrient balances and the timing and placement of fertilizers, will continue and accelerate in the future.

Projections for fertilizer consumption have been derived on the basis of the relationship between yields and fertilizer application rates that existed during 1995/97. Data on fertilizer use by crop and fertilizer application rates (kg of fertilizer per ha) are available for all major countries and crops, accounting for 97 percent of global fertilizer use in 1995/97 (FAO/IFA/IFDC, 1999 and Harris, 1997). This relationship is estimated on a cross-section basis for the crops for which data are available and is assumed to hold also over time as yields increase (see Daberkov *et al.*, 1999). It provides a basis for estimating future fertilizer application rates required to obtain the projected increase in yields for most of the crops covered in this study. It implicitly assumes that improvements in nutrient use efficiency will continue to occur as embodied in the relationship between yields and fertilizer application rates (fertilizer response coefficients) estimated for 1995/97. For some crop categories such as citrus, vegetables, fruit and "other cereals", fertilizer consumption growth is assumed to be equal to the growth in crop production: i.e. for these crops, the base year input-output relationship between fertilizer use and crop production is assumed to remain constant over the projection period. To account fully for all fertilizer consumption, including its use for crops not covered in this study, fertilizer applications on fodder crops were assumed to grow at the same rate as projected growth for livestock (meat and milk) production, and fertilizer applications on "other crops" is at the average rate for all crops covered in the study.

The overall result, aggregated over all crops, is that fertilizer consumption will increase by 1.0 percent p.a., rising from 138 million tonnes in 1997/99 to 188 million tonnes in 2030 (Table 4.14). This is much slower than in the past for the reasons explained below. Wheat, rice and maize, which together at present account for over half of global fertilizer use, will continue to do so, at least until 2030. By 2015 maize will rival wheat as the top fertilizer user because of the projected increase in maize demand for feeding purposes in developing countries (see Chapter 3). Fertilizer applications to oilseeds (soybeans and rapeseed) are expected to grow fastest.

North America, western Europe, East and South Asia accounted for over 80 percent of all fertilizer use in 1997/99. Growth in fertilizer use in the industrial countries, especially in western Europe, is expected to lag significantly behind growth in other regions of the world (Table 4.15). The maturing of fertilizer markets during the 1980s in North America and western Europe, two of the major fertilizer consuming regions of the world, account for much of the projected slowdown in fertilizer consumption growth. In the more recent past, changes in agricultural policies, in particular reductions in support measures, contributed to a slowdown or even decline in fertilizer use in this group of countries. Increasing awareness of and concern about the environmental impacts of fertilizer use are also likely to hold back future growth in fertilizer use (see Chapter 12).

Over the past few decades, the use of mineral fertilizers has been growing rapidly in developing countries starting, of course, from a low base (Table 4.15). This has been particularly so in East and South Asia following the introduction of high-yielding varieties. East Asia (mainly China) is likely to continue to dwarf the fertilizer consumption of the other developing regions. For sub-Saharan Africa, above average growth rates are foreseen, starting from a very low base, but fertilizer consumption per hectare is expected to remain at a relatively low level. The latter probably reflects large areas with no fertilizer use at all, combined with small areas of commercial farming with high levels of fertilizer use, and could be seen as a sign of nutrient mining (see also Henao and Baanante, 1999).

Average fertilizer productivity, as measured by kg of product obtained per kg of nutrient, shows considerable variation across countries. This reflects a host of factors such as differences in agro-ecological resources (soil, terrain and climate), in management practices and skills and in economic incentives. Fertilizer productivity is also strongly related to soil moisture availability. For example, irrigated wheat production in Zimbabwe and Saudi Arabia shows a ratio of 40 kg wheat per kg fertilizer nutrient at yield levels of 4.5 tonnes/ha. Similar yields in Norway and the Czech Republic require twice as large fertilizer application rates, reflecting a considerably different agro-ecological resource base. Furthermore, a high yield/fertilizer ratio may also indicate that fertilizer use is not widespread among farmers (e.g. wheat in Russia, Ethiopia and Algeria), or that high yields are obtained with nutrients other than mineral fertilizer (e.g. manure is estimated to provide almost half of all external nutrient inputs in the EU). Notwithstanding this variability, in many cases the scope for raising fertilizer productivity is substantial. The degree to which such productivity gains will be pursued depends to a great extent on economic incentives.

The projected slowdown in the growth of fertilizer consumption is due mainly to the expected slowdown in crop production growth (Table 4.1). The reasons for this have been explained in Chapter 3. Again, this is not a sudden change but a gradual process already under way for some time, as illustrated by the annual growth rates for the last ten years (1989-99) shown in Table 4.15. In some cases it would even represent a "recovery" as compared with recent developments. As mentioned, fertilizer is most productive in the absence of moisture constraints, i.e. when applied to irrigated crops. For this reason, the expected slowdown in irrigation expansion (Section 4.4.1) will also slow the growth of fertilizer consumption. The continuing trend to increase fertilizer use efficiency, partly driven by new techniques such as biotechnology and precision agriculture, will also reduce mineral fertilizer needs per unit of crop output. There is an increasing concern about the negative environmental impact of high rates of mineral fertilizer use. Finally there is the spread of organic agriculture, and the increasing availability of non-mineral nutrient sources such as manure; recycled human, industrial and agricultural waste; and crop by-products. All these factors will tend to reduce growth in fertilizer consumption.

4.6.2 Farm power

Human labour, draught animals and engine-driven machinery are an integral part of the agricultural production process. They provide the motive power for land clearance and preparation, for planting, fertilizing, weeding and irrigation, and for harvesting, transport and processing. This section focuses on the use of power for primary tillage. Land preparation represents one of the most significant uses of power. Since land preparation is power intensive (as opposed to control intensive), it is usually one of the first operations to benefit from mechanization (Rijk, 1989). Hence any change in the use of different power sources for land cultivation may act as an indicator for similar changes in other parts of the production process.

Regional estimates of the relative contributions of different power sources to land cultivation have been developed from estimates initially generated at the country level. On the basis of existing data and expert opinion, individual countries were classified into one of six farm power categories according to the proportion of area cultivated by different power sources, at present and projected to 2030. The categories range from those where hand power predominates, through those where draught animals are the main source of power, to those where most land is cultivated by tractors. The figures were subsequently aggregated to estimate the harvested area cultivated by different power sources for each region (see Box 4.5 for details of the methodology).

Overall results. It was estimated that in 1997/99, in developing countries as a whole, the proportion of land cultivated by each of the three power sources was broadly similar. Of the total harvested area in developing countries (excluding China), 35 percent was prepared by hand, 30 percent by draught animals and 35 percent by tractors (Table 4.16). By 2030, 55 percent of the harvested area is expected to be tilled by tractors. Hand power will account for approximately 25 percent of the harvested area and draught animal power (DAP) for approximately 20 percent.

Box 4.5 Methodology to estimate farm power category

Individual countries were classified by expert opinion into one of six farm power categories according to the proportion of area cultivated by different power sources. The six categories identified are given in the table below. The percentages of the area cultivated by different power sources are indicative only and refer to harvested land, which represents the actual area cultivated in any year, taking into account multiple cropping and short-term fallow. Upper and lower limits were set for the area cultivated by each power source (bottom row in the table).

Farm power category at country level		Percentage of area cultivated by each power source		
		Hand	Draught animals	Tractors
A =	humans are the predominant source of power, with modest contributions from draught animals and tractors	>80	< 20	<5
B =	significant use is made of draught animals, although humans are still the most important power source	45-80	20-40	<20
C =	draught animals are the principal power source	15-45	>40	<20
D =	significant use is made of motorized power, including both two-wheel and four-wheel tractors	20-50	15-30	20-50
E =	tractors are the dominant power source	<5	<25	50-80
F =	fully motorized	<10	<10	>80
Minimum and maximum percentage		5-90	5-70	2-90

Where possible, country classifications were verified against existing data. Sources included a number of country farm power assessment studies commissioned by the Agricultural Engineering Branch of FAO and published reports. The classifications from various sources proved to be fairly consistent. The categories were converted into the physical area cultivated by each power source by multiplying the estimated percentage figure for each power source with the data for harvested area for the base year 1997/99 and the projected area for 2030. The country figures were subsequently aggregated to estimate the harvested area cultivated by different power sources for subregions and regions.

This approach has several advantages. First, it highlights the role of humans as a source of power in those parts of the world where they are responsible for much of the land preparation. This is essential for understanding concerns arising from future projections about the size and composition of the agricultural labour force, particularly in countries where a sizeable share of the population is expected to be affected by HIV/AIDS within the next 20 years. The significance of humans as a power source can be easily overlooked if their contribution is expressed solely as a percentage of the total power input, rather than the area they cultivate. Second, when projecting future combinations of farm power inputs at the country level, account can be taken, albeit by way of expert judgement, of the developments in the overall economy and in the agricultural sector, competing claims on resource use, and opportunities for substitution between power sources. Third, this approach is independent of estimates of inventories of draught animals and tractors, data for which are often unreliable or not readily available. The number of tractors and draught animals working in agriculture may vary considerably from published data. These numbers depend on several unknown variables, such as the working life, the proportion working in agriculture as opposed to off-farm activities, and the proportion in an operational state. Finally, the process of converting different power sources into a common power equivalent is a process fraught with difficulties and inaccuracies.

Nevertheless, there are also several limitations associated with the methodology. First, only one farm power category is selected to represent the power use for land cultivation within an entire country. This overlooks the diversity that exists inside many countries, particularly when the use of a specific power source is highly influenced by soil and terrain constraints, by cropping patterns, or is highly differentiated between commercial/estate and smallholder sectors. A second limitation is the use of a single average percentage figure for each power source to convert categories to harvested area, rather than actual percentages.

Table 4.16 Proportion of area cultivated by different power sources, 1997/99 and 2030

Region		Percentage of area cultivated by different power sources		
		Hand	Draught animal	Tractor
All developing countries	1997/99	35	30	35
	2030	25	20	55
Sub-Saharan Africa	1997/99	65	25	10
	2030	45	30	25
Near East/North Africa	1997/99	20	20	60
	2030	10	15	75
Latin America and the Caribbean	1997/99	25	25	50
	2030	15	15	70
South Asia	1997/99	30	35	35
	2030	15	15	70
East Asia	1997/99	40	40	20
	2030	25	25	50

Notes:Figures have been rounded to the nearest 5 percent. China has been excluded from the analysis because its size and diversity made it impossible to estimate a single farm power category for the country.

There are marked regional differences in the relative contributions of the power sources, both at present and in the future. Tractors are already a significant source of power in the Near East/North Africa region and in Latin America and the Caribbean: approximately half of the harvested area is currently prepared by tractor in these regions. This is expected to rise to at least 70 percent of harvested area by 2030. Draught animals are at present relatively important sources of power in the rice and mixed farming systems of South and East Asia, accounting for over one-third of harvested area. However, the shift to motorized power by 2030 will be substantial. The area cultivated by tractors will rise in South Asia from 35 percent of harvested area to 70 percent, and in East Asia (excluding China) from 20 percent to over 50 percent. This increase in area cultivated by tractor arises from two factors: an increase in total harvested area at the country level, combined with a reduction in the area cultivated by humans and draught animals as a result of substitution between power sources.

In contrast, humans are and will continue to be the main power source in sub-Saharan Africa. Almost two-thirds of the harvested area is prepared by hand at present and although this will fall to 50 percent by 2030, the physical area involved will remain broadly constant. The area cultivated by draught animals and tractors is expected to increase (both in physical area and proportional terms) but they will not offset the dominance of hand power.

When countries are classified by farm power category, some common characteristics can be observed:

■ *Countries in which humans are the predominant power source.* The agricultural sector typically employs two-thirds of the workforce and generates over one-third of the GDP. During the 1990s many of these economies were almost static with annual average growth rates in total GDP of less than 1 percent and income per head below US$500.

■ *Countries in which draught animals are a significant or predominant source of power.* The principal difference between the hand power and draught animal power countries is in terms of land use. The intensity of cultivation on both rainfed and irrigated land is higher and more area is under irrigation. Indeed, in all regions the highest cropping intensities occur in DAP countries. This suggests there are no labour displacement effects associated with the use of draught animals.

■ *Countries in which tractors are a significant or predominant source of power.* High levels of tractorization are generally associated with relatively well-developed economies and the production of cash crops. Non-agricultural revenues can facilitate their adoption. The increased use of tractors is consistently associated with expanding the area under irrigation, but cultivating it less intensively than in countries using hand or animal power. In these countries agriculture is no longer the dominant sector, employing less than half of the workforce and generating less than one quarter of GDP. Their economies are more buoyant and incomes per capita are at least three times as high as those in either hand power or DAP countries in the same region. The rate of growth in the agricultural workforce is small and in some countries the absolute number of people working in agriculture has started to fall. This is often considered to be one of the more significant turning points in the process of economic development.

Forces changing the composition of farm power inputs. The stimulus to change the composition of farm power inputs will come from either changes in the demand for farm power or from supply-side changes, or both. Any increase in total agricultural output (be it from area expansion, an increase in cropping intensity or an increase in yield) requires additional power, if not for technology application then for handling and processing increased volumes. Similarly, land improvements (such as terracing, drainage or irrigation structures), soil conservation and water harvesting techniques frequently place additional demands on the power resource.

In response, farmers can either increase their inputs of farm power or increase the productivity of existing inputs through the use of improved tools and equipment. Alternatively, adopting different practices or changing cropping patterns may reduce power requirements. For example, the use of no-till and direct seeding practices eliminates the need for conventional land preparation and tillage; broadcasting rice overcomes the labour-intensive activity of transplanting seedlings, and the use of draught animal power, benevolent herbicides or no-tillage with continuous soil cover (see Chapter 11) can overcome labour bottlenecks associated with weeding.

Motivations to mechanize may also arise from supply-side changes in the availability and productivity of farm power inputs, as well as a wish to reduce the drudgery of farm work. The health, nutritional status and age of the workforce affect the productivity of labour. The availability of household members for farm work is influenced by other claims on their time, such as household tasks, schooling and opportunities for off-farm work. The household composition also changes through rural-urban migration or the death of key household members. The productivity of draught animals is affected by their health and nutrition, the training of animals, operator skills and availability of appropriate implements. Productive and sustainable use of motorized inputs is dependent on operator skills, appropriate equipment and an infrastructure capable of providing timely and cost-effective access to repair and maintenance services.

The changing composition of farm power inputs will also have an impact on the division of agricultural tasks among household members. The range of tasks performed by different members of the household varies according to sex, age, culture, ethnicity and religion (see Box 4.6). It also varies according to the specific crop or livestock, sources of power input and equipment used.

Patterns of mechanization up to 2030. Most of the changes in farm power categories during the next 30 years are expected to occur in countries that already make significant use of tractors. By 2030, tractors will be the dominant source of power for land preparation in southern Africa, North Africa/Near East, South Asia, East Asia and Latin America and the Caribbean. Southeast Asia is also expected to shift from draught animals to making greater use of tractors. The reasons underlying these shifts are explained below.

In a few countries, it is expected that the present composition of farm power inputs is not sustainable. In eastern Africa, for example, the number of draught animals has been decimated in some areas through livestock disease and cattle rustling, thereby removing a principal power source from certain farming systems. The sustainability of tractor-based systems is highly dependent on the profitability of agriculture and an infrastructure capable of providing timely access to fuel and inputs for repairs and maintenance. The

Box 4.6 Gender roles and the feminization of agriculture

Women are key players in both cash and subsistence agriculture. Their daily workload is characterized by long hours, typically 12- to 14-hour days, with little seasonal variation. The use of their time tends to be fragmented, mixing farm work with household duties and other off-farm activities. The range of agricultural tasks performed by different household members varies according to sex, age, culture, ethnicity and religion. It also varies according to the specific crop or livestock activities, sources of power input and equipment used. Hence gender roles differ markedly, not only between regions and countries, but also within countries between neighbouring communities.

In sub-Saharan Africa, women contribute between 60 and 80 percent of the labour in food production. While there are significant variations in gender roles, women overall play a major role in planting, weeding, application of fertilizers and pesticides, harvesting, threshing, food processing, transporting and marketing. Men are largely responsible for clearing and preparing the land, and ploughing. They also participate along-side women in many of the other activities. In many countries men are responsible for large livestock and women for smaller animals, such as poultry, sheep and goats. Women are usually most active in collecting natural products, such as wild foods, fodder and fuelwood. Men are usually associated with the use of draught animals or tractors. However, with appropriate training and implements, women also prove to be very effective operators of mechanized inputs.

In Asia, women account for 35 to 60 percent of the agricultural labour force. Women and men often play complementary roles with a division of labour similar to that found in sub-Saharan Africa. In many Asian countries women are also very active with livestock, collecting fodder, preparing buffaloes for ploughing, feeding and cleaning other cattle, and milking. In Southeast Asia they play a major role in rice production, particularly in sowing, transplanting, harvesting and processing. Women also supply a significant amount of labour to tea, rubber and fruit plantations.

In the Near East, women contribute up to 50 percent of the agricultural workforce. They are mainly responsible for the more time-consuming and labour-intensive tasks that are carried out manually or with the use of simple tools. In Latin America and the Caribbean, the rural population has been decreasing in recent decades. Women are mainly engaged in subsistence farming, particularly horticulture, poultry and raising small livestock for home consumption.

Gender roles are dynamic, responding to changing economic, social and cultural forces. The rural exodus in search of income-earning opportunities outside agriculture, usually dominated by men, has resulted in increasing numbers of female-headed households and the "feminization" of agriculture. Similar patterns arise from the death of male heads of household. Women are being left to carry out agricultural work on their own, changing the traditional pattern of farming and the division of tasks among household members. For example, women in female-headed households without recourse to adult male labour may clear and prepare land, including ploughing with oxen (tasks which traditionally would have been performed by men).

The feminization of agriculture, most pronounced in sub-Saharan Africa but also a growing phenomenon in other parts of the world, has significant implications for the development of agriculture. The needs and priorities of both rural women and men must be taken into account in any initiative to support and strengthen the sector.

Source: Based on FAO (1998a).

failure of government-based initiatives often results from introducing a level of mechanization that is inappropriate for the state of economic development and political stability. As a result, in the absence of further government interventions, it is expected that some countries will revert to increasing the use of hand or draught animal power during the next 30 years.

Persistence of hand and animal power in sub-Saharan Africa. Human labour is the most significant power source throughout sub-Saharan Africa.

The human contribution is most pronounced in Central and western Africa where it accounts for 85 and 70 percent of harvested area, respectively. These areas include the forest-based farming systems of Central Africa, characterized by shifting cultivation and the gathering of forest products, the root crop system stretching across West Africa, Central Africa and parts of East Africa, and the cash tree crop system in West Africa (FAO, 2001d). A relatively high proportion of rainfed land is under cultivation (45 percent of the potential area) but it is not used intensively as reflected in a relatively low cropping intensity of 60 percent. The presence of trees, stumps and shrubs makes it difficult to use ploughs without considerable investment of time and effort in land clearance (Boserup, 1965; Pingali, Bigot and Binswanger, 1987). Moreover, the incidence of tsetse fly (which breeds in tropical forests and forest margins) makes the area unsuitable for many types of draught animals. There is very little irrigated land.

Draught animals (predominantly work oxen) are concentrated on rainfed land in the cereal-based farming systems in the northern parts of West Africa, throughout the maize mixed systems of eastern Africa and the highland mixed systems of Ethiopia. Countries making significant use of tractors are scattered throughout the region.

Two-thirds of the countries in sub-Saharan Africa are not expected to change their power category by 2030. Although there will be some movement in the relative contributions of hand, draught animal and tractor power to land preparation, much of the region will continue to be cultivated using hand and animal power. All countries that are expected to change either from hand power to draught animal power, or from DAP to tractors, will experience lower population growth rates, higher incomes per head and higher income growth rates than those countries with the same farm power category at present that are not expected to change.

The process of urbanization in this region will provide some stimulus to switch power sources, as it not only draws labour away from the agricultural sector but also has implications for wage levels and the composition of the remaining labour force. Typically the young, able-bodied, educated and skilled migrate. The shift to urbanization is most pronounced in countries switching from DAP to tractors or already using tractors as the dominant power source. In these countries, a growth of almost 30 percent in the proportion of the population living in urban areas is expected by 2030, twice the rate of urbanization expected in countries not switching power source or shifting to DAP. Countries that will continue to use draught animals as a significant source of power will remain predominantly rural.

Another factor driving the process of change in eastern and southern Africa will be the impact of HIV/AIDS on the workforce (Box 4.7). Those countries that are expected to switch from hand power to DAP are projected to lose almost 20 percent of their agricultural labour to AIDS by 2020, that is, more than twice as much as those countries continuing to use hand power. Similarly, those shifting from DAP to tractors are expected to experience higher losses in their labour force (12 percent by 2020) than countries continuing with DAP. Some of the highest losses (16 percent by 2020) are projected for countries already making significant use of tractors. Thus the impact of HIV/AIDS will make it vital for many countries to change their source of farm power in order to overcome serious labour shortages at critical times of the farming year.

Increasing use of tractors in the Near East/North Africa and Asia. The development of regional markets and strong links with Europe are expected to be important engines of growth for North African countries. Oil wealth will continue to underpin development in the Near East. Economic development will be coupled with continued growth in non-agricultural employment and the migration of people from the land to urban areas. By 2030, over 75 percent of the population in the Near East/North Africa region will be living in urban areas. The option of using tractors becomes more viable with increasing costs of labour and increasing shortage of land for fodder production for draught animals.

Prospects for *mechanization in Asia* are based on projections of buoyant economies and high rates of growth in income per capita, but the process of urbanization would not appear to be so significant. More than half of the population in South Asia will continue to be based in rural areas by 2030. Single-axle tractors will be an increasingly important form of farm power in irrigated farming systems, which are suited to their use. The process of mechanization will be facilitated in this region by proximity to sources of manufacture, namely India (the world's largest manufacturer of tractors) and China (a source of low-cost power tillers).

Stable use of tractors in Latin America and the Caribbean. Projections for economic growth in Latin America and the Caribbean are on a par with other regions and per capita incomes are among the highest in the developing world. However, almost half of the countries in the region are not expected

Box 4.7 Household vulnerability to the loss of human and draught animal power

Households reliant on human power, and draught animals to a lesser extent, are extremely vulnerable to the loss of their principal power source. More than 15 countries in sub-Saharan Africa are projected to lose at least 5 percent of their workforce to HIV/AIDS by 2020. The pandemic will impact heavily in the agricultural sector where losses will typically account for at least 10 percent of the workforce and, in at least five countries, more than 20 percent. In a region where people are a significant, and often the dominant, source of power for both household and farm activities, this loss of labour will have a dramatic impact on rural livelihoods.

HIV/AIDS usually strikes at the heart of the household, killing women and men in their economic prime. Not only do households lose key family members but they also lose time spent by other household members caring for the sick. The situation is exacerbated by urban dwellers returning to their villages to be cared for when they become ill, thereby placing further strain on rural households. In addition to the immediate emotional, physical and financial stresses, the remaining family members have to take on the long-term care of orphan children. In some cultures, widows also have to cope with the threat of property-grabbing by the relatives of the deceased.

In parts of eastern and southern Africa, the vulnerability of rural livelihoods has been worsened by the decimation of the DAP base caused by the switch from hardy local breeds to cross-breeds, coupled with the failure to carry out regular healthcare practices and increased livestock susceptibility to disease (such as East Coast fever). Cattle rustling is also a threat, particularly in areas close to international borders. In the absence of alternative power sources, such as tractor hire, households have reverted to hand power. Areas under culti-vation have fallen significantly and households that once were food self-sufficient and producers of surplus for sale, now regularly experience food shortages. Household transport has become more problematic and the opportunity to earn additional income from hiring out draught animals has also disappeared.

Food insecurity, arising from the inability to produce or purchase sufficient food for the household throughout the year, is a persistent characteristic of subsistence agriculture. Short-term coping strategies include reducing the number of meals eaten per day, with very poor households spending up to two days between meals, or by switching to less nutritious foods. The poor may gather and sell natural products (such as wild fruits, mushrooms, tubers, firewood and grass thatch) or beg for food. Households may also engage in off-farm activi-ties, trading and making handicrafts, or rely on remittances from family members living elsewhere. Some survival strategies, such as the sale of assets to buy food, taking out loans to purchase inputs, or hiring out family labour to work on other farms, invariably place the household at greater risk in subsequent seasons. Longer-term adaptive strategies to overcome labour shortages include reallocating tasks between household members, using labour-saving technologies and switching to less labour-intensive cropping patterns and practices.

to change farm power categories during the next 30 years. Several countries are at the limits of tech-nical change in terms of farm power. Much of their agricultural sector is already fully mechanized and any further expansion in the use of tractors will be largely constrained by topographical features, notably the Amazon basin and the mountainous regions of the Andes. In other countries (such as El Salvador, Guatemala, Mexico and Paraguay) the shift towards no-till farming and conservation agri-culture may reduce or eliminate the need for the increased use of tractors. For a few countries, economic conditions are determining factors and stagnant incomes in the smallholder sector inhibit any increase in the use of tractors.

Livestock production

5.1 Introduction

Livestock production is the world's largest user of land, either directly through grazing or indirectly through consumption of fodder and feedgrains. Globally, livestock production currently accounts for some 40 percent of the gross value of agricultural production. In industrial countries this share is more than half. In developing countries, where it accounts for one-third, its share is rising quickly; livestock production is increasing rapidly as a result of growth in population and incomes and changes in lifestyles and dietary habits.

Growth in the livestock sector has consistently exceeded that of the crop sector. The total demand for animal products in developing countries is expected to more than double by 2030. By contrast, demand for animal products in the industrial world has been growing at low rates, and livestock production in this group of countries is expected to grow only slowly over the projection period (see Table 5.1).

Satisfying increasing and changing demands for animal food products, while at the same time sustaining the natural resource base (soil, water, air and biodiversity), is one of the major challenges facing world agriculture today. Global agriculture as a whole will be increasingly driven by trends in the livestock subsector, many of which are already apparent:

- An increasing proportion of livestock production will originate in warm, humid and more disease-prone environments.
- There will be a change in livestock production practices, from a local multipurpose activity to a more intensive, market-oriented and increasingly integrated process.
- Pressure on, and competition for, common property resources such as grazing and water resources will increase.
- There will be more large-scale industrial production, located close to urban centres, with associated environmental and public health risks.
- Pigs and poultry will increase in importance compared with ruminants.
- There will be a substantial rise in the use of cereal-based feeds.

Meeting these challenges raises crucial global and national public policy issues that must be addressed. Broadly, these encompass equity and poverty alleviation, the environment and natural resource management, and public health and food safety.

The main purpose of this chapter is to discuss the factors and perspective issues underlying the projections for livestock production as presented in Chapter 3. Developments in demand for livestock products will be briefly summarized in Section 5.2 and those for production of livestock commodities in Section 5.3. This latter section also discusses issues in increasing livestock productivity and expected changes in livestock production systems. Some selected livestock-related issues expected to take on greater importance over the projection period are dealt with in Section 5.4.

5.2 Consumption of livestock products

As incomes increase, demand for greater food variety grows. Demand for higher-value and quality foods such as meat, eggs and milk rises, compared with food of plant origin such as cereals. These changes in consumption, together with sizeable population growth, have led to large increases in the total demand for animal products in many developing countries, and this trend will continue.

The rising share of animal products in the diet is evident in developing countries. Even though calories derived from cereals have increased in absolute terms, as a share of total calories they continue to fall, from 60 percent in 1961/63 to an expected 50 percent in 2030. Similarly, the contribution of other traditional staples (potatoes, sweet potatoes, cassava, plantains and other roots) fell from second largest contributor to dietary calories (10 percent) in 1961/63 to lowest (6.2 percent) by 1997/99. By then, animal products had become the second major source of calories (10.6 percent) in developing countries.

In industrial countries, cereals contribute significantly less as a share of calories consumed – around 34 percent – while the contribution of animal products has remained stable at around 23 percent. In developing countries the per capita consumption of animal products is still less than a third of that in industrial countries, so there remains a significant potential to increase the contribution of animal products to the diet, both in absolute and percentage terms.

In industrial countries, the consumption of animal proteins increased in the 1960s and 1970s from 44 to 55 g/capita/day. After this, animal protein consumption remained fairly stable. In developing countries, however, although the level of consumption of animal proteins increased steadily from 9 g/capita/day in 1961/63 to 20 g/capita/day in 1997/99, there is still significant potential for increases.

Between 1997/99 and 2030, annual meat consumption in developing countries is projected to increase from 25.5 to 37 kg per person, compared with an increase from 88 to 100 kg in industrial countries. Consumption of milk and dairy products will rise from 45 kg/ person/p.a. to 66 kg in developing countries, and from 212 to 221 kg in industrial countries. For eggs, consumption will grow from 6.5 to 8.9 kg in developing countries and from 13.5 to 13.8 kg in industrial countries.

Wide regional and country differences are also evident in the quantity and type of animal products consumed, reflecting traditional preferences based on availability, relative prices and religious and taste preferences. Some of the more important aspects include the following:

■ In *sub-Saharan Africa*, low consumption levels of animal products have changed little over the last 30 years. These contribute about 5 percent to per capita calorie consumption, about half the percentage of the developing countries as a whole and a fifth of that of the industrial countries. Milk contribution to total calories and protein per capita has remained constant in recent years, indicating an increase in total milk availability equivalent to population increases. Only minor increases in consumption are projected.

■ In the *Near East and North Africa*, the contribution of animal products to the relatively high calorie consumption is small, just 8.7 percent in 1997/99. The contribution of animal products (primarily poultry meat and milk) to total calorie intake will increase to 11.4 percent by 2030.

■ In *Latin America and the Caribbean* (excluding Brazil), consumption of animal products (meat) historically has been higher than in other developing country groups and is predicted to increase further. Currently, animal products provide 16.6 percent of the dietary energy, but meat consumption per capita is still only about 60 percent of that of industrial countries, implying that there is scope for further growth.

■ *Brazil* is unusual in terms of its large and increasing dietary contribution of animal products (18.8 percent in 1997/99). The gap between Brazil and the rest of Latin America is expected to widen. Meat consumption per capita, at present over three-quarters the level of industrial countries, is projected to reach the level of the latter by 2030. Per capita milk consumption, at present just over half the level of industrial countries, is projected to reach three-quarters of the industrial countries' consumption by 2030.

- In *South Asia* (excluding India), there has been a slow but steady growth in animal product consumption. This increase is mostly the result of an increase in the contribution of milk, already high at a per capita level 50 percent above the average for developing countries, together with an increase in the contribution of poultry meat. The contribution of eggs is well below the developing country average.

- In *India*, the relative contribution of animal products to diets is predicted to increase up to 2030 largely as a result of increases in the consumption of milk and milk products.

- In *East Asia* (excluding China) there is also a steady increase in the contribution of animal products to the diet. However, unlike South Asia, this increase is a result of the contribution of meat, predominantly pork.

- In *China*, the projected rapid rise in the contribution of animal products to dietary energy from 15 to 20 percent between 1997/99 and 2030 will be mainly on account of a substantial increase in the contribution of pork and poultry. Per capita consumption of milk is very low and projected to remain so (rising from 7 kg p.a. in 1997/99 to 14 kg in 2030). By contrast, egg consumption in China is very high – more than double the average for developing countries and even above the industrial country average – and will rise from 15 kg/ person/p.a. in 1997/99 to 20 kg in 2030.

5.3 Production

Chapter 3 details the production and trade projections for livestock commodities. This section will look more at the implications. The location of production and processing activities is increasingly determined by factors such as the availability, quality and cost of inputs and proximity to markets. With the expected expansion in demand for animal products, traditional mixed farming practices alone will no longer be capable of meeting requirements.

The supply of animal products can be increased by raising the number of animals, and by improving productivity and processing and marketing efficiency, and by adopting various combinations of these factors. In most regions, land availability limits the expansion of livestock numbers in extensive production systems, so the bulk of the production increase will come from increased productivity, through intensification and wider adoption of better production and marketing technologies.

The strongest structural trend in livestock production has been the growth of intensive, vertically integrated, intensive establishments close to large urban centres, particularly for pig and poultry meat production in East Asia and Latin America, and broiler production in South Asia. Similar trends are apparent in dairy and beef production, albeit to a lesser degree. In East Asia the growth in demand for feedgrains associated with these production systems has been met by increased imports, effectively substituting imports of livestock products by imports of feedgrains.

In framing public policies, governments and other stakeholders are confronted with important trade-offs. For example, many developing countries favour industrial livestock production in order to provide affordable animal protein to urban populations (nutritional and public health benefits). Yet this may occur at the expense of diminishing the market opportunities and competitiveness of small rural producers. Similarly, stricter food safety regulations to enhance public health constitute barriers that often prevent poor farmers from entering formal markets.

5.3.1 Livestock production and productivity

The projections of total demand for livestock products were reached by taking the projections of food demand for direct human consumption, and adding on other demand components, such as the use of livestock products for non-food industrial uses, milk consumed by offspring and eggs for hatching and wastage. Production projections for each country were arrived at by making a detailed analysis of past trade in livestock products, including trade in inputs such as cereals and oilseeds used for animal feed.

The overall results are presented in Table 5.1 (unlike Table 3.11 these are aggregated over all livestock products covered in this study, including meat, milk and eggs). The main feature is a gradual slowdown in the growth of global livestock production. This will be made up of slow growth in the industrial countries, a recovery in the transition countries and a pronounced slowdown from rapid to moderate growth rates in the developing countries. This latter trend is heavily dominated by the expected slowdown in two major countries that experienced fast growth in livestock production in the past, China and Brazil. Excluding these two countries, the slowdown is much more gradual. In individual countries, developments are much more varied, with accelerating livestock production growth in several countries.

Table 5.2 (a more detailed expansion of part of Table 3.11) gives details of the livestock production data and projections for the six livestock products covered in this study, underlying the overall results presented in Table 5.1. In developing countries there has been a continued increase in production. Annual growth rates for the six commodities ranged from 3.7 to 9.4 percent for the period 1989-99. By

contrast, over the same period production in the transition countries actually fell, and grew only slightly in the industrial countries.

By 1997/99, the share of the developing countries in world meat production was 53 percent and in milk production 39 percent, compared with 40 and 28 percent only ten years earlier. This was in part a result of the collapse of production in the transition countries, but it is an ongoing trend even in the absence of this phenomenon. Annual growth of meat production in developing countries to 2030 is projected at 2.4 percent and of milk at 2.5 percent. This would raise developing countries' share in world meat production by 2030 to 66 percent (247 million tonnes) and in milk production to 55 percent (484 million tonnes).

The growth in white meat (pork and poultry) production in developing countries between 1989 to 1999 has been remarkable – more than double the growth of red meat (cattle, sheep and goats). There are, however, major regional differences. Growth in poultry production has been spectacular in East Asia (11.7 percent p.a.) and South Asia (7.2 percent p.a.) and reflects the rapid intensification of the poultry industry in these regions. Latin

Table 5.1	Annual growth rates of total livestock production					
	1969-99	1979-99	1989-99	1997/99 -2015	2015 -2030	1997/99 -2030
				Percentage		
World	2.2	2.1	2.0	1.7	1.5	1.6
excl. China	1.7	1.3	0.8	1.6	1.5	1.5
Developing countries	4.6	5.0	5.5	2.6	2.1	2.4
excl. China	3.5	3.5	3.6	2.8	2.5	2.7
excl. China and Brazil	3.3	3.3	3.3	2.9	2.6	2.8
Sub-Saharan Africa	2.4	2.0	2.1	3.2	3.3	3.2
Latin America and the Caribbean	3.1	3.0	3.7	2.4	1.9	2.1
excl. Brazil	2.3	2.1	2.7	2.4	2.1	2.3
Near East/North Africa	3.4	3.4	3.4	2.9	2.6	2.7
South Asia	4.2	4.5	4.1	3.3	2.8	3.1
East Asia	7.2	8.0	8.2	2.3	1.6	2.0
excl. China	4.8	4.7	3.7	3.0	2.7	2.8
Industrial countries	1.2	1.0	1.2	0.7	0.4	0.6
Transition countries	-0.1	-1.8	-5.7	0.5	0.6	0.5

Note: Total livestock production was derived by aggregating four meats, milk and eggs at 1989/91 international commodity prices used to construct the FAO indices of agricultural production.

Table 5.2   Livestock production by commodity: past and projected

	1967/69	1987/89	1997/99	2015	2030	1969 -1999	1989 -1999	1995/97 -2015	2015 -2030
	\multicolumn{5}{c}{Million tonnes}					% p.a.		
Total meat									
World	92	166	218	300	376	2.9	2.7	1.9	1.5
excl. China	84	142	162	218	277	2.1	1.3	1.8	1.6
Developing countries	28	66	116	181	247	5.2	5.9	2.7	2.1
excl. China	21	41	60	98	147	3.8	3.9	3.0	2.7
excl. China and Brazil	18	34	47	79	123	3.5	3.3	3.1	2.9
Sub-Saharan Africa	3	4	5	9	16	2.3	2.2	3.3	3.5
Latin America and the Caribbean	10	19	28	43	58	3.5	4.5	2.6	2.1
excl. Brazil	7	11	15	24	33	2.5	3.1	2.7	2.3
Near East/North Africa	2	5	7	13	19	4.4	3.8	3.5	2.9
South Asia	3	5	7	13	23	3.7	2.8	3.6	3.9
East Asia	10	33	69	103	131	7.1	7.6	2.4	1.6
excl. China	3	8	13	21	32	5.1	4.1	3.0	2.8
Industrial countries	46	71	85	99	107	1.9	1.8	0.9	0.5
Transition countries	17	29	17	20	22	0.0	-6.4	0.8	0.8
Bovine meat									
World	38.0	53.7	58.7	74.0	88.4	1.4	0.8	1.4	1.2
Developing countries	11.8	19.3	28.0	41.2	55.0	3.0	3.8	2.3	2.0
excl. China	11.7	18.4	23.2	33.5	44.1	2.5	2.2	2.2	1.8
excl. China and Brazil	10.0	14.4	17.3	25.2	34.1	2.0	1.5	2.3	2.0
Sub-Saharan Africa	1.6	2.2	2.6	4.3	6.7	1.5	1.7	3.0	3.0
Latin America and the Caribbean	6.8	10.4	13.1	18.2	22.5	2.5	2.1	1.9	1.4
excl. Brazil	5.1	6.5	7.2	9.9	12.5	1.4	0.4	1.9	1.6
Near East/North Africa	0.7	1.3	1.8	2.8	4.1	3.2	3.4	2.4	2.6
South Asia	1.7	3.1	4.0	5.7	7.4	3.1	2.3	2.1	1.7
East Asia	1.0	2.3	6.4	10.1	14.4	6.4	11.5	2.7	2.4
excl. China	0.8	1.4	1.6	2.5	3.5	2.1	2.3	2.6	2.2
Industrial countries	19.1	23.8	25.0	26.6	26.5	0.6	0.6	0.4	0.0
Transition countries	7.0	10.6	5.7	6.3	6.9	-0.3	-7.5	0.5	0.6
Ovine meat									
World	6.6	9.1	10.8	15.3	20.1	1.9	1.4	2.1	1.8
Developing countries	3.0	5.0	7.4	11.2	15.4	3.4	3.7	2.5	2.1
Sub-Saharan Africa	0.6	0.9	1.3	2.2	3.4	2.8	3.5	3.1	3.0
Near East/North Africa	0.9	1.5	1.8	2.6	3.5	2.3	1.9	2.2	2.0
South Asia	0.6	1.1	1.3	2.1	3.1	3.5	1.4	2.6	2.6
East Asia	0.4	1.1	2.5	3.8	4.8	7.0	8.1	2.6	1.5
Industrial countries	2.4	2.8	2.7	3.1	3.5	0.6	-0.8	0.9	0.8
Transition countries	1.3	1.3	0.8	0.9	1.1	-1.0	-6.4	1.3	1.1

Table 5.2 *cont.*   Livestock production by commodity: past and projected

Pig meat									
World	34.1	66.3	86.5	110.2	124.5	3.2	2.7	1.4	0.8
excl. China	28.1	46.2	48.1	57.9	66.2	1.7	0.4	1.1	0.9
Developing countries	9.7	28.0	49.3	69.5	82.8	6.1	5.7	2.0	1.2
excl. China	3.8	7.9	10.9	17.2	24.5	3.7	3.4	2.7	2.4
Latin America and the Caribbean	1.8	3.0	3.9	6.0	7.8	2.1	3.9	2.5	1.8
excl. Brazil	1.1	1.9	2.3	3.4	4.4	1.7	2.8	2.3	1.8
East Asia	7.6	24.2	44.3	61.6	71.9	6.8	6.0	2.0	1.0
excl. China	1.6	4.0	5.9	9.3	13.6	5.1	3.3	2.8	2.5
Industrial countries	16.6	26.0	29.3	32.3	33.1	1.8	1.4	0.6	0.2
Transition countries	7.7	12.3	7.9	8.4	8.6	-0.1	-5.3	0.4	0.1
Poultry meat									
World	12.9	37.2	61.8	100.6	143.3	5.2	5.4	2.9	2.4
excl. China	12.1	34.6	51.2	81.4	117.5	4.8	4.1	2.8	2.5
Developing countries	3.3	13.2	31.3	59.1	93.5	7.9	9.4	3.8	3.1
excl. China	2.5	10.6	20.7	39.9	67.7	7.4	7.2	4.0	3.6
excl. China and Brazil	2.2	8.6	15.6	31.9	56.4	6.9	6.4	4.3	3.9
Sub-Saharan Africa	0.3	0.7	0.9	1.9	4.1	3.8	2.6	4.3	5.1
Latin America and the Caribbean	1.0	4.7	10.5	18.2	27.3	7.8	9.0	3.3	2.7
excl. Brazil	0.7	2.7	5.4	10.2	16.0	6.7	8.4	3.8	3.0
Near East/North Africa	0.4	2.1	3.2	7.1	11.6	7.7	5.2	4.7	3.3
South Asia	0.2	0.5	1.1	3.9	10.6	7.7	7.2	7.9	6.9
East Asia	1.5	5.3	15.5	27.9	39.9	8.5	11.7	3.5	2.4
excl. China	0.7	2.6	4.9	8.7	14.1	7.3	6.1	3.4	3.2
Industrial countries	8.1	18.8	27.7	37.5	44.1	4.0	3.9	1.8	1.1
Transition countries	1.5	5.2	2.9	4.1	5.7	1.6	-6.7	2.0	2.3
Milk (whole milk eq.)									
World	387	528	562	715	874	1.3	0.6	1.4	1.3
Developing countries	78	149	219	346	484	3.6	4.1	2.7	2.3
excl. China and Brazil	69	128	189	301	425	3.5	4.1	2.8	2.3
Sub-Saharan Africa	8	13	16	26	39	2.7	1.9	3.0	2.8
Latin America and the Caribbean	24	40	57	81	105	2.6	3.9	2.1	1.8
excl. Brazil	17	26	36	52	69	2.2	4.0	2.1	1.9
Near East/North Africa	14	21	28	41	56	2.3	3.1	2.2	2.1
South Asia	30	65	104	174	250	4.5	4.9	3.1	2.4
East Asia	3	10	15	25	34	6.9	4.5	2.9	2.2
excl. China	1	4	5	8	12	7.3	3.2	3.0	2.4
Industrial countries	199	236	246	269	286	0.7	0.5	0.5	0.4
Transition countries	110	144	97	100	104	-0.3	-4.6	0.2	0.2
Eggs									
World	18.7	35.6	51.7	70.4	89.9	3.4	4.2	1.8	1.6
Developing countries	4.9	16.2	33.7	50.7	69.0	7.0	8.0	2.4	2.1
excl. China	3.2	9.5	13.5	24.6	37.8	5.0	3.4	3.6	2.9
Sub-Saharan Africa	0.3	0.7	0.9	1.8	3.4	3.7	2.6	4.0	4.1
Latin America and the Caribbean	1.2	3.6	4.6	7.3	10.4	4.5	2.5	2.8	2.3
Near East/North Africa	0.4	1.5	2.2	3.6	5.3	6.0	4.1	3.0	2.6
South Asia	0.3	1.4	2.2	5.7	9.9	6.3	4.7	5.8	3.7
East Asia	2.6	9.1	23.8	32.1	40.0	8.3	10.7	1.8	1.5
excl. China	0.9	2.4	3.6	6.0	8.8	5.0	3.5	3.0	2.6
Industrial countries	10.7	12.8	13.7	14.8	15.5	0.6	0.9	0.5	0.3
Transition countries	3.1	6.5	4.3	5.0	5.5	0.7	-4.7	0.8	0.7

America saw annual growth rates of 9 percent. Yet in sub-Saharan Africa the annual growth rate was only 2.6 percent. Red meat accounted for almost 37 percent of total meat production in the developing countries in the late 1980s, but declined to 31 percent in 1997/99 and this proportion is expected to decline further.

Egg production increased in developing countries during the last ten years (1989-99) with similar regional differences. Annual growth rates for East Asia, South Asia and sub-Saharan Africa were 10.7, 4.7 and 2.6 percent, respectively. Latin America saw a growth rate of 2.5 percent p.a. and the industrialized countries 0.9 percent p.a., while in the transition countries production fell by 4.7 percent p.a. Milk production in developing countries grew at 4.1 percent p.a. over the same period, with the highest annual growth found in South Asia (4.9 percent) and the lowest in sub-Saharan Africa (1.9 percent). Milk production in industrial and transition countries followed the same trend as egg production.

Productivity can be measured by the amount of meat or milk produced per animal per year. More sophisticated productivity analyses – based on unit of output per unit of biomass or feed input or based on financial flows – are much more difficult to undertake. For example, in high-income countries there is a growing demand for free-range meat at premium prices, and this might still allow farmers to make higher net returns despite lower carcass weight and lower offtake rates.

Increased production can be achieved by a combination of expansion in animal numbers and increased productivity. Higher productivity is a compound of higher offtake rates (shorter production cycles by, for example, faster fattening), and higher carcass weight or milk or egg yields. The projections show that the increase in livestock numbers will remain significant, but less so than in the past. Higher carcass weights will play a more important role in beef production, while higher offtake rates (shorter production cycles) will be more important in pig and poultry meat production.

There are problems in getting reliable data for offtake rates and carcass weights. To circumvent these, meat production can be compared directly with herd sizes. For example, over the last decade (1989-99), beef production in developing countries increased by 3.8 percent p.a., while cattle numbers increased by only 1.3 percent (Table 5.3), implying an annual productivity improvement of 2.5 percent. Small ruminant production increased by 3.7 percent p.a. while flock size increased by only 1.5 percent, suggesting a 1.2 percent annual productivity improvement.

There are substantial differences between regions and countries, however. In sub-Saharan Africa the increase in cattle numbers was greater than the growth in production, indicating a decline in meat productivity. In Asia, where land is scarce, growth in herd size for cattle and buffaloes was much lower than the growth in output, indicating that intensification and increased productivity were relatively more important. Increases in productivity were also responsible for the increases in white meat and egg production.

Meat or milk output per animal remains higher in industrial countries than in developing ones. For example, in 1997/99 the yield of beef per animal (carcass weight) in developing countries was 163 kg compared with 284 kg in industrial countries, while average milk yields were 1.1 and 5.9 tonnes p.a., respectively. Pork and poultry productivity levels are more similar across regions, reflecting the greater ease of transfer and adoption of production techniques.

5.3.2 Production systems

The changes in global demand for animal products and the increasing pressures on resources have important implications for the principal production systems found in developing countries.

Grazing systems. A quarter of the world's land is used for grazing, and extensive pasture provides 30 percent of total beef production and 23 percent for mutton (FAO, 1996e). In developing countries, extensive grazing systems have typically increased production by herd expansion rather than by substantial increases in productivity. However, globally the market share from these extensive systems is declining relative to other production systems. The availability of rangelands is decreasing, through arable land encroachment, land degradation, conflict and so on. Hence the scope for further increasing herd numbers in these systems remains limited.

Crop-livestock production systems. In developing countries, most ruminant livestock are found in

TABLE 5.3 Meat production: number of animals and carcass weight

	Number of animals (millions)				Number of animals (% p.a.)			Carcass weight (kg/animal)		
	1967/69	1987/89	1997/99	2030	1969 -1999	1989 -1999	1997/99 -2030	1967/ 69	1997/ 99	2030
World										
Cattle and buffaloes	1 189	1 418	1 497	1 858	0.8	0.5	0.7	174	198	211
Sheep and goats	1 444	1 708	1 749	2 309	0.9	-0.1	0.9	14	14	17
Pigs	566	838	873	1 062	1.4	0.3	0.6	65	78	84
Poultry	5 585	10 731	15 067	24 804	3.8	3.4	1.6	1.3	1.6	1.8
Developing countries										
Cattle and buffaloes	799	1 013	1 156	1 522	1.3	1.3	0.9	150	163	188
Sheep and goats	862	1 121	1 323	1 856	1.6	1.5	1.1	13	13	16
Pigs	297	493	581	761	2.2	1.6	0.8	49	73	82
Poultry	2 512	6 168	10 544	19 193	5.6	5.5	1.9	1.2	1.4	1.8
Sub-Saharan Africa										
Cattle and buffaloes	130	159	200	285	1.5	2.4	1.1	137	130	157
Sheep and goats	182	269	346	501	2.4	2.6	1.2	12	12	17
Pigs	6	13	18	27	4.5	2.3	1.4	45	47	63
Poultry	313	555	720	1 459	3.1	2.4	2.2	0.9	1.0	1.4
Latin America and the Caribbean										
Cattle and buffaloes	219	317	350	483	1.6	0.9	1.0	191	211	230
Sheep and goats	152	145	119	145	-0.5	-2.5	0.6	15	13	16
Pigs	63	74	76	108	0.6	0.1	1.1	65	72	83
Poultry	558	1 248	2 075	3 815	4.5	5.7	1.9	1.2	1.5	1.9
Near East/North Africa										
Cattle and buffaloes	37	37	39	62	0.0	0.7	1.5	107	158	194
Sheep and goats	205	241	256	350	0.9	0.5	1.0	14	16	20
Poultry	215	722	1 101	2 135	6.3	4.9	2.1	1.1	1.1	1.6
South Asia										
Cattle and buffaloes	293	348	384	424	1.0	1.0	0.3	95	121	151
Sheep and goats	148	241	289	405	2.5	1.7	1.1	11	12	15
Pigs	6	12	17	23	3.4	3.6	1.0	35	35	55
Poultry	232	472	717	2 256	4.4	4.7	3.6	0.9	0.9	1.6
East Asia										
Cattle and buffaloes	121	153	183	268	1.8	2.0	1.2	147	144	176
Sheep and goats	174	226	312	455	2.0	3.1	1.2	12	13	15
Pigs	221	393	470	602	2.4	1.7	0.8	47	75	83
Poultry	1 195	3 171	5 930	9 529	6.5	6.1	1.5	1.3	1.6	1.8
Industrial countries										
Cattle and buffaloes	263	253	254	243	-0.5	0.2	-0.1	212	284	308
Sheep and goats	397	394	341	358	-0.1	-2.2	0.2	16	17	20
Pigs	172	206	210	220	0.7	0.4	0.1	75	85	89
Poultry	2 167	2 941	3 612	4 325	1.8	2.2	0.6	1.4	1.8	2.1
Transition countries										
Cattle and buffaloes	127	152	87	94	-1.0	-6.4	0.2	144	155	170
Sheep and goats	185	193	85	95	-1.9	-9.3	0.3	14	15	18
Pigs	97	139	81	82	-0.5	-6.2	0.0	77	82	84
Poultry	906	1 622	920	1 287	0.4	-6.9	1.1	1.3	1.4	1.6

mixed farming systems. These are estimated to provide over 65 percent of beef, 69 percent of mutton and 92 percent of cow milk (FAO, 1996e). The complementarity between crop and livestock production is well known. Crops and crop residues provide feed, while livestock provide animal traction, manure, food, a form of savings or collateral, income diversification and risk reduction. Although short-cycle species, such as chickens and pigs, are often very important for household food security and immediate cash needs, only ruminants can convert highly fibrous material and forages with little or no alternative use into valuable products. An estimated 250 million work animals provide draft power for cultivation of about half the total cropland in developing countries.

Intensive industrial livestock production systems. The trend towards intensification is most pronounced in Asia, where there is a shortage of land but an abundance of relatively cheap labour. This has encouraged small-scale intensive systems such as "cut and carry" and stall feeding, which have higher labour but lower land requirements. Increasing access to capital has allowed for investment in machinery, housing and inputs such as improved breeds, concentrate feeds and veterinary drugs. This has resulted in improved productivity and has accounted for faster growth in production of monogastric animals such as pigs and chickens than of ruminants. The consequence has been a reduction in the value of livestock's alternative uses, as the value of its food products becomes relatively more important.

In sub-Saharan Africa, semi-intensive and intensive dairying has developed close to urban centres and, where agro-ecological conditions permit, on the basis of cultivated fodder and agro-industrial by-products. In Latin America, intensive poultry production and, to some extent, dairying have developed partly in response to the high level of urbanization and a resumption of economic growth in the 1990s.

Large-scale and vertically integrated intensive industrial poultry and pig production systems have increased significantly in the developing world, particularly in East Asia. They make use of improved genetic material and sophisticated feeding systems, and require highly skilled technical and business management. They are also dependent on inputs

of high-energy and protein-rich feeds and animal health prophylactics, and consume considerable amounts of fossil fuel, both directly and indirectly. The wholesale transfer of these types of production systems has been facilitated by the relative ease and speed with which the required infrastructure and equipment can be operationalized in so called "turnkey" operations. In recent years, industrial livestock production grew at twice the annual rate of the more traditional, mixed farming systems (4.3 against 2.2 percent), and at more than six times the annual growth rate of production based on grazing (0.7 percent; FAO, 1996e). The major expansion in industrial systems has been in the production of pigs and poultry since they have short reproductive cycles and are more efficient than ruminants in converting feed concentrates (cereals) into meat. Industrial enterprises now account for 74 of the world's total poultry production, 40 percent of pig meat and 68 percent of eggs (FAO, 1996e).

5.4 Major perspective issues and possible policy responses

In this section a number of selected livestock-related issues that are expected to increase in importance over the projection period will be discussed.

5.4.1 Livestock, economic development and poverty alleviation

As discussed in Chapter 8, a key challenge for development and poverty alleviation is the identification and promotion of broad-based income opportunities that may lead to significant pro-poor growth. The livestock sector appears to present a major opportunity to enhance the livelihoods of a large portion of the world's poor.

Livestock ownership currently supports and sustains the livelihoods of an estimated 675 million rural poor (Livestock in Development [LID], 1999). These people depend on livestock fully or partially for income and subsistence. Livestock can provide a steady stream of food and income, and help to raise whole-farm productivity. They are often the only livelihood option available to the landless, as they allow the exploitation of common property resources for private gain. In addition, livestock are often the only means of asset accumulation and risk diversification that can prevent the rural poor in marginal areas from sliding into poverty. Recent statistics show that an estimated 70 percent of the poor are women, for whom livestock represent one of the most important assets and sources of income (DFID, 2000a). Livestock ownership also tends to increase consumption of animal products and create employment opportunities.

In spite of the trend towards increasing scales of production and vertical integration, the greater part of the food consumed in developing countries is still produced by semi-subsistence farmers. The projected growth in the demand for animal products therefore offers opportunities for the rural poor since they already have a significant stake in livestock production. Unfortunately, until now the large majority of the rural poor have not been able to take advantage of these opportunities. Thus far, the main beneficiaries have been processors and traders, middle-class urban consumers, and a relatively small number of large producers in high-potential areas with good access to markets.

A number of handicaps prevent the poor from taking advantage of the available development potential. These are summarized below.

Financial and technical barriers. Financial barriers prevent small farmers from intensifying their production. The investment required often exceeds their capital wealth. Policies and institutions must facilitate forms of targeted small- to medium-scale credit, based among others on the strengthening of property rights, to ensure the poor's future involvement in increasing livestock production and processing. Technical barriers prevent small producers from supplying a safe and relatively uniform product to the market. The lack of appropriate infrastructure to preserve perishable products affects the negotiating power of small production units, particularly if they are distant from consumption centres. Technical barriers exist in the form of sanitary requirements (including animal welfare) as a prerequisite to trade. Perceived or real livestock disease incidence may exclude groups of farmers or whole countries from international, regional and local markets. Policies and institutions must facilitate access to technologies, goods and services, and encourage the establishment of product standards and safety norms that do not exclude smaller producers, yet do not compromise public health.

A combination of higher production and higher transaction costs can make small producers *uncompetitive* and limit their access to markets. They do not benefit from the economies of scale available to large-scale units, so their production costs are usually higher, outweighing any cost advantages from the discounted value of family labour. Hidden and overt subsidies to facilitate the supply of cheap animal products to the cities may impact on small-scale producers, public health and the environment, but there is a lack of objective data to assess this.

Transaction costs can be prohibitively high for small-scale producers because of the small quantities of marketable product and the absence of adequate physical and market infrastructures in remote areas. Transaction costs are also increased where producers lack negotiating power or access to market information, and remain dependent on intermediaries. Public policies are needed to develop market infrastructures, including appropriate information systems enabling small-scale producers to make informed marketing decisions. Producers' associations or cooperatives enable producers to benefit from economies of scale by reducing transaction costs.

Reducing *risks* and mitigating their effect on poor livestock-dependent people are prerequisites for a sustainable reduction in poverty. Small-scale production is associated with both market and production risks. Market risks include price fluctuations of both inputs and products and are often associated with a weak negotiating position. While subsistence farming often has sound risk-coping mechanisms, many small-scale producers lack the assets or strategies to sustain full exposure to market risks. If the poor are to participate fully in the market, safety nets are needed to cope with the economic shocks invariably present in free markets.

Production risks arise from resource degradation, extreme weather events such as droughts and floods, and disease outbreaks. Both small-scale and intensive livestock production systems are at risk from the ravages of epidemic diseases and droughts, but the poor are particularly vulnerable to these types of shocks because of their limited assets and the lack of insurance schemes. Public and private services in disaster-prone poor countries almost invariably lack the capacity to plan for such risks, or to respond in a timely manner. Building up such response capacity of communities and institutions is important, and drought and disease preparedness strategies need to be an integral part of public policy.

Development interventions in the livestock sector generally have not been very successful. Many livestock development projects have not succeeded because of inappropriate technologies and failure to deliver services to poor farmers. However, even in cases where the technologies were appropriately targeted and the focus was distinctly pro-poor, many technical projects have failed to improve the livelihoods of the poor substantially. Clearly, an enabling institutional and political environment is indispensable if interventions and strategies are to focus on the poor in a sustainable way (LID, 1999; IFAD, 2001).

5.4.2 Livestock health, welfare and nutrition

Animal health. Infectious and parasitic diseases of livestock remain important constraints to more productive and profitable livestock production in many developing regions. Diseases reduce farm incomes directly and indirectly: directly, by causing considerable losses in production and stock as well as forcing farmers to spend money and labour on their control; and indirectly by the consequent restrictions on exports.

Infectious diseases such as rinderpest, foot-and-mouth disease, contagious bovine pleuropneunomia, classical and African swine fever and peste des petits ruminants are still major threats to livestock production in developing countries. Through increased movements of livestock, livestock products and people, they also threaten production in industrial countries.

The global eradication of rinderpest by 2010 remains an achievable goal. However, the other epizootic diseases can be brought under control only gradually through intensive, internationally coordinated animal health programmes. There has been a major shift away from countrywide eradication programmes towards more flexible control strategies, where interventions are focused on areas offering the highest returns. Risk analysis and animal health economics help determine where disease control investment will have the greatest impact and benefit. Because of the large externalities of outbreaks of these diseases, management of their control remains a public sector responsibility. However, many developing countries do not have effective veterinary institutions capable of the task and public funding for disease control has been on the decline over the last decades.

Among the parasitic diseases, trypanosomosis (sleeping sickness) poses an enormous constraint to cattle production in most of the humid and subhumid zones of Africa. Flexible combinations of aerial spraying, adhesive pyrethroid insecticides, impregnated screens and traps and sterile insect techniques (SIT), supported by use of trypanocidal drugs, hold the promise of gradually recovering infested areas for mixed farming. These strategies will provide much greater benefits than the increased livestock output alone. They will raise crop output, allow the introduction of mixed farming, and improve human welfare by preventing sleeping sickness, providing higher and more stable incomes, and improving nutrition.

From a production viewpoint, helminthosis and tick-borne diseases are particularly important. Helminths (worms), while rarely fatal, can seriously affect productivity and profitability. Although helminths can be effectively controlled, parasite resistance to drugs through the inappropriate use of antihelmintics is a growing problem. Ticks have the capacity to transmit diseases, notably East Coast fever in eastern and southern African countries. But the cost of traditional dipping with acaracides for tick control is becoming prohibitive, and raises environmental concerns about disposal of the waste chemical. As production systems become more intensive, diseases affecting reproductive performance, and nutritional imbalances will also assume greater importance.

Intensification of livestock production is thus going to face growing constraints both from epidemic and endemic disease agents. In many parts of China, for example, high pig and poultry densities close to high concentrations of human population are breeding grounds for existing and emerging diseases, such as avian influenza and nipah virus in pigs. Foot-and-mouth and other diseases are endemic in India, with its very high cattle density. In the Near East, diseases such as bluetongue and Rift Valley fever occur in the large populations of sheep. Some livestock diseases (zoonoses) also cross from animals to humans, directly or through an intermediate host (e.g. new influenza strains through mosquitoes).

In industrial countries, centuries of systematic eradication efforts have eliminated some livestock diseases and drastically reduced the incidence of others. The maintenance of this situation is becoming more and more difficult with the ever-increasing intensity of international travel and the growing long-distance trade in animals and livestock products. Veterinary barriers are part of the response, such as the screwworm control programme that started in the United States and Mexico, and is now protecting almost all of Central America as well, or the foot-and-mouth disease control programme that is protecting Australia and Indonesia.

Biotechnology offers the promise of solving some of the technical constraints through improved prevention, diagnosis and treatment of animal disease. Genomics, for example, may well contribute to the development of new generations of vaccines using recombinant antigens to pathological agents. A far wider range of effective, economic vaccines that are easy to use and do not require a cold chain can be expected in the future. The development of cost-effective, robust pen-side diagnostics will enhance the veterinary services offered in developing countries. However, technological advances must be matched by enhanced epidemiological and logistical capacities, and by greatly improved coordination of all institutions involved in animal disease control from local up to international level.

Animal welfare. Unregulated intensification of livestock production is associated with animal management practices, such as space, light and movement limitations, which do not allow the animals to express their natural behaviour. Such practices are increasingly disliked in more affluent societies. Similar reservations are also expressed about animal transportation over large distances, and about certain feeding and medication practices. Genetic selection for increased weight in broilers has also been linked to animal health problems: skeletal and circulatory systems taxed by the rate of muscle formation, leading to increased rates of heart failure and broken limbs. Particularly in the industrial world, such concerns are likely to have an increasing influence on production systems over

the projected period. This trend is already reflected in EU regulations regarding the minimum cage sizes for battery hens.

Feed quantity and quality. A large majority of the world's livestock, particularly ruminants in the pastoral and low-input mixed farming systems, suffer from either permanent or seasonal nutritional stress. Finding better ways to use fibrous plant material is a high priority. A better understanding of how the rumen functions has led to proven techniques for treating crop residues and other low-quality roughage, and further developments can be expected. Improving the capacity of the rumen to digest high-fibre diets could dramatically improve the prospects of ruminant production in areas with easy access to roughage with low feed quality.

Critical dry season feed shortages can be alleviated by techniques for cultivating, collecting, storing and conserving fodder, forage and residues. There is considerable potential for incorporating a wide range of high-quality products (e.g. leguminous feeds) into animal rations. Strategic feeding and supplementation with key ingredients, such as minerals or nitrogen, can improve the overall utilization of low-quality diets. The feed industry has introduced a range of enzyme additives to enhance the nutritive value and quality of feeds. For example, silage additives containing cellulase and hemi-cellulase enzymes and bacterial inoculates have been shown to increase digestibility and improve preservation.

Plant genomics and phytochemistry have identified antinutritional factors (ANFs) in plants, such as phytic acid in maize. The transfer of a detoxifying (dehalogenase) gene from the soil bacterium Morexella to rumen bacteria has been successful under experimental conditions, opening the possibility of enriching the ruminant ecosystems with microbes with improved ability to detoxify ANFs. Microbial and ionophoric feed additives have been shown to increase feed conversion efficiency by up to 7.5 percent, although the results remain variable and the precise mechanisms and complex interaction of breed, micro-organism and animal are yet to be fully understood.

For pigs and poultry, improving feed conversion efficiency will be crucial to profitability. It will also reduce dependency on feedgrains where appropriate. Over the past decade, feed conversion rates for pigs and poultry have improved by 30 to 50 percent, in part through breeding and in part through the addition of enzymes to feeds. Still, in monogastrics only 25 to 35 percent of the nutrients consumed are captured in the final products. Further understanding of digestive physiology and biochemistry can be expected to improve feed utilization.

5.4.3 Livestock and trade

Expanding demand for livestock products, particularly in developing countries, has translated into rapid growth in trade in meat products over the past decade (see Chapter 3). This dynamic growth has been facilitated by a changing policy environment that has reduced market barriers to trade. It has also been favoured by technical factors, such as increasing specialization of production and processing operations as well as advances in transportation and cold-chain technology. The growing complexity of global meat markets, driven by heightened consumer-related demands about product type, quality and safety will increasingly dictate the patterns of trade. The competitiveness of meat exporters will increasingly hinge on their ability to respond to rapidly changing consumer preferences, and to the myriad of international regulations related to food safety and animal health standards.

One of the factors shaping future developments in the global meat trade is the continuing specialization of production and processing. Increasing production specialization, driven by consumer preferences for specific types and cuts of meat, will result in countries trading different types of meat cuts. For example, the United States imports manufacturing-grade beef and exports high-quality grain-fed beef to China. China in turn imports chicken feet and wings and exports higher-value processed cuts to Japan, in some cases using cheaper imported cuts from other countries as an input. Increasing product differentiation in response to varying preferences in different markets will accelerate in the future as consumers become more sophisticated and demanding, leading to a more complex and diverse meat economy.

The complexity of the global trading system for meat products is expected to increase further in response to consumers' preoccupation about the ways in which meat is produced and sold. Another factor pushing towards increasing complexity is the increased risk of international spread of animal diseases, zoonoses and food-borne infections associated with increased trade in livestock and livestock products. With greater scrutiny of meat production systems and the "hoof to plate" approach to food safety and quality, there is a risk of proliferation of divergent food standards, sanitary assurances and certification procedures. This would disadvantage meat-exporting developing countries, which would face higher costs in adhering to these changing requirements.

In response to food safety and animal disease concerns, countries have encouraged initiatives towards the establishment of science-based food safety regulations in terms of the sanitary and phytosanitary standards (SPS) under the Uruguay Round global trade agreements. The Agreement on SPS aims to eliminate the use of unjustified, unscientific regulations to restrict trade. Under the new rules, countries maintain the right to set their desired level of protection or "acceptable risk" of food safety and animal and plant health, but are obliged to adhere either to international standards or, when setting their own standards, to provide scientific backing for the latter. This right has been used and put to the test mostly by industrial countries. However, national standards, and some international standards, may be the subject of difficult and contentious trade disputes in the future. They will also exert increasing pressure towards improved delivery of veterinary and public health services in both industrial and developing countries.

The SPS notification process should provide a mechanism for introducing increased transparency into the system of notification, consultation and resolution of trade disputes, as also will moves to harmonize measures based on international standards of the Codex Alimentarius Commission and the International Office of Epizootics (OIE). Perhaps the success of the SPS Committee can be measured by the fact that to date, only one livestock/meat case has involved the formal dispute settlement body. In 1999, the World Trade Organization Hormone Case resolved a long-standing dispute between the EU and the United States, supported by Canada. Since 1989, the EU had banned imports of red meat from animals treated with six different growth promotants.

WTO arbitrators determined that the United States and Canada were entitled to suspend tariff concessions for certain EU products, in compensation for the EU not opening its market to beef treated with hormones.

Trade in livestock products has also been facilitated by the spread of cold-chain facilities and by lower costs for constructing, operating and maintaining this equipment. This has made transportation of fresh livestock products technically feasible and accessible to a much larger group of countries and greatly reduced the costs of transportation. Previously, such developments may have provided an incentive for agribusinesses to move their production to countries where labour, environmental and public health regulations are less stringent. However, consumer concerns and international regulations are no longer conducive to such shifts. Stricter food sanitation regulations and labelling and certification requirements demand controls and investment in infrastructure to a level that may be difficult to achieve in most developing countries in the near future. Additional factors are the costs and risks related to the transport of animal feed versus transport of livestock products. Shipment of animal feeds, in livestock product equivalent units, is still significantly cheaper than the shipment of livestock products. In addition, livestock feeds are far less perishable and thus run less risk during transport than fresh livestock products.

Technical advances in packaging, transportation and information will continue to be among the key determinants in meat trading patterns. They will allow meat shippers to deliver frozen, fresh and chilled meat products to buyers thousands of kilometres away, with no substantial loss in quality. Loading and unloading have always accounted for a relatively large share of total transportation costs. The use of containers, however, has radically reduced these front- and back-end costs. Shorter delivery times, improvements in pre- and post-shipment handling activities, and advances in technologies extending shelf-life, including packaging, will continue to reduce the transaction costs of trade.

Improvements in transportation systems inside developing countries will influence their consumption and trading patterns. As infrastructure improves, advances in intermodalism – moving goods by linking together two or more modes of transportation – will make it easier to move imported perishable meat products within the country. This will be accompanied by advances in controlled atmosphere (CA) technologies, added refinements and cold-chain facilities that extend the shelf-life of meat, thus expanding the types of meat that can be shipped in refrigerated containers without spoilage.

Continuing liberalization of meat markets as well as closer regional integration (in particular the accession of Eastern European countries to the EU) will focus attention on individual countries' comparative advantage and international competitiveness. The cost of capital and labour as well as industries' ability to respond to consumer requirements in meat processing will increasingly dictate the patterns of meat trade. The rapid pace in technology transfer, along with cross-border investments in meat production and processing, may accelerate trading changes together with increased product diversity.

Pressures on the competitiveness of traditional developed country meat exporters have stemmed from the increasing stringency of environmental regulations targeting animal feeding operations and waste management. In the Netherlands, it is estimated that strict regulations on phosphate quotas, regulations on waste treatments, restrictions on storage and field applications and, more recently, direct output controls are costing up to US$4.05/hog (Metcalfe, 2001). In the United States, it has been estimated that waste management costs vary from US$0.40 to US$3.20 per hog, or 1 to 8 percent of total hog production costs. Application of these regulations results in higher producer costs and a loss of competitiveness for specific industries, particularly those in the EU.

Social, health and environmental issues related to animal production are being raised more explicitly in WTO negotiations. The competitive playing field for animal industries will change if these issues are formally incorporated as "trade issues," and related international regulations are harmonized (as has happened with animal diseases, zoonoses and food safety). In addition, decisions on how to incorporate animal welfare concerns, traceability and meat labelling will affect exporter competitiveness. Specifically, it will disadvantage developing country exporters who, by the nature of their smaller-scale export operations, will have higher compliance costs.

5.4.4 Intensification: public health risks and consumer choices

Food safety and emerging zoonoses. Access to safe and healthy food products is an important public good. Animal products, especially animal fat, are linked with human health risks, but some of these risks are associated only with overconsumption. At low to moderate intakes, meat, milk and egg products are beneficial and provide essential amino acids, minerals and vitamins. Indeed, an increase in consumption of animal products in developing countries will help to combat some forms of undernutrition.

Growing densities of livestock, particularly in peri-urban and urban areas, import of feedstuffs from distant areas, and shifts in dietary habits have raised concerns about diseases, microbial contamination of food and general food safety.

Changes in production systems, changing feeding practices and the safety of animal feed may increase the risk and change the pattern of disease transmission. The recent upsurge of human cysticercosis in eastern and southern Africa following the expansion in pork production is an example of how a zoonotic disease may become a significant risk when production systems change without accompanying changes in veterinary regulations and enforcement.

Certain livestock diseases (zoonoses) can also affect humans, e.g. brucellosis and tuberculosis, and new zoonoses may originate from livestock populations (e.g. nipah and avian flu). The potential dangers are clearly demonstrated by the emergence of BSE (bovine spongiform encephalopathy) in cattle and its ramifications for human health (variant Creutzfeld-Jakob disease) and the livestock industry. It is estimated that new diseases have been detected at the rate of one a year over the last 30 years.

Meat, milk and eggs are perishable products and susceptible to contamination by microbes. Some of these, such as salmonella and *Escherichia coli*, reside in the intestinal tract of livestock. Thus inappropriate handling, slaughter hygiene, processing or preservation throughout the food chain can result in contamination and propagation of microbes and create serious health risks for consumers. Although many microbial contaminants are of little or no effect if products are prepared appropriately, contamination is the leading reason for ever-increasing sanitary standards imposed on livestock processing and transport. Stricter regulations – combined with a growing number of affluent consumers prepared to pay a premium for organic or free-range livestock products – are likely to have a lasting effect on production methods, particularly in the industrial countries. Changes will include the application of pre-harvest food safety programmes through the Hazard Analysis Critical Control Point (HACCP) concept at farm level, from breeding to the slaughterhouse, and pathogen reduction or elimination programmes along the production chain, such as the elimination of *Salmonella enteritidis* from breeder flocks.

Food safety also concerns biological and chemical contaminants. Aflatoxins, for example, are of major importance in humid and warm environments, and drug residues are another major category of contaminants.

In order to improve food safety from feed production to the supermarket shelf, basic food quality control systems are evolving into quality assurance systems, and these in turn are moving towards total quality management (TQM) systems. The costs of compliance with these systems are high and will lead to further concentration and integration of the food chain.

Antimicrobials and hormones. Antimicrobials are widely employed in the livestock sector, and not just for therapeutic purposes. It is common practice, in modern farming systems with high animal density, to supplement feed with subtherapeutic doses of antimicrobials so as to enhance growth rates. Constant exposure to antimicrobials, however, promotes the development of resistant microbes. Resistance to antimicrobials in farm animal pathogens can be passed on to bacteria of humans through the exchange of genetic material between micro-organisms. This can increase public health costs by necessitating the use of more expensive drugs for treatment, and longer hospital stays. The problem of antibiotic resistance is compounded by the fact that no truly novel antibiotics have been developed over the last decade. Most antimicrobial resistance in human pathogens stems from inappropriate use of these drugs in human medicine, but the use of antimicrobials in the livestock sector plays a contributing role.

The World Health Organization (WHO) has recently called for a ban on the practice of giving healthy animals low doses of antibiotics to improve their productivity, and the EU has implemented a partial ban on six antibiotics that are also used in the treatment of humans. Although such moves are strongly opposed by livestock producers, they are likely to gain momentum, particularly in the light of consumer demands. In 1998, the Danish poultry industry decided voluntarily to discontinue the use of all antimicrobial growth promoters, despite concerns that this would result in decreased productivity and increased mortality. Contrary to expectations, mortality was not affected and the feed-conversion ratio increased only marginally. Similar steps have been taken by the United Kingdom's largest poultry producer.

Hormones that increase feed conversion efficiency are used in many parts of the world, particularly in the beef and pig industry. No negative impacts on human health as a result of their correct application have been scientifically proven. However, the EU, partly in response to consumer pressure, has taken a strict stand on the use of hormones in livestock production. This has led to major trade disputes with the United States. These examples show that consumer concerns will increasingly influence not only the quality of the end product, but also the ways in which it is produced.

Livestock biotechnology. As defined by the Convention on Biological Diversity (CBD), biotechnology is "any technological application that uses biological systems, living organisms, or derivatives thereof, to make or modify products or processes for specific use". Much of the debate to date has been about biotechnology applications to crops, but many aspects, including intellectual property rights and biodiversity, are also relevant for animals.

Biotechnology with respect to animals may be classified into two groups: reproductive interventions, including artificial insemination (AI), embryo transfer (ET), in vitro maturation, fertilization and sexing, and cloning; and DNA-based genetic interventions, including marker-assisted selection, DNA vaccines, GM marker vaccines, recombinant vaccines and the development of GM livestock.

AI is an established technology particularly in the commercial dairy sector in all industrial countries. In future, widespread use of AI is likely to occur in the more favoured production environments of developing countries, where the demand for milk provides an economic incentive for its introduction. ET allows cows of high genetic potential to produce a much larger number of calves than with normal reproduction. But this technology is currently limited to only a small part of commercial herds and breed improvement programmes in some industrial countries. This and other advanced reproductive technologies are unlikely to spread widely in developing countries within the foreseeable future.

Recent advances in cloning of mammalian cells could potentially have a very large impact, particularly for dairy cattle in industrial countries. There is, however, the danger of further narrowing the genetic base and decreasing disease resilience. Cloning is a wasteful process (only 2 to 5 percent of attempts to clone animals currently succeed) and cloned animals often develop serious health problems. Cloning is thus an area where a number of complex ethical and scientific issues still have to be resolved.

The rapid increase in the understanding of the genetic make-up of animals is likely to have a major impact on animal breeding. Genetic improvement could be accelerated by direct identification of genes affecting important traits for economic performance or disease resistance, together with the use of neutral markers. This technology is being developed for a large range of traits, and could provide a short cut to genetic development in developing countries, not only for productive traits but also for adaptation to adverse environmental conditions such as climate and diseases.

Existing technologies – including even the more basic biotechnologies available as listed above – are underutilized in many developing countries. This is a result of a number of factors, including underinvestment in education, health, financial, communication and transport infrastructure that would reduce the costs and risks of input supply and product marketing (Thompson, 1997). These limitations are likely to hold back even more the application of the new sophisticated technologies, so that the technology gap between industrial and developing countries may grow rather than shrink.

Over 80 percent of research activities in biotechnology are conducted by large private companies for commercial exploitation to meet the requirements of developed markets and large-scale commercial producers (Persley and Lantin, 2000). They are thus unlikely to be very suitable for the conditions of small-scale farmers in tropical regions and this may lead to increasing inequality of income and wealth within countries. In addition, legal registration requirements can act as barriers to commercial product introduction, giving relative advantage to large, mostly international corporations that have sufficient institutional infrastructure and financial resources to meet intensive registration requirements.

GM livestock and feeds. Estimates of the current number of GM animals being produced are difficult to establish. The vast majority have been used for biomedical research, and have been engineered to produce different products such as human proteins in their milk. Researchers are currently planning to modify chickens to express certain products in their eggs that may have medicinal use.

GM cattle, sheep, pigs and chickens are also being produced for eventual direct human consumption, although this is still largely in the experimental phase. For example, genes are being transferred to livestock with the aim of increasing the endogenous production of growth hormones to speed up growth rates. So far, however, this process has had negative impacts on animal welfare.

The main risks to animal and human health and the environment from GM livestock arise from pleiotropy, i.e. unexpected changes in form and function resulting from gene insertion, especially when these take place without sufficient testing before widespread release.

GM animals and their products themselves are not at present used for human consumption. But products are on the market from livestock fed with GM crops (e.g. corn and soybean feeds used in poultry meat and egg production and cottonseed cake in dairy production). In 2001, the Indian Government commissioned field trials of transgenic cotton in various agroclimatic regions of the country to assess its environmental safety. Parallel nutritional studies were carried out in buffaloes and cows to determine whether transgenic cottonseed and oil have any effect on animal health, milk production and quality with regard to human health and toxicity.

Little impact of GM food and feed has been seen so far in international markets, but some biotechnology-related changes in international trade are expected. Currently, importers wanting to obtain GM-free food to satisfy strong consumer demand have had the choice between importing feed from GM-free regions, or importing segregated non-GM products from GM-producing regions. Market segmentation in food products is taking place, in the EU and Japan particularly, where GM-free soya food products are receiving price premiums above products made from mixed GM-free and GM soya products. So far this market development is reported to have had little impact on the pattern of global trade in foods, as GM-free products can be obtained without significant extra cost. However, consumers are only now beginning to demand products from animals fed on GM-free diets, and this trend may have a significant impact on international bulk-based feed handling and commodity trade flows.

5.4.5 Livestock, environment and animal genetic diversity

There can be substantial environmental benefits from keeping livestock in mixed farming systems. Historically, mixed crop-livestock systems have formed the basis for agricultural intensification. In these systems, livestock not only accelerate nutrient turnover, but also provide a mechanism to import and concentrate nutrients, which is key to intensification. Livestock can also perform important functions in landscape and ecosystem maintenance.

However, many contentious issues remain to be addressed. Widespread poverty-led land degradation, and associated biodiversity losses, are occurring in semi-arid and humid environments because of increased population pressure, ill-defined access to land and other resources, and poor access to markets and financial services. Degradation reinforces poverty by reducing the productivity of shared resources and by increasing vulnerability.

Pollution of land, water and air from intensive livestock production and processing in both industrial and developing countries has raised awareness of the associated environmental problems. This pollution can directly damage the environment, and it can also become a major vehicle for disease transmission. Animals are also associated with global warming. Directly and indirectly, domesticated livestock produce greenhouse gases: carbon dioxide (CO_2) and methane (CH_4) together with small quantities of ozone (O_3) and nitrous oxide (N_2O). These issues are covered in detail in Chapter 12. Examples of other livestock-related environmental impacts are summarized below.

Land degradation is the result of a complex interaction involving livestock movement and density, land tenure, crop encroachment and fuelwood collection. It is particularly evident in the semi-arid lands of Africa and the Indian subcontinent. Changing land tenure, conflicts, and settlement and incentive policies have undermined traditional land use practices and exacerbated the situation in many cases. Earlier reports of widespread irreversible degradation may have exaggerated the extent of the problem, but there is no reason for complacency and concerted action is needed.

Deforestation. Commercial ranching has been responsible for the destruction of vast areas of rain forest, with a serious loss of biodiversity. The problem arose from policies that promoted inappropriate ranching and was largely, but not exclusively, confined to Central and South America.

Animal genetic diversity. Intensification is associated with the risk of losing farm animal genetic diversity through a reduction of the gene pool. Almost 5 000 identified breeds and strains of farm animals have been identified. Currently some 600 breeds are in danger of extinction, and further erosion of many traditional and locally adapted breeds is expected.

Over the past decades, several new breeds of farm animal species have been successfully developed. Such animals produce more meat, milk or eggs, as long as they receive ample quantities of high-quality feed and are protected from harsh weather, pests, diseases and other kinds of environmental stress. But the consequence has been an increasing dependency on a narrowing genetic resource base, facilitated by biotechnologies such as AI, which allows easy transfer of genetic material across international borders.

The wholesale transfer of breeds suited to high-input production systems from industrial to developing countries is one of the greatest threats to domestic animal diversity. Development policies favour such introductions, which generally work against the survival of local breeds. It is estimated,

for example, that while 4000 of the world's remaining breeds are still popular with farmers, only about 400 are the subject of genetic improvement programmes, almost all of them in industrial countries.

Maintaining animal genetic diversity allows farmers to select stock or develop breeds in response to environmental change, disease threats, consumer demands and changing market and societal needs. Genetic diversity thus represents a storehouse of largely untested potential. Breeds that utilize low-value feeds, survive in harsh environments or have tolerance to or resistance against specific diseases may offer large future benefits. Wild relatives of common breeds, in particular, may contain valuable but as yet unknown qualities that could be useful now or in the future (FAO, 1999d). The characterization, conservation and use of tropical animal breeds are vital if livestock production is to respond to changing production environments worldwide. Adapted livestock are more resistant to disease and environmental challenges, and can remain productive without the need for high-value inputs, and thus increase farm income and contribute to poverty alleviation.

Considerable potential exists for the within-breed improvement of locally adapted breeds, given that most indigenous populations have not been subject to heavy selection pressure for particular traits. Such programmes promise significant returns, but are necessarily long-term in nature and require substantial commitment and investment.

5.5 Concluding remarks

Increasing population, urbanization and disposable incomes in developing countries are fuelling a strong growth in demand for animal food products. This will have a major impact on the location and organization of livestock production. Changes in the latter will in turn strongly impinge on animal and human health, the livelihoods of the poor and the environment. The following consequences of these trends are expected:

■ a major increase in the share of the developing countries in world livestock production and consumption;

- a gradual substitution of cereals and other basic foods by meat and milk in the diets of developing countries;
- a change, varying in speed between regions, from multiple production objectives to more specialized intensive meat, milk and egg production within an integrated global food and feed market;
- rapid technological change and a shift to more industrial production and processing;
- a rapid rise in the use of cereal-based animal feeds;
- greater pressures on fragile extensive pastoral areas and on land with very high population densities and close to urban centres; and
- increasing livestock disease hazards, with public health consequences, particularly in areas with high livestock densities near human population centres.

The future holds both opportunities and serious pitfalls for animal production in developing countries. There is a danger that livestock production and processing will become dominated by integrated large-scale commercial operations, displacing small-scale livestock farmers and thus exacerbating rural poverty and malnutrition.

Uncontrolled expansion of highly intensive animal production could have major environmental consequences. On the other hand, if correctly managed, a dynamic livestock sector could prove catalytic in stimulating rural economies. However, if it is to take on this role, it will require proactive policies from the private and public sector, such as:

- the removal of policy distortions that artificially increase economies of scale and disadvantage small-scale producers;
- building of institutional and infrastructural capacities to allow small-scale rural producers to compete and integrate successfully within the developing livestock industry;
- a conducive environment, through public sector investment where necessary, to allow producers to increase production through improved efficiency and productivity; and
- effective reduction of environmental, animal and human health threats.

Without such proactive development policies, the impact of increased demand for livestock products on food security and safety, environmental protection and poverty reduction will be far less favourable.

6.1 Introduction

Globally, forestry, like other sectors, will continue to face numerous challenges and changes resulting in an increasingly intense scrutiny of the contribution of forests to sustainable development. The general trend during the past 20 years has been an intensification of demand pressures on a diminishing forest resource base. The demand for forest products and services has increased in response to increase in population and income.

Historically, production of wood and wood products has been the main objective of forest management and other functions were not explicitly accounted for. However, there has increasingly been a shift away from this towards assigning a higher priority to the environmental and social functions of forests. Conflicting demands and differing views of the relative importance of various goods and services provided by forests and alternative land uses will increasingly have to be reconciled. The complexity of predicting future developments in forestry is increased by demands for a more equitable distribution of the benefits from forests, for safeguarding the rights of forest dwellers and indigenous people, and for ensuring widespread participation in decision-making.

The predecessor to this study, *World agriculture: towards 2010*, (Alexandratos, 1995), provided a general indication of the principal changes occurring in the forestry sector. Many of the changes outlined in the previous study still persist, but a number of new issues are also emerging, reinforcing or shifting the direction of development of forestry at all levels. A brief comparison between the conclusions and projections of the two studies is useful, both to demonstrate changes in fundamental structures in the forestry sector since the previous study, and also to highlight conceptual changes in outlook studies. Since the 1995 study, FAO has markedly strengthened its capacity in forestry sector outlook work through a variety of studies including a *Global Forest Products Outlook Study*, a *Global Fibre Supply Model* and a range of regional outlook studies, commencing with an *Asia-Pacific Forestry Sector Outlook Study* and a *Forestry Outlook Study for Africa* (in progress). At the same time, baseline forestry data have been updated and improved through the *Global Forest Resources Assessment 2000* (FAO, 2001c).

Probably the greatest conceptual shift has been the development, throughout the 1990s, of a clear consensus that there is no impending "global forest crisis". In part, this recognizes that previous projections of consumption of wood products have not adequately taken into account all relevant factors. Thus, for example, the 1995 study noted "a nearly tripling of consumption (of forest products) in developing countries over the next two decades ... In the developed countries, the growth in consumption of forest products is projected to be lower than the growth of their economies, with consumption somewhat less than doubling over the next 20 years". These projections implied an industrial roundwood consumption of around 2.7 billion m^3 by 2010. More recent projections suggest global consumption of roundwood will be markedly lower, reaching only 2.4 billion m^3 by 2030. Similarly, estimates of fuelwood consumption are projected to be lower than the 1995 forecast, while the supply situation looks more promising. There is evidence of real improvements in forest management and harvesting and processing technologies, increases in plantation establishment and a better appreciation of the role of trees outside forests in wood supplies. At the same time, the pressure for agricultural expansion – often achieved at the expense of forest land – is expected to be less during the next three decades, suggesting that a more sustainable future may be at hand for the forestry sector.

This chapter revisits some of the issues dealt with in the 1995 study, discusses new factors that may affect the development of the sector, and outlines a broad scenario to illustrate likely developments during the next three decades. The following sections consider, in turn: the forces shaping developments in the forestry sector; the principal changes expected over the period to 2015 and 2030; and major perspectives and issues relating to forestry, including possible policy responses. A concluding section summarizes likely scenarios for developments in the forestry sector to 2015 and 2030.

6.2 The presente state of forests

Forest area. One of FAO's major forestry programmes is the periodic Forest Resources Assessment (FRA), which produces a statistical snapshot of the global status of forest resources once in a decade. The most recent of these, the *Global Forest Resources Assessment 2000* (FRA 2000; see FAO, 2001c) estimates the global forest area in 2000 at about 3 870 million ha or about 30 percent of the land area. Tropical and subtropical forests comprise 48 percent of the world's forests, while temperate and boreal forests account for the remaining 52 percent. Natural forests are estimated to constitute about 95 percent of global forests, while plantation forests constitute around 5 percent. Developing countries account for 2123 million ha (55 percent) of the world's forests, 1850 million ha of which in tropical developing countries. FRA 2000 revealed a net annual decline of the forest area worldwide of about 9.4 million ha between 1990 and 2000. Annual forest clearance was estimated at 14.6 million ha, and annual forest area increase at 5.2 million ha. Nearly all forest loss is occurring in the tropics.

Table 6.1 Global forestry at a glance, 2000

	Total forest				Annual cover change	
	Area	Share of land area	Area per capita	Plantation Area	1990-2000	
	'000 ha	%	ha	'000 ha	'000 ha	%
Africa	649 866	21.8	0.85	8 036	- 5 262	- 0.78
Asia	547 793	17.8	0.15	115 847	- 364	- 0.07
Oceania	197 623	23.3	6.58	2 848	- 365	- 0.18
North and Central America	549 304	25.7	1.15	17 533	- 570	- 0.10
South America	885 618	50.5	2.60	10 455	- 3 711	- 0.41
Europe	1 039 251	46.0	1.43	32 015	881	0.08
World	3 869 455	29.6	0.65	186 733	- 9 391	- 0.22

Source: FAO (2001c).

Supply of wood and wood products. Supplies of wood are not drawn on all global forests but from a significantly smaller area. FAO's Global Fibre Supply Model (see Bull, Mabee and Scharpenberg, 1998) estimates that only around 48 percent of global forests are available for wood supply. The remainder is either legally protected or physically inaccessible, or otherwise uneconomic for wood supply.

The greater part of global wood production is burned as fuel; fuelwood at present accounts for over 1.7 billion m^3 or around 55 percent of the annual global wood harvest. The vast majority of wood burning occurs in developing countries where wood is often the most important source of energy. Annual global production of industrial roundwood currently amounts to slightly over 1.5 billion m^3, around 45 percent of the global wood harvest. Two-thirds of industrial wood products are consumed in developed countries, which account for less than one-fifth of the global population. Interestingly, annual per capita wood consumption in tropical and temperate and boreal regions is approximately equal (just over 0.5 m^3). In temperate and boreal countries, however, more than 75 percent of wood consumption is in the form of industrial wood products while, conversely, in tropical countries more than 85 percent of wood consumption is in the form of fuelwood.

Forest-dependent people. It is estimated that around 450 million people, or about 8 percent of the global population, live in forest ecosystems. Around 350 million of the world's poorest people are entirely dependent on forest ecosystems for their livelihoods. These people are often marginalized in terms of opportunities for development and, as such, are dependent on forests for their livelihood. Reducing forest pressures created by expansion of agriculture, grazing and fuelwood gathering depends on more general economic and social development to provide alternative income-earning opportunities. While development can reduce the pressures emanating from rural poverty, it also generates increased demands for a number of forest products. Direct and indirect impact of this on forests will largely depend on technological changes in production and the pattern of distribution of and access to forest products.

Conservation of biodiversity and other values. The global area of forests in formally protected areas is estimated to be 479 million ha, or 12.4 percent of total forest area (FAO, 2001c). On the surface, this shows considerable progress towards the recommendation of the International Union for the Conservation of Nature and Natural Resources (IUCN) of having at least 10 percent of forest area in protected areas. Significant challenges still remain, however, in ensuring balanced representation of all forest types, and in ensuring effective protection of forests. At present, many countries still fall well short of having 10 percent of their forests protected, while in many other countries the representation of protected forest types is badly skewed. Furthermore, in many developing countries, protection is afforded to protected forests on paper only, while in reality they remain subject to substantial encroachment including logging, deliberate burning and other forms of clearing or degradation.

6.3 Forces shaping forestry and areas of change

Statements about what is likely to happen in the future necessarily include elements of speculation, increasingly so the longer the projection period. As explained in Chapter 1 (see also Appendix 2 on the approach followed in this study), the objective is to sketch out the "most likely developments" to 2015 and 2030 in the forestry sector, taking into account foreseeable changes. Projections are not trend extrapolations – they incorporate an evaluation of factors, already under way or expected to become important in the future, which would cause deviations from past trends.

In many countries forestry is a low-priority sector and what happens to forests and forestry is to a great extent decided by changes outside the sector. The multifunctionality of forests, arising from the diverse values assigned to them, adds to the complexity of management. The development and future of individual forests, and of forests as a whole, will largely depend on the relative importance assigned to each of these values and on the strength of political advocacy for each value. The key questions in developing an outlook for forestry over the next three decades are the factors that will have an impact on each of these demands and how policy environments will evolve to alleviate and reconcile demand pressures.

Increasing demands for forest products arise from growing populations, increasing prosperity as a result of economic development and changes in consumer preferences. In this latter category, the most significant changes in preferences include the explicit recognition of the value of the environmental and social dimensions of forestry. Trends in each of these central, shaping pressures are examined below.

Demographic changes. Population growth, together with agricultural expansion and agricultural development programmes, is a major cause of forest cover changes (FAO, 2001c). Population growth results in an increase in demand for food, which is met through intensification and arable land expansion. As discussed in Chapter 4, a net increase of 120 million ha is projected in developing countries to expand crop production over the period to 2030. Part of this is expected to come from forest clearance. Population growth will also increase demand for wood and wood products, including fuelwood. Changes in population distribution, especially urbanization, will change the demand for fuelwood, especially charcoal, when incomes are insufficient to procure alternative sources of energy. Depletion of wood resources from areas supplying urban centres has been a major problem in several countries and this is expected to persist during the next three decades, unless alternative sources of energy are more widely used.

The overall impact of demographic changes on forests will vary between countries. In most developed countries where population has stabilized, deforestation has been arrested. Indeed the extent of forest cover is on the increase, primarily because of reversion of agriculture areas to forests. The number of countries where this process could happen is expected to increase.

Economic changes. Rising incomes and changes in income distribution are major factors affecting forestry. With rising incomes, demand for forest products increases, especially for processed items (e.g. panel products, furniture, printing and writing paper, and so on). In countries where control of access to forests is weak, increased demand for forest products is likely to exacerbate

problems of poor forest management and excessive (including illegal) logging. In some countries, however, increasing prosperity may actually reduce some of the poverty-driven pressures on forests by providing alternative sources of livelihood. Growth of the industrial and services sectors reduces direct dependence on land.

Forests have often served as an engine for economic growth. In the early stages of economic development many countries have harvested forests as a means of capital accumulation. Most developed countries have gone through a phase of forest depletion during the early years of industrial development and many developing countries are currently moving through this phase. The challenge for developing countries that rely on forestry as a significant earner of foreign exchange is to maintain a healthy rate of development without excessively depleting forests.

The scenario that may emerge by 2015 and 2030 could be as follows:

■ More countries in Asia and South America would have made significant progress in terms of economic growth, leading to significantly reduced dependence on land.

■ Some countries that have depended on forests to generate investment funds would have diversified sufficiently, reducing their dependence on forests and increasing their willingness to adopt more sustainable management practices. However, there will still be several countries in Asia, Africa and South America where economic development and growth of the non-farm sectors would be inadequate to facilitate such a major shift from land-based activities. Deforestation and its attendant problems are likely to persist in these countries.

■ In several countries economic progress would enable substantial increases in investment in large infrastructure projects, often resulting in deforestation. This is particularly the case in a number of countries in Asia and South America. Forest clearance for infrastructure development (such as large irrigation and hydroelectric projects) and expansion of area under industrial crops could counterbalance the reduction in forest clearance for arable cropping.

■ Even in Asian and Latin American countries that make significant economic progress, if large areas of deprivation persist, people will continue to depend on land and other natural resources, exerting direct and indirect pressure on forests.

■ While consumption of forest products is expected to stabilize in the developed economies, an increase in consumption of all wood products is expected in most developing countries.

■ The demand for forest services such as recreation and ecotourism is expected to increase.

Political, social and institutional changes. The last two decades have witnessed major changes in the openness, transparency and decentralization of governance. Local communities and groups have achieved an increasing say in the management of natural resources.

More pluralistic institutional arrangements for resource management are expected to emerge and strengthen. Several countries have revised their forest policies and changed legislation, enabling the participation of the private sector in forest management, including community groups and farmers. Participatory approaches such as joint forest management and management by forest-user groups are becoming more acceptable, although the area managed under such arrangements is still very small. These efforts are expected to gain momentum during the next three decades.

Forestry departments in many developing countries could change their structure and functions. Their primary responsibility will shift from direct management of forests to policy development and other regulatory functions. Actual resource management will largely be the responsibility of the private sector, including farmers and local communities. The proliferation of information (and the potential for misinterpretation) requires proper interpretation and improved public access to information. This will be another important emerging function of the public sector forestry agencies.

Forests, environment and global initiatives. The growing concern about the environment encompasses a wide range of forest functions, including protection of watersheds, conservation of biodiversity and the perceived role of forests as carbon sinks. As water becomes a more critical resource, the role of forests in flood control, in rainfall interception and in controlling erosion will assume even

greater importance, requiring better knowledge and quantification.

These concerns have become a key driving force in forestry initiatives following the 1992 UN Conference on Environment and Development (UNCED). UNCED Agenda 21, in particular Chapter 11, has formed the basis for several initiatives. Three international, legally binding conventions, the Convention on Biodiversity (CBD), the UN Framework Convention on Climate Change (UNFCCC), and the Convention to Combat Desertification in those Countries experiencing Serious Drought and/or Desertification, particularly in Africa (CCD), are of direct relevance to forestry. They have highlighted the importance of forests in providing a range of goods and environmental and social services, and have helped to widen the focus of international attention from forests proper to trees in fragile ecosystems such as drylands and mountains.

National and international non-governmental organizations (NGOs) and advocacy groups have brought forestry and environmental issues to the forefront of public debate. NGOs and civil society will continue to have a pioneering role in influencing forestry policies and programmes. Since UNCED, countries have attempted to make forestry policy more consistent with the concepts of sustainability and environmental soundness. The Ad hoc Intergovernmental Panel on Forests (IPF), the Intergovernmental Forum on Forests (IFF) and the United Nations Forum on Forests (UNFF) have helped to articulate the policy, institutional, social, environmental and technological issues relating to forest management, although commitment to a legally binding mechanism for conserving and managing forests still remains elusive.

These developments have not only generated pressures against the conversion of forests to alternative land uses but they have also resulted in a closer scrutiny of forestry practices, including logging of natural forests and plantation management. These concerns are expected to dominate the debates at local, national and global level and would have the following implications:

■ increasing concern for protecting the environment, leading to a better recognition of the global, national and local significance of forests; improved mechanisms to determine the trade-

Table 6.2   Trade in forest products as a proportion of production, 2000

Region	Industrial roundwood			Sawnwood			Wood-based panels			Pulp			Paper and paperboard		
	P mln m³	E m³	S %	P mln m³	E m³	S %	P mln m³	E m³	S %	P mln mt	E mt	S %	P mln mt	E mt	S %
Africa	69	6	9	8	2	25	2	1	50	3	1	33	3	1	33
Asia	229	11	5	54	7	13	47	15	32	22	2	9	95	12	13
Oceania	47	9	19	8	2	25	4	1	25	3	1	33	4	1	25
North and Central America	624	15	2	194	57	29	60	14	23	86	18	21	112	25	22
South America	153	2	1	30	5	17	11	3	27	11	5	45	10	1	10
Europe	464	71	15	128	56	44	65	23	35	47	11	23	100	58	58
World	1587	114	7	421	128	30	189	56	30	171	38	22	323	98	30

Note: P=production; E=exports; S=exports as share of production (in percentage).
Source: FAOSTAT.

offs between competing objectives; and appropriate arrangements to share costs and benefits;

■ recognition of the rights of indigenous and forest-dependent people and an increasing concern to protect their livelihood;

■ integration of tree cultivation with other land uses, to reduce environmental problems and to enhance income opportunities; and

■ global and regional initiatives to enhance the role of forests in mitigating climate change, especially their potential as carbon sinks.

International trade. A significant proportion of global production enters international trade, reflecting the difference between the geographic distribution of industrial roundwood production and processing capacity, and of demand for forest products. Table 6.2 shows that the export share of more processed products is significantly higher than that of less-processed products, reflecting most countries' efforts to capture foreign exchange earnings through higher value-added forest processing. For example, the export share of sawnwood and wood-based panels is much higher than that of its raw material, namely industrial roundwood, and likewise for paper and wood pulp.

The largest forest product trade flows are intraregional. Europe, North America and Asia combined account for 90 percent of the value of forest product exports and around 95 percent of imports. Europe accounts for almost a half of the global trade in forest products, and 80 percent of European trade is between European countries. Similarly, trade between the United States and Canada accounts for around 70 percent of North and Central American trade, while Japan is a partner in around 35 percent of trade in the Asia-Pacific region.

As further trade liberalization takes place and barriers to trade diminish, there could be significant changes in the composition, volume and direction of trade during the next three decades. Countries will continue to develop domestic wood-processing facilities, and thus the share in trade of processed and semi-processed products will increase relative to that of unprocessed products. Economic growth in Asia is creating new opportunities for trade of processed and unprocessed wood and this could accelerate the shift in the direction of trade already evident. Increased demand from Asia coupled with restrictions or outright bans on logging of natural forests in some of the Asian countries have to some extent contributed to unsustainable (often illegal) logging in Africa and Latin America.

Developments in science and technology. Developments in science and technology and their wider diffusion could lead to the emergence of more knowledge-based operations and management. The ability to take advantage of natural processes will increase considerably, leading to more benign approaches to resource management. Access to knowledge and information will improve, and even societies that have historically been isolated will have access to information. Of particular interest to forestry are the following probable developments:

- Better understanding of ecological processes will improve the ability to fine-tune human interventions, minimizing the adverse effects of several forestry practices.
- Application of improved remote sensing technologies will allow the state of resource changes to be assessed on a real-time basis and to be available to wider groups of people.
- Improved knowledge of the chemical and genetic make-up of a large number of species will widen the scope for their use, leading to the emergence of new processes and products.
- Significant improvements in processing technologies will reduce raw material requirements (for example, see Abramovitz and Mattoon, 1999) and environmental pollution.
- Increased use of tree breeding and application of vegetative propagation technologies will ensure availability of high-quality planting materials, especially for establishment of large-scale plantations.
- Energy sources will probably shift, with wind and solar power becoming more important and widely accessible. This may have a significant impact on forestry, especially on the use of fuelwood.

Tree breeding is increasingly drawing on a range of biotechnological tools. These fall broadly into three categories: (i) biotechnologies based on molecular markers that help quantify genetic diversity and identify promising genotypes and genes for subsequent use in breeding programmes; (ii) technologies that enhance vegetative propagation; and (iii) genetic modification. While the first two techniques are relatively commonly used to support existing tree-breeding programmes, genetic modification is still at an experimental stage. To be of practical value, genetic modification must offer unique features that cannot be economically delivered through selection or more conventional breeding and that are capable of offsetting the costs and time involved. All new technologies must be tested and proved to be environmentally

safe before they can be used on a large scale. Important traits in forest trees, such as growth rate, adaptability to harsh environmental conditions and stem and wood quality, are governed by an array of genes and may not lend themselves readily to genetic modification. Should genetic modification be pursued on a large scale in forestry, the characteristics most likely to be targeted are simple inherited traits, such as modified lignin content, insect and virus resistance and herbicide tolerance. There are, to date, no reported large-scale or commercial plantations of genetically modified trees.

Although changes up to 2015 may be slow, by 2030 developments in science and technology are expected to have more far-reaching impacts. But the pace of adoption of technological change will vary enormously between and within countries. This will depend on investment in human resources, on providing favourable institutional arrangements, etc. In the short term, disparities in access to and ability to use knowledge may widen. Large areas of underdevelopment may persist in a number of countries, limiting the participation of a sizeable section of people in knowledge and technology-based activities.

Globalization and its impacts. Economic liberalization and increasing interdependence will provide new opportunities, but will also mean that perturbations in one sector or country could easily affect other sectors and countries. Environmental interdependence such as in global climate change or the transboundary effects of deforestation and other related changes, could affect economic interdependence. The implications of economic and environmental integration on forestry could be far-reaching. Of particular importance could be the following factors:

- The movement of technologies, processes and products across countries will increasingly be based on comparative (competitive) advantages. Traditional forest-based enterprises that fail to take advantage of the emerging opportunities will fade out and be replaced by more efficient producers.
- Regional and international trade agreements will continue to alter the pattern of trade in forest products. While tariff barriers may become less important, several non-tariff barriers, such as those relating to certification

and ecolabelling, could become critical in determining access to certain markets (Bourke and Leitch, 1998).

- Global, national and local pressure groups will play an increasing role in shaping forest management, each group trying to push their agenda based on the group's interests and objectives.

However, economic integration is not necessarily a benign process conferring benefits to all at the same time. Disparities in access to knowledge, resources and markets could widen the economic gap between countries, and could also accentuate resource-use conflicts within countries.

6.4 Probable changes up to 2015 and 2030

In view of these driving forces, what will be the changes in forestry during the next three decades? Because of the complexity of factors and their interaction, it is difficult to provide a clear indication. However, we can indicate the broad direction of probable changes.

Forest and tree resources. The key issue for forest and tree resources is progress towards wider adoption of sustainable forest management (SFM). There is some evidence of progress, although scattered and disparate. In some areas, excellent forest management complies with criteria agreed upon for SFM. Elsewhere, mainly in the tropics, substantial tracts of forests remain unmanaged and are frequently severely degraded by careless logging, burning or other destructive practices. Initiatives to remedy the situation include the development and implementation of criteria and indicators for SFM, and subsequent adjustments to forest management practices; establishment of certification processes; implementation of pilot projects and operations aimed at reduced impact logging; and extension of improved silvicultural systems.

All these initiatives are helping efforts to manage forests to meet a range of long-term objectives and demands. In most developed countries, forest stocks have been increasing for a number of years, both because management of existing forests has been intensified (including the establishment of forest plantations) and because significant land areas of marginal farmland have been allowed to regenerate as secondary natural forest. The challenge for these countries is to continue to improve forest management to encompass the new dimensions of sustainability. Conversely, in a number of developing countries, even to arrest deforestation and forest degradation would mean considerable progress.

Shift in the source of wood supplies. During the next three decades, the shift away from dependence on natural forests for wood production is expected to strengthen. Increasingly wood is being obtained from intensively managed plantations, thus reducing the pressure on natural forests and woodlands and enabling the latter to fulfil other functions. Between 1980 and 1990 the area under forest plantations increased from about 18 million ha to nearly 44 million ha. By 2000 it had increased further to a total of 187 million ha. Although plantations account for only 5 percent of forest cover, they have become a significant source of roundwood supply.

There is also an increase in tree cultivation on farms and other land outside forests. In several countries (e.g. in Bangladesh, India, Kenya and South Africa) trees from farms are an important source of industrial roundwood and fuelwood. Often such efforts are supported by industries through provision of technical assistance and assured prices. Stable or increasing prices and secure tenure are needed to encourage such efforts. As farmers adopt more integrated land uses with trees as an important component, an increasing proportion of local wood supplies is expected to come from trees outside forests.

Consumption of industrial roundwood. Consumption of industrial roundwood during the next 30 years is expected to move modestly upwards. The most recent projections of FAO's Global Forest Products Model estimate that global consumption of industrial roundwood in 2030 will be around 2 400 million m^3, an increase of around 60 percent on current consumption. Much of this increase will result from an increase in population and income in developing countries, leading to an increase in per capita consumption of wood (and particularly paper) products. Offsetting these trends will be changes in consumer preferences, changes in techniques of forest management, more efficient utilization of wood and residues, and the development of new products.

Current indications are that the real prices of most forest products are unlikely to increase significantly in the long term. There may be short-term fluctuations, but these will be evened out through changes in supplies and shifts in sources of supplies, including substitution by non-wood products. Similarly, the spread of SFM may reduce supply and increase costs in the short term, but it is unlikely to affect supplies in the long term. In fact, by making supplies sustainable, it is expected to stabilize long-term supplies and thus help maintain price stability.

A marked change will take place in the share of roundwood grown on plantations. Industrial roundwood production from plantations is expected to at least double during the next 30 years, from the current 400 to around 800 million m³. By 2030 plantations will supply about a third of all industrial roundwood.

Much of the growth in demand for wood during this period will, consequently, be met from increased plantation supplies, and this will necessitate some technological changes to match wood qualities with consumer demands. In parallel, significant changes in the type of timber produced in natural forests are likely to induce shifts in the relative prices of forest products, leading to shifts in demand. For example, there is an increasing scarcity of large tropical logs, which will necessarily constrain the production of hardwood plywoods and large-dimension sawn timber. Consequently, demand for softwood, plywood or other reconsti-

tuted panels may increase. Similarly, solid wood products made from tropical timbers are likely to become premium products leading to a shift in demand towards cheaper wood-veneered products.

Changes in forest-based industries. Changes in prices and demand, together with the development of new technologies and changes in resource availability, are affecting the types and volumes of forest products. The most striking change in the manufacturing of forest products during the past 30 years has been the growth in production of wood-based panels and paper products as compared with sawn timber. Since 1970, global production of sawn timber has remained largely static, while production and consumption of wood-based panels have more than doubled and production of paper and paperboard has almost tripled.

As Table 6.3 shows, output levels of processed wood products (wood-based panels and paper products) have been increasing, even though levels of industrial roundwood production have remained static or declined. This has been made possible by improvements in processing efficiencies. This trend encompasses the effects of utilizing recycled paper and wood but, more important, the development of products that minimize waste by making fuller use of processing residues. A generation ago, solid wood furniture was the norm, but today much utilitarian furniture is made from fibreboard with a veneer covering or from cheap and abundant softwoods such as plantation-grown pines. Similarly, there has been considerable expansion in the types and range of paper products. The development of a broader range of reconstituted paper and panel

Table 6.3   Average annual production of selected forest products

Five-year period	Industrial roundwood mln m³	Sawnwood mln m³	Wood-based panels mln m³	Wood pulp mln mt	Paper and paperboard mln mt	Recovered paper mln mt
1971-75	1 318	428	87	102	131	35
1976-80	1 415	452	102	125	170	51
1981-85	1 457	445	104	136	193	61
1986-90	1 660	503	124	155	240	83
1991-95	1 501	437	130	162	282	104
1996-00	1 523	419	165	171	323	134

Source: FAOSTAT.

and board products has also allowed much greater use of processing residues, enhancing the overall efficiency of wood use.

One of the key changes that can be expected over the period to 2015 is the rationalization of industry capacities. At present there is a significant global overcapacity in pulp and paper facilities, panel plants and sawmills, as evidenced by relatively low returns to forestry sector investments. Much of this overcapacity is the result of distortions in investment decisions created by government incentives (including low stumpage prices, direct investment incentives and tariff protection). Obsolete processing technology and inappropriate location of processing have increased the uncompetitiveness of forest-processing industries in many countries.

A number of factors are forcing greater competition in forest product markets. Trade liberalization is an obvious one. The pressures created by wood scarcity, widespread adoption of market-based economies and greater environmental scrutiny are other factors altering the structure and location of forest-based industries. The period to 2015 and 2030 will certainly witness a much greater reliance on competitive advantages in determining patterns of forestry industry investment.

Wood as a source of energy. As noted above, an estimated 55 percent of global wood production is used as fuelwood. Tropical countries account for more than 80 percent of global fuelwood consumption. There is, however, considerable regional and national variation. Asia and Africa together consume more than 75 percent of global fuelwood. The Regional Wood Energy Development Programme (RWEDP) in Asia estimates that wood supplies 18 percent of all energy used in its member countries. In Africa wood is still the most important source of household energy, used mainly for cooking, although cottage industries (such as food drying and brick-making) also consume significant volumes in some countries.

Fuelwood consumption is mainly determined by income levels, availability of wood supplies and availability of alternatives. As consumers become wealthier they prefer other forms of conventional energy, particularly electricity and liquid fuels. Therefore, while overall consumption of fuelwood is increasing, per capita consumption is declining. In most countries there is a surplus of fuelwood, although there is often localized scarcity, especially in areas adjacent to urban centres, which are sometimes subject to rapid deforestation for charcoal production. In some cases, increasing electricity and gas prices (for example because of the privatization of utilities) have resulted in a switch back to fuelwood.

Dramatic changes in fuelwood consumption are unlikely over the next 15 years. Access to alternative fuels will become easier, but the majority of current wood-using communities are likely to be still burning wood in 2015. The shift towards alternative fuels may accelerate beyond 2015, depending on developments of infrastructure and on improvements in the efficiency and cost-effectiveness of generating energy.

Role of forests in mitigating climate change. Increasing concern over CO_2 emissions and their potential impacts on global climate has focused attention on the role of forests in regulating atmospheric concentrations of carbon. Forests serve as carbon reservoirs by storing large amounts of carbon in trees, understorey vegetation, litter and soil. Newly planted forests, or degraded forest areas that are allowed to regenerate, act as carbon sinks by absorbing and storing carbon as their biomass increases. Conversely, when forests are cleared or degraded, their sink potential is reduced and they can become a substantial source of CO_2 emissions. Therefore projects that focus on afforestation, improved forest management and forest protection can reduce net volumes of atmospheric carbon dioxide and thus play an important role in mitigating climate change.

Estimates of the carbon sink capacity of forests suggest a potentially highly significant role. Overall, forests contain just over half of the carbon in terrestrial vegetation and soil, amounting to 1 200 Gt of carbon. Boreal forests account for more carbon than any other terrestrial ecosystem (26 percent), while tropical and temperate forests account for 20 and 7 percent, respectively. The Intergovernmental Panel on Climate Change (IPCC) has estimated that, globally, carbon sequestration from reduced deforestation, forest regeneration and plantation development could equal 12 to 15 percent of the total carbon dioxide emissions expected to be generated by fossil fuels between 1995 and 2050.

The Kyoto Protocol is the key framework for international agreement on stabilization of greenhouse gas concentrations in the atmosphere. Article 3.3 of the Kyoto Protocol clearly states that "the net change in greenhouse gas emissions by sources and removals by sinks resulting from direct human-induced land-use change and forestry activities, limited to afforestation, reforestation and deforestation since 1990, measured as verifiable changes in carbon stocks, shall be used to meet the commitments under this Article of each party". However, the protocol is unclear about the extent to which forests will be included in the principal mechanisms for regulating carbon emissions. There remains considerable debate over rules for incorporating forestry projects in emission reductions under the clean development mechanism (CDM).[1] Some efforts are under way to develop trading in carbon emission permits and some CDM pilot projects are under implementation, but there is still uncertainty regarding the future potential of these arrangements.

Non-wood forest products (NWFPs). Forests produce many other products besides wood. The vast majority of these NWFPs are subsistence goods, used by the household that collects them, or traded only in local markets (see Table 6.4). There are, however, a number of products that have been developed commercially and an estimated 150 NWFPs are traded internationally. Most are traded in relatively small quantities and supplies are typically unstable, with varying quality and unreliable supply chains. There are, however, a number of exceptions to these general trends including, for example, rattan, bamboo and gum Arabic.

As mass-produced consumer goods become increasingly accessible to even very remote communities, the dependence on inferior NWFPs has declined. The disintegration of many traditional livelihood systems has exacerbated this trend, as has the overexploitation of many NWFP stocks. In the past, the commercialization of NWFPs has been viewed as a potential development path for forest dwellers in less developed countries. Increasingly, however, there is recognition that natural systems for the production of NWFPs are not intensive enough and generally too unreliable to provide a basis for large-scale industrial development. Development of NWFP programmes is also hindered by lack of focus, institutional and policy neglect, lack of research and development funding by NWFP producers and lack of capital. The future is likely to see only a handful of key species being cultivated on a commercial scale no longer controlled by forest dwellers. Thus, the opportunities for poor forest-based communities to profit substantially from NWFP development are likely to remain very limited.

Protected areas, biodiversity and environmental protection. IUCN has established a goal of having 10 percent of each country's land area under some form of protected status. Ideally, this should

Table 6.4   Classification of non-wood forest products (NWFPs) by end uses

End use	NWFP examples
Food products and additives	Wild meat, edible nuts, fruits, honey, bamboo shoots, birds' nests, oilseeds, mushrooms, palm sugar and starch, spices, culinary herbs, food colorants, gums, caterpillars and insects, fungi
Ornamental plants and parts of plants	Wild orchids, bulbs, cycads, palms, tree ferns, succulent plants, carnivorous plants
Animals and animal products	Plumes, pelts, cage birds, butterflies, lac, cochineal dye, cocoons, beeswax, snake venom
Non-wood construction materials	Bamboo, rattan, grass, palm, leaves, bark fibres
Bio-organic chemicals	Phytopharmaceuticals, aromatic chemicals and flavours, fragrances, agrochemicals/insecticides, biodiesel, tans, colours, dyes

Source: FAO (1998c).

[1] The clean development mechanism enables developed countries (with national emission reduction commitments) to undertake emission reduction projects in less developed countries, and to include the reductions thus achieved in their national commitments.

include a representative sample of all ecofloristic zones, so that 10 percent of all types of natural forests would be protected. At present, it is estimated that some 80 countries have formally attained the IUCN 10 percent goal, but around 100 countries currently still have less than 5 percent of their national land area under protection. Overall, it is estimated that more than 30 thousand protected areas have been established and that these cover about 3.7 percent of the global land area.

Working in collaboration with FAO's Global Forest Resources Assessment 2000, the World Conservation Monitoring Centre (WCMC) reported that the number and extent of protected areas grew steadily throughout the latter part of the twentieth century. However, much work needs to be done to harmonize national and international data, including data from different agencies in the same country. In addition, the interpretation of the concept of protected areas often differs substantially among countries, making aggregation at the global level unreliable.

Although the number and extent of protected areas have increased markedly in most regions since the 1960s, in many countries there will be diminishing opportunities to establish protected areas in their traditional form. In countries where conservation efforts fall below the IUCN target, there are often already intense land use pressures, or strong conflicts between economic and environmental objectives. Consequently, there will be a need to develop innovative means to combine conservation objectives with SFM and the creation of sustainable livelihoods. For the next 30 years, therefore, it is expected that the total land area under strict protection will increase only moderately, but greater areas are likely to be placed under SFM, in which conservation concerns are among the criteria.

Biodiversity is important at the level of ecosystems, species, populations, individuals and genes. There is a need for better management to safeguard biodiversity at all these interacting levels. While some losses in biodiversity over time are inevitable through both natural and human-induced causes, diversity can be conserved and managed through a wide range of actions, from the establishment of nature reserves and managed resource areas, to the inclusion of conservation

considerations in improvement and breeding strategies of species under intensive use. The key to success will lie in the development of programmes that include a strong element of active gene management and that harmonize conservation and sustainable use of forest genetic resources within a mosaic of land use options. The sustainability of programmes will depend on genuine efforts to meet the needs and aspirations of all interested parties, and will require close and continuing collaboration, dialogue and involvement of stakeholders in planning and executing related programmes.

Some environmental threats will continue to require protective actions involving forestry. For example, combating desertification will remain a priority for many sub-Saharan countries, as well as throughout the Near East, and parts of North and South Asia. Watershed protection and flood control will continue to be crucial in countries with extensive steep lands and in their downstream neighbours. Weaknesses in institutional arrangements to mobilize concerted action will remain the principal constraint, exacerbated by the lack of financial resources, and difficulties in identifying beneficiaries and their willingness and ability to pay.

Forest-based recreation and ecotourism. Ecotourism is a rapidly growing function of forests. It is defined as ecologically sustainable tourism that fosters employment and cultural understanding, appreciation and conservation. More generally, it comprises a symbiotic relationship between the environment and tourism, and provides an economic incentive for conservation.

Ecotourism comprises a relatively small part (about 7 percent) of the world's travel and tourism industry. Nonetheless, it constitutes a significant source of foreign exchange earnings in some countries. A major question mark is the extent to which it can serve as a means of financing conservation. The market for forest- and conservation-based holidays is limited, and often a few sites with exceptional tourism values attract the vast majority of visitors, such as Yellowstone National Park in the United States, or the Khumbu region in Nepal. Paradoxically, this places extreme pressures on these exceptional sites. Nonetheless, in many instances, forests contribute to overall landscapes that are highly valued for tourism or recreational purposes, for example in Switzerland or the Canadian Rocky Mountains.

Changing consumer preferences are the key determinant of the future growth of ecotourism. In developed countries, where leisure activities are most extensive, recreation markets are becoming more competitive. It is thus quite conceivable that actual forest-based recreation may decline over the next 30 years, even though forests are likely to be increasingly accessible. If increased recreation does occur, it is most likely to be where significant other services are provided. In developing countries, forest recreation is likely to increase together with increased affluence and the development of infrastructure to facilitate access to forests.

Food security and poverty alleviation. There is increasing discussion about what forests and forestry could do to alleviate the widespread prevalence of poverty and food insecurity. In many countries, forests and forest products are an important source of livelihood for rural communities. Traditional approaches to forest management by public sector agencies or the private sector have failed to consider this aspect. Forest-based informal activities such as firewood collection, charcoal production, pit sawing and collection and trade of NWFPs, including medicinal plants, will remain important means of livelihood for a large number of people. In recent years poverty alleviation has become the focal area of attention for international organizations (see Chapter 8), and the role of forests and forestry in this can be expected to be reconsidered.

6.5 Major perspective issues in world forestry

Emerging markets and their impacts on the direction and volume of trade. One of the principal questions relating to the future development of industrial forestry is the extent to which new markets will emerge, and the effects they may have on directions and volumes of trade. Several demand growth poles may emerge over the coming 30 years, with significant impacts on global trade. The largest potential markets are in the populous countries of Asia, particularly China and India. At present these two countries, with around 40 percent of world population, consume less than 10 percent of the world's industrial wood. Other developing countries such as Indonesia, Brazil, Bangladesh and Nigeria all have large populations, but relatively low per capita rates of industrial wood consumption. If developing countries like these significantly raise their per capita consumption of industrial wood products, there will be a significant shift in patterns of trade.

Table 6.5 shows that in Sweden and the United States, per capita consumption of wood products is currently at least tenfold that of the developing countries shown. Given that most of the world's most populous developing countries will continue to experience per capita incomes far below levels in developed countries, per capita consumption of wood products will also continue to lag behind.

The transition countries form another important emerging market. Throughout the 1990s most

Table 6.5   Apparent consumption of wood products (per thousand persons)

	Industrial roundwood (m³)		Sawnwood (m³)		Wood-based panels (m³)		Wood pulp (mt)		Paper and paperboard (mt)	
	1978	1999	1978	1999	1978	1999	1978	1999	1978	1999
China	168	177	40	31	2	30	3	10	13	71
Brazil	384	502	113	101	17	14	15	27	23	42
India	28	27	14	18	0	0	1	2	2	4
Indonesia	64	152	19	9	3	6	1	7	3	24
Nigeria	91	88	29	18	3	2	0	0	2	3
Sweden	5 562	6 963	522	463	151	128	573	901	180	206
United States	1 470	1 546	529	584	153	187	196	216	272	350

Source: FAOSTAT.

of these countries have been in serious recession. During the next 15 years, there seems little doubt that major wood producers such as Russia are likely to regain much of their pre-transition wood supply capacity. At the same time, wood product consumption in Eastern European countries is also likely to rise, following increasing consumer affluence. This improved outlook suggests that marked changes in patterns of trade, particularly in European and North Asian markets, might be expected.

Impact of globalization. Continuing globalization has an impact on markets for forestry and forest products in four ways: further trade liberalization in forest product markets; standardization in production processes, product quality and regulatory environments; an increasing role for foreign investment and investment by multinational corporations; and wider spread of environmental advocacy and concerns.

During the past decades there has been significant liberalization of trade in forest products. The liberalization process culminated in the General Agreement on Tariffs and Trade (GATT) Uruguay Round, during which the majority of tariff barriers for forest products were reduced to fairly moderate levels or completely removed. Similarly, the past two decades have seen considerable standardization in market and regulatory requirements. Most countries have adopted trade, investment and even environmental policies, which have reduced national insulation and promoted a greater level of standardization. Thus, while there is still room for further progress towards free trade in forest products and standardization in markets, changes in trade conditions during the next 30 years are likely to be less important than other factors in determining the shape of forestry.

The globalization of environmental advocacy really came of age at UNCED in 1992. UNCED witnessed the birth of global acceptance of sustainable development, and acceptance of SFM as the core paradigm for forestry development. At the same time, throughout the 1990s environmental advocacy has firmly embraced the concept of transnational environmentalism. Issues such as global warming and the role of forests in carbon sequestration, the development of multicountry criteria and indicators of SFM, and international forest certification processes are all indicative of growing concerns about the environmental implications of forestry. Perhaps even more important implications are those related to the globalization of environmental concerns. The scrutiny and interventions of environmental NGOs and development agencies have imposed greater accountability on multinationals and have forced governments to write and enforce environmental legislation and regulations. Governmental decisions relating to forests are now subjected to wider scrutiny at local, national and global levels, and there are increasing pressures to make forestry practices compliant with widely acceptable environmental standards.

There is a long history of foreign direct investment (FDI) in forestry operations. Most FDI in forests and forest-processing operations is undertaken by multinational companies. FDI is seen as a means of developing national economies by generating employment, income and foreign exchange earnings. Education, training and technology transfer also accrue to the "host" country. FDI in forestry operations, particularly those relating to forest purchases or harvesting concessions, is often controversial. Forests are often seen as strategic national assets, and there has been considerable resistance in many countries to relinquishing control of forests to foreign interests. Similarly, multinational companies operating in countries with weak regulations have been responsible for much destructive forest harvesting. Multinational companies are often perceived to be less accountable for environmental damage than domestic companies or government logging agencies.

Privatization of forests remains a significant vehicle for promoting FDI in forests although, with the exception of a few countries, it has not progressed very far. It seems likely that privatization will eventually progress in a number of countries, although this process is likely to be slow and is more likely to find favour in developed countries. Eastern Europe, where there are efforts to transfer formerly centrally controlled forest assets to private ownership, may be an exception to this rule.

Impact of developments in science on forestry practices. In simple terms, a central objective of most forestry science is to improve some part of the processes that convert sunlight into cellulose or of those that deliver forest goods and services. Traditionally, forestry research focused on making the production process faster, cheaper and better. Such research focused on seedling production, tree growing, harvesting and wood processing as well as all the intermediate steps. The bulk of forest science still operates within these parameters, and over the coming 30 years we should expect improvements across the broad range of the production process, making wood production faster and delivering cheaper and better forest products.

An equally important new trend is to incorporate social and environmental objectives in forest science, management and production. In a number of countries paper production is required to include a minimum proportion of recycled material. Similarly, intensive forest harvesting covering large areas is increasingly considered unacceptable by the public, even though in many instances it is (in the short term) the cheapest means of harvesting and may mimic nature better than alternative systems.

There are also important changes in the technology supporting the monitoring of forests. The increased ease in collecting information (through satellite imagery, the Geographic Information System [GIS], etc.), collating it (database technology), and disseminating it (e-mail and the Internet) means that countries' forest management

practices are under far greater international scrutiny. Thus there is far greater leverage to ensure that forestry practices are sustainable.

Deforestation and forest degradation. Deforestation refers to change of forest land cover where tree crown cover is reduced to less than 10 percent. Deforestation has the highest profile of all environmental challenges confronting forestry. It has an impact on the availability of forest goods and services by reducing the area of forest. High rates of deforestation (and forest degradation) may also increase pressure on forestry policy-makers to place forest areas in strictly protected areas or to prevent harvesting in other ways (e.g. through logging bans). This may increase pressure on accessible areas and further reduce production that can be harvested from remaining forest areas.

There is an identifiable relationship between deforestation and levels of industrial development. In subsistence economies with slowly growing populations, forests may absorb the impact of demands for wood and other forest products. The advent of widespread economic development and rapid population growth has generally seen a period of rapid deforestation in industrializing countries. At a certain level

Figure 6.1   Percentage of deforestation by region, 1990-2000

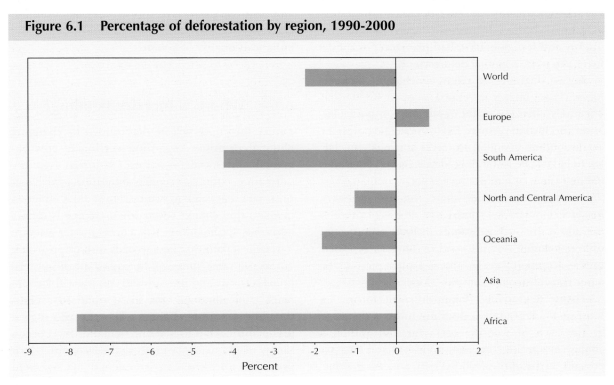

Source: FAO (2001c).

of development, however, further land clearing and forest harvesting become uneconomic, and may also generate increasing criticism related to neglect of social and environmental dimensions of forest management. Thus there exists an apparent curve of changing forest use, which leads from subsistence, through heavy exploitation, to harmonization of harvesting with protection and conservation, as well as social needs and aspirations.

Figure 6.1 shows evidence of this development path. In most European countries, for example, the exploitive period is long past, and forest areas are now gradually expanding. Most developed countries in other regions, including Japan, Australia, New Zealand, Canada and the United States also exhibit either increasing or static total areas of forest.

On a regional basis, deforestation is most rapid in Africa and in South America. It also remains a serious problem in many countries in Asia, Oceania and Central America. Nonetheless, the experience of developed countries suggests that a reversal in deforestation is imminent in a number of developing countries. In fact, the FAO *Global Forest Resources Assessment 2000* shows that in the past decade, developing countries such as Bangladesh, Viet Nam, Algeria and the Gambia have reversed deforestation, and are expanding net forest area. For more than a decade, developing countries such as China, India, the Libyan Arab Jamahariya, Turkey and Uruguay have had plantation establishment in excess of annual deforestation.

Forest degradation refers to changes between forest classes (e.g. from closed to open forest) that negatively affect the stand or site and, in particular, lower production capacity. Extensive logging, without corresponding attention to forest management, has led to large-scale forest degradation, especially where the extraction of mature trees is not accompanied by their regeneration, and where the use of heavy machinery causes soil compaction or loss of productive forest area. Other causes include overgrazing, overexploitation (for firewood or timber), repeated fires and attacks by insects, diseases, plant parasites or other natural sources such as cyclones.

Forest degradation takes different forms. In most cases, degradation does not imply a decrease in the area of woody vegetation, but rather a change in species composition, loss of soil fertility, erosion, reduction in biodiversity and fragmentation of forest areas. One of the most easily identifi-able and important indicators of degradation is depletion of standing biomass, particularly mature trees. Forests may be degraded from virgin primary status to secondary forests, to scrub or other wooded land and then perhaps cleared and reafforested as plantation. Broad measures of forest area and deforestation do not capture the changes in values that these shifts encompass. A shift to principles of SFM and active monitoring of indicators of SFM are likely to help document and eventually arrest forest degradation.

Current policy directions will undoubtedly result in a decrease in net deforestation in many countries during the next decade. The rate of forest degradation is also likely to decrease as more attention is paid to sustainable harvesting and utilization practices. In Asia, for example, countries such as Thailand and the Philippines have imposed bans on natural forest harvesting. Although illegal logging and shifting cultivation continue, it should be expected that ultimately the profits of plantation establishment and natural regeneration will outweigh those of defying the bans. Improved policies, particularly those encouraging sustainable forest management, should ensure incremental improvement in forest management on a global basis, which will gradually shift the balance of forestry degradation and clearance to stasis, and then to reforestation and enhanced quality of forests.

6.6 Where is forestry heading?

Future directions will be determined by changing national priorities regarding economic, environmental and social objectives for forestry. At the same time, countries remain subject to a variety of pressures that regularly compromise their ability to invoke rapid change, even when change is clearly desirable. Each country will, consequently, attempt to choose a path that best accords with its priorities, but at the same time will be cognizant of international scrutiny and pressures to comply with benchmark environmental and social standards. Thus, the most important changes will occur with regard to the approaches to meeting demands for a broad range of goods and services, the ways in which pressures for better environmental management are addressed, and the means by which increasing demands for social equity are satisfied. Each of these key dimensions is discussed below.

Through the 1980s and into the early 1990s, global wood supply projections regularly showed a pessimistic slant. Forecasts showed rapid deforestation, increasing populations and economic development, with an apparent prospect of ever-increasing demand for wood and wood products to be provided from ever-diminishing forest areas. There is, however, a general consensus between recent outlook studies that this rather gloomy picture was overstated. While forest areas continue to decline globally and demands increase, it can also be argued that we have greater wood supply potential than at any time in history.

The key to increasing wood supplies lies in innovation and technological development. In the past 30 years, as concerns for forests and wood supplies have mounted, so too have the number of adaptive measures designed to intensify forest management and wood production. These trends are already evident in an acceleration of plantation establishment, introduction of reduced impact logging techniques, development of reconstituted wood products, increasing emphasis on recycling systems, and introduction of techniques to promote SFM.

The most recent forestry outlook studies indicate that trade and other market mechanisms will ensure that, at a global level, wood supplies will be adequate for the foreseeable future. However, localized shortages will occur in many countries. Wood will also remain scarce for impoverished or otherwise marginalized people with little purchasing power. Providing better livelihoods and living conditions for these people remains a fundamental challenge in many countries. In some countries these problems are being addressed by implementing systems of community-based forest management, or by improving regulation of and rights to forest access.

The question as to whether there is enough wood is likely to be less important in the future. Rather, the key questions will be: Where should it come from? Who will produce it? And how should it be produced? It seems clear that plantations will play a relatively greater role in providing wood. Similarly, trees planted outside the forest, around the farm or household or on boundaries, roadsides and embankments will continue to be an extremely important and underestimated source of wood

supplies. Changing patterns of resource availability will have profound implications for forestry administrations in many countries, requiring changes in technology, processing capacity and location of processing industries.

The future for many forest services appears comparatively bright. Environmentally based services such as watershed protection, carbon sequestration, contribution to water quality and maintenance of biodiversity and genetic resources are in the mainstream of environmental concerns. Environmental services of forests have the advantage that their benefits extend outside the forest and consequently affect a more affluent constituency – and therefore more easily attract funding. Other services that are largely used by the rural poor are less likely to attract attention, now or in the future.

In the past, the areas of fuelwood and NWFPs have been generally marginalized in forest policy-making. While the future is likely to see greater attention paid to these products, there is little evidence of rapid progress in their adoption into the policy mainstream. In many instances there is an assumption that economic development will eliminate reliance on these products, and thus they tend to be treated as less important. While there is doubtless some truth in this viewpoint, the reality is that vast numbers of people will remain dependent on these products and therefore they merit attention by policy-makers.

The overriding influence on forestry's future direction will be the evolution and adoption of principles of SFM. Implementation of SFM is likely to involve a range of measures that will substantially alter the ways forests are managed in the future. The two major components of SFM are likely to be increases in the area of forest within protected area networks, and changes in forest harvesting and management to increase the social and environmental acceptability of production forestry.

The most significant change will be the broadening of forest management objectives to incorporate not only wood production, but also environmental and social values. Initially this is likely to encompass significant increases in protected areas. In the immediate future, increasing the area of forest in legally protected areas should be feasible without too much impact on future wood supplies. Further ahead, identifying additional areas for full conservation will become more difficult. It is likely that, by 2030, new conservation areas will encompass less stringent preservation aspects, and will adopt multiple-use dimensions within a framework of sustainable management. At present, too many conservation areas are afforded protection only in name. In future, it will be important to ensure that there is adequate funding to manage and protect such areas properly.

Eventually all types of land uses, including natural forests and plantations outside the legally protected areas, will incorporate these objectives. By 2030, environmental protection is likely to have become fully integrated into SFM (not just "boxed-in" in designated areas). The accessibility of forests for recreational use will also have improved markedly.

Deforestation is expected to slow down in a majority of countries, and efforts to reverse forest degradation and improve the quality of forests, including enhancing biodiversity and restoring natural processes, will have gathered enormous momentum. It is difficult to project the magnitude and rate of change. There will be notable inter-country and interregional differences. In Asia, for example, rates of agricultural expansion into forest land are likely to decline, largely because forests most suitable for arable cultivation have already been converted. Slow economic growth, continued dependence of a large share of the population on subsistence farming and the availability of forest land suitable for conversion to arable cultivation are likely, however, to lead to continued deforestation in sub-Saharan Africa and, to some extent, in South America.

The likelihood is that moves to collaborative forest management will make only steady progress over the medium term after which, depending on observable success, such moves may accelerate. A crucial requirement is the strengthening of forestry institutions, which are generally under-resourced and weak. Forestry institutions also need to be proactive in intersectoral discussions, in recognition of the fact that many of the most important forces shaping the future of forestry are largely outside the direct influence of the sector. Ultimately, the importance of forestry will depend on how effectively it can be integrated into the activities of all other sectors and to what extent the public as a whole will be convinced about forestry's current and future role in promoting sustainable development.

World fisheries

7.1 Introduction

The future of world fisheries[1] will be affected by fluctuations in the abundance of fish and living marine resources, along with factors such as technical developments, scientific discoveries, climate change and economic progress. But the dominant factor will be the critical issue of how capture fishing and aquaculture activities are managed.

The chapter starts by providing a summary of the status of world fisheries at the end of the 1990s, covering the main trends in supply (the status of stocks and fish production), demand (food consumption and other uses of fisheries) and employment. There follows a discussion of fisheries over the period to 2030.

Clearly, trying to depict world fisheries three decades from now is subject to many uncertainties. The discussion in this chapter is necessarily based on assumptions and analyses that include an element of subjectivity. The broad picture might be useful as a sketch of possible developments in production and consumption, even though fisheries in particular parts of the world may not conform to this picture.

7.2 World fisheries at the end of the 1990s

7.2.1 Fisheries production and stocks at the end of the 1990s

Capture fisheries. During the 1990s, the production of capture fisheries fluctuated between 80 and 85 million tonnes p.a. for the marine sector. Fish from the Pacific Ocean dominated world capture fisheries, accounting for almost two-thirds of total world supplies in 1999.

During the same period, the production from fisheries in inland waters expanded slightly, increasing from 6.4 million tonnes in 1990 to nearly 8.3 million tonnes in 1999. However, inland fishery catches are believed to be greatly under-reported (FAO, 1999e). This is because of the dispersed and informal nature of many inland fisheries, and because many inland fish are bartered, sold or consumed locally without entering into the formal economy. For example, inland fisheries in Brazil, Ghana and several Southeast Asian countries may be twice to six times as productive, in terms of catch actually taken, as officially reported.

---

[1] The use of the term "fisheries" is in its broadest sense. It includes capture fisheries and aquaculture and sometimes associated handling, preservation, processing, marketing and trading activities, but excludes fishing for, and trade in, ornamental fish.

In several countries efforts are under way to improve the data on inland fisheries and on the range of services inland waters provide. One preliminary result is that it appears to be more profitable to use Brazilian floodplains for fisheries than to use cleared floodplains for cattle grazing.

The gross volume of marine catches has fluctuated and does not show any definite trend. But the species composition of the catch has changed. High-value species – bottom-dwelling species (demersal) and large surface-dwelling species (pelagics) – are gradually being substituted by shorter-lived pelagic and schooling fish. FAO studies indicate many causes for this shift. These include the thinning out of (overfished) top predators; increases in natural production of small pelagics through nutrient enrichment of coastal areas, enclosed and semi-enclosed seas; and changes in fishing strategy and technology. The main force underlying such changes, however, is the change in harvesting costs as fishing technology has advanced and as various stocks have been depleted, and the impacts of these changing costs on operations.

During the second half of the twentieth century, capture technologies evolved rapidly. In the 1950s the invention of gear handling devices, such as the power block for purse seines, resulted in major improvements in the efficiency of fishing methods such as trawling, gillnetting, purse seining and lining. The development and introduction of synthetic fibres in fishing gear during the 1960s significantly increased the efficiency of such gears, thereby lowering that cost component of production. Also, new vessels were able to remain for longer periods in the fishing grounds and the introduction of freezing on board helped increase the autonomy of vessels.

Modern fishing fleets have also seen continued growth in efficiency over the last decades, thanks to improvements in navigation and fish-locating equipment, increased knowledge of fish behaviour during the capture process, and development of gear monitoring instruments. The growth of efficiency has contributed to increased exploitation rates and overexploitation in some fisheries, as well as to overcapacity in others.

Status of fish stocks. In order to help organize its data, FAO distinguishes 16 statistical regions for the world's fishing areas. Figure 7.1 shows available

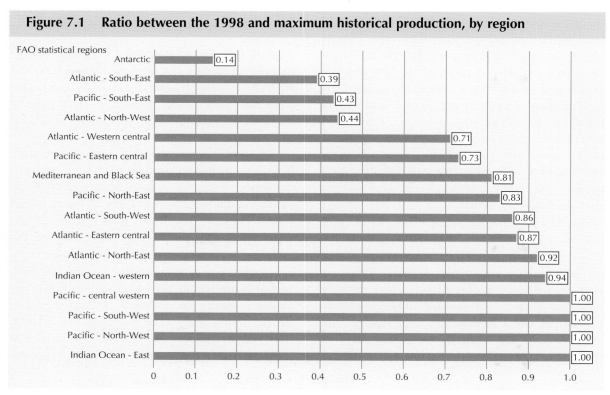

Figure 7.1   Ratio between the 1998 and maximum historical production, by region

Source: FAO (2000b).

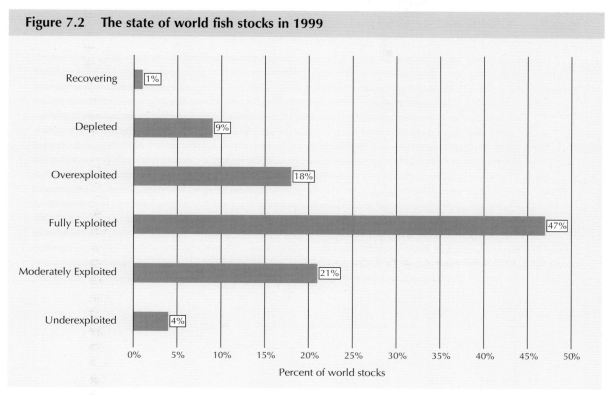

Figure 7.2   The state of world fish stocks in 1999

Source: FAO (2000b).

data for these FAO statistical regions[2] of the world's oceans in 1998. Four regions (the Eastern Indian Ocean and the Northwest, Southwest and Western Central Pacific Oceans) were at their maximum historical level of production. Eight regions (the Western Indian Ocean, the Mediterranean and Black Sea, the Northeast and Eastern Central Pacific, and the Eastern Central, Western Central, Southwest and Southeast Atlantic) were slightly below maximum historical levels of production. The remaining four regions (the Antarctic, the Southeast and Northwest Atlantic and the Southeast Pacific) were well below maximum historical levels of production. While there may be some natural oscillations in productivity, the main factor responsible for the declines in production

levels in most of the regions is overfishing.

Another useful indicator is the state of stocks of various fishery species (Figure 7.2). For the 590 "stock" items for which FAO had some information, 149 were in an unknown state. Among the 441 for which data were available, 9 percent were depleted,[3] 18 percent were overfished,[4] 1 percent recovering, 47 percent appeared fully exploited, 21 percent were moderately exploited, and only 4 percent were classified as underexploited, i.e. they could sustain catches higher than current levels.

A more general assessment of the condition of world stocks depends on how one categorizes the stocks that at present are exploited close to the maximum sustainable yield (MSY) level. If the fully exploited stocks are accepted as being in good

---

[2]   The Pacific is divided into the northwest (PNW), northeast (PNE), eastern central (PEC), western central (PCW), southwest (PSW) and southeast (PSE) regions; the Indian Ocean into the eastern (IE) and western (IW) regions; the Atlantic into the northeast (ANE), northwest (ANW), eastern central (AEC), western central (ACW), southwest (ASW) and southeast (ASE) regions. The last two regions are the Mediterranean and Black Sea (MBS),and the Antarctic (ANT).

[3]   A "depleted" stock is one that is driven by fishing to a very low size (compared with historical levels), with dramatically reduced spawning biomass and reproductive capacity. Such stocks require rebuilding. The recovery time will depend on the current condition of the stock, the level of protection afforded to the stock, and environmental conditions.

[4]   A stock is considered "overfished" when exploited beyond the point at which the volume is considered "too low" to ensure safe reproduction. The term "overfished" is often used when biomass is estimated to be below a specified biological reference level such as the maximum sustainable yield (MSY).

Figure 7.3   Condition of stocks by FAO statistical region

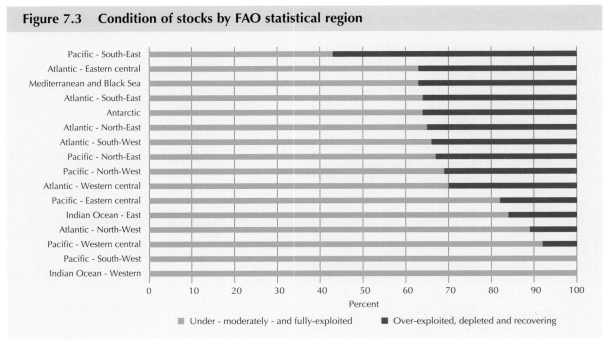

Source: FAO (2000b).

condition because they comply with the United Nations Convention on the Law of the Sea (UNCLOS) requirement of being at or above the MSY level of abundance, then 72 percent of the stocks appear exploited at sustainable levels. However, this still implies that 28 percent of stocks are overfished or even depleted and therefore need rebuilding.

Similar conclusions arise by analysing the stock data by FAO statistical region. Comparing the total of fully, under- and moderately exploited stocks with the total of overexploited, depleted and recovering stocks, the data suggest that in each region more than 50 percent of stocks are being sustainably exploited, except in the Pacific Southeast (Figure 7.3).

Aquaculture. Until the 1950s, aquaculture involved a limited number of species: oysters, mussels, carp, trout and shrimps. This was largely because these were the species for which large volumes of young/spat could be easily obtained for production purposes, and such culture systems were generally extensive. Since the 1950s, scientists gradually solved the problem of artificial reproduction for different carp (e.g. common, Chinese and Indian carp), salmonids, oysters and shrimps, making it

possible to stabilize and expand production of these species.

The bulk of aquaculture production comes from species that feed low on the food chain. Among the finfishes, carp, which generally feed on plankton and plants, accounted for nearly 70 percent of world production of cultured finfishes in 1999. Commanding low to moderate prices, such fish tend not to be traded internationally. Aquaculture currently provides a considerable share of the production of several high-priced species such as salmon, marine shrimps and oysters.

In general, efforts to culture other finfish and crustaceans implied protracted and costly research and development efforts. This has hampered greater aquaculture-based production of many of these types of species. As a result, even by the end of the 1990s, a relatively small number of species (less than 40) accounted for 90 percent of the volume of world aquaculture production.

During the 1990s, aquaculture production increased at the rapid rate of approximately 10 percent p.a., rising from a 1990 level of 13 percent of world fisheries production (excluding aquatic plants) to 26 percent nine years later. Currently, aquaculture production already far exceeds the year 2010 production level projected as recently as in 1995 (Alexandratos, 1995).

Aquaculture's share of fish used in food consumption has risen in volume terms from 19 percent in 1990 to 34 percent in 1999. In value terms, this contribution is 39 percent – the ex-farm value of aquaculture (fish and shellfish) production was estimated in 1999 at approximately US$48 billion, compared with an estimated landed value of capture fisheries at approximately US$76 billion.

The growth of aquaculture has been unevenly distributed around the globe. Production has been heavily concentrated in Asia both in volume and value terms. At the beginning of the 1990s, Asia provided 83 percent of the world's total volume of aquaculture production, and by 1999 that share had risen to 89 percent. Most of the increase was attributable to China, whose share of the world's aquaculture production increased from 50 percent in 1990 to 68 percent in 1999.

There is a growing perception that much of aquaculture production leads to environmental damage. This is not correct for many aquaculture activities. Molluscs, seaweed and most forms of carp culture, for example, have only a very limited impact on the aquatic environment, and these species account for more than half of world aquaculture production.

However, it is generally recognized that the early rapid expansion of shrimp farming in countries around the Bay of Bengal, elsewhere in Asia, and in Latin America caused considerable environmental and social damage. This was in large part a result of unsuitable procedures for site selection, pond construction and pond management. Conversely, aquaculture itself has increasingly been negatively affected by coastal development, habitat loss, and non-point-source pollution, issues that are usually beyond the direct jurisdiction or control of fisheries management authorities.

7.2.2 Demand for fishery products at the end of the 1990s

Demand for fishery products consists of two components: direct consumption of fish as food and fish used for animal feeding. Increasingly, fishery resources are also in demand as a source of leisure activities.

Direct consumption of fish as food. Globally, fish provide about 16 percent of animal proteins consumed and are a valuable source of minerals and essential fatty acids. Regionally, these consumption figures vary, with fish accounting for an average of 30 percent of consumed animal proteins in Asia, approximately 20 percent in Africa, and around 10 percent in Latin America and the Caribbean.[5]

During the 1990s, global apparent consumption of fish increased.[6] In 1990, apparent direct food consumption of fish, crustaceans and molluscs amounted to 71.2 million tonnes (live-weight equivalent). This figure increased to 97.2 million tonnes in 1999, an increase of more than a third over a period of only nine years, faster than the growth of world population. Thus average apparent consumption of fish increased from 13.4 kg per capita in 1990 to 16.3 kg per capita in 1999.

This development was heavily dominated by events in China, which emerged as the world's largest fish producer during this period.[7] Growth in Chinese fish production was so rapid that it overshadowed developments elsewhere. In fact, excluding China, the apparent consumption per person in the rest of the world actually declined from 14.4 kg in 1990 to 13.1 kg in 1999.

However, it is important to note that such global figures mask the very wide differences among countries in the level of fish used in food consumption. This ranges from less than 1 kg per person p.a. in some countries to over 100 kg per person in others. Moreover, in recent years the form in which fish has been consumed has changed. The volume of fishery products marketed in their fresh state has increased not only in absolute terms, but also as a

---

[5] In discussing food uses of fish, the term "fish" refers to fish, crustaceans and molluscs, excluding aquatic mammals and aquatic plants.

[6] FAO does not obtain data from countries on consumption. Thus, when the term "consumption" is used it refers to "apparent consumption" which is the equivalent of domestic fish production plus fish imports minus exports and non-food uses of fish.

[7] The accuracy of the Chinese figures for fish production and the utilization of fish for food and feed has been called into question by some scientists. FAO and China are examining this issue.

Box 7.1 Biodiversity and fisheries

Over 1 100 species of fishes, molluscs and crustaceans directly contribute to production of the world's major fisheries. There are many additional species contributing to smaller-scale fisheries. In aquaculture, although the majority of production comes from a few species, there are over 300 species, which do contribute.

Biodiversity as defined in the Convention on Biological Diversity – namely, the variability within species, among species, and of ecosystems, and the ecological complexes of which they are part – is a basic element necessary for sustainable fisheries and aquaculture. The biodiversity of natural populations provides the resource base for commercial fisheries and the means to adapt to environmental changes and fishing pressure, and the biodiversity of farmed fish allows for continued breed improvement to meet production and changing market demands.

Like most human activities that alter resources and the environment in which they are found, commercial fishing and aquaculture activities can have negative impacts on the environment and, hence, on biodiversity. For example, in capture fisheries practices such as the use of poison and dynamite are extremely harmful (especially in coral reefs) and have been generally banned. Similarly, the use of non-selective fishing gears can have negative impacts on biodiversity, particularly if the amounts of species harvested and impacted are unsustainable. This happens, for example, where there is a high bycatch of relatively scarce species, high discard rates, and where only high-valued catch is kept. There is growing concern that the practice of bottom trawling can have negative impacts on bottom-dwelling flora and fauna in sensitive areas such as coral reefs, seagrass beds, etc. Similarly, increased production of so-called reduction fisheries (i.e. fish being reduced to fishmeal and oil) have been increasingly criticized for their likely impact on the food chains of important aquatic species and birds.

Fishery administrations and Regional Fishery Management Bodies (RFMBs) are responding to stakeholders' concerns and are working to address the impacts of these activities through better regulations to minimize or mitigate their impacts on biological diversity. One approach has been the development and use of more selective fishing gears such as those that reduce the capture of marine mammals, undersized target species and unwanted bycatch. Additional efforts are being made to ensure that commercial exploitation does not result in the eradication of species or seriously harm genetic diversity.

Nonetheless, some currently targeted species such as the white abalone, the barn-door skate and large coral reef fishes are at risk. Genetic diversity and the related ability for species to adapt have been lost by intense fishing on susceptible populations such as spawning aggregations of the orange roughy in New Zealand, thereby eliminating this large, slowly reproducing species. In response, some stakeholder groups are working, through the Convention on International Trade in Endangered Species of Wild Fauna and Flora (CITES), to prohibit international trade of commercially caught species that may be at risk.

Inappropriate fisheries and aquaculture activities are but some of the threats to aquatic biodiversity. Additional threats, often more severe, include pollution, loss of habitat, habitat degradation and global warming. Unfortunately, aquatic biodiversity usually faces several such threats simultaneously, making the solution to such problems less amenable to relatively simple or quick remedies.

---

share of all uses of fish. In 1999 slightly more than a third of all fish was marketed fresh against only one fifth in 1987.

Fish used for animal feed. During the 1990s, between 28 and 33 million tonnes of fish were used each year for the production of fishmeal and oil. Almost all of this was landed by marine capture fisheries. When the capture of pelagic fish off the west coast of South America contracted as a result of the El Niño phenomenon, so did the production of fishmeal in the world: in 1998 only 23.9 million tonnes of fish were reduced to fishmeal and oil. By 1999

this figure increased again to 30 million tonnes or almost 24 percent of total world catch of fish, representing a return to a more normal level as a result of the recovery of fishing in South America.

Fishing as a source of leisure activities. Two types of users are increasingly using fisheries for leisure activities: active users such as recreational and sports fishers, and so-called non-consumptive and passive users such as tourists, sport divers and individuals who want to enjoy the environmental values of pristine marine environments (including fisheries). In many parts of the world, the number of stakeholders who want to use fisheries for leisure purposes exceeds the number of commercial fishers. Because all stakeholders, commercial or otherwise, have valid claims for access to living aquatic resources and have a stake in influencing the use of world fisheries, the emergence of this group is an important development. Because of their numbers and the fact that they are increasingly well represented and organized, these stakeholders in affluent developed economies have a growing influence on how fisheries can be used, including the use of fisheries for food production.

This new category of stakeholders has developed various strategies for effecting its objectives. In some instances, stakeholders have moved to prevent or otherwise modify exploitation of some valued species, either because the exploitation of these species had the potential to directly threaten species (e.g. marine mammals) or because of the effects capture fisheries could have on associated species, e.g. seabirds or seals.[8] In other instances, they have asserted particular policy perspectives regarding the appropriateness of exploitation of certain environments or species and have worked to establish parks or protected areas in which certain fisheries activities are prohibited or limited. In some cases such efforts have resulted in fishing methods being significantly modified, restricted and even prohibited as, for example, under the 1989 UN Resolution 44/225 on Large-Scale Pelagic Driftnet Fishing and its Impacts on the Living Resources of the World's Oceans and Seas and the related 1989 Wellington Convention for the Prohibition of Fishing with Large Driftnets (and Protocols).

7.2.3 Trade in fishery products at the end of the 1990s

The gradual opening up of markets during the last decade has expanded international trade in fish and fish products in both volume and geographic coverage. Nearly 40 percent of all fishery production is now internationally traded. Around 80 percent of fish for human consumption ends up in three main markets (Japan, the United States and the EU).

Exporting countries increasingly see trade in fish and fish products as a powerful vehicle for earning foreign exchange, especially through the export of high-valued species. For example, in Indonesia the government is aiming for approximately three-quarters of all foreign exchange earned from fish and crustacean exports to come from high-valued cultured species (Purwanto, 1999). Globally, gross earnings from fish exports by developing countries grew rapidly from US$5.2 billion in 1985 to US$15.6 billion in 1999, a level that raised fish export earnings to a level on a par with the total value of exports of such commodities as bananas, coffee, cocoa and rubber.

International trade is particularly important in fishmeal and oil. Some 65 percent of the volume of world fishmeal production enters international trade. This reflects the fact that fishmeal is produced primarily in South America but is used as animal feed in feedlots in more wealthy economies elsewhere in the world. This growth in trade also reflects the global growth of the aquaculture sector, where fishmeal and oil are key feed ingredients.

The increase in trade in fish and fish products in the 1990s has not been without difficulty. Stringent standards for fish imports have been introduced, mostly in the markets of developed countries. Approaches such as HACCP methodology aim to ensure that consumers receive safe and hygienic quality food products. This has forced exporters to choose between finding new output markets, or introducing systematic procedures to meet the requirements of the markets to which they want to export. One effect has been that some countries saw their exports halted until they could introduce HACCP procedures.

7.2.4 Employment in fisheries

Employment policies in the fisheries sector are seldom explicitly elaborated. Where they are explicit, they frequently concern the participation of local fishers as crew in a joint venture or on foreign flagged, licensed vessels fishing in the national exclusive economic zones (EEZs).

---

8   Stakeholder groups were key forces driving international efforts to set up the International Plan of Action for Reducing Incidental Catch of Seabirds in Longline Fisheries, the International Plan of Action for the Conservation and Management of Sharks, and the International Plan of Action for the Management of Fishing Capacity.

Despite this, other developments in marine policy have had an impact on fisheries employment. For example, the introduction of the EEZs early in the 1980s affected fisheries employment. Employment in long-distance fishing fleets changed when countries lost access to fishing grounds. This affected both those that already existed, and those developed during the 1950s and 1960s by countries such as Japan, the former Soviet Union and several countries in Europe. Some countries, particularly Spain and Portugal, relocated much of their activities to other nations' fishing grounds through fishing agreements. Other countries, particularly developing countries in Asia, expanded employment in fisheries, in part through effective occupation of their own fishing grounds in their newly created EEZs, but also by engaging in high sea fisheries. Still other countries worked directly to reduce the numbers of those employed in high seas and other fisheries. For example, from 1980 to 1995, the number of active Japanese marine fishers was reduced from some 457 to 278 thousand.

Although information on fisheries employment is only fragmentary, the available data show that during the 1990s employment in capture fisheries and aquaculture continued to contract in the developed economies and to expand in the developing countries. Already in the early 1990s, more than 90 percent of those fully employed in the fishery sector were employed by either developing economies or economies in transition (FAO, 1997d). At the same time, fisheries employment in developed economies fell in capture fisheries, in part compensated by the slowly expanding employment in aquaculture. Overall, growth in employment in fisheries is slowing down significantly.

When fisheries are used as a means for providing social safety nets or for last-resort employment opportunities, such employment policies are rarely explicitly enunciated. For example, there has been an increasing tendency to allocate inshore areas with fish resources to small-scale fishers. This can be seen as an indirect way of preserving local employment opportunities and a source of food for local fishing communities. However, such policies rarely provide legal rules and mechanisms for controlling the inflow of fishers from other areas or communities. As a result, unrestricted catches by unregulated numbers of fishers mean that inshore fisheries are increasingly unable to render the benefits they were intended to provide, particularly when there are problems in other sectors that drive people into fisheries as an activity of last resort.

7.3 Plausible developments in world fisheries

Developments in the global fisheries picture will to a great extent be determined by developments in fisheries management, which will have to face problems such as the increasingly contentious issue of access rights allocation. In this section, developments in production, demand and trade will be discussed, leading up to a concluding discussion of management issues in fisheries.

7.3.1 Developments in fisheries production

Overall, production from wild capture fisheries is approaching its biologically sustainable limit for many fish species. The bulk of further production increases will therefore have to come from aquaculture. However, aquaculture is at present also constrained to the extent that it relies for feeding purposes on fishmeal and fish oil that come from capture fisheries. This constraint may be overcome when cost-effective substitutes for fishmeal and fish oil emerge.

Beyond these constraints, future fish production increases will depend on further development of post-harvest utilization technologies, and on successful resolution of various negative externalities such as habitat destruction, bycatch and pollution from intensive aquaculture.

The future for the *aquaculture* sector varies by region. In Asia aquaculture is long practised and well established. It is expected that aquaculture will continue to grow and supply important segments of fish markets in Asia, increasing its share in total fish production.

The aquaculture industry will respond to the growing demand for fish by increasing production of already domesticated species, by domesticating additional species, and by expanding marine ranching. In doing so, the industry will need to resolve issues such as how to maintain genetic diversity of domesticated species, how to develop technologies for new species, how to

Box 7.2 The use of genetically modified organisms (GMOs) in aquaculture

Traditional animal breeding, chromosome-set manipulation and hybridization have already made significant contributions to aquaculture production. Their contribution is expected to increase as aquatic species become more domesticated and as breeding and genetic technology continue to improve.

A relatively new form of genetic modification involves moving genes between species, thus producing "transgenics". Experimental and pilot projects on transgenic organisms have demonstrated that commercially important traits can be greatly improved. Growth hormone genes from a variety of fishes have been put into other species such as carp, catfish, tilapia and salmon to improve growth rates by increasing the production of the fishes' own growth hormone. A gene that produces an antifreeze protein in the Arctic flounder, thus allowing it to survive in subfreezing waters, has been put into Atlantic salmon in the hope of increasing this species' tolerance of cold waters.

At the end of the 1990s, no commercial aquaculture producer was marketing transgenic aquaculture species for human consumption. Consumer resistance and concerns about the protection of intellectual property rights are the main reasons for not marketing the animals at present being developed in research programmes. There is concern that new genetic technologies, and specifically transgenic technology, are poorly understood and that they may pose risks to the environment and human health. These concerns will be addressed, *inter alia*, in the biosafety protocols on the international movement of GMOs now being developed under the Convention on Biological Diversity. The EU has issued directives for the trade, transport and release of GMOs.

Many resource managers, aquaculturists and scientists believe that GMOs and transgenics will eventually become accepted in aquaculture and will contribute to increased production. Clear and accurate information on the benefits and risks of GMOs, and policy guidelines for their responsible use, are urgently needed in order to allow the field to progress.

resolve a looming shortage of feed for cultured fish and shrimps, and how to deal with pollution issues.

Aquaculture is an ancient activity that was usually carried out in harmony with other uses of the environment, but few of the old ways of raising aquatic animals and plants still prevail. Modernized technologies have resulted in substantial reductions in production costs for several species, but they have also created negative environmental externalities. Aquaculture will be increasingly subject to regulations ensuring judicious use of culture technologies and site selection. The development of technologies, and a reduction in the costs of dealing with the undesirable environmental effects of aquaculture installations and intensive feed regimes, will allow these problems to be much reduced, if not overcome. Environmental considerations may also lead to the shift of aquaculture installations away from the coastal zone into offshore and submergible installations at sea, and in densely populated areas inland, into intensive recirculating systems.

There is concern in the aquaculture industry that the limited availability of feeds could constrain the expansion of aquaculture production. Aquaculturists using semi-intensive and intensive culture systems for fish and marine shrimps are particularly dependent on feeds that incorporate fish oils. Fishmeal producers expect that within a decade or so, the aquaculture industry will use up to 75 to 80 percent of all fish oil produced, and about half of the available white fishmeal. Although considerable research efforts have been undertaken, a satisfactory replacement for fish oil has still to be found. The problem of finding new fish feeds will soon become more pressing, given that semi-intensive and intensive culture systems for shrimps, marine fish and increasingly for carp and other freshwater fishes will expand.

Despite boom-and-bust cycles of overexpansion, market gluts and falling prices in northern Europe and North America, it is likely that the culture of salmon, sea bass, bream and turbot will still continue to expand, particularly in Europe. But the rate of expansion is likely to decrease and stabilize, probably at a few percent per year in terms of the volume of production. The rate of expansion will depend on the intensity of market promotion efforts, particularly in non-traditional markets.

However, there are cultured species such as carp, tilapias and catfishes that have much larger "natural" markets. For example, carp account for about 70 percent of world cultured production of finfish although they are produced in only a few countries (China, India and Indonesia account for more than 90 percent of world production; FAO, 2001e). It is plausible to expect continued growth in traditional markets. Total production may well double by 2015, with growth slowing thereafter. If non-Asian markets could be found for Indian and Asian carp, production could grow more quickly.

After several decades of promotion, the culture of tilapia (mostly *Oreochromis nilotica*) took off at the end of the 1990s, not in its natural home of Africa, but in Asia, where it has found ready markets. It is also finding markets in Europe and North America. It seems most likely that tilapia culture will expand in Asia, intensify in Latin America and take off in Africa. The sector has been growing in volume by over 10 percent p.a. at the end of the 1990s, and it is likely to maintain that rate at least for the next two decades. Tilapia is a relatively inexpensive fish, and it can be used as an ingredient in processed products. By 2015, it is plausible that there will be a production volume in the order of 3 to 5 million tonnes p.a., increasing to possibly twice that amount by 2030.

Crustaceans account for about 5 percent of world production of fish and shellfish. While the culture of crustaceans has progressed steadily in volume, this has not always been the case for individual species. Given the growing demand for this luxury product and an increased ability to deal with disease (a severe problem in the culture of tropical marine shrimps), it would seem reasonable to expect that the culture of present domesticated crustaceans will expand as it did during the last few years of the 1990s.

Molluscs, foremost among them oysters, account for about one-third of world production of fish and shellfish. Recently, the production of mussels has stagnated. However, efforts are under way for these products also to find new markets and these are likely to be successful. Production of mussels can be expected to grow through new offshore technologies. Thus mollusc production is likely to resume its pace of growth.

Because of the high costs and long periods involved in achieving controlled reproduction and survival of the offspring of newly domesticated species, most recent domestications have been achieved in developed countries, in particular along the shores of the North Atlantic. This is likely to continue in the future. Production of species such as char, cod, halibut, coryphena and possibly large tunas and turbot is likely to grow and eventually stabilize. The level at which this will happen is largely determined by production costs vis-à-vis those in capture fisheries. For halibut, cod and tunas, which lately saw relatively high volumes in capture fisheries, cultured production will eventually be quite high. For Atlantic cod it may be several millions of tonnes. The technology is being developed for some types of halibut, production of which may increase quickly and reach hundreds of thousands of tonnes before the industry reaches serious market problems. Capture fisheries of char and turbot amount to only a few thousand tonnes per year. So, in terms of volume and number of producers, growth of aquaculture may be slow for these species.

Among the species mentioned above, Atlantic cod is perhaps the most interesting. Once this species has been domesticated and a commercially viable culture technology develops, the impact could be considerable not only on codfish fisheries and their markets but on the white fish industry as a whole. It took about 15 years for the Atlantic salmon industry to reach a yearly production of half a million tonnes and, as noted above, the marketing challenge was considerable. It is likely to take less time for the cod culture industry to reach that figure. It will start from a much better market position. Total capture fishery production of Atlantic cod is about 1.3 million tonnes p.a. Atlantic cod has a market also in southern Europe and in Latin America, and the species can be sold in many product forms. Thus, assuming that commercially viable technology for the culture of cod is developed soon, by 2015 cultured cod production could be at least 1 to 2 million tonnes p.a.

At present, of the major fishing nations, only Japan is engaged in sea (or ocean) ranching on a large scale. Over 80 species are being ranched or researched for eventual stocking in marine, brackish and fresh waters (Bartley, 1999) and, as a result, the population has increased for a number of species. Many other countries are also active in the enhancement and ranching of aquatic species, but most of them stock only a few marine species. Salmonids are the most widely stocked group of fish.

The future of sea ranching will depend on a number of crucial issues. One of these is how to deal with the externalities of ocean ranching and the potential ecosystem-related imbalances that it could create (Arnason, 2001). Another is the ownership of the released animals. As long as ownership is not legally defined and technically impossible to defend, the private sector and governments will be reluctant to spend resources on maintaining ranching programmes. It is possible that new technology will permit the identification of released animals and their separation on capture. But, before sea ranching can contribute significantly to world fish supplies it must be integrated into an overall fishery management approach that includes habitat protection.

In inland fisheries, enhancement of fish stocks has been practised successfully in Asia, particularly in China. There is considerable potential for stocking and harvesting fish in both permanent and seasonal freshwater bodies in other parts of the world. The obstacle to rapid development of inland fish stocking is not technical, but rather financial and institutional. Most small freshwater bodies have a large number of users with long-standing traditional usage rights. Enhancement practices must be compatible with these rights and include mechanisms for perpetuating them. No doubt the practice of stocking fish in such bodies will spread, but the spread will be slow in Africa, and is not likely to receive priority in Latin America (with some notable exceptions). In Asia the practice will continue.

In summary, it is unrealistic to expect any sudden change or marked increase in the rate of aquaculture growth worldwide over the next decade or so, unless significant market shocks destabilize the production of other protein sources that currently serve as substitutes for fish in general and aquacultured fish in particular. Nevertheless, it seems possible that the growth in aquacultured fish and shellfish production will continue, at least until 2015, at close to the fairly high growth rates recorded recently (5 to 7 percent p.a.).

Wild *capture fisheries* production in the longer term will be influenced by advances in harvesting technologies and marketing developments. Estimates of total sustainable production still refer to the figure of approximately 100 million tonnes p.a.

(National Research Council, 1999). This is higher than the annual catches of 80 to 85 million tonnes of the 1990s because it assumes efficient utilization of the stocks in healthy ecosystems where critical habitats have been conserved. Moreover, this estimated potential yield includes large quantities of living marine resources which thus far have been little exploited. Of these the best known are krill, mesopelagic fish and oceanic squids. The major factor affecting wild capture production is whether fisheries management can ensure that fish are harvested in a sustainable and economically efficient manner.

Clearly, technical developments will continue to increase the efficiency of fishing. Navigational technologies, electronic fish-finding devices, freezing technologies and gear improvements have already increased fishing productivity substantially. As international awareness increases about various fishing practices, more selective fishing gears that can reduce the catch of non-target species are being developed and spread. In addition, social pressures will continue to drive innovations to make use of unwanted but unavoidable catch. Nonetheless, technical gains will probably not fully compensate for management strategies where these are increasingly economically inefficient.

Market changes will also drive changes in technology. There is a growing trend to consume fresh fish instead of processed fish products. Increased access to markets, lower costs of the technology for delivering fresh fish, and price premiums paid for fresh fish are creating incentives to make this shift. The advent of labelling schemes further supports market changes. For example, labelling products by country of origin, and use of ecolabels recognizing sustainably harvested fish enable producers to differentiate their products and take advantage of changing consumer preferences. However, these developments will not necessarily alter total production, just the form in which fish are produced for food.

The increasing competition of fish substitutes will force wild capture fisheries to deliver products that enjoy price premiums. For example, at the beginning of the 1990s, imports of farmed tilapia (fillets) and supply of pond-cultured catfish were believed to depress prices of white fish in the United States market. Similar situations have been observed in south European countries with respect

Box 7.3 An aquaculture scenario for Africa

Africa's per capita fish consumption is relatively low (6.8 kg in 1999) and there is therefore ample scope for increases in per capita fish demand. There are however several reasons to expect at best a stagnation from now until 2015. Indeed it is not clear at present how per capita fish supply in Africa could be maintained even at its present low level.

Assuming that exports and imports remain at the levels of 1999, fish supply would have to expand by 46 percent over the period to 2015, and by 92 percent to 2030, simply to maintain present per capita supplies. It is not immediately obvious where the fish will come from. Local wild stocks are close to being fully exploited, both in inland and marine waters. Thus, it is unlikely that fishers can respond with higher production to an increase in demand. Export demand is likely to be supplied first, in response to prices that are higher than those affordable by African consumers. Imports of small pelagic species could grow, but it is all a matter of price. Supply for domestic consumption will probably decline and prices edge upwards. Consumers will therefore look for alternatives, which they are likely to find mostly in vegetables and, to a lesser degree, in red meats and poultry.

Recently aquaculture production has expanded rapidly in Africa, but it is essentially Egypt that has accounted for the expansion. In 1999 Egypt accounted for 80 percent of total African production, estimated at 284 thousand tonnes. Production in Egypt consisted of tilapia (46 percent), carp (33 percent) and mullet (19 percent). Conceivably, tilapia and carp could be sold in the rest of Africa, but this is not likely to happen for two reasons. Egypt depends on imports to keep up fish consumption and the average consumer is wealthier than most of the potential importers in the rest of Africa. Thus, future increases in aquaculture production of carp and tilapia are likely to enter the Egyptian market rather than being exported to other African markets. For Nigeria, the second largest aquaculture producer in Africa, the situation is similar.

Elsewhere in Africa aquaculture for local consumption is still small. The contribution to total fish supplies is less than 1 percent of apparent supply. Thus, even in the best of circumstances it will take at least a decade and probably longer before African aquaculture will have a significant impact on African fish supplies.

During the latter part of the projection period the supply situation might change somewhat in Africa as commercial aquaculture is likely to be introduced, initially mainly to supply export markets. Also, the purchasing power of the urban population could be large enough to absorb some of the production of local aquaculture entrepreneurs. African governments will have to put a number of remedial policies in place. These should include, but not be limited to, facilitating fish imports, stimulating aquaculture and the production of short-cycle terrestrial animals (in substitution for fish).

to European sea bass and sea bream. In 1991 some 3 800 tonnes of European sea bass came from culture, or just under 40 percent of the total supply. By 1999 aquaculturists produced more than 41 000 tonnes of the species, accounting for 85 percent of the total supply, and sea bass prices dropped. Similarly, the sheer volumes of cultured European sea bass, Channel catfish and Atlantic salmon that are now available have affected prices for these and other substitutable food sources.

7.3.2 Developments in demand for fishery products

Fish food products. Growth in food demand for fish will not be uniform, and there will be significant differences between regions and countries. As discussed in Chapters 2 and 3, per capita consump-

tion of meat could by 2030 reach 37 and 89 kg per person in the developing and developed countries respectively. Delgado *et al.* (1999) suggest that this increase in meat consumption is expected to occur in parallel with a small decline in the real world market prices of both meat and grains. If these projections prove correct, fish consumption could still increase in developing countries where meat is not a common substitute for fish. Although consumption of poultry and pig meat is expected to continue to increase in the developed countries, this is unlikely to have a major effect on trends in fish consumption in these countries, where demand for fish is more influenced by variety and health considerations than for reasons of economy.

The world average per capita fish consumption grew from 9.4 kg in 1961/63 to 16.3 kg in 1999, or by over 70 percent. A study for FAO (Ye, 1999), taking into account per capita income growth and trend factors, estimates that per capita fish consumption could reach 22.5 kg by 2030, an increase of almost 40 percent over 1999 consumption. Total world demand would then reach 186 million tonnes in 2030, or almost 90 million tonnes more than in 1999. However, the plausible developments in capture fisheries and aquaculture discussed above may limit the growth in fish supply. In view of this, total demand is more likely to lie in the range of 150-160 million tonnes by 2030, or between 19 and 20 kg per capita.

Ye (1999) suggests that per capita consumption in North America, Europe and Oceania will increase rapidly until 2015. Consumption will continue to increase thereafter, but at a slightly slower rate, reaching between 30 and 35 kg per capita in 2030.[9] However, in terms of total consumption, this increase is relatively small because populations in these regions are expected to grow only marginally. In Japan, the market is considered to be saturated, and it is likely that the total volume of fish consumption will decline slightly.

The picture in Asia is mixed. The 1999 average per capita consumption in the region was 17.4 kg.

China's per capita consumption was much higher (25.6 kg). Demand is expected to continue to increase so that per capita demand in some countries in this region could reach about 40 kg by 2030. China's very rapid growth in per capita fish consumption will continue, as elsewhere in Asia, although this rate will begin to slow down over the next 30 years.

Since South Americans are large consumers of red meats, and are likely to remain so, per capita fish consumption is relatively low (9.6 kg in 1999), especially compared with North and Central America where per capita consumption is in the order of 16.5 kg. Furthermore, although per capita consumption in South America could grow slowly over the next 30 years, it is not likely to exceed 15 kg by 2030.

Demand for fish for animal feeding purposes. Small pelagic species and macro-algae have traditionally been used for animal feeding purposes and also as fertilizers. The use of fish and algae as fertilizer has almost ceased but vast quantities of small pelagic species and small volumes of fish offal are still regularly converted into fishmeal and oil, the bulk of

Box 7.4 Krill as a source of human food and animal feed

Krill is a large group of zooplankton found in all oceans, with the Antarctic stocks being the best known. Krill is the major source of food for large whales in southern oceans. It is estimated that the annual volume of these stocks is in the order of 300 to 400 million tonnes – four to five times the total present marine harvest of capture fisheries. Krill has been moderately harvested in the past, reaching a peak catch in 1982 (approximately 500 thousand tonnes). Annual catches dropped to about 100 thousand tonnes at the end of the 1990s.

Krill are small crustacean organisms dispersed in the water column. At present they can be captured only with fishing methods that filter water. To date, pelagic trawls with small meshes in the body of the trawl have been the most successful technology. Further improvement of krill capture technology is possible. Most likely this will involve increasing the volume of water filtered by the gear through use of stronger and thinner twine. Also the use of multiple trawls will increase efficiency. Development of technology permitting the easy spotting of the densest concentrations of krill would be a major advance in krill capturing.

Krill can be used for human food, as animal feed, and as a source of medical products. Such products have been developed and proven to have a market. In order to ensure quality, most processing of krill is done on board the fishing vessels. In the industry it is expected that more krill-based products will be developed, which is likely to spur future demand.

Japan, Poland and Ukraine have been the most active countries in exploiting Antarctic krill, and it is likely that they will continue this fishery. It is probable that international companies will also take an increased interest in this resource, and contribute to expansion. If catching efficiency improves significantly, krill resources in other oceans may also be exploited.

---

9   The preliminary long-term projections based on FAO Food Balance Sheets (FAO, 2002b) suggest similar results, with the rate of increase in per capita consumption declining and per capita consumption in North America, Europe and Oceania reaching between 24 and 33 kg by 2030.

which is used as feed ingredient. Fish used as raw material in the manufacture of fishmeal accounted in 2000 for about a quarter of total fish production (30 out of 130 million tonnes). As discussed in Chapters 3 and 5, livestock production will continue to increase rapidly over the period to 2030, and with it demand for animal feeds, hence also for fishmeal and fish oil. To this should be added a growing demand for both fishmeal and oil used in feeds for intensive aquaculture systems.

Thus far, except for small volumes of offal, all raw material for fishmeal and oil has been supplied by capture fisheries. In all likelihood this will continue to be the situation over the projection period. However, the competition for pelagic fish, a limited and highly fluctuating resource, will become more intense, and it is unlikely that the fishmeal industry will have the economic means to continue to purchase the 30 million tonnes of fish p.a. for reduction. The industry will need to find other raw materials, the most plausible of which could be zooplankton or krill (see Box 7.4). Finally, the onboard processing of unavoidable bycatch and offal into fishmeal may expand.

Demand for fisheries in leisure activities. Recreational and sports fishers are becoming more numerous than commercial fishers in North America, northern Europe and Oceania, and the number of leisure fishers will continue to grow around the world. These stakeholders will increasingly obtain guaranteed access to particular stocks and fishing grounds, sometimes regulated by managing fisheries agencies and sometimes by purchasing access rights from commercial fishers. Their impact will be felt in both freshwater and marine fisheries. Indeed by 2030 the economic importance of recreational and sports fisheries could easily be greater than that of commercial fisheries for several species and waterbodies. This is a plausible scenario not only for developed countries, but also for "big game fishing" and similar activities occurring in tropical areas and on high seas. If access is reallocated in particular fisheries there will be an impact on revenues of the commercial fishing industry, even if the volumes captured in leisure fisheries are likely to remain comparatively small.

There is also a growing demand for "non-consumptive use" of living aquatic resources in developed countries. This demand for access to pristine and exotic resources has found indirect support in a number of United Nations conventions and agreements. The influence of non-consumptive user groups has also increased in less developed countries. It seems likely that demand for non-consumptive uses will continue to grow gradually. However, this is not expected to result in any significant reduction of commercial capture fisheries, particularly in developing countries, unless switching to such activities creates alternative livelihood opportunities that address local food security issues.

7.3.3 Developments in fishery trade.

In coming decades, the world's trading system is likely to be further liberalized under WTO auspices. Trade barriers for fisheries products will eventually be lowered or even dismantled. In anticipation of these developments, fish processors are investing in low-cost countries (e.g. China and the Baltic States) to take advantage of cost differences. Regional specialization in fish production means that some fish-processing industries will probably remain in industrialized countries, even if they have to rely on buying raw material from distant suppliers.

The expansion of international markets can be considered as useful but also problematic. For many communities, especially in developing countries, fish play a very important role in food security and constitute an important part of the diet. However, if fish producers or processors take advantage of access to distant markets for their fish in order to receive a higher net income, the local availability of fish may be reduced (see Box 7.4). Thus, increased trade can provide a mechanism for economic development, but local communities also need to change to accommodate it.

Carp production in China and India is very important but is not expected to expand greatly beyond existing markets in Asia, largely because the demand is limited by regional consumer preferences. In Africa and Latin America the situation is somewhat different in the sense that aquaculture constitutes a very small share of local fish supplies. In these regions abrupt increases of aquaculture fish production for local markets are not impossible, but at present seem unlikely. Thus, if a sudden increase in aquaculture production were to take place, it would most likely be directed at

Box 7.5 International agreements

The Code of Conduct for Responsible Fisheries (CCRF) was developed to address the management of all aspects of capture fisheries and aquaculture, fish processing and trade, including environmental issues and fishing on the high seas. The 1993 FAO Compliance Agreement (the Agreement to Promote Compliance with International Conservation and Management Measures by Fishing Vessels on the High Seas), which addresses fishing on the high seas, is an integral part of the CCRF.

The application of the precautionary approach is a fundamental concept in the CCRF. It requires cautious aquatic resource management when information is uncertain, unreliable or inadequate, and should lead to reduced risks to the ecosystem. The precautionary approach has been enshrined in several other international initiatives, including the UN Convention on Biological Diversity and the UN Fish Stocks Agreement (i.e. the Agreement for the Implementation of the Provisions of the United Nations Convention of the Law of the Sea of 10 December 1982, relating to the Conservation and Management of Straddling Fish Stocks and Highly Migratory Fish Stocks).

foreign markets. The direct effect on supplies in local markets would likely be small, with little competitive impact on local capture fisheries producing for local consumption.

7.3.4 Developments in fishery management

The management issues that are receiving increasing attention in the fisheries sector fall under the headings of maintenance of biodiversity, ecologically sustainable development and, more recently, ecosystem management. These are important considerations for many stakeholders, all of whom are increasingly vocal about what constitutes appropriate practices and acceptable uses of aquatic living resources.

However, there is a largely unaddressed, yet fundamental issue underlying these concerns. It is that in spite of their intrinsic renewable nature, capture fisheries resources are essentially finite. And, as finite resources, they can only be exploited so much in a given period, beyond which overexploitation will cause them to produce less in the longer term, if not entirely collapse. This simple fact means that total wild capture fisheries production cannot increase infinitely, that these exhaustible resources have to be treated with care, and that what is produced must be shared among those engaged in harvesting. The latter can either be done on an ad hoc free-for-all basis which probably is ultimately unsustainable, through adminis-

trative processes, or through the use of rights-based management systems.[10]

The single most important development for the future of wild capture production is how best to contain total catches at sustainable levels while allocating catches. To date, with only few exceptions, the matter of allocating resources among producers and other stakeholders has been addressed only indirectly, and not very successfully.

To achieve sustainable and equitable use, most fisheries stakeholders and administrators will eventually have to recognize the power of using instruments that create and enhance incentives, rather than administrative decisions. Fisheries management strategies based on explicit and well-defined access rights will have to become more common, or otherwise management entities will be overwhelmed by allocation conflicts. To facilitate decision-making, management responsibilities will need to be devolved to fishing interests and communities, with increased participation by stakeholder groups. As part of this, frameworks that control access to capture fisheries need to be strengthened, and created where they do not exist. This will not be a matter only for local or artisanal fisheries – the governance of high seas fisheries will also need to be more explicitly regulated.

In developing countries, the need to design management strategies that can cope with exit and entry issues in small-scale and artisanal fisheries will become more pressing. Although some countries

---

[10] The development of rights-based management systems to create and enhance incentives for sustainability does not equate to or preclude existing cultural, social and traditional arrangements in fishing communities. In fact, if well designed, rights-based management systems will reinforce and strengthen community arrangements while also strengthening rights for exploiting fisheries resources.

Box 7.6 Resource sharing

The sustainable harvest of natural resources in high demand is generally possible only if those resources are subject to explicit and well-defined property rights or rigorous administrative processes to regulate and enforce harvest levels. By contrast, wild fish stocks are often characterized by weak property rights, inadequate management regulations and ineffective enforcement. This situation has made it possible to exploit capture fish stocks beyond sustainable levels. In some instances, particularly for slow-growing demersal species, stocks have been reduced so seriously that the commercial fisheries activities causing the decline cease for lack of fish to catch.

Rights-based management systems create positive incentives for the rights holders and clarify their responsibilities. However, the allocation of rights where they have not explicitly existed before is a complicated matter. Most living aquatic resources are hidden from our view, fluctuate in their abundance, span jurisdictional boundaries and can occur jointly. One consequence of these characteristics is that creating or setting up rights systems can be extremely contentious, complex and expensive. This is particularly true in situations where policies have led to excess fishing capacity and overfishing and where the introduction of rights-based management includes concurrent reductions in fishing capacity (e.g. vessel buyout programmes and redeployment programmes for fishers). Finally, the initial enforcement of newly created rights can be difficult. The result is that rights-based management systems are not being rapidly adopted and implemented. Progress is slow, and excess fishing capacity continues to exist in many fisheries. The problem is severe and occurs in most parts of the world, including developing countries.

Because fisheries resources are becoming increasingly scarce, conflicts over the allocation and sharing of these resources are becoming more frequent. Most conflicts over fisheries resources arise when the resource is (or is perceived to be) so scarce that sharing it becomes difficult. When rights are well defined, understood and observed, allocation conflicts tend to be minimized. However, when rights to the use of a stock are not well defined, understood or upheld, divergent assumptions about what rights users may have often result in conflicts over scarce fisheries resources. Conflicts can be minimized by clarifying roles, responsibilities and the general management of a fishery, by following risk-based decision-making strategies and using conflict mitigation processes.

(for example, the Philippines and Sri Lanka) are working to confront this problem, the vast majority of countries are still struggling with how to do so. Not being able to cope with issues of mass entry and exit could have severe implications for the food security of the people involved. Eventually, it will become politically and socially imperative that alternative social safety nets are developed so that fisheries are not the sector of last resort.

Sustainability issues are also important in aquaculture. The early experience with modern semi-intensive and intensive methods showed that access to and use of water, seed, feed, etc. was perhaps not as low cost as initially envisioned. During the 1990s aquaculturists had to accept that access to sites on land and in water and feeding practices became increasingly regulated to reduce the negative externalities of some aquaculture technologies (e.g. intensive pond or cage culture). In many regions restrictions on species transfers, site selection and feed use have become quite severe, and this has increased production costs. Where profit margins are large, some of these regulatory changes have altered, but not eliminated, aquaculture production. However, in many developing countries the application of such regulatory concerns is difficult. There is little incentive to comply unless the consumers demand it and are prepared to pay a higher price for fish products. In addition, the very limited government budgets and the large number of aquaculture producers make compliance difficult to verify and enforce.

7.4 Concluding remarks

The preceding sections discussed many problems related to the fisheries sector, including how to increase productivity; how to maintain capture fisheries at sustainable levels; how to rebuild over-exploited stocks; how to ensure adequate feed supplies to aquaculture; how to deal with aquaculture-induced environmental problems; how to guarantee user rights to various groups with interests in inland and marine fisheries; and how to strengthen the role of fisheries as a source of employment and foreign exchange earnings.

Views on what would be the appropriate policy responses to such problems differ considerably among the various stakeholder groups and jurisdictions (be they local, national, regional or global) and this makes it difficult to reach agreements on how to solve common concerns in fisheries. Moreover, the global picture is oversimplified as it masks the many possible local or regional specific situations and characteristics.

The most likely scenario to 2030 is that the management of and production from wild capture fisheries will continue much as it is today, and that aquaculture production will continue to grow. Fish will remain an important source of food and employment. The ongoing development and expansion of aquaculture and mariculture will likely continue to be the main source of increased fisheries production. Wild capture fisheries will also certainly remain an important source of food, but will be subject to production constraints.

It remains to be seen whether fisheries-related activities will be managed so as to allow expression of the full productive, economic and social potential that both aquaculture and wild capture fisheries have to offer. There are signs that the management of fisheries can address allocation and sharing issues. But to do so, some major policy and management challenges must be met. Adequate administrative processes and fisheries management capabilities must be developed to cope with issues of allocation of finite resources, in the absence of fully defined and explicit property rights in both the wild capture and aquaculture arenas. Perverse incentives must be removed, and incentive-enhancing management strategies must be created at local, national and international levels. And environmental, cultural and socio-economic concerns related to fisheries must be adequately addressed.

Agriculture in poverty alleviation and economic development

8.1 Introduction

This chapter deals with some of the most contentious issues in the field of economic development. The first section sets the stage with a discussion of the importance attached to poverty reduction by the international community. World leaders have repeatedly proclaimed their commitment to poverty reduction at various summits. In addition, a set of internationally agreed development targets and indicators for measuring progress have been accepted by the Development Assistance Committee (DAC) of the Organization for Economic Co-operation and Development (OECD) and the United Nations General Assembly. In general these targets relate to various aspects of poverty. Sections 8.2 and 8.3 describe the international poverty targets and lay out the main features of the different strategies proposed by international organizations for achieving pro-poor growth.

Few people would quarrel with the proposition that the reduction of hunger should receive priority in the fight against poverty, since hunger is surely the most dire manifestation of poverty. What is often forgotten is that fighting hunger is itself essential for fighting poverty: hunger reduces productivity, increases susceptibility to illness, reduces cognitive ability in children and reduces the willingness to undertake risky investments, thus affecting the rate of economic growth. Hunger also transmits itself from one generation to the next. Evidence of these effects is presented in Section 8.4. Fighting hunger is thus one of the most effective ways of promoting pro-poor economic growth.

Section 8.5 takes up the issue of promoting pro-poor growth. The central theme of this section is that poverty is still disproportionately concentrated among the rural population who still rely heavily on agriculture and on rural off-farm activities for their livelihood strategies. This observation forms the basis for the main argument of Section 8.5, that the best hope of achieving pro-poor growth lies in the interaction between agricultural development and the growth of rural off-farm activities, provided certain conditions are met.

8.2 Internationally agreed poverty reduction targets

8.2.1 Setting the targets

Starting in the early 1990s the international community has adopted a series of targets for the reduction

Table 8.1   Millennium Development Goals (MDGs)

Goal	Target
1. Eradicate extreme poverty and hunger	1. Halve, between 1990 and 2015, the proportion of people whose income is less than US$1 a day
	2. Halve, between 1990 and 2015, the proportion of people who suffer from hunger
2. Achieve universal primary education	3. Ensure that, by 2015, children everywhere, boys and girls alike, will be able to complete a full course of primary schooling
3. Promote gender equality and empower women	4. Eliminate gender disparity in primary and secondary education preferably by 2005 and in all levels of education no later than 2015
4. Reduce child mortality	5. Reduce by two-thirds, between 1990 and 2015, the under-five mortality rate
5. Improve maternal health	6. Reduce by three-quarters, between 1990 and 2015, the maternal mortality rate
6. Combat HIV/AIDS, malaria and other diseases	7. Have halted by 2015, and begun to reverse, the spread of HIV/AIDS
	8. Have halted by 2015, and begun to reverse, the incidence of malaria and other major diseases
7. Ensure environmental sustainability	9. Integrate the principles of sustainable development into country policies and programmes and reverse the loss of environmental resources
	10. Halve, by 2015, the proportion of people without sustainable access to safe drinking-water
	11. By 2020, to have achieved a significant improvement in the lives of at least 100 million slum dwellers
8. Develop a global partnership for development	12. Develop further an open, rule-based, predictable, non-discriminatory trading and financial system
	13. Address the special needs of the LDCs
	14. Address the special needs of landlocked countries and small island developing states
	15. Deal comprehensively with the debt problems of developing countries through national and international measures in order to make debt sustainable in the long term
	16. In cooperation with developing countries, develop and implement strategies for decent and productive work for youth
	17. In cooperation with pharmaceutical companies, provide access to affordable, essential drugs in developing countries
	18. In cooperation with the private sector, make available the benefits of new technologies, especially in information and communication

Source: Adapted from UN (2001b).

Box 8.1 The problems with international poverty data

A discussion on the quality of poverty statistics is beyond the scope of this chapter. Serious doubts do however exist as to the significance of the widely used global data on poverty (namely the World Bank's US$1 a day poverty line) that cannot be ignored. A few major objections are reported here, to warn the reader about important caveats that should be borne in mind concerning the global numbers in this section. For a fuller discussion of the shortcomings behind the international poverty lines the interested reader is referred to Deaton (2001) and Srinivasan (2000), on which the following list is based.

■ Poverty is sensitive to temporal and spatial variations (due partly to cultural norms) that are ignored in international poverty lines.

■ International poverty measures, even for the same country, are highly sensitive to the way purchasing power parity (PPP) indexes are calculated. Introducing a different PPP data set causes the 1993 poverty headcount for sub-Saharan Africa to jump from 38.1 to 48.7 percent (Deaton, 2001, p. 5).

■ PPP data are available only for a limited, albeit increasing, set of countries and years, and are applied to the other countries by extrapolation.

■ The price indexes used in poverty calculations have several flaws: (i) they are often national averages, not those actually faced by the poor; (ii) they are seldom updated; (iii) they may have an "urban bias", in the sense that price data collection is often more accurate in urban than in rural areas; and (iv) a rural versus urban breakdown is not always available.

■ Survey data also have flaws linked to: (i) coverage (publicly provided goods are not captured); (ii) different recall periods across countries, which generate data that are not comparable; and (iii) inconsistencies in the way income survey data are transformed to make them comparable with expenditure survey data.

■ There is often a discrepancy between national account data and survey findings (for instance in India). This makes it impossible to draw any meaningful inference about the relationship between variables measured on the basis of surveys (such as poverty) and variables measured on the basis of national accounts (such as GDP growth).

of poverty in its various dimensions. Goals have been set for reduction of poverty and hunger; for school enrolment; gender equality; reduction of infant, child and maternal mortality; access to reproductive health services; and the adoption of national strategies for sustainable development.

In 1995, the Copenhagen declaration issued at the UN World Summit on Sustainable Development (WSSD) established the general goal of eradicating poverty and eliminating hunger and malnutrition, but no date was specified.[1] The following year the members of the DAC of OECD adopted the goal of halving the proportion of people living in extreme poverty by 2015, initially with no specific base, later specifying 1990 as the base year (OECD, 1996). Also in 1996, the WFS set the goal of halving the number of undernourished by 2015.

Other UN conferences convened in the 1990s agreed on targets for education; child health, labour and nutrition; reproductive health and infant and maternal mortality; education; gender equality; employment; housing and access to safe water and sanitation; and the environment. International targets have also been agreed for political and civil rights, crime prevention and drug control. The UN Millennium Declaration, adopted in September 2000, consolidated most of these goals.

In 2001, in order to harmonize reporting on progress towards the goals of the Millennium Declaration and other International Development Goals (IDGs), the UN General Assembly adopted a set of eight Millennium Development Goals (MDGs), 18 targets (Table 8.1) and 48 indicators. The MDG on undernourishment differs from that agreed upon at the WFS. The former is about halving the proportion of people who suffer from hunger, while the latter has the more ambitious aim of reducing the number of undernourished

---

[1] The Copenhagen declaration is frequently claimed to have set the target of halving poverty by 2015. However, no such target is found in the text of the declaration or the annexed plan of action, which only talk of "eradicating absolute poverty by a target date to be specified by each country in its national context" (UN, 1995).

by half by 2015. There is some discrepancy among the goals in the choice of base year: 1990/92 for the WFS goal, 1990 for the MDG on halving hunger.

Does it make sense to have internationally agreed development goals? The short answer is yes.

It is true that internationally agreed development goals, and the indicators used to monitor progress towards their achievement, are prone to endless critiques. Maxwell (1999) and Gaiha (2001) wrote specifically on the shortcomings of the IDGs, while Deaton (2001) and Srinivasan (2000) are two recent examples of many contributions to the problems of measuring poverty, and particularly of arriving at internationally comparable poverty statistics (see Box 8.1).

Nevertheless, even critics acknowledge that "simple targets have real value, and that politicians have a perfectly legitimate need to find simple messages that will galvanize public opinion" (Maxwell, 1999); or that "they are ... useful in drawing attention to pervasive deprivation in developing countries, and to the need for a determined and coordinated effort on the part of the development community to reduce it substantially in the not-too-distant future" (Gaiha, 2001).

Such targets and indicators should not be seen as accurate measures of progress, or as finely tuned criteria to guide policy and to prioritize spending at the international, national or subnational levels. They should be judged and used for what they really are: crude targets and indicators to raise international consciousness among the public and within the development community, and to assess roughly whether progress is being made. However, seeing indicators as a means to focus attention and action can only be meaningful as a step towards some kind of priority setting.

8.2.2 Measuring progress

Thus far the discussion has focused on putting order in the sometimes confusing web of internationally agreed development commitments. In this section, the discussion will be centred on the actual trends towards the achievement of some of these goals and the prospects up to 2015.

Table 8.2 presents the latest World Bank estimates and projections of poverty (less than US\$1 per day). The data show that progress towards the 2015 target is being made. At the global level, poverty has declined both in absolute numbers (if only marginally) and in relative terms. Table 8.2 also presents the baseline scenario for the future as developed by the World Bank (2001c). These forecasts suggest that the world as a whole is roughly on track to reach by 2015 the MDG of halving the proportion of poor people.

The global data hide huge discrepancies between and within regions. East Asia may achieve

Table 8.2   World Bank estimates and projections of poverty

	1990	1999	2015	1990	1999	2015
	Million persons			% of population		
Developing countries	1 269	1 134	749	32.0	24.6	13.2
Sub-Saharan Africa	242	300	345	47.7	46.7	39.3
Near East/North Africa	6	7	6	2.4	2.3	1.5
Latin America and the Caribbean	74	77	60	16.8	15.1	9.7
South Asia	495	490	279	44.0	36.9	16.7
East Asia	452	260	59	27.6	14.2	2.8
Transition countries	7	17	4	1.6	3.6	0.8
Total	1 276	1 151	753	29.0	22.7	12.3
idem excl. China	916	936	700	28.1	24.5	14.8

Source: Adapted from World Bank (2001c), Table 1.8. The definition of regions is not always identical to that used in this study, e.g. Turkey is not included in the developing Near East/North Africa and South Africa is included in developing sub-Saharan Africa.

the target early in the new millennium. South Asia made considerable progress in percentage terms during the 1990s, and achieving the goal of halving US$1/day poverty is deemed feasible.

The goals seem much more challenging in the other regions. In sub-Saharan Africa poverty has in fact increased between 1990 and 1999. The available projections suggest that for sub-Saharan Africa the MDG poverty goal may be beyond reach. Very little progress will be achieved unless performance is significantly enhanced in the near future, and the absolute number of poor may in fact rise considerably. Should this scenario materialize, close to half the world's poor will live in sub-Saharan Africa in 2015.

Latin America and the Near East/North Africa have made only marginal progress (in relative terms) during the 1990s. If the forecasts are accurate they should at least come within reach of the poverty reduction target.

The transition countries in Eastern Europe and Central Asia present a different picture. A big surge in poverty occurred in the region after 1990 (the base year for the target). Most of these countries were then on the brink of a recession after the collapse of the centrally planned regimes and the beginning of the transition towards market economies. The first years of the transition recorded substantial increases in poverty rates across the region (Milanovic, 1998) that have since been only partially reversed. Achieving the target would require faster poverty reduction than at present.

To summarize, the poverty projections imply that:

- the goal of halving by 2015 the proportion of people living in poverty from that prevailing in 1990 may be achieved – the proportion falls from 29.0 percent in 1990 to 12.3 percent in 2015;
- the absolute numbers in poverty may not be halved, as they decline from 1.27 billion in 1990 to 0.75 billion in 2015;
- much of the decline results from prospective developments in East Asia and South Asia;
- in contrast, sub-Saharan Africa's absolute numbers in poverty kept increasing in the 1990s

and are projected to continue to do so until 2015.

The World Bank baseline scenario reported in Table 8.2 is based on the assumption of somewhat higher economic growth than in the past. In fact, the World Bank in its penultimate (2001) report stressed that "if policies are inadequate to achieve more than the slow growth of the 1990s, then the number of people living in extreme poverty would remain near current levels for the next 15 years" (World Bank, 2001a, p. 42).

Shane *et al.* (2000) use a world general equilibrium model to run alternative poverty scenarios to 2015.[2] The broad trends yielded by their projections are generally consistent with those produced by the World Bank. Two aspects of their results are of particular interest. First, they point to the sensitivity of the projections to the income distribution parameters for China. If China's income distribution is forecast to deteriorate significantly over the time horizon considered, the decrease in the number of world poor would amount to a mere 6 percent (against 20 percent in their baseline). Second, their simulations of different types of growth patterns show that growth in agricultural productivity is the most promising avenue to achieve poverty alleviation. In their model, growth based on increases in total factor productivity (TFP) in agriculture is more poverty reducing than either general TFP growth or labour productivity growth.

Developments in a few large countries that host the majority of the world's poor (China, India, Indonesia, Brazil, Bangladesh and Nigeria) are crucial in determining global numbers. It is therefore also interesting to look at prospects for individual countries. In a recent study, Demery and Walton (1999) show that if per capita growth continues along the path recorded during the 1990-95 period the poverty target will be met, as poverty would be reduced by more than half in large countries such as India, China, Indonesia and Brazil. However, 22 out of their sample of 36 countries will fall short of the poverty reduction target.[3]

---

[2] They actually refer to food insecurity rather than poverty, but define food-insecure people as those under a certain income threshold, which is the common definition of poverty.

[3] Demery and Walton (1999) also build two alternative scenarios to account for a change in policy regimes. Not surprisingly, they arrive at the result that "better" policies would lead to significantly faster poverty reduction.

It is also recognized that where income is very unequally distributed, the poor fail to connect to the growth process (Timmer, 1997). According to some estimates, high-inequality countries would need twice as much growth as low-inequality countries to meet the poverty target (Hammer, Healey and Naschold, 2000).

A word of caution is needed about such perspective analyses. While the data definitely point to the feasibility of achieving the target for poverty reduction at the global level, even the pessimistic scenarios assume higher growth rates than those historically recorded. When past rather than forecast GDP growth rates are used to predict future trends, future poverty reduction will fall short of the 2015 target. The only exception is the Demery and Walton (1999) study, which is based on a smaller sample of countries.

Hence, such perspective analyses do not leave room for unqualified optimism. Business as usual will not result in achieving the targets: stronger commitment and faster results are needed. This is especially true for the two regions that account for more than half of the world's poor. Unless current trends are reversed, South Asia and sub-Saharan Africa face much less rosy prospects and a possible increase in the number of poor.

8.3 The main international strategies for poverty reduction: a summary assessment

The ranking of poverty reduction on the international development agenda has probably never been so high. With the symbolic landmark of the new millennium, several multilateral and bilateral donors and agencies felt the urgency to take stock of their own and others' experience with poverty alleviation programmes, and to come up with new strategic proposals for poverty reduction in the years to come. The years 2000-2001 have seen a flurry of strategic poverty reduction documents and reports.

This section briefly reviews these reports, focusing in particular on those by the World Bank, the International Fund for Agricultural Development (IFAD) and the United Nations Development Programme (UNDP). We shall consider only selected aspects: the definition of poverty; the role of growth; the role of inequality and redistribution; the role of agricultural growth; and empowerment and participation. The aim is not to provide a complete review, nor a comparative evaluation of these reports, but to highlight some major elements of consensus and divergence that are of particular relevance to this chapter.

8.3.1 Definition and measurement issues: the multiple dimensions of poverty

Reflecting the recent evolution of thinking about poverty (Kanbur and Squire, 1999), all agencies now define poverty as having multiple dimensions. Poverty is not seen as simply lack of income or consumption: it includes deprivation in health, education, nutrition, security, power and more. It is also widely accepted that these dimensions of deprivation interact with and reinforce each other.

Some differences persist in the emphasis each agency places, implicitly or explicitly, on the various dimensions of poverty, and on the consequences that are drawn in the strategic documents concerning measures of poverty, and implications for policy and action.

The World Bank's *World Development Report 2000/1* broadens the World Bank's earlier definition of poverty, which included income, education and health dimensions (World Bank, 1990) to include vulnerability and exposure to risk, and voicelessness and powerlessness (World Bank, 2000b, p. 15).

Much of the initial part of the *World Development Report 2000/1* deals with the problems of measuring poverty in its multiple dimensions. When it turns to policy discussion, however, poverty is identified mainly in terms of income poverty. This reflects the difficulty of building a good poverty indicator that is both multidimensional and robust to cross-country comparisons. The problems of constructing such an indicator are introduced in the report itself (World Bank, 2000b, p. 22).

IFAD's *Rural Poverty Report 2001* (IFAD, 2001) discusses these problems at greater length. Three categories of poverty measures are defined (see p. 19-20): a scalar approach (using a single indicator for a single dimension, e.g. income or consumption); a multidimensional-indexed approach (several indicators are combined into a single indicator); and a vector approach in which several indicators are used to classify people according to each indicator. The IFAD report explicitly chooses one indicator (the scalar absolute poverty line fixed over time) which it finds best suited for its purpose of international comparison.

UNDP's approach is different. The multidimensional definition of poverty is key to their *human poverty* approach, which focuses "not on what people do or do not have, but on what they can or cannot do" (UNDP, 2000b, p. 22). This approach is reflected in their choice of a poverty indicator. UNDP's human poverty index combines "deprivation in a long and healthy life" (measured by the percentage of people expected not to survive to age 40), "deprivation in knowledge" (measured by adult illiteracy), and "deprivation in economic provisioning, from private and public income" (measured by lack of access to health services and safe water, and percentage of children under five who are underweight).

All indexes have advantages and disadvantages. Some may ignore relevant information (e.g. income poverty measures). Others may attribute somewhat arbitrary weights to the different dimensions of poverty (e.g. the human poverty index). Others again may not be suitable for comparisons across different groups (e.g. the multidimensional vector approach).

Of course there is no such thing as a perfect poverty indicator. Rough poverty indicators such as the US$1/day poverty line are usually thought to provide an adequate approximation to the other dimensions of poverty at the national and regional level of aggregation (see, for instance, Kanbur and Squire, 1999). More detailed information on the other dimensions will normally be needed when the poverty measure is supposed to inform policy programmes at the local level, or policies to target some specific non-income objective.

It is important to stress that the multidimensional definition of poverty adds further complications to the task of measuring poverty. Poverty measurement issues form a substantial part of the existing literature on poverty. The generalized acceptance of a multidimensional definition is likely to give new impetus to this debate, as it did in the past following the introduction of UNDP's human development index (HDI).

Such a debate may not develop an "ultimate" indicator of poverty (i.e. one that is both multidimensional and robust to international comparisons). Its key contribution may rather be to develop a consistent framework for choosing appropriate indicators for each particular need, in research as well as in policy-making.

8.3.2 Strategies for poverty reduction: the quest for pro-poor growth

Starting from similar definitions of poverty, most of the documents reviewed here broadly agree on the basic determinants of poverty, and on the issues that well-conceived poverty alleviation policies should tackle. Most strategies now include topics such as institutions and access to markets, human and social capital, empowerment, decentralization, democracy, accountability and governance, international finance and trade, technology, the environment, social policies and aid.

Although differences of emphasis still remain, it is tempting to say that a new consensus has emerged on a poverty reduction recipe. This can be summarized as follows: foster broad-based, pro-poor economic growth, favouring poor people's access to markets and assets, basic education, health and sanitation services, safe water, and power/decision-making. Such access is seen as an end in itself (in that it addresses some of the dimensions of poverty directly, a "welfare" effect), and as a means for poverty reduction through increased economic growth (an "efficiency" effect).

Economic growth is considered central in all the strategic documents surveyed. This is neither new nor surprising. What is new, though, is the range of issues that are now required for achieving sustained growth. In the World Bank report (World Bank, 2000b, p. 49) divergences in growth rates are said to depend on the interaction among a range of factors including initial conditions, policy choices, external shocks and even "no small measure of good luck". The World Bank document also discusses the roles of education, life expectancy, population growth, trade openness and sound macro policies, institutions, ethnic fragmentation, geography and environmental degradation as determinants of growth and its sustainability.

IFAD's approach, partly because of the nature of the Organization and its mandate, calls much

more specifically for labour-intensive, food production based growth. IFAD's approach also explicitly
mentions transient and chronic poverty as different
problems requiring quite different approaches to
analysis and action (a point that does not appear to
receive the same attention by the World Bank). The
focus on labour-intensive growth is also emphasized in IFAD's report but not in that of the World
Bank. This is a surprising omission, since it was key
to the World Bank's strategy in its previous poverty
report (World Bank, 1990).

The UNDP strategy is different in many
respects. The UNDP document does not deny the
role of economic growth in poverty reduction, but
it focuses more on the need to include poverty alleviation and empowerment targets and policies in
growth strategies explicitly and consistently, rather
than treating them as largely separate issues.

All agencies seem to accept the qualification
that, to be effective in reducing poverty, growth
needs to be "pro-poor" or "broad-based". While
this may appear to be part of the new consensus,
different participants in the debate may have
different understandings of the meaning and
implications of these terms, which are rarely articulated in full. To a large extent such understanding
may depend on how the roles of inequality and
redistribution measures are viewed. It may also
depend on two other features that figure prominently in all documents: the role of the agricultural
and rural sector in growth and poverty alleviation,
and the role and importance of the empowerment
and participation of the poor. In what follows these
three aspects are briefly reviewed.

8.3.3 Strategies for poverty reduction: the role of inequality

The World Bank's recipe focuses on win-win measures that are both growth enhancing and equalizing, such as improving the poor's access to land
(especially through market-based schemes), basic
education and health. However, the World Bank's
position on policy measures that may involve a
trade-off between a little less but more equal
growth is not spelled out clearly.

The IFAD strategy stresses even more strongly
the need for redistributive measures for effective
poverty reduction. It points to the overlooked
success of many traditional land reform
programmes (with beneficial efficiency effects), and
to the need to combine redistribution of land assets
with redistribution of water and education assets,
and provision of complementary services.

UNDP's main criticism of traditional antipoverty
strategies is that they did not properly link policies
to promote economic growth with social policies
addressing poverty concerns. Although its flagship
report lacks specific policy suggestions on the
issue, UNDP clearly affirms that "many poverty
programmes do not adequately address inequality".
In particular, it stresses that they "rarely deal with
inequality in the distribution of land, which
continues to be the most important asset of the
rural poor", and that "in some regions with high
inequality ... economic growth cannot be accelerated enough to overcome the handicap of too
much income directed to the rich" (UNDP, 2000b,
p. 42-3).

8.3.4 Strategies for poverty reduction: does agriculture matter?

The role of agricultural growth in alleviating
poverty has attracted wide attention. This is hardly
surprising as most of the world's poor still live in
rural areas. This section focuses only on how the
reports reviewed see the role of agricultural growth
in poverty reduction. A more extensive analysis of
this role is provided in Section 8.5 below.

IFAD's approach explores the issue in greater
detail, and also takes a more distinct position. It
points in particular to food staples production as
the subsector with the greatest potential for rural
poverty reduction. IFAD's argument is based essentially on (i) the large share of food (and particularly
staple food) in the total consumption of the poor;
(ii) the large share of calories the poor derive from
staple food consumption; and (iii) the large share of
income the poor derive from staple food production. The other strategic documents reviewed do
not share this focus on staple food production.[4]

The World Bank report asserts that growth in
the agricultural sector may benefit the poor dispro-

---

4 Although IFAD's report (IFAD, 2001) does not deny the importance of cash crop production, off-farm income and income diversification in general
for the welfare of the rural poor, a "critical role" is assigned to staples (see, for instance, p. 3).

portionately and may thus be associated with declining inequality. The World Bank and UNDP share the view that agricultural growth has a particular role to play in poverty alleviation, but their argument is based largely on the more familiar observation that this is one of the sectors from which the poor derive much of their income. The World Bank recommends removing the remaining biases against agriculture, and seems to put a lot of faith in this type of intervention when it states that "the winners (of past reforms) are often those in rural areas" (World Bank, 2000b, p. 66).

The *UN World Economic and Social Survey 2000* (UN, 2000) makes agricultural growth a central issue for "escaping the poverty trap". It states that agricultural growth can contribute strongly to poverty reduction, mainly because of its demand linkages, and because agricultural and related activities tend to be more labour intensive and less import intensive than manufacturing activities.

8.3.5 Two new issues: empowerment and participation

Issues such as growth, inequality and the role of agriculture have always featured prominently in the development debate. What is new in the current wave of poverty reports is the focus on empowerment and participation. All three agencies stress the importance of empowerment of the poor, and encourage their participation in the poverty reduction process. A significant feature in the World Bank's approach is the inclusion of "voicelessness and powerlessness" among the main dimensions of poverty (World Bank, 2000b, p. 15). This perspective leads to treating empowerment as an end of poverty alleviation strategies, rather than primarily as a means.

IFAD's perception is very similar to the World Bank's. "Lack of power" enters IFAD's definition of poverty (IFAD, 2001, p. 2). Empowerment and participation are seen as essential means for effective poverty reduction in the rural sector.

UNDP has a similar understanding of the matter in that it includes "[not] participating fully in the life of the community" as one of the dimensions of poverty and "empowering the poor" as one of the key elements of the "new generation of poverty programmes" it calls for. Unlike the World Bank and IFAD, however, UNDP puts more stress on issues such as governance, fighting corruption and decentralization than on empowerment *per se*. The "means" aspect of empowerment is emphasized more than the "goal" aspect. Empowerment is, for instance, seen as an essential tool for achieving more effective targeting in poverty alleviation programmes.[5]

8.3.6 From vision to action: putting the strategies to work

An evaluation of the merits of these strategies is far beyond the scope of this chapter. A few general remarks are however possible.

First, the concepts of empowerment and participation. As a technique for field-level development work, participation has a long and well-established history of successes. It assumes different and new meanings when it is linked to the concept of empowerment, and when it is viewed as one of the main ends of poverty alleviation efforts. However, such meanings are not always fully spelled out in terms of operational concepts and policy guidance. If empowerment and participation are to serve as powerful tools in the quest for poverty reduction, considerable efforts will be needed to make them workable concepts.

By recognizing the multiple dimensions of poverty, and by opening to a wealth of new concepts, the recent wave of strategies reviewed here run the risk of generating an all-inclusive framework that may lack focus. Hence it may not serve what should be one of the main tasks of a poverty report, that is, helping to outline the priorities for concerted international action. It is certainly true that poverty is a multidimensional concept, but it does not automatically follow that each of its dimensions may relate to a specific set of causes and, hence, of policies.

The most welcome conclusion of this brief review is that both poverty and inequality are universally placed at the top of the international development policy agenda. It was not so a decade ago. Still, continued efforts are needed to enhance

---

[5] The United Kingdom's Department for International Development (DFID) seems to have a similar position, viewing the role of the "voices of the poor" as an element largely embodied in democracy and good governance, rather than a key dimension in itself (DFID, 2000b).

our understanding of what works in terms of poverty reduction and increased food security. In particular, the focus should be partly shifted from the pursuit of win-win policies, towards policy options that involve trade-offs, with which policy-makers are more often confronted. Growth is certainly good for poverty, but what else is needed? What kind of growth is more poverty reducing? How can the related goals of poverty reduction and enhanced food security be best pursued? The rest of this chapter explores some of these questions.

8.4 Micro and macro evidence on the impact of undernourishment

This section presents arguments in favour of a focus on fighting hunger as a complement to fighting poverty. Hunger is a violation of a basic human right, and also imposes significant economic costs on society. Hence its reduction and ultimate eradication are the most urgent tasks facing national governments, civil society and the international community.

The plan of this section is as follows. The next subsection documents the serious economic costs of widespread hunger for individuals and nations. This is followed by a discussion of the relationship between household income growth and nutrition, showing that income growth alone, at the rates observed in the recent past, cannot be expected to remove hunger in a reasonably short period. It follows that direct public interventions are also required. However, rapid income growth is needed, since it is a necessary condition for poverty and hunger alleviation, and would make it possible to finance public action against hunger. In Section 8.5 it is argued that agricultural development, coupled with the development of rural non-farm (RNF) activities, is one of the most effective means of promoting income growth.

8.4.1 The economic impact of hunger

It is necessary to define some frequently used terms. The definitions used here correspond to those used in FAO (1999a). The term "undernourishment" is used to describe the status of persons whose food intake does not provide enough calories to meet their physiological requirements on a continuing basis.[6]

An alternative approach is to assess nutritional status through the physiological outcomes of poor nutrition. The term "undernutrition" is used here to describe the status of persons whose heights and weights lie below the lower limits of the ranges established for healthy people. It is critical to note that poor anthropometric status is the outcome not only of insufficient food intake but also of sickness spells. Infectious diseases tend to raise nutritional needs and lower the capacity to absorb nutrients. Food intakes that are adequate for a healthy person may be inadequate for someone in poor health, leading to weight loss in adults and children and growth retardation in growing children. Thus anthropometric measures incorporate information about food consumption as well as health inputs.

Among adults a commonly used measure is the body mass index (BMI), defined as the ratio of bodyweight in kg to the square of height in metres. A BMI in the range of 18.5 to 25 is considered to be healthy for adults, as recommended by FAO, WHO and others. BMI can clearly vary over an adult's lifetime, but physical stature is determined by the time an individual reaches adulthood.

A natural question at this point is whether it makes sense to collect information on both food intake and stature. The answer is yes, because data on heights and weights are expensive to collect and for that reason are not collected on a regular basis. This makes it hard to use them for regular monitoring.[7] Data on dietary energy supply (DES), using FAO's methodology, do have the merit of being relatively easy to collect and can be used for monitoring purposes. Another reason for collecting both types of information is that they are needed to guide public policy. It is not usually clear simply by looking at a country's undernutrition record whether this results from inadequate food intakes or from frequent sickness spells, despite calorie intake levels that would otherwise be sufficient. The

---

6 See Box 2.1 in Chapter 2 for a discussion of the data and criteria used to measure undernourishment.

7 On the other hand, since anthropometric data do not change rapidly (except during wars or famines) it is not necessary to collect these data more often than once every five years.

implications for public policy are quite different in these two cases. In the former case the aim should be to increase food intakes, while in the latter the emphasis should be on public health, sanitation and the provision of clean drinking-water, etc.

What then are the economic costs of under-nourishment and undernutrition? First, at the most basic level a person requires adequate nutrition in order to perform labour. If this nutrition is not forthcoming, or if the person lives in an unhealthy environment, the result is poor nutritional status and the person's ability to do sustained work is reduced. Furthermore, if the person is shorter or has a smaller body frame as a result of past nutritional deprivation, he or she may lack the strength to perform physically demanding but also better rewarded tasks. Thus one would expect to find poor nutritional status, as measured by height, associated with lower wages and earnings.

Second, there is evidence that poor nutritional status leaves people more susceptible to illness. Thus a vicious cycle may exist whereby inadequate food intakes combined with frequent sickness spells result in poor nutritional status, which in turn creates an increased susceptibility to illness. Evidence on this point is presented below.

Third, there is a risk of intergenerational trans-mission of poor nutritional status. For example, women who suffer from poor nutrition are more likely to give birth to underweight babies. These babies thus start out with a nutritional handicap.

Fourth, there is evidence that poor nutrition is associated with poor school performance in children of school age. At its simplest, this is expressed as "a hungry child cannot learn". This would not necessarily imply any impairment in the child's cognitive ability, but merely that because of hunger the child is listless or tired and inattentive and cannot participate in learning activities. Unfortunately, it may also be the case that cognitive ability itself is impaired as a result of prolonged and severe malnutrition. In either case, the upshot is that children do poorly at school, thereby damaging their future economic prospects.

Fifth, people who live on the edge of starvation can be expected to follow a policy of safety first with respect to investments. They will avoid taking risks since the consequences for short-term survival of a downward fluctuation in income will be cata-strophic. But less risky investments also tend to have lower rewards. Once again, the tendency is for poor nutrition to be associated with lower income.

Finally, there is some evidence that the macro-economic performance of the whole economy may suffer as a result of the cumulative impact of these effects. It has been shown recently that the overall effect may be to reduce a country's rate of economic growth.

Nutrition and productivity. What is the mechanism by which poor nutrition affects productivity? Dasgupta (1997, p. 15) explains that a person's physical work capacity can be measured by his or her maximal oxygen uptake. The higher its value, the greater the capacity of the body to convert energy in the tissues into work. Here is the crux of the matter: clinical tests suggest that the maximal oxygen uptake per unit of muscle cell mass is more or less constant in well-nourished and mildly undernourished people. Since lean body mass is related to muscle cell mass, it follows that a higher BMI implies a higher maximal oxygen uptake and hence greater work capacity. Also, if two people have the same BMI, the taller of the two has more lean body mass, and hence higher maximal oxygen uptake and work capacity. Studies also suggest that maximal oxygen uptake depends on the concentration of haemoglobin in the blood. Since that depends on iron intakes, the connection between iron-deficiency anaemia and low productivity is also explained.

Empirical studies are now available on the impact of poor nutrition on individual productivity and wages in ten developing countries in Asia, Africa and Latin America. The evidence convinc-ingly bears out the hypotheses advanced above. Table 8.3 below summarizes the main features of these studies.

It is useful to distinguish between the effect of current undernutrition, as expressed in BMI, from the crystallized effect of past undernutrition, as expressed in adult height.

As far as the former is concerned, all the studies referred to above found that increased BMI had a significant impact on output and wages. For example, Croppenstedt and Muller (2000) found that in rural Ethiopia an increase of 1 percent in BMI increased farm output by about 2.3 percent and wages by 2.7 percent. Thomas and Strauss (1997) found that a 1 percent increase in BMI in

Table 8.3   Summary of studies on the productivity impact of poor nutrition

Study authors	Country	Group studied	Main findings
Croppenstedt and Muller (2000)	Ethiopia	Rural households, mainly agricultural	Output and wages rise with BMI and WfH. Adult height has positive impact on wages
Bhargava* (1997)	Rwanda	Rural households, mainly agricultural	BMI, energy intake have positive impact on time spent on heavy activities for men, but not women
Strauss (1986)	Sierra Leone	Rural households, mainly agricultural	Calorie intake has positive impact on productivity
Satyanarayana et al. (1977)	India	Indian factory workers	WfH is significant determinant of productivity
Deolalikar (1988)	India	Southern Indian agricultural workers	Significant effect of WfH on farm output and wages
Alderman et al. (1996)	Pakistan	Rural households, mainly agricultural	Adult height is significant determinant of rural wages
Haddad and Bouis (1991)	Philippines	Sugar-cane growers	Adult height is significant determinant of rural wage
Strauss and Thomas (1998)	Brazil and United States	Adult male population, Brazil, United States	Adult stature, BMI have positive impact on wages in Brazil. Only stature has positive effect on wages in the United States
Thomas and Strauss (1997)	Brazil	Urban population sample	BMI, adult height have strong, positive impacts on market wages
Spurr (1990)	Colombia	Sugar-cane cutters and loaders	Weight, height are significant determinants of productivity
Immink et al. (1984)	Guatemala	Coffee and sugar-cane growers	Adult height has positive impact on productivity

Notes: BMI=body mass index, i.e. weight (kg)/square of height (metre). WfH=weight for height.

*   This study is not directly relevant since it focuses on time allocation decisions. However, it could be argued that *ceteris paribus* the ability to spend more time on heavy activities enhances one's productivity and earning capacity in agriculture.

their sample from urban Brazil was associated with a 2.2 percent increase in wages.

Another possibility is to measure current nutrition through calorie intakes. Here also Strauss (1986) and Thomas and Strauss (1997) reported significant impacts of increased calorie consumption on farm output and wages. For example, the latter study found that an increase of 1 percent in calorie intakes increased wages by about 1.6 percent at calorie intake levels of around 1 700 calories per day, but that this effect ceased to operate

after calorie consumption levels reached around 1 950 calories per day.

The role of micronutrient deficiencies in reducing work capacity has also received increased attention lately. Horton (1999) states that "Studies suggest that iron deficiency anaemia is associated with a 17 percent loss of productivity in heavy manual labour, and 5 percent in light blue-collar work (studies cited in Ross and Horton [1998])".

Strauss and Thomas (1998) present a succinct and illuminating review of the impact of adult stature and BMI on productivity through an analysis of two data sets from the United States and Brazil. They found that adult stature is positively correlated with wages in both countries, but the effect is strong in Brazil and weak in the United States. However, stature is also positively correlated with education. The suspicion naturally arises as to whether the seeming effect of stature on wages is simply a reflection of the fact that taller people are also better educated. Since it is widely accepted that better education does lead to higher wages, perhaps that is the underlying cause of the dependence of wages on stature. However, Strauss and Thomas (1998) show that this cannot be the explanation since in Brazil the impact of stature on wages was strong even in those adults who had no education.

Low stature and low BMI are also associated with lower labour force participation – not only do people with lower stature or BMIs earn less, but they are less likely to be in a position to earn wages at all. The probable reason for this is that people with low BMIs and low stature are also more likely to fall sick.

This evidence supports the hypothesis that higher stature and BMI are associated with higher wages because of their impact on maximal oxygen-carrying capacity, not because they are proxies for otherwise unobserved qualities that are attractive to employers. Even if one were to argue that stature captures other unobserved investments in human capital in childhood, it is hard to explain the finding that people with no education are likely to earn higher wages if they are taller. Since manual jobs done by uneducated people typically involve heavy labour and do not require much initiative, but rather a willingness to carry out instructions, it is difficult to see what employers could be looking for in a tall person other than sheer physical strength. More evidence for this hypothesis is provided by the finding that the impact of higher stature on wages is weaker in the United States, where mechanization is more prevalent, than in Brazil where mechanization is less prevalent and thus physical strength matters more. Also note that even in Brazil, the better educated, who presumably do more sedentary labour, cannot expect to get higher wages if their BMI is 26 or higher, while in the United States, obesity actually lowers one's chances of getting a higher wage.

For adults, the arrow of causation seems to point from stature to wages and not the other way round. Stature does not vary over a person's adult lifetime, so when a correlation between stature and wages is found, it can safely be assumed that a change in wages cannot cause stature to vary, yet variation in stature can cause wages to vary.

The implications of these findings are profound. The loss of income to those suffering from undernutrition can be large. Thus it seems that in Brazil people with BMIs of 26 earn wages that are considerably higher than wages earned by those with a BMI of 22 (both BMIs are well within the range of good nutritional status). Furthermore, people with BMIs of 26 are far more likely to find work than people with BMIs of 22.

Nutrition and health. Inadequate consumption of protein and energy as well as deficiencies in key micronutrients such as iodine, vitamin A and iron are key factors in the morbidity and mortality of children and adults. It is estimated that 55 percent of the nearly 12 million deaths each year among under five-year-old children in the developing world are associated with malnutrition (UNICEF, 1998). Similarly, it has been estimated that 45 percent of all deaths in developing economies in 1985 can be attributed to infectious and parasitic diseases such as diarrhoea and malaria, while these diseases account for about 4.5 percent of all deaths in industrial countries (Strauss and Thomas, 1998, p. 767). Based on research on the European past, Fogel (1994) finds that improvements in stature and BMI explained "over 80 percent of the decline in mortality rates in England, France and Sweden between the last quarter of the eighteenth century and the third quarter of the nineteenth".

Modern evidence from a number of Asian countries is presented in Horton (1999). As many as 2.8 million children and close to 300 000 women die needless deaths every year because of malnutrition in these countries. Also noteworthy is the fact that anaemia is responsible for 20 to 25 percent of maternal deaths in most of these countries. This last observation points to the importance of micronutrient deficiencies in malnutrition. Iron deficiency is also associated with malaria, intestinal parasitic infestations and chronic infections. Chronic iodine deficiency causes goitre in adults and children, besides having an impact on mental health. The importance of subclinical vitamin A deficiency in child mortality has only recently been recognized through meta-analysis of clinical studies (Horton, 1999, p. 249). The relative risk of mortality for a child with subclinical vitamin A deficiency is 1.75 times that for a child who does not suffer from this deficiency. Thus in the Asian region, if about 10 percent of children suffer from subclinical deficiency, a conservative estimate, then 300 000 child deaths could be prevented through a successful vitamin A supplementation programme.

Nutrition and school performance. Considering the importance of nutrition in human development, there is a relative dearth of studies focusing on the role of the different malnutrition aspects on cognitive achievement among children in developing countries. Nevertheless, there is sufficient empirical evidence to indicate that early childhood nutrition plays a key role in cognitive achievement, learning capacity and ultimately household welfare. Available studies have shown that low birth weight, protein energy malnutrition in childhood, childhood iron-deficiency anaemia and iodine deficiency (e.g. being born to a mother with goitre) are all linked to cognitive deficiencies and the effects are more or less irreversible by the time a child is ready to go to school (Horton, 1999, p. 249). Childhood anaemia is associated with a decrease in score of one standard deviation in cognitive tests. Children are most vulnerable to malnutrition *in utero* and before they are three years of age as growth rates are fastest and they are most dependent on others for care.

Nutrition and macroeconomic performance. Horton (1999) provides a rough measure of the overall economic costs of malnutrition as a percentage of GDP for selected Asian countries. The author presents evidence for India, Pakistan and Viet Nam on the losses of adult productivity as a proportion of GDP resulting from stunting, iodine deficiency and iron deficiency. Estimates are also presented for losses, including childhood cognitive impairment associated with iron deficiency, for Bangladesh, India and Pakistan. When a significant proportion of the population is undernourished, potential rates of GDP growth can be curtailed. Adult productivity losses arising

from the combined effect of stunting, iodine deficiency and iron deficiency are equivalent to 2 to 4 percent of GDP every year in these countries. These are very large totals indeed. It should be noted that these estimates were produced under conservative assumptions.

Thus there is evidence that there are considerable losses at the national level from malnutrition and that these losses accumulate over time. A recent FAO study (Arcand, 2001) indicates a strong relationship between economic growth and nutritional factors, as measured either by the prevalence of food inadequacy (PFI) or the dietary energy supply (DES) per capita. The impact of nutrition on economic growth appears to be strong and to operate directly, through the impact of nutrition on labour productivity, and indirectly through improvements in life expectancy.

Based on historical studies of single countries, Fogel shows that improvements in nutrition and health explain half the economic growth in the United Kingdom and France in the eighteenth and nineteenth centuries (Fogel, 1994). Using an accounting approach with concepts from demography, nutrition and health sciences, he stresses the physiological contribution to economic growth over the long term. Reductions in the incidence of infectious diseases, together with changes in diet, clothing and shelter, increased the efficiency with which food energy was converted into work output and translated into higher economic growth.

8.4.2 Income growth and hunger

Not only is undernutrition an unacceptable violation of human rights, it also imposes a heavy economic burden on nations.

Given that the reduction of undernutrition is vital, how can it best be tackled? Can income growth among poor households take care of the problem on its own, or does it need a helping hand in the form of direct public interventions? Smith and Haddad (1999) used data from 63 countries in five regions, covering 88 percent of the developing world's population over the period from 1970-95, to analyse the determinants of child malnutrition, as measured by the percentage of underweight children under five. One of their principal findings is that growth in per capita national income contributed to half the reduction in child malnutrition over this period.

However, Alderman *et al.* (2001) show that the WFS target is unlikely to be met without robust income growth, and not through income growth alone. They assert that "a combination of growth and specific nutrition programs will be needed".

It is reasonable to say that while income growth has a substantial impact on undernutrition, taken alone it will not take care of the problem. The reasons for this are as follows. Nutritional status is the outcome of food intakes as well as health inputs. Therefore the solution to undernutrition is increased intakes of calories and micronutrients or better health and sanitation, safe drinking-water, etc. or both. Private income growth is not guaranteed to improve nutrition for several reasons. First, household income growth does not necessarily lead to increased calorie intakes. Second, some inputs into nutrition are public goods. Better health requires public investments. Third, since private investments in nutrition have a long-term payoff, private capital markets are unlikely to finance this investment if collateral cannot be provided. Fourth, parents are likely to underinvest in nutrition of girls, particularly in those countries in Asia where sons are more highly valued.

Although income growth, certainly at low levels of per capita income, will lead to growth in calorie consumption, the magnitude of this effect is unclear. A vast number of studies have attempted to measure the elasticity of demand for calories,[8] i.e. the percentage increase in calorie consumption associated with a 1 percent increase in income. In a seminal study, Reutlinger and Selowsky (1976) came up with estimates of the income elasticity of demand for calories that ranged from 0.15 to 0.30 for households at the calorie requirement level. Subsequent studies produced elasticity estimates ranging from 1.2 to as low as 0.01.

There are at least two reasons why this range is so large. One is the degree of aggregation. Some degree of aggregation is unavoidable in any household survey. For example, all cereals may be aggregated into one food group. If incomes rise, consumers are likely to substitute within the broad group to foods with a lower or higher calorie content than the average for the group. But the income elasticity of demand for calories is calculated under the assumption that there is no substitution within the food group. Hence it may be biased upwards or downwards depending on the foods to which consumers switch and the degree of aggregation. Another reason is the use of different functional forms and econometric estimation methods.

Some of the early calorie elasticity estimates tended to be on the high side – elasticities of 0.5 were not uncommon. With better recognition of the problems involved in estimating these elasticities, and improvements in survey techniques and econometric estimation techniques, elasticity estimates have generally decreased in size. In recent years, with the important exception of Subramanian and Deaton (1996) who obtained an estimate of about 0.45, most researchers have obtained low to very low elasticities, in the range of 0.01 to 0.15.

Behrman and Deolalikar (1989) provide an explanation for the finding that these elasticities are low. Their hypothesis is that there is a strong demand for more variety in foodstuffs and that this demand manifests itself even at relatively low income levels. This hypothesis was tested on the data set used for the University of Pennsylvania International Comparison Programme (ICP) project, which had data on prices, quantities and purchasing power parity incomes from 34 countries for 1975 and 60 countries for 1980. Nine food groups were covered, i.e. the degree of aggregation was quite high. As food budgets increased from very low levels, there was a very pronounced increase in the demand for food variety (Behrman and Deolalikar, 1989, p. 671). An important implication of this finding is that since the elasticity of substitution is higher among poor households, any increase in food prices will cause the poor to curtail their food consumption more dramatically than the rich. Hence food consumption by the poor will respond strongly to food subsidies that are sharply targeted on them.

8.5 Agricultural and rural non-farm growth

Pro-poor income growth is thus a necessary but often insufficient condition to reduce hunger within a reasonable timespan. Without direct

---

8 See Bouis (1994) and Strauss and Thomas (1995) for details.

public measures to alleviate the most pressing and transient problems, income growth will only gradually solve the problem of hunger. But to finance direct public measures, income growth is needed. The question then becomes, under what circumstances is income growth pro-poor? This section attempts to answer the question.

A consensus has now emerged that the structure of growth matters for its impact on poverty and on human development generally. The 1996 UNDP *Human Development Report* shows clearly that economic growth, as measured by growth in per capita GDP, is associated with better human development. The relationship is quite strong: countries that achieved higher per capita GDP growth rates over the period from 1960 to 1992 also generally achieved higher values of the HDI, restricted to those components that do not rise automatically with income.

The report also shows that for the same growth rate, some countries managed to improve the HDI more than others. For example, why did Egypt not succeed in increasing its HDI strongly despite enjoying fast growth in per capita GDP? Why did other countries with the same per capita GDP growth rate, such as Lesotho, Indonesia and Malaysia, do much better at improving their HDI performance?

The answer lies in the fact that economic growth, reduction in poverty and inequality reduction are all outcomes of the same deeper processes (Srinivasan, 2000, makes this point forcefully). If these are such as to increase the returns to the assets possessed by the poor then economic growth and poverty reduction will be seen to go together. On the other hand if the process favours assets possessed by the wealthy then they will not. Hence the sectoral composition of growth is important; it matters greatly for poverty and hunger alleviation, in which sector overall economic growth originates.

This section argues that economic growth originating in agriculture, when coupled with growth in RNF incomes, is likely to be strongly poverty reducing, provided that it does not occur against a backdrop of extreme inequality in asset ownership, especially of land. Timmer (1997) found that in countries with highly skewed income distribution, growth reaches the poor with difficulty, whether it originates from increases in agricultural or non-

agricultural productivity. According to some estimates, high-inequality countries would need twice as much growth as low-inequality countries to achieve the same reduction in poverty levels (Hammer, Healey and Naschold, 2000).

Why does economic growth originating in the agricultural sector matter for poverty reduction? The majority of the world's poor still live in rural areas and depend crucially on agriculture for their livelihoods. Hence an increase in agricultural productivity should raise incomes in agriculture. This alone will not necessarily help the poor. The next step is to ask where the extra income is spent. There is some evidence that in many countries this income increment is largely spent on goods such as the services of merchants, artisans, mechanics, etc. and on simple agricultural and household goods. The defining characteristic of most of these goods and services is that they are effectively non-tradable. Furthermore, these commodities generally require low inputs of capital and skills to supply and are ideally suited to the capabilities of the rural poor. But because they are effectively non-tradable, their growth is constrained by the growth of demand in the local rural market, which is stagnant. Hence, if this barrier could be removed, these commodities could grow and help the poor escape poverty. That is precisely what the extra income from agricultural growth does: it creates demand for these locally non-tradable goods, provided this extra income is not hoarded or spent outside but is spent locally, which is more likely in a society of smallholders than in one of large landlords. If all goes well, a virtuous cycle could be created, with agricultural and RNF income growing and helping each other to grow. Four important pieces of evidence are needed to validate this hypothesis. First, incremental budget shares of non-tradables out of agricultural income have to be large; second, income from non-tradables should be important for the poor; third, poverty reduction should follow agricultural growth with a lag; and finally, high initial inequality will short-circuit this process.

The argument presented above will now be discussed in more detail. The majority of the world's poor still live in rural areas and depend crucially on agriculture for their livelihoods (IFAD, 2001). It seems probable then that raising the profitability of agriculture will be helpful to the poor. This involves taking steps to increase agricultural productivity per hectare, or encouraging a switch to higher-valued crops.

The initial impact of increased profitability in farming is to raise the incomes of those who own land. This in itself may help reduce poverty if the poor also own some land and participate in the productivity increase, but obviously not if the very poor do not generally own land. But there may also be an increase in demand for labour because agriculture itself, and the construction of the infrastructure needed to support agricultural development, are both very labour intensive. Those who were earlier unemployed may thereby find work while those already employed may find themselves working more hours. Either way their incomes go up. However, for poverty reduction, it is not the initial rise in incomes that matters. What does matter is what incomes are spent on.

It is well known that, as incomes rise, the proportion spent on staples declines, while the proportion spent on superior foods such as superior grains, vegetables, fruit, milk, meat, etc. increases. These commodities are effectively non-tradable and more likely to be purchased locally, because they are bulky or perishable. At the same time the proportion of income spent on the services of merchants, artisans, mechanics, etc. is likely

to go up. This is partly because agricultural growth creates a demand for agricultural implements, but also because rural consumers start to demand goods such as bicycles, which need repairs, or start to eat outside the home so creating a demand for food stalls. Services are by definition also non-tradable. Finally, there is a third category of effectively non-tradable goods comprising simple agricultural inputs such as hoes, rakes, spades and so forth, which may be bought and sold locally, but which do not have much of a market outside rural areas.

The combined effect of these patterns of rural spending can be large. Using household consumption data from 1980s surveys in Burkina Faso, the Niger, Senegal and Zambia (with additional data from Zimbabwe), Delgado, Hopkins and Kelly (1998) show that the share of additional income spent on non-tradables ranges from 32 percent in Senegal to 67 percent in Burkina Faso and Zambia. This spending had multiplier effects that were also calculated. The combined impact on household incomes turns out to be surprisingly large. For example, in Burkina Faso, a US$1 increase in income from farm tradables led to an increase of US$1.88 in income from non-tradables, while in Zambia a US$1 increase led to an increase of US$1.57 in income from non-tradables.

Table 8.4 Non-farm shares in total rural income and employment

Regions and subregions	Non-farm income share		Non-farm employment share		Average[2] per capita GNP 1995 (US$)
	Mean[1] (%)	Coefficient of variation	Mean[1] (%)	Coefficient of variation	
Sub-Saharan Africa	42	0.45	...	...	726
Eastern/southern Africa	45	0.47	...	...	932
West Africa	36	0.36	...	...	313
Asia	32	0.33	44	0.32	1 847
East Asia	35	0.19	44	0.29	2 889
South Asia	29	0.52	43	0.40	388
Latin America	40	0.20	25	0.33	2 499

[1] Mean refers to the mean over the case studies considered for each region and subregion.

[2] Average per capita GNP is calculated as the simple average over the countries covered by the case studies and is based on estimates from the World Bank, World Development Report 1997.

Note: The income shares represent the share of non-farm income in total income of households that are mainly farm households (and the rural landless). The employment shares represent the share of households in the rural population (both in rural areas and small rural towns) with non-farm activity as the primary occupation.

Source: Special chapter on rural non-farm activities in FAO (1998d).

Table 8.5   Income sources in rural India by expenditure quintile, 1994

| | Per capita consumption expenditure quintile | | | | | |
	All	Bottom	2nd	3rd	4th	Top
	(%)					
Total farm income	55	38	38	45	50	65
Total off-farm income	43	60	59	53	46	33
Wages	14	43	36	25	14	4
Agricultural wages	8	28	21	13	8	2
Non-agricultural wages	6	16	15	10	6	2
Self-employment	12	11	17	16	15	8
Regular employment	17	4	7	12	19	21
Other income	3	2	2	3	3	3

Source: Lanjouw and Shariff (2001).
Note: Subtotals do not necessarily add up to totals because of rounding.

The next step is to show that income from non-tradables looms large in the incomes of the poor. Since livestock can easily be raised at little cost on smallholdings, small animals such as sheep and goats are often kept by the poor, and livestock income is generally of more importance to them[9] (see Adams and He, 1995, for an example). Services such as running a food stall or setting up a simple repair shop do not require much in the way of either skills or capital. Neither does the manufacture of simple agricultural implements. Hence it is precisely in the provision of these goods and services that the poor can find gainful employment and thus raise their incomes. The labour required to supply these goods and services does not have to be withdrawn from some other activity, since there is often a great deal of unemployment, disguised or open in the rural areas of developing countries. Tables 8.4 to 8.6 provide evidence as to the importance of RNF incomes to the poor.

From Table 8.4 it is clear that the mean share of non-farm income in household income is nowhere less than about 30 percent and is as high as 45 percent in eastern/southern Africa. Shares in employment are equally high, ranging from 25 percent in Latin America to almost 45 percent in parts of Asia. Table 8.5 provides evidence that in one large poor country, India, the share of income from non-farm sources is highest, at 60 percent, for households in the bottom expenditure quintile.

It declines as income increases, down to only 33 percent for the richest quintile. A broadly similar trend holds in Mexico, as Table 8.6 shows, although households are classified by the amount of land farmed, rather than expenditure as in the case of India. To sum up, non-farm income is important to all rural households, but is particularly important to poor rural households.

Hence, for the poor, the RNF sector offers a relatively easy escape route from poverty. But anyone who thinks of supplying these goods and services runs into a demand bottleneck. Because they are effectively non-tradable in most circumstances, they can only be sold locally. There is not much local demand for them in a stagnant rural economy and, until the economy is created, there is no point in expanding output. But if agricultural productivity and hence the incomes of those who own land can be increased – and if they spend this extra income on goods and services provided by the RNF sector – then the bottleneck to the RNF sector's expansion can be cleared and it can grow and provide important benefits for the poor. Even landless agricultural labourers and others not directly employed in this sector benefit since their power to bargain for higher wages goes up if alternative sources of employment are available.

A critical implication is that the impact on poverty occurs with a lag. Growth in agricultural income will not initially reduce poverty and may

9   Another consideration is that livestock are often kept on marginal or degraded land that would otherwise not contribute to income.

Table 8.6   Sources of income in the Mexican *ejido* sector by farm size, 1997

| | All | Farm size in rainfed equivalent hectares | | | | |
		<2	2-5	5-10	10-18	>=18
Number of households	928	131	244	239	179	135
Shares in total income				(%)		
Total farm income	45.1	22.9	28.1	41.8	50.3	62.0
Total off-farm income	54.9	77.1	71.9	58.2	48.7	38.0
Wages	25.6	40.3	36.9	30.4	18.2	11.1
Agricultural wages	6.7	10.0	8.5	4.2	5.7	2.2
Non-agricultural wages	18.9	30.3	28.4	26.2	12.5	8.9
Self-employment	8.4	17.1	14.2	4.6	12.1	6.8
Remittances	6.5	2.6	5.4	8.9	6.0	6.0
Other income	14.4	17.1	15.3	14.3	13.3	14.1

Source: de Janvry and Sadoulet (2000).

not at first have any impact even on the wages of unskilled agricultural labour. It is only later, after incomes have been generated in the RNF sector, that poverty should begin to decline. Once it does, however, it should decline very quickly.

Good econometric evidence of a positive relationship between agricultural growth and poverty alleviation is available from India, which has had a long period of sustained agricultural growth starting from the early 1970s. The most detailed study is by Datt and Ravallion (1998), who relate differences in poverty reduction to differences in agricultural growth rates for different Indian states. Since macroeconomic, trade, sectoral and social policies apply to the whole of India and so are all held fixed, the "pure" effect of agricultural growth on poverty reduction can be isolated.

The main point of the Datt and Ravallion (1998) paper is the following. From the early 1970s, when growth in agricultural yields in India became strong, poverty as measured by the squared poverty gap index began to decline. The squared poverty gap index does not merely count the number of people whose incomes are below the poverty line. It also measures how far below the poverty line their incomes are, and gives progressively higher weights to incomes the further they are below the poverty line. Not only did the number of people in poverty decline, as measured

by the headcount index, but poverty also became less severe, i.e. the consumption of the poorest of the poor also increased. The claim that agricultural yield growth bypassed the poorest cannot be supported on the basis of this finding.

What were the channels through which agricultural growth helped the poor? An important finding is that rural wages increased, but with a lag: "Higher average farm-yields benefited poor people both directly and via higher real wages ... The benefits to the poor from agricultural growth were not confined to those near the poverty line" (Datt and Ravallion, 1998).

The fact that wages do respond to agricultural growth but with a lag is a critical piece of evidence, showing that time is required for the RNF sector to grow after the initial impetus from growth in agriculture. When the RNF grows, the demand for labour goes up. Agricultural workers find that their bargaining power has gone up and they can start demanding higher wages. Therefore agricultural growth should cause wages to go up, but with a lag.

Thus far the discussion has concentrated on how the process works if everything goes well. Under what conditions would the process not benefit the poor?

Agricultural growth puts money initially into the hands of those who own land. Its impact on poverty depends on whether this income is spent on goods and services that are supplied locally, or on imports. The poor will not benefit if it is not spent locally, on goods and services provided by the RNF sector. But this is what may happen when there are marked inequalities in landownership and the initial increase in agricultural income is concentrated in a few hands. Wealthy landowners may have metropolitan tastes and the wealth to indulge them, and will be unlikely to patronize local suppliers. There may be gains to the poor arising out of extra agricultural employment and possibly lower food prices, but the add-on effect on local employment and industry arising out of expenditure by farmers on locally made products will be lacking.

Bautista's (1995) case study in the Philippines illuminates these issues. He points out that over the period 1965-80, crop production in the Philippines grew at a rate of 5.2 percent p.a. and livestock at a rate of 6.4 percent, among the highest growth rates in Asia. The growth of crop production was evenly shared between rice and non-traditional export crops. These high growth rates were at least partly a result of a sevenfold increase in real government expenditure on agriculture, the bulk of which was devoted to irrigation that took half of all agricultural investment by 1980. This was at the cost of investments in rural roads whose share dropped to barely 2 percent of agricultural expenditure. At the same time, human development was exceptionally good in the Philippines, with rates of literacy, infant mortality and life expectancy all either better or comparable with its neighbours in Southeast Asia. Despite all this, there was no significant reduction in poverty.

The primary reason was that the income gains from agricultural growth were highly concentrated. First, where rice farmers were concerned, only those who had access to irrigation could benefit. Despite all the investment in irrigation, only 18 percent of arable land was irrigated by 1980. Second, subsidies on credit and fertilizers were pocketed by large farmers who also enjoyed better access to infrastructure. Large farmers also enjoyed implicit subsidies – through low tariffs, an overvalued exchange rate and a low interest rate – on imported farm machinery that displaced landless agricultural labourers. The consequences were clear: "Income gains were concentrated in the already more affluent segment of the rural population. As a result, rural consumption favored capital-intensive products and imported goods rather than labor-intensive, locally produced goods ... Accordingly the rate of labor absorption in both agriculture and industry was very low, and given the rapid expansion of the labor force, it prevented real wage rates from moving upward. As a result ... the incidence of poverty increased over the period" (Bautista, 1995, p. 144).

A similar situation can arise in countries where governments place legal and administrative hurdles in the path of smallholders who wish to grow commercial crops. The cultivation of these crops is then in the hands of wealthy farmers who are likely to spend any increments in their income on imported goods while spending little on locally produced goods. Allowing small farmers to share in the profits from commercial crop cultivation would have increased the likelihood of these profits being spent locally, thus creating income-earning opportunities for others. But if this is not the government's policy, the result is that the agriculture sector registers growth but this growth has little or no impact on poverty.

Thus, agricultural growth provides opportunities for the poor to increase their incomes. Whether the poor can seize these opportunities depends on their education and health, on their access to credit and savings services, and on whether they are excluded by social custom or government fiat from income-earning activities (such as women shut out from credit markets). Measures to increase the capital available to the poor (human, financial, physical, natural and social) are therefore likely to pay big dividends in terms of their ability to lift themselves out of poverty.

To conclude, the key point is that growth in agricultural incomes, by creating demand for the output of the RNF non-tradable sector, makes it possible for that sector to grow. Since the capital and skill requirements of the sector are well suited to the capabilities of the poor, its rapid growth can help eliminate poverty. Thus agricultural growth ultimately reduces poverty and does so with a lag. But this benign process cannot work if there are marked initial inequalities in the agricultural sector since these act to prevent agricultural incomes from being spent locally and therefore do not create the multipliers needed.

Agricultural trade, trade policies and the global food system

9.1 Introduction

For centuries countries have relied on trade in agricultural and food commodities to supplement and complement their domestic production. The uneven distribution of land resources and the influence of climatic zones on the ability to raise plants and animals have led to trade between and within continents. Historical patterns of settlement and colonization contributed to the definition of trade patterns and to the emergence of an infrastructure to support such trade. More recently, transnational firms with global production and distribution systems have taken over from post-colonial trade structures as a paradigm for the organization of world agricultural trade. Changes in consumer taste have encouraged the emergence of global markets and added to the significance of trade. Few countries could survive the elimination of agricultural trade without a considerable drop in national income, and none could do so without considerable reduction in consumer choice and well-being.

The projections presented in Chapter 3 suggest that the role of trade in meeting global food demand will further increase over the next 30 years. Developed countries will provide a growing share of developing countries' food needs, and in return will continue to import larger quantities of other agri-

cultural products, notably tropical beverages, rubber and fibres. However, the developing countries are not a homogenous trading bloc. While the group as a whole will increase its net exports of tropical products and import more and more temperate-zone commodities, within the group there will remain important net exporters of temperate-zone commodities.

Like all projections, the global trade outlook presented in this chapter is based on numerous assumptions about the likely evolution of policies that will affect trade flows, as well as basic trends in income, population and productivity. A principal premise of the quantitative projections is the continuation of existing policies related to the support and protection of agriculture, including policy changes that will be implemented in the future, for example, the EU's Everything but Arms Initiative (EBA) that foresees a liberal import regime for rice and sugar from the least developed countries (LDCs) in the future. If policies differ substantially in the future, so will the outcomes. If, for instance, the reform process that began under the Uruguay Round Agreement on Agriculture were to achieve a fundamental reform of the sector, and if there were significant reductions in production-enhancing subsidies and protection in industrial countries, this could have an impact on predicted trading patterns. And if policy reforms extended beyond the developed countries and led to the removal of the remaining bias against agriculture in the policy of several developing countries, this could mobilize resources to enhance productivity and stimulate development of the rural economy.[1] As a consequence, much of the qualitative discussion in the chapter is an attempt to indicate how policies might develop over the next three decades.

This chapter begins with a discussion (Section 9.2) of the evolution of patterns in global agricultural trade. The analysis reviews the changing share of agricultural trade relative to manufactures in world trade and identifies the rapidly changing role of agricultural trade in the developing countries. This is followed by a discussion of the agricultural trade balances for developing countries by major commodity categories. It shows changes in the net trade balances of these product categories. It also provides an overview of complementarity and competition in global agricultural trade and how, where, and to what extent policies have affected the current trading patterns between developing and developed countries.

Section 9.3 examines the trade policy environment, focusing on the Agreement on Agriculture. It assesses the progress made to create a "fair and market-oriented" trading system for agriculture, and looks at other trade policy developments in agriculture, particularly the role of preferential and regional trade agreements. Section 9.4 addresses the prospects for and likely impacts of further reforms in the current and future rounds of negotiations, with particular attention to the ways in which the new emphasis on non-trade concerns will influence trade liberalization. This is followed by an analysis of how different policy options could affect and alter the baseline projections to 2030. The focus is on the stake of developing countries in international agricultural trade policy reform. The chapter concludes (Section 9.5) by looking at the emerging issues that will affect the overall trade policy environment and how they may influence agricultural trade in the coming three decades.

9.2 Long-term trends in the pattern of global agricultural trade

9.2.1 From agricultural exports to manufactures exports

The last 50 years have witnessed an impressive growth in international trade. The volume of global merchandise trade has increased 17-fold, more than three times faster than the growth in world economic output.

A number of factors contributed to this growth. Average import tariffs on manufactures, for instance, fell from 40 to 4 percent over the four decades of trade negotiations under the General Agreement on Tariffs and Trade (GATT) (Abreu, 1996). Non-policy factors have also been important, including a reduction in transport costs and new transportation facilities as well as cheaper and more efficient communications. Moreover, growth in manufactures trade has been spurred by the rapid expansion of intra-industry and intrafirm trade, exploiting a division of labour within companies operating across various countries or continents. Much of this trade is an exchange of components or semi-processed products. Finally, manufactures benefited from a virtuous circle whereby the gains from trade translated into higher incomes and in turn fuelled growth in trade.

Agricultural trade has also grown during the last 50 years, but only at about the rate of global economic output. Notable among the factors that contributed to this relatively slow growth in trade was the failure to include agriculture fully in the multilateral trade negotiations under GATT that were so successful in reducing industrial tariffs. As a result, agricultural tariffs are as high now, on average, as industrial tariffs were in 1950. The effects of high border protection have been compounded by domestic support policies in many developed countries and in some developing countries by policies that promoted import substitution at the expense of international trade.

---

[1] As discussed in Chapter 10, many countries ran policies that were implicitly or explicitly biased against their agricultural sector. While some of this bias appears to have been removed in the course of domestic policy reforms and structural adjustment programmes, the patterns of government investment still favour urban areas in many countries.

Growth in agricultural exports from developing countries was also held back by the limited absorption capacity of their export markets. The major part of their agricultural exports was destined for largely saturated developed country markets without much responsiveness in demand. Tropical products such as coffee, cocoa, tea or rubber were most severely affected by these limitations. Rising output from developing countries met inelastic demand in developed countries and resulted in a persistent downward pressure on prices. In fact, lower prices cancelled out much of the gains in export volumes with the result that export earnings increased only moderately.

Moreover there has been very little intra-industry or intrafirm trade in food and agricultural products. This reflects mainly the nature of agricultural trade, which is often largely determined by different agro-ecological conditions. But intra-industry trade has also been held back by traditional trade and investment barriers that made international sourcing more difficult than for manufactures. When and where these barriers have declined, the exchange of processed and semi-processed agricultural products has increased considerably and brought about levels of intra-industry/intrafirm trade close to levels observed for non-agricultural products (see Chapter 10). Much of this trade has been stimulated by the activities of global food companies and traders, but has also involved retailers and small food exporters exploiting niche markets.

The result of the rapid export growth in manufactures and slow growth in agricultural exports was a dramatic decline in the relative importance of agricultural exports. The share of developing countries' agricultural exports in their overall exports fell from nearly 50 percent at the beginning of the 1960s to barely more than 5 percent by 2000[2] (Figure 9.1). Even for the group of the 49 LDCs, where agriculture is often the largest sector of the economy, the share of agricultural exports declined from more than 65 percent in the early 1960s to less than 15 percent by 2000 (Figure 9.1). Whereas the low shares for developing countries are, among many other factors, also a reflection of protectionist policies in OECD countries, OECD policies have probably contributed to

Figure 9.1 The agricultural trade balance and share of agricultural exports

Agricultural net exports from developing countries have been declining...

...and even more so the importance of agriculture in total merchandise exports

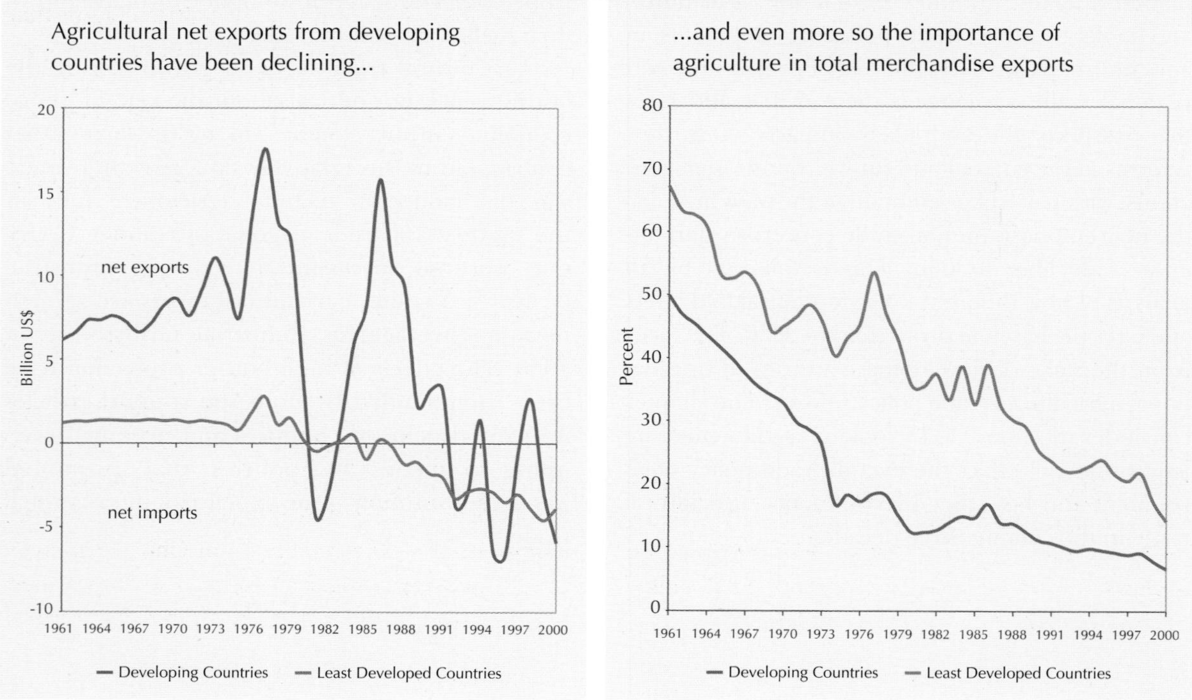

2 Both for agriculture and total merchandise trade, gross trade data include intradeveloping country trade.

Figure 9.2   Least developed countries (LDCs) have become major net importers of agricultural products

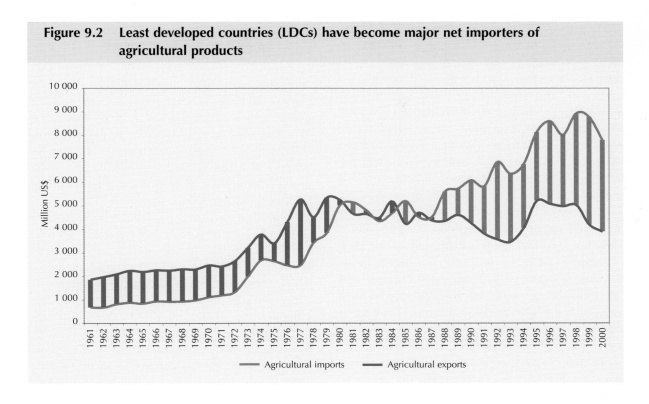

keeping shares high for the LDCs. The numerous preference agreements in which they participate offer the LDCs higher export prices and encourage them to export more than otherwise.

9.2.2 From net exporters to net importers of agricultural commodities

Together with the overall decline in the share of agriculture in international trade, the structure of agricultural trade has changed markedly. One manifestation of this change is the balance in food trade between developed and developing countries. In 1961/63 developing countries as a whole had an overall agricultural trade surplus of US$6.7 billion, but this gradually disappeared so that by the end of the 1990s trade was broadly in balance, with periodic minor surpluses and deficits. The outlook to 2030 suggests that the agricultural trade deficit of developing countries will widen markedly, reaching an overall net import level of US$31 billion. Net imports of food will increase to about US$50 billion (in US$ of 1997/99, for details see Table 9.1).

The 49 LDCs have been in the forefront of this shift: their agricultural trade deficit has increased so rapidly that, already by the end of the 1990s, imports were more than twice as high as exports (Figure 9.2). The outlook to 2030 suggests that this trend will show no sign of abating. The agricultural trade deficit of the group of LDCs is expected to widen further and will increase overall by a factor of four over the next 30 years.[3]

As already discussed in Chapter 3, the evolution of the overall net agricultural trade balance *per se* may not necessarily represent a deterioration of the overall economic situation in a developing country. For some countries, a growing agricultural trade deficit may simply reflect rapid overall development. This is, for instance, the case of countries such as the Republic of Korea in which the growing agricultural deficit has gone hand in hand with high rates of overall development and growing food consumption. Likewise, rising imports of vegetable oil in China primarily reflect an improved ability to meet domestic food needs through imports. A declining agricultural

---

[3] The increase in the trade deficit is expressed in constant international dollars. For the commodities covered in this study, it is projected to widen from about US$4 billion in 1997/99 to US$8.5 billion in 2015 and US$16.6 billion in 2030, respectively. The country composition of the group of LDCs has been assumed to remain unchanged.

trade balance is, however, a negative developmental outcome in countries that continue to depend to a high degree on export earnings from agriculture or that have to divert scarce foreign exchange resources (eventually building up unsustainable foreign debt) to pay for growing food imports. It is an even more negative indicator where such food imports are not associated with rising food consumption per capita but are necessary just to sustain minimum levels of food consumption.

Commodity dependence. Notwithstanding the declining importance of agricultural exports for developing countries as a whole, some developing countries still rely heavily on agricultural exports for their foreign exchange earnings. In more than 40 developing countries, the proceeds from exports of a single agricultural commodity such as coffee, cocoa or sugar account for more than 20 percent of total merchandise export revenue and more than 50 percent of total agricultural export revenue (Figure 9.3). In Burundi, for example, coffee exports alone accounted for 75 percent of the country's foreign exchange earnings in 1997/99. Half of these countries are located in sub-Saharan Africa, and three-quarters are LDCs and/or small islands. The heavy reliance on a few crops is often a reflection of the fact that many of these economies are very small.

Dependence on a few commodities brought about numerous problems for these countries during the period of low prices in 1999/2001. Particularly low world prices for coffee and sugar reduced their overall foreign exchange availability, lowered rural wages, increased rural poverty and thus underlined the importance of undistorted agricultural trade for overall economic development. The sluggish demand for primary agricultural commodities and the recurring conditions of boom and bust in their exports created problems for commodity-dependent economies. Unstable commodity prices and export earnings are known to make development planning more difficult and to generate adverse short-term effects on income, investment and employment.

9.2.3 Projected shifts in the agricultural trade balance of developing countries

The outlook is that developing countries will become significant net importers, with a trade deficit of almost US$35 billion by 2030 (Table 9.1).[4] The following section focuses on the structural changes within agricultural trade by commodity group, to identify the main factors that have brought about these shifts in trade patterns and the likelihood of their continuation.

The first category includes temperate-zone commodities (wheat, coarse grains and livestock products) of which the developed countries produce the bulk of world-exportable surpluses. These commodities are also highly protected in many countries and the world market price is often influenced by heavily subsidized export surpluses. Developing countries are already significant importers of these commodities and the projections suggest that by 2030 their net imports will have increased further, e.g. more than 2.5-fold for wheat and coarse grains (Table 3.5) and almost fivefold for meat (Table 3.14).

One of the most important changes that affected the overall agricultural trade balance of developing countries was the rapid growth in imports of temperate-zone commodities. The net imports in this product category increased by a factor of 13 over the last 40 years, rising from a minor deficit of US$1.7 billion in 1961/63 to US$24 billion in 1997/99 (Table 9.1). These figures are in current US$.

A number of factors contributed to this shift. First, developing countries found it increasingly difficult to compete with subsidized surpluses of temperate-zone commodities disposed of by OECD countries. These subsidies hindered exports from all developing countries, although some countries such as Argentina, Brazil and Uruguay managed to remain net exporters. Subsidies and subsequent surplus disposal put downward pressure on international prices, and held back export volumes and earnings for temperate-zone commodities. Second, the developing countries themselves added to the growing trade deficit by taxing their own agricul-

---

[4]   All projected trade balances in Table 9.1 are expressed in constant (1997/99) US$, but the historical data are in current US$. Therefore, the rates of increase of the historical period cannot be compared with those for the projections.

Figure 9.3 Dependence on agricultural export earnings by commodity, 1997/99

Share of export earnings in total mechandise exports (percentage)

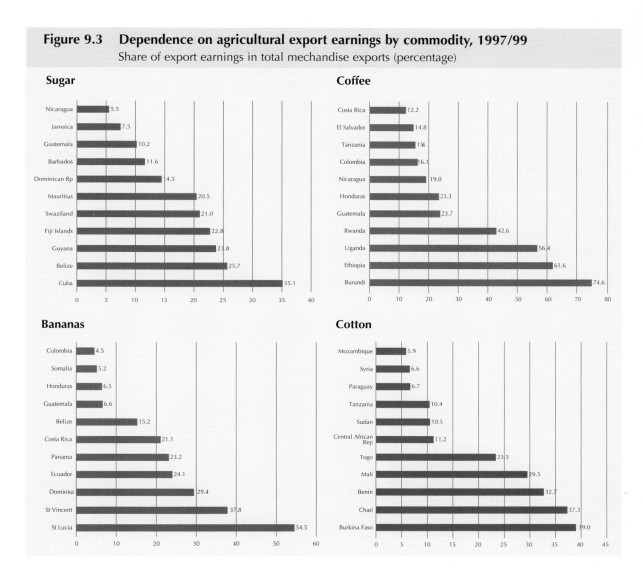

ture directly through low procurement prices and quotas or indirectly through overvalued exchange rates and high protection rates for non-agricultural goods. In many cases, these policies had been in place before OECD surplus disposal policies became a significant factor in world agricultural trade. Third, overall economic development contributed to higher imports of temperate-zone commodities. With higher incomes and more people to feed, demand for temperate-zone commodities in developing countries increased fast, too fast for domestic production to keep pace. The rise in import requirements was particularly

pronounced in countries where agro-ecological constraints restrained agricultural production growth and where urbanization and income growth brought about rapid increases in demand (notably in the Near East/North Africa region).[5]

Within the group of temperate-zone commodities, there is a discernible shift in trade from cereals to livestock products, notably to meats. This is the result of various mutually reinforcing developments that took place in parallel in developed and developing countries. In developed countries, technological and organizational progress in livestock production (vertical integration, etc.)

[5] The rapid overall increase in temperate-zone commodities over the last 40 years masks the fact that import growth varied considerably during this period. Imports rose particularly fast during the 1970s and early 1980s when the oil boom afforded many developing countries, particularly in North Africa and the Near East, with the foreign exchange earnings needed to increase imports of products such as cereals, meats and dairy products, for which domestic production capacity was limited. Together with the decline in oil prices, however, growth in food imports slowed down to considerably lower rates in the late 1980s and 1990s.

Table 9.1 Trade flows between developing and developed countries

Commodity category	Net trade of developing countries (negative values denote net imports)					Cumulative increase	OECD support
	1961/63	1979/81	1997/99	2015[1]	2030[1]	1997/99 -2030	PSE 1998/00
	Billion US$ (current)			Billion US$ (in US$ of 1997/99)		Percentage	Billion US$
Total agriculture	6.68	3.87	-0.23	-17.6	-34.6		258.57
Total food	1.14	-11.52	-11.25	-30.7	-50.1	+345	n.a.
1. Temperate-zone	-1.72	-18.17	-24.23	-43.8	-61.5	+154	134.22
Cereals (excluding rice)	-1.57	-14.25	-17.40	-31.9	-44.6	+156	40.09
Wheat	-1.53	-10.45	-10.30	-17.3	-23.5	+128	18.13
Coarse grains	-0.04	-3.80	-7.10	-14.7	-21.1	+195	21.97
Meat	0.22	-0.56	-1.18	-3.4	-5.8	+389	49.16
Ruminant	0.27	0.14	-0.93	-2.5	-4.6	+395	32.30
Non-ruminant	-0.06	-0.71	-0.25	-0.8	-1.2	+372	16.87
Milk	-0.37	-3.36	-5.65	-8.4	-11.1	+97	44.97
2. Competing	3.13	4.29	6.20	6.3	5.9	-4	111.28
Rice	-0.07	-1.44	-0.39	-0.5	-0.7	+82	26.38
Vegetable oils and oilseeds	0.81	0.52	-0.57	-0.6	-0.6	+17	5.47
Fruit, vegetables and citrus	0.24	1.67	8.40	9.7	11.2	+33	57.44 [3]
Sugar	1.02	3.83	1.30	1.3	0.9	-30	6.73
Tobacco	0.20	0.07	1.26	0.9	0.6	-55	1.92 [3]
Cotton lint	0.91	-0.13	-3.46	-4.2	-5.0	+46	6.81 [3]
Pulses	0.02	-0.23	-0.34	-0.3	-0.4	+14	6.53 [3]
3. Tropical	3.83	17.55	19.16	22.8	26.0	+36	0.92 [3]
Bananas	0.28	1.00	2.64	3.5	4.0	+53	0.32 [3]
Coffee	1.78	9.49	9.77	11.1	12.4	+27	0.28 [3]
Cocoa	0.48	3.30	2.82	3.6	4.2	+49	0.03 [3]
Tea	0.48	0.85	1.39	1.5	1.7	+20	0.29 [3]
Rubber	0.89	2.91	2.54	3.1	3.7	+45	0.01 [3]
4. All other commodities	1.46	0.20	-1.36	-3.0 [2]	-5.0 [2]	+267	11.15 [3]
Other study commodities	0.36	0.83	0.21	0.2	0.2	+10	n.a.
Commodities not covered in this study	1.10	-0.63	-1.57	n.a.	n.a.	n.a.	n.a.

Notes:
[1] Based on projected growth in quantities (as in Chapter 3), applied to the 1997/99 trade values from FAOSTAT, rounded numbers; the projected trade balances in values are implicitly expressed in constant US$ of 1997/99, while the historical values are in current US$. It follows that the implied rates of change over time are not comparable between past and future. For such comparisons refer to Chapter 3.

[2] "Guesstimates".

[3] Pro-rated according to shares in values of production.

n. a.=not available.

outpaced productivity gains in cereal production. This made it more profitable to convert cereals into meat domestically and export meats rather than cereals (OECD, 1998). In developing countries, income growth, particularly in Asia, has contributed to a shift in consumption patterns from grain-based diets to livestock-based diets with rapidly rising per capita consumption levels (Chapter 2). Moreover, alongside overall economic development and urbanization, some developing countries that recorded higher income growth also created the infrastructure facilities (e.g. cold chains) that were needed to handle livestock product imports.

The long-term outlook presented in Chapter 3 suggests that the shift from cereals to the meat trade will continue. While cereals (excluding rice) will still account for most of the increase in absolute terms, meats are expected to exhibit the strongest relative growth. Developing countries' imports of livestock products will increase very rapidly, although starting from relatively low trade levels in the base year 1997/99. Also developing countries' overall imports of temperate-zone products will continue to rise, albeit at a lower pace. By the year 2030, developing countries are expected to increase their net imports of temperate-zone commodities to about US$61 billion (in 1997/99 US$, Table 9.1; for changes in quantities see Chapter 3, Tables 3.5, 3.14).

One of the crucial issues in this context is whether and to what extent developing countries would be able to expand their production and exports if policy distortions, particularly those imposed by OECD countries, were to be removed. Would the removal of distortions change developing countries' net trade position? Numerous studies suggest that this is unlikely to be the case. Instead, there is wide agreement that the limiting effect of agro-ecological constraints often outweighs the effects of policy distortions. In fact, a removal of OECD subsidies would largely result in a shift in market shares from subsidized to unsubsidized producers within the group of OECD countries. Only a few developing countries with additional production capacity for temperate-zone commodities, such as Argentina, Brazil, Uruguay or Thailand would expand their net export positions (see, for instance, FAO, 2000e, p. 27-28). The majority of developing countries, however, would not shift from net importers to net exporters.

Where the responsiveness of agricultural supply is particularly low (as in many LDCs) and where non-agricultural protection remains high, countries will experience higher food import bills and a deterioration of their terms of trade.

The second category includes primarily *competing commodities*, i.e. those commodities that are produced in both north and south, even though they may originate from different primary products (sugar from beets or cane, oil from several oilcrops), or commodities for which competition is limited to certain parts of the year (fruit and vegetables). Overall, there is considerable competition for export shares in these markets. Subsidies in OECD countries often offset the comparative advantage of producers in developing countries.

Developing countries' export interests are adversely affected by OECD policy distortions affecting competing products. Many developing countries have a comparative advantage in producing these commodities, either because their production is labour intensive (fruit and vegetables) and/or because they are strongly favoured by the agro-ecological conditions of tropical or subtropical regions (tropical fruit, sugar and rice). Developing countries' net exports in this product category amounted to about US$6 billion in 1997/99, about twice as much as in the early 1960s (in current US$, Table 9.1). At a net export level of US$8.4 billion, fruit, vegetables and citrus accounted for the largest portion under this heading, and exhibited the highest growth over the past 40 years.

Yet export growth might have been even more rapid had it not been for policy distortions, particularly OECD subsidies that totalled about US$111 billion in 1998/2000 (Table 9.1). Fruit and vegetables, together with rice, accounted for nearly three-quarters of this OECD subsidy. Developing countries, at least in aggregate, are likely to benefit from a cut in OECD subsidies and an increase in access to developed countries' markets.

The third category encompasses *tropical commodities* that are mainly produced in developing countries, but primarily consumed in OECD countries. These are mostly tropical products such as coffee, cocoa or rubber for which developing countries have been increasing output substantially over the past decades. Developed countries' import markets for these commodities have become increasingly saturated. Demand has become

Figure 9.4   Coffee exports, world and Viet Nam

Coffee exports: Revenues grow slower than volumes for the world as a whole...

but also for new, dynamic exporters like Viet Nam

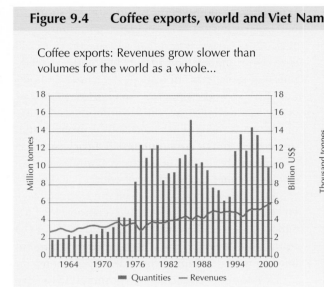

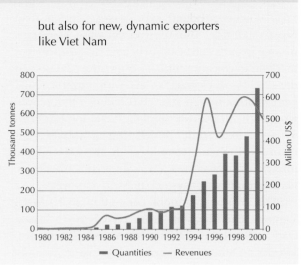

inelastic, and prices are subject to a secular decline. Since developed countries do not produce these commodities in significant quantities, they do not support or protect these markets.

Developing countries have been rather successful in expanding production and exports of tropical products. Overall, net exports of tropical products have increased by a factor of five, from about US$3.8 billion in 1961/63 to about US$19.2 billion by the late 1990s at current prices (Table 9.1). Export growth will continue over the next 30 years, and in 2030 could be higher by some 36 percent (in volume terms).

The single most important commodity in this group is coffee. Net exports of coffee account for about half of total net exports of tropical products. They increased by a factor of 5.5 since the early 1960s (in current prices), largely matching the growth of overall net exports of tropical products. Production of coffee (green beans) increased by 60 percent from 1961/63 to 1997/99 and, as domestic consumption in developing countries remained low, much of the additional produce went into export markets. Coffee exports rose rapidly over much of the 1960s and 1970s. But as developed countries' markets for coffee became increasingly saturated, growth in export volumes was no longer matched by a similar growth in export revenues. From 1974/76 to 1998/2000, volumes of coffee exports doubled while export revenues increased by merely 15 percent (Figure 9.4).

In addition to these demand-related reasons for a slowdown in export growth, growth in production and trade was also affected by policies. In developing countries, the export taxes of large exporting countries resulted in periods of attractive prices and lured more competitive countries into these markets. In developed countries, tariff escalation on more processed products kept a lid on exports of processed coffee and thus contributed to the slow growth in export earnings. Over and above tariff escalation, OECD trade and support policies play no significant role in these markets. Overall OECD countries spent less than US$1 billion, i.e. less than 0.5 percent of total producer support to subsidize the production of these commodities in 1998/2000 (Table 9.1).

The country that made the most substantive inroads into the international coffee market was Viet Nam. Over the 1990s, the country's coffee exports rose by a factor of ten in terms of quantities and a factor of five in terms of revenues (Figure 9.4). Together with rapidly rising rice production and exports, Viet Nam's coffee boom provided a boost to the country's overall rural economy. Fertilizer use increased by about 50 percent and agricultural GDP grew by about 4.5 percent p.a. The coffee boom also boosted food production and thus contributed to a reduction in hunger and poverty. Rural poverty fell from 66 percent in 1993 to 45 percent in 1998 (World Bank, 2001c) and from 1990/92 to 1997/99, the share of undernourished persons declined by 8 percentage points or 4 million people (FAO, 2001a).

The coffee production and export boom in Viet Nam, however, also augmented the downward pressure on international prices. By the end of 2001, international prices for green coffee had declined to a 30-year low. Low coffee prices had serious impacts for many rural poor in those areas where production was growing less rapidly. Wages for pickers, for instance, fell along with prices for coffee and reached levels below US$1-2 per day in many African and Latin American countries. Oxfam has documented such impacts on wages and rural poverty in a number of case studies for the United Republic of Tanzania and Mexico (Oxfam, 2001). Moreover, the low international prices also affected average coffee qualities. It rendered some of the previous efforts to create high-quality coffee segments unprofitable (for instance in Colombia) and, as much of the production in the newly emerging coffee producers such as Viet Nam was of inferior quality, it lowered both average prices and average qualities.

The effects of the coffee boom confirm a number of general problems that plague producers and markets for tropical products. First, where rapidly rising supply is primarily exported to largely saturated markets, prices tend to decline rapidly and sharply without necessarily resulting in the expected contraction in production. The asymmetry in the supply response is because many of these crops are perennials, grown on a multiyear basis with a relatively large share of total investment requirements locked in at the beginning of the investment periods. To increase the returns on the fixed investment, investors have to reduce the variable costs of production. In plantation production systems, this means lower wages for rural workers (pickers); in smallholder systems, it means a considerable decline in family incomes. Second, while the promotion of production in new low-cost environments (Asia) can be an interesting agricultural development strategy in a given country, the same policy pursued simultaneously in many countries can nullify or reverse the advantage for the original beneficiary. Markets for tropical agricultural products are particularly prone to this *fallacy of composition* problem (see also Section 3.6 in Chapter 3 for a discussion of problems with tropical commodities).

9.3 The trade policy environment for agriculture

The environment in which agricultural trade policy is formed includes both the domestic political balance between competing ideas and interests and the international climate for trade rules and negotiations. This section examines current issues in the trade policy environment for agriculture and the extent to which a continuation of present trends is likely.

9.3.1 Global trends in trade policy reform

At the completion of the Uruguay Round of multilateral trade negotiations, the Agreement on Agriculture was hailed as an important first step towards the fundamental reform of the international trading system for agriculture. Since then, however, many countries have been disappointed by the modest benefits deriving from it. Indeed, some observers argue that the Agreement on Agriculture may have "institutionalized" the production- and trade-distorting policies of the OECD countries without addressing the fundamental concerns of developing countries (Green and Priyadarshi, 2002). Negotiations on the continuation of the reform process began in March 2000, as mandated under Article 20 of the Agreement on Agriculture, but the progress that can be achieved will depend, in part, on the experience of World Trade Organization (WTO) Members with the reform process thus far.

The Doha Round of multilateral trade negotiations. Article 20 of the Agreement on Agriculture declares Members' "long-term objective of substantial progressive reductions in support and protection resulting in fundamental reform" and pledges to continue the reform process, "taking into account:
■ the experience ... from implementing the reduction commitments;
■ the effects of the reduction commitments on world trade in agriculture;
■ non-trade concerns, Special and Differential Treatment (SDT) to developing country Members, the objective to establish a fair and market-oriented agricultural trading system; and other objectives and concerns mentioned in the Agreement; and

■ what further commitments are necessary to achieve the above-mentioned long-term objectives."

The negotiations on agriculture were subsumed in a broader round of multilateral trade negotiations launched by WTO at its Fourth Ministerial Conference held in Doha, Qatar in November 2001 (WTO, 2001e).[6] The round will consider agriculture as part of a "single undertaking", and the negotiations are scheduled to conclude not later than 1 January 2005. Modalities for further commitments on agriculture are to be established no later than 31 March 2003 and comprehensive draft schedules based on these modalities should be submitted no later than the date of the Fifth Session of the WTO Ministerial Conference to be held in September 2003 in Cancún, Mexico.

During the first phase of the negotiations on agriculture, which ended in March 2001, WTO Members submitted their proposals and other papers reflecting their positions and concerns. A total of 44 negotiating proposals were submitted, sponsored by 125 countries either individually or in groups. The proposals address the range of issues mandated for the negotiations, including the three pillars of the Agreement on Agriculture (market access, domestic support and export competition), as well as the cross-cutting issues of SDT for various groups of countries and non-trade concerns ranging from food security to animal welfare. They span a broad range of possible outcomes, from the rapid elimination of all trade-distorting barriers and subsidies to a slowing or even reversal of the reforms undertaken in the Uruguay Round.

The Ministers made a commitment to undertake negotiations on agriculture that, without prejudging the outcome, would aim at substantial improvements in market access, the reduction of all forms of export subsidies with a view to phasing them out, and substantial reductions in trade-distorting domestic support. It was agreed that SDT should be applied to developing countries to reflect their development needs, including food security and rural development. Non-trade concerns will also be taken into account in the negotiations. Given the wide divergence of the initial positions, radical reform seems unlikely in the near term. Nevertheless, successive rounds of gradual liberalization implemented over the next 30 years could result in "fundamental reform of the sector" within the timespan of this study.

Market access. Bound tariffs remain high in the OECD countries, especially for temperate-zone basic food commodities and products that compete with them. Developing countries typically also have high bound tariffs on agricultural goods, at 68 percent on average, but their applied tariffs are often much lower. Some countries bound tariffs for certain sensitive staple products at very low levels, often because they had previously relied on other means to control imports, such as quota exemptions under GATT Article XII for balance of payments purposes (Green and Priyadarshi, 2002). Many agricultural goods do, however, enter industrial countries duty free or at low tariffs. OECD tariffs on unprocessed tropical products are typically quite low, but tariff escalation is a problem in some of these commodities. In the EU and Japan, for example, bound tariffs on coffee escalate from zero percent on green beans to 7.5 and 12 percent, respectively, on roasted beans; while in India, this escalation goes from 100 to 150 percent (WTO, 1994).

Tariff schedules also reflect the prevalence of tariff rate quotas (TRQs),[7] with both in-quota and over-quota tariffs listed, and the proliferation of complex duties that combine elements of specific and ad valorem duties (UNCTAD, 1999b). Thirty-eight WTO Members have tariff quota commitments in their schedules, for a total of 1 379 individual quotas. But TRQs have done little to improve market access. Many countries allocated the quotas to traditional suppliers and counted pre-existing preferential access quotas as part of their minimum access commitments.[8] New access

---

[6] The subject areas for negotiations will include the following: agriculture, services, market access for non-agricultural products, aspects of TRIPS, antidumping, subsidies and countervailing measures, dispute settlement, trade and environment, trade and investment, trade and competition policy, government procurement and trade facilitation.

[7] A TRQ is a specified quantity that can be imported at tariff rates (in-quota rates) well below the "normal" (over-quota) ones. The TRQ quantities and in-quota rates should be set at levels allowing for imports no smaller than the market access opportunities that existed in the Agreement on Agriculture base year 1986-88 or in any case to allow for a "minimum access", whichever is higher.

[8] Examples are sugar imports in the EU and sugar and beef imports in the United States.

volumes created by TRQs were typically less than 2 percent of world trade, and TRQ utilization rates or fill rates – the amount of trade that actually takes place relative to the TRQ level – have averaged about 60-65 percent (see OECD, 2001g). Unlike simple tariffs, TRQs generate market rents that may be captured by various groups (producer, exporting government, importing government, trader) depending on the administrative mechanism and the degree of market competition (ABARE, 1999). Thus, even if new market access is created, the producer may not capture the benefits, and there may be vested interests arguing against further expansion of these quotas.

Thirty-eight Members have the right to invoke special agricultural safeguards (SSGs) on a total of 6 072 individual tariff items. During the implementation of the Agreement on Agriculture, a total of 649 tariff items have been subject to special safeguard notification, with more than half of the price-based actions being taken by the United States, and more than half of the volume-based actions taken by the EU (WTO, 2000a).[9] Most developing countries do not have access to SSGs because they did not use the tariffication procedure. Their only practical means of stabilizing domestic markets in the face of import surges or fluctuating world prices is by varying their applied tariffs (within their bound rates), although such "price bands" could yet be challenged under the rules of WTO.

Domestic support. The Agreement on Agriculture included rules and disciplines on domestic support in recognition of the potential these policies have to distort trade. The specific provisions aim to ease trade conflicts between developed countries, to reform policies that resulted in overproduction in the past, and to ensure that commitments on market access and export competition are not undermined through domestic support measures.

Although the Agreement on Agriculture began the process of disciplining domestic support measures, the rules thus far have done little to restrain the subsidies provided by OECD countries. Furthermore, the types of policies available to developing countries under the Agreement may not be appropriate to the conditions of their agricultural sectors or sufficient to enable them to overcome the handicaps they face in international markets (Green and Priyadarshi, 2002).

The major portion of domestic support expenditures is provided by three WTO Members: the EU, Japan and the United States (Table 9.2). Although several OECD countries have reformulated their agricultural policies towards less distorting instruments, the overall level of support to OECD agriculture has not declined since the Agreement on Agriculture came into force. Support to agricultural producers in OECD countries, as measured by the producer support estimate (PSE)[10] increased relative to the Uruguay Round base period to US$258 billion in 1998-2000 (OECD, 2001e). The aggregate measurement of support (AMS) limits have not been constraining thus far: few developed countries used more than 80 percent of their commitment level in 1996. Only four WTO Members have reported that they are using or have used blue box[11] policies (the EU, Japan, Norway and the Slovak Republic).

The maximum permitted domestic support in most developing countries[12] is set by the de minimis level and the Agreement on Agriculture provisions for product-specific support. This is because only a few developing countries reported AMS figures and only 12 set them at levels above the de minimis. In theory, domestic support could total 20 percent of the total value of agricultural production (10 percent product-specific plus 10 percent non-product-specific support). In fact, very few developing countries have provided support to agriculture (exempt and non-exempt) in excess of 2-3 percent of the value of production, and many have reduced support since the Agreement on Agriculture base period because of budgetary and administrative constraints. Indeed some countries still tax the agricultural sector or specific commodities, although they are not

---

[9] Two product categories were responsible for more than half of all SSG actions: dairy and sugar and sugar confectionery.

[10] Defined by the OECD as the annual monetary value of gross transfers from consumers and taxpayers to support agricultural producers, measured at farmgate level, arising from policy measures that support agriculture, regardless of their nature, objectives or impacts on farm production or income. The PSE includes the effects of border measures and thus captures both support and protection.

[11] Direct payments under production-limiting programmes. They belong to the "blue box", i.e. they are not subject to reduction commitments if they meet certain criteria.

[12] For support measures included in the AMS and subject to reduction commitments.

Table 9.2   Domestic support expenditures 1996, US$ million

| Member | AMS | % of AMS commitment used | Measures exempt from reduction commitments | | | | Total expenditures |
			De minimis	Blue box	Green box	SDT	
Australia	113	26	2	0	740	-	855
Brazil	0	0	363	0	2 600	269	3 232
EU	61 264	67	915	25 848	26 598	-	114 625
India (1995)	-23 847	-31	5 956	0	2 196	254	8 406
Japan	29 562	72	331	0	25 020	-	54 913
Kenya							0
Korea, Rep.	2 446	93	427	0	6 443	38	9 354
Morocco							0
New Zealand	0	0	0	0	151	-	151
Norway	1 633	79	0	638	520	-	2 791
Pakistan	-193	-	-	-	440	-	247
South Africa	451	82	203	0	544	-	1 198
Switzerland	2 962	74	0	0	2 128	-	5 090
United States	5 898	26	1 153	0	51 246	-	58 297

Source: WTO (2001a) and FAO (2000e).

allowed to offset such negative product-specific supports with positive non-product-specific support.[13]

Export competition. The third main pillar of the Agreement on Agriculture dealt with export competition. Although the original GATT 1947 prohibited export subsidies in most sectors, an exception had been made for primary products, including agriculture. The Agreement on Agriculture sought to redress this omission by establishing disciplines on the use of export subsidies.

Only eight of the 25 WTO Members that have export subsidy commitments are developing countries, as defined by WTO (Brazil, Colombia, Cyprus, Indonesia, Mexico, South Africa, Turkey and Uruguay), and only one (Colombia) reported using export subsidies in 1998. The majority of direct export subsidies are used by the EU, which in 1998 accounted for more than 90 percent of all direct export subsidies under the Agreement on Agriculture (Table 9.3).

The terms of the Agreement on Agriculture also commit Members to negotiate disciplines on the use of export credit guarantees or food aid shipments, which might be used to circumvent the disciplines on direct subsidies. The United States is by far the biggest user of officially supported export credit guarantees, having provided 46 percent of all export credits used from 1995 to 1998. Three other exporters (Australia, the EU and Canada) were responsible for almost all the remainder. The subsidy element in export credits is fairly small compared with direct export subsidies (OECD, 2000b).

Food aid has also received attention as a possible means of circumventing disciplines on export subsidies. The Agreement on Agriculture stipulates only that food aid should be provided in accordance with FAO Principles on Surplus Disposal and, to the extent possible, fully in grant form. While most food aid is made in grant form, some donors provide food aid in kind, and there has been an apparent countercyclical relationship

---

[13] Many developing countries reported that all their support to agriculture met the green box and/or SDT criteria for exemption, without specifying the measures or providing budgetary outlays. The conformity of these policies with the Agreement on Agriculture will remain uncertain unless assessed, perhaps in the context of dispute settlement.

Table 9.3	Export subsidy use (million US$)	
Member	**1998**	**% of commitment**
Australia	1	6
Brazil	0	0
Canada	0	0
Colombia	23	22
EU	5 843	69
Indonesia	0	0
New Zealand	0	0
Norway	77	65
South Africa	3	28
Switzerland	292	65

Source: WTO (2001a).

between such aid and international commodity prices, suggesting that aid is increased when prices are low and withdrawn when prices rise.

Non-trade concerns. The Agreement on Agriculture allows significant scope for governments to pursue important "non-trade" concerns – such as food security, environmental protection, rural development and poverty alleviation – through the use of domestic support measures that are exempt under the green box (Agreement on Agriculture, Annex 2), the de minimis or SDT provisions for developing countries (Agreement on Agriculture, Article 6). Furthermore, Article 20 of the Agreement requires that non-trade concerns be taken into account in the mandated negotiations on the continuation of the reform process in agriculture.

However, many countries feel that the Agreement on Agriculture does not provide sufficient policy flexibility for the pursuit of their non-trade concerns. They argue that agriculture has specific characteristics not shared by other sectors, and should not be subject to the same types of discipline on the use of subsidies and border protection.

Thirty-eight countries submitted a note in the ongoing WTO negotiations in which they address the "specific and multifunctional" characteristics of agriculture such as its contribution to rural development, food security, environment and cultural diversity (WTO, 2000b).[14] Several formal negotiating proposals have elaborated on these themes. Norway, for example, used the non-trade concerns argument to justify its proposal that certain basic food commodities be exempt from further market access commitments and that production-distorting supports for commodities destined for the domestic market be subject to less stringent AMS disciplines (WTO, 2001a). Similarly, Japan uses the multifunctionality of agriculture to justify its call for discretion in setting tariffs and providing domestic support (WTO, 2000c).

Most WTO Members have agreed that agriculture is multifunctional, although several have argued that it is not unique in this regard. The Members have also agreed that each country has the right to pursue its non-trade concerns. The question being debated in the WTO is whether "trade-distorting" subsidies, or other subsidies not currently exempt from disciplines, are needed in order to help agriculture perform its many roles.

Where Members differ sharply is in their views regarding the appropriate policy response to non-trade concerns. Many countries, particularly the members of the Cairns Group, have argued that non-trade concerns do not justify the use of production- or trade-distorting support and protection. South Africa, for example, insists that the non-trade concerns of some countries should not become trade concerns for others (WTO, 2000d). Furthermore, many developing countries have sought to distinguish their concerns regarding food security and sustainable development from the non-trade concerns of developed countries. India in particular has argued that the food security and livelihood concerns of low-income countries with large agrarian populations "should not be confused or equated with the non-trade concerns advocated under 'Multifunctionality of Agriculture' by a few developed countries" (WTO, 2001b).

---

[14] The economic debate surrounding the multifunctionality of agriculture is generally framed in terms of public goods and externalities. Economists generally agree that agriculture can be a source of public goods. These are goods that are non-rival (one individual's consumption of the good does not reduce the quantity available to others) and non-exclusive (once the good becomes available it is not possible to prevent someone from consuming it). The non-rival and non-exclusive nature of public goods means that they are not properly priced in a market system. Agriculture may also be a source of positive externalities. Externalities impose social costs or generate social benefits jointly with the production of a good, beyond the private costs or benefits deriving from the good. Positive externalities are consumable but are not priced in the market. For public goods and positive externalities associated with agriculture, it can be argued that producers should be rewarded in order to ensure a sufficient supply of the desired goods (Blandford, 2001).

Developing countries and the Agreement on Agriculture. Clearly, the Agreement on Agriculture does not in itself help developing countries to strengthen their agriculture. It expects essentially the same type of commitments by developing countries, including the least developed, as by developed countries. Measures for SDT for developing countries were generally in the form of lower reduction commitments or longer implementation periods that did not recognize the fundamental differences between the agricultural sectors of developed and developing countries. Many SDT provisions were of little use to developing countries. A longer time period to reduce subsidies on domestic support or exports is meaningless for a country that does not provide such subsidies. But, by binding subsidies at their de minimis or base levels, the future ability of developing countries to invest in the agricultural sector may have been constrained.

Another issue of interest to many developing countries is the full implementation of the Marrakesh Decision on Measures Concerning the Possible Negative Effects of the Reform Programme on Least Developed and Net Food-Importing Developing Countries. This issue will be addressed in the current negotiations on agriculture. Trade Ministers have approved the recommendations of the Committee on Agriculture that the delivery of food aid to Least Developed Countries (LDCs) be fully in grant form and, to the extent possible, that levels of food aid be maintained during periods of rising world prices. The Ministers have also endorsed the recommendation that a multilateral facility be explored to assist LDCs and Net Food-Importing Developing Countries (NFIDCs) with short-term difficulties in financing normal levels of commercial imports, including the feasibility of creating a revolving fund (FAO, 2001j).

9.3.2 Regional and preferential agreements

Although the WTO rules apply generally to international trade, almost all WTO Members also participate in regional trade agreements. Moreover, many developing countries are granted preferential access into industrial markets for tropical products. The terms under which trade takes place within these agreements is important to the development of agriculture and the future of the food system. Thus a discussion of the long-term outlook for agri-

culture must include the role of regional trade agreements (RTAs) and the future of non-reciprocal preferential trade agreements (PTAs).

Regional trade agreements. The past decade has seen a proliferation of regional agreements involving agriculture, and this trend is likely to continue and intensify, particularly if the WTO multilateral negotiations stall. At the same time, PTAs that have so far characterized much of the trade between developing and developed countries are set to change significantly.

Nearly all WTO Members are parties to at least one RTA, some belonging to ten or more. Since 1948, more than 225 RTAs have been notified to GATT/WTO, of which about 150 are considered active, and more than two-thirds of these have been formed since the WTO entered into force in 1995 (WTO, 2001d). About 60 percent of these new agreements were formed by countries in Europe: among countries of Central and Eastern Europe, between those countries and the EU or between the EU and countries in other regions such as North Africa. In addition, many others, particularly in Africa, have been declared but not yet notified to WTO.

Countries have many reasons for joining RTAs. The economic benefits can be significant if lower-cost imports from the partner country displace higher-cost domestic goods (trade creation) or if access to a larger market allows producers to achieve greater economies of scale. Regional agreements may also stimulate foreign investment and technology transfer among members. RTAs may provide a forum in which trade liberalization can be pursued at a different pace, faster or slower, than in the multilateral system. But the benefits may extend beyond pure economics. For example, regional integration may be a useful strategy for improving regional security, managing immigration flows or locking in domestic reforms.

Against these potential gains, regionalism is not without costs. The classic trade diversion effect occurs when lower-cost imports from non-members are displaced by higher-cost products from a member, raising consumer costs and exacerbating inefficient production patterns. In addition to diverting trade away from efficient suppliers, complicated rules of origin and sourcing requirements may create difficulties for members themselves. Other costs are largely related to the administrative burdens associated with negotiating and operating RTAs. These can be particularly severe for small countries with scarce negotiating capacity and for countries involved in numerous agreements.

Is regionalism a threat to the multilateral system or a response to its failure? As noted above, about two-thirds of the RTAs that are currently in force have been notified since the creation of WTO, and a large share of them are association agreements with the EU. Since agriculture is typically given special treatment in these agreements (i.e. less integration and longer transition times than other sectors), it is difficult to argue that they are responding to a failure of the Agreement on Agriculture to liberalize fast enough. For many of these RTAs, it is likely that the agricultural sector was less important than other sectors and that non-economic considerations have greater importance than the potential economic benefits.

Other RTAs, such as the North American Free Trade Agreement (NAFTA) among Canada, Mexico and the United States, and the Mercado Común del Sur (Mercosur) among Argentina, Brazil, Paraguay and Uruguay, have liberalized agriculture at a considerably faster pace than in the Agreement on Agriculture, albeit more slowly than other sectors. Some of the agreements that are currently being negotiated, such as the Free Trade Area of the Americas and the Caribbean, and the African Union, envision a more aggressive pace of agricultural liberalization and integration than the multilateral system seems capable of delivering. If the WTO negotiations stall, it seems likely that these countries would put greater efforts into their regional integration plans.

Bilateral, regional and plurilateral institutions also play a role in global food regulation (Josling, Roberts and Orden, 2002). There is the potential for bilateral or regional standards to become *de facto* international norms without the full participation of all trading countries. When nations try to harmonize their respective technical regulations to permit the free movement of goods within a region, their external trading partners may face new technical requirements for gaining entry to the unified market. These external regulatory changes, or even proposed regulatory changes, can lead to market disruptions for the private sector, which in turn can produce trade conflicts for the public sector to resolve. New regional trade alliances, as well as the enlargement and deeper integration of older alliances, have been factors in the increase in technical barriers brought to the attention of policy-makers by exporters.

Regional agreements can, however, also point the way towards global standards. An early example of the regional "test bed" function was the provision in the United States-Canada Free Trade Agreement for the mutual recognition of testing and inspection facilities for livestock destined for sales across the mutual border. An early version of what eventually became the Sanitary and Phytosanitary (SPS) Agreement was inserted into NAFTA.[15] The three partners agreed to incorporate the use of scientific evidence in domestic regulations and to set up a mechanism for examining cases where one country (typically the exporter of a particular product) considered that the importer was using sanitary and phytosanitary regulations to protect domestic production. The main impetus for such a provision was the view that market access negotiated under NAFTA could be severely jeopardized if arbitrary health and safety rules were to be allowed to go unchallenged.

A more recent example of a bilateral agreement to handle food regulatory issues is that between Australia and New Zealand. These two countries regulate food safety jointly through the Australia New Zealand Food Authority (ANZFA) and administer standards through the Australia New Zealand Food Standards Council (ANZFSC). Such a close working agreement is made easier, and more necessary, by the existence of a fairly complete free trade area between the two countries, the Australia-New Zealand Closer Economic Relations Agreement (CER). As two exporters of food products, relatively remote from other landmasses, both countries are anxious to keep out plant pests and animal diseases where possible and to maintain their reputation as suppliers of disease-free, quality foods (Josling, Roberts and Orden, 2002).

---

[15] The negotiations were progressing in parallel in the period 1991-93, with the NAFTA negotiations a little ahead. However, much of the work in the GATT negotiations on the SPS Agreement had been conducted in a subcommittee of the agriculture negotiating group and had made progress before the NAFTA process was initiated.

At the other end of the spectrum of bilateral approaches to food regulations is the attempt to use the transatlantic partnership between the United States and the EU. Several agreements of potential significance have been negotiated, although not without considerable time and effort. These tend to be agreements for the mutual recognition of testing and certification, rather than the recognition of the standards of the transatlantic partner. On the most contentious issue, however, that of the introduction of genetically modified (GM) foods, fundamental differences in approach still remain unsolved (Patterson and Josling, 2001).

Among the more ambitious regional initiatives in the area of plurilateral coordination of food regulations is the one discussed in the Asia-Pacific Economic Cooperation Council (APEC) process. APEC debated establishing an APEC Food System, which would have included both food safety and trade liberalization instruments. The food safety part of the programme has yielded a framework for mutual recognition agreements (MRAs) in the area of conformity assessments. Building on the SPS Agreement and the CODEX guidelines for such MRAs, the APEC Food MRA system allows countries to negotiate agreements that will facilitate trade in foodstuffs in the Asia-Pacific region. Whether this is the start of a more substantial cooperation in the food regulations area will in part determine the attraction of such plurilateral schemes.

Preferential agreements. Developing countries often depend on PTAs for access to the protected developed country markets in Europe, North America and Japan, particularly so for agricultural exports. But some of the existing and proposed preferential schemes may be incompatible with the terms of WTO or with regional integration agreements. Even when compatibility is not a concern, the value of these preferences is likely to shrink as general tariffs decline. Thus, many of the developing countries that currently depend on trade preferences may face difficult adjustments in the coming decades (Tangermann, 2001). Four non-reciprocal preferential arrangements are of particular relevance: the Generalized System of Preferences (GSP) under WTO, the ACP-EU Cotonou Agreement, the United States Trade and Development Act of 2000, and the Everything But Arms Initiative to provide duty-

free and quota-free market access to the EU for the products of LDCs.

Generalized System of Preferences. The broadest of the existing non-reciprocal preferential arrangements is the GSP. Within the GSP, the preference-granting country has unilateral discretion over product coverage, preference margins and beneficiary countries. Many developing countries have complained that the product coverage of GSP schemes is not consistent from year to year, making it difficult for them to attract the necessary investment to develop the supply-side capacity to take advantage of the preferences. Others have noted that the GSP schemes granted by different countries vary considerably, making it difficult for exporters to know what tariffs their products will face in various markets.

Several proposals in the WTO negotiations on agriculture address these problems, calling for preferences to be made more transparent, stable and reliable. Some countries have argued that, in addition to the two main categories of countries that currently receive preferences under GSP (developing and least developed countries), other subgroups of countries should be eligible for non-reciprocal preferences. These subgroups comprise countries that are particularly disadvantaged in global markets by factors such as their size or geographical remoteness.

ACP-EU Cotonou Agreement. The EU historically granted trade preferences to the ACP countries under the Lomé Conventions that, in their later years, operated under a waiver from the GATT/WTO principle of non-discrimination that expired in 2000. The Lomé preferences have been extended for seven years under the ACP-EU Cotonou Agreement (which has also been given a WTO waiver), during which time the parties will negotiate new WTO-compatible trade arrangements, to come into force on 1 January 2008. These negotiations are envisaged to result in regional economic partnership agreements (REPAs), in effect, fully reciprocal free-trade areas with the EU. Agriculture will play a major role in these new agreements.

As the REPAs will have to be fully consistent with GATT Article XXIV and include "substantially all trade", it will be more difficult for the EU to restrict access for agricultural products from REPA partners, although some quantitative constraints can presumably continue in cases where domestic production is also curtailed. One would expect pressure from developing countries that are not signatories to the Cotonou Agreement to make sure that those countries negotiating REPAs do not perpetuate preferential access for particular commodities that would be inconsistent with WTO rules. But, on the other hand, within the REPAs there may well be scope for agricultural policies that benefit ACP countries as they continue to supply the EU market.

United States Trade and Development Act of 2000. The US Trade and Development Act of 2000 extends certain trade benefits to selected groups of developing countries, including those covered by the Caribbean Basin Initiative and the Africa Growth and Opportunity Act. The US Trade and Development Act of 2000 is somewhat less comprehensive than the Cotonou Agreement, and some of its provisions could prove problematic in the context of regional economic agreements among developing countries. The main difficulty will likely relate to eligibility requirements and rules of origin.

For the African initiative, eligibility for the programme is to be determined annually by the United States Government on the basis of numerous criteria spelled out in the legislation. Rules of origin are quite detailed and strict, and penalties for transshipment are severe. Generally speaking, eligible products may receive preferential treatment only if they are produced almost entirely in one or more beneficiary countries or the United States. Origin rules for certain textile products, for example, allow no more than 7 percent of the fibre content by weight to originate outside the United States or one of the beneficiary countries, while for other products, no more than 25 percent of the final value may originate from outside. No provision is made for the regional agreements that may exist between eligible and non-eligible countries (OAU/AEC, 2000).

Everything But Arms Initiative (EBA) and LDCs. The conditions of access for LDCs appear to be one area where trade liberalization is making some headway. Agricultural products are included in the schemes that have been discussed in this regard. In 2001 the EU granted duty-free and quota-free market access, under the EBA, to all products originating from the 49 LDCs except armaments and, after 2006-2009, bananas, rice and sugar. Since the EU announcement, some other countries have declared their intention to extend similar preferential access for LDCs. Several proposals have been tabled in the WTO agriculture negotiations that would make this a permanent obligation of developed country members, a commitment that was taken up by all WTO Members in the Doha Ministerial Statement. These preferences could mark a significant improvement in the market access enjoyed by LDCs, although some have questioned the degree to which developing countries as a group would benefit, since the economic gains for LDCs under this initiative would come largely at the expense of other developing countries, some of them only slightly better off than the beneficiaries. Of course, this holds for all preferential schemes that cover some, but not all, developing countries (including those mentioned above).

9.4 Towards freer trade in agriculture: what is important from a 30-year perspective?

As discussed earlier, both market and policy factors have affected the evolution of volumes and patterns of trade. While some indications of the relative importance of the two factors have been given, no quantitative assessment has been provided. This section will attempt to distinguish policy factors from non-policy factors and give some guidance as to what would happen if trade policy distortion were to be removed. The removal of these policy distortions will again affect trade patterns and bring about benefits and costs for all countries, particularly those that are participating in the reform process. This gives rise to questions such as: How important are policy and non-policy factors in developing and developed countries? What will happen if some, or all, distortions are removed? What differential impacts can be expected from specific reform packages? What would help or harm developing countries and what should be the priority areas for developing countries in the trade liberalization process?

9.4.1 How significant are the expected overall benefits from freer trade?

The basic case for multilateral trade liberalization rests on the potential for large global welfare gains.

Estimates for the magnitude of these welfare gains vary considerably depending on the base year for the model simulations and the comprehensiveness of the reforms over sectors and participating countries. A recent study (Brown, Deardorff and Stern, 2001) estimated global cumulative welfare gains from comprehensive trade liberalization (agriculture, manufactures, services) at about US$1 900 billion over a period of ten years.[16] The World Bank estimates that global gains from comprehensive trade reform could amount to about US$830 billion and that low- and middle-income countries would be able to share in benefits of about US$540 billion. All estimates include both static and various forms of dynamic gains[17] (World Bank, 2001c).

The likely impacts of further liberalization in the agricultural sector have been the subject of several detailed analyses. While such analyses differ significantly in model structure and assumptions, they agree roughly in terms of the likely magnitude of impacts on agricultural commodity prices, trade volumes and economic welfare. In general, the results suggest that the expected benefits of agricultural trade liberalization are less important for developing countries than for developed countries.

ABARE, for example, has analysed the impacts of agricultural trade liberalization using a computable general equilibrium (CGE) model (ABARE, 2001). This analysis found that a further 50 percent reduction in agricultural support levels alone would create a static US$53 billion increase in global GDP in 2010. Some US$40 billion of this would accrue to developed countries. Full liberalization of agriculture and manufacturing would boost global GDP by US$94 billion, with developing countries capturing most of the increment and about half the total (because they have relatively high levels of industrial protection). If dynamic gains are incorporated, global GDP would increase by US$123 billion relative to the base case, with more than half these gains going to developing countries (because they have greater potential for productivity improvements). In the agricultural sector, the gains for developing countries would be greatest in those that are either producing or are capable of producing the commodities that are currently most heavily supported in the developed countries such as livestock products, grains, oilseeds, sugar, fruit and vegetables.

The results of a study by Anderson (Anderson *et al.*, 2000) are summarized in Table 9.4. The study suggests that if all agricultural protection and trade barriers were removed globally, the world as a whole could expect an annual welfare gain of about US$165 billion (in constant 1995 US$). The gains could be significantly higher if reforms were extended to include freer trade in services and manufacturing as well as a liberalization of investment flows. But the results also show that the benefits of freer trade in agriculture would largely accrue to the developed countries. In fact, if reforms are limited to the developed countries, more than 90 percent of the additional welfare gains will remain within the group of high-income (OECD) countries. This reflects largely the fact that subsidies and other distortions in OECD agriculture are extraordinarily high. Their removal would create an increase in welfare for consumers in OECD countries (through lower consumption prices) and an increase in welfare for producers in non-subsidizing OECD countries (through higher farm prices).

Table 9.4 **Welfare gains from agricultural trade liberalization (per year, in 1995 US$)**

Liberalizing region	Benefiting region	Welfare gain (billion US$)
High income	High income	110.5
	Low income	11.6
	Total	122.1
Low income	High income	11.2
	Low income	31.4
	Total	42.6
All countries	High income	121.7
	Low income	43.0
	Total	164.7

Source: Anderson *et al.* (2000).

---

[16] Serious doubts have been raised about the plausibility of these estimates. For example, an analysis by Dorman suggests that the estimates only measure the benefits, without fully accounting for the costs associated with the reallocation of resources, etc. (Dorman, 2001).

[17] The dynamic gains often account for the major part of all welfare gains. Estimates for these gains are particularly high when they are based on additional productivity gains that are assumed to emerge when firms start to penetrate world markets and they are forced to adopt new technologies. In addition, firms can benefit from scale economies and a larger market. The assumed underlying relationship between openness and productivity growth applied in these models is econometrically estimated. It should be noted that even the authors of these studies underline that "much more work needs to be done in this area" (World Bank, 2001c, p. 167).

USDA has analysed the potential impacts of further liberalization of agriculture (USDA, 2001d). This report uses a combination of static and dynamic CGE models and other analytical tools to assess the economic costs of the current distortions in global agricultural markets and the potential effects of trade liberalization on global welfare and on various countries, commodities and economic agents. For a complete removal of subsidies and tariff barriers, the USDA study assesses global welfare gains at US$56 billion annually. This total entails both static (US$31 billion) and dynamic welfare gains (US$25 billion). The static benefit would accrue almost entirely to developed countries (US$28.5 billion of the total of US$31 billion) while developing countries are expected to share in a larger part of the dynamic gains (US$21 billion out of the total of US$25 billion).[18] The USDA study also finds that world commodity prices would be an estimated 12 percent higher on average, with the biggest gains for livestock and products, wheat, other grains and sugar. There are interaction effects among policy categories, but about half of the impact on prices would derive from the elimination of tariffs and other border measures, about a third from elimination of domestic supports, and the remainder from the elimination of export subsidies.

Despite the evidence of aggregate welfare gains, the USDA study found that some countries may lose from agricultural trade liberalization. Within countries, there will be both winners and losers as resources are reallocated according to their comparative advantage. Exporting countries stand to gain from improved terms of trade, relative to those they would obtain in the absence of reform. Importers may benefit from improved domestic resource allocation but losses in consumer welfare outweigh gains in producer welfare in exporting countries. Also, net exporters with preferential arrangements are likely to lose, as well as food importers where there is no potential improvement in domestic efficiency to offset the effect of higher world prices. In the case of the former, it is questionable, however, whether the dependency fostered by preferential agreements is of long-term benefit.

9.4.2 What if all OECD countries dismantle their agricultural subsidies?

The results from the various impact studies provide useful information about the general changes that are likely to emerge from trade liberalization for global agricultural markets. All of these results refer to baseline scenarios that describe a situation or an outlook without policy reforms. These baseline scenarios may differ in many important aspects from the projections presented in this study (Chapter 3). Below, the results of a policy reform scenario will be discussed, which takes the baseline scenario projections of this study as a starting-point.[19]

The current levels of farm support form the starting-point for two different policy reform scenarios. In a first step, all market price support is phased out in equal annual steps over a period of 30 years.[20] These price support measures are commonly regarded as the most distorting kinds of subsidies and form a subset of the so-called amber box measures of the Uruguay Round. They stimulate production in a direct and immediate way (Figure 9.5, 2nd diagram). In a second step, the gradual elimination of price support is accompanied by a complete phase-out of all non-price-related subsidies. This second scenario reflects a comprehensive removal of agricultural policy interventions in all OECD countries, i.e. a removal of subsidies to the tune of US$266 billion (OECD, 2001e). Both the gradual elimination of market price support and the phase-out of other subsidies are implemented at the level of individual commodity markets and countries or country groups (Figure 9.5, 3rd diagram).

For countries where producer support estimates are not available, i.e. non-OECD and non-transition economies,[21] domestic and international prices are linked via simple price transmission equations that translate, at varying degrees,

---

[18] These additional gains are assumed to emerge from "increased savings and investment as policy distortions are removed, and from the opportunities for increased productivity that are linked to more open economies" (USDA, 2001d, p. 6). This means that the gains would only be forthcoming if developing countries embark on domestic policy reform as well.

[19] A detailed description of the underlying model is provided in Schmidhuber and Britz (2002).

[20] No allowance is made for possible de minimis provisions that would afford individual countries a subsidy limit of up to 5 and 10 percent of the value of production for developed and developing countries, respectively.

[21] The scenario also assumes the removal of some US$2 billion of agricultural subsidies (OECD, 2001f) in economies in transition.

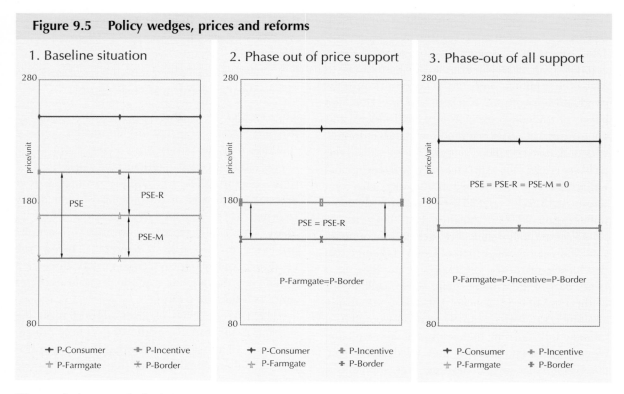

PSE: Producer support estimate
PSE-M: Market price support of the PSE
PSE-R: Non-price related support (PSE minus PSE-M)

changes in international prices into changes in domestic prices. These varying degrees of price transmission are encapsulated in a price transmission elasticity, which represents both tariff-based protection and "natural" protection. The elasticity can range from 0 to 1, where a value of 1 represents full transmission of price signals from the world market, while a value of 0 denotes complete insulation. In the scenario runs, these elasticities are increased year by year. In the first scenario, all price transmission elasticities are gradually increased to reach a value of 0.8 by 2030, wherever they are below this value in the baseline projections. In the second scenario, all price transmission elasticities are gradually increased from 0.8 to a level of 1 by 2030 which, together with a complete elimination of support policies in OECD and transition countries, represents the comprehensive policy reform scenario.

Impacts by commodity group. The most significant changes are expected to occur for temperate-zone commodities that account for the major portion of OECD policy distortions. OECD countries would

also be most affected by these policy reforms. There would be a shift in market shares from currently highly protected and supported producers to countries that have relatively liberal agricultural policy regimes. In general, production in Japan, Norway, Switzerland and, to a lesser extent, the EU would decrease and production in Australia, New Zealand, the United States and Canada would increase.

Some developing countries would also gain. The main beneficiaries would be Argentina (wheat, maize and beef) and Brazil (poultry). The majority of developing countries would reduce somewhat their imports of temperate-zone commodities but the price effects that emerge from international markets are mostly too small to change the net trade picture significantly, either for the majority of individual developing countries or for developing countries as a whole. This reflects their low supply responsiveness for temperate-zone commodities, particularly compared with developed countries. It is also a reflection of low and declining demand elasticities, which are assumed to fall with rising income levels (for details, see Schmidhuber and Britz, 2002).

Developing countries are estimated to gain more from OECD policy reform for competing products. OECD producer support for competing products accounts for about 40 percent of total PSE support. Many developing countries could step up their production of these commodities and increase exports. Examples are Thailand (rice and sugar), China (fruit and vegetables), Brazil (sugar), Malaysia, Indonesia and Argentina (vegetable oils), Zimbabwe (tobacco) and Pakistan (cotton). However, the majority of developing countries would remain net importers: they would import lower volumes at higher prices.

Prices and markets for tropical products would not be affected substantially. There is no significant production in OECD countries and, hence, no producer support to be removed. Developing countries might be able to reap more significant gains if OECD policy reforms were extended to include a reduction in protection of processed products (tariff escalation) or the abolition of commodity-specific consumer taxes.

Impacts by country group. The impacts of OECD policy reform would be felt most strongly in those OECD countries where producer support has been highest. Consumers would gain significantly from lower prices while producers would reduce output and lose market share. OECD producers in countries where support is small (e.g. beef and dairy in Australia or New Zealand) would benefit and gain market share at the expense of producers in protected markets. This outcome is consistent with the results from the above-mentioned USDA and ABARE studies that suggest that the major part of all welfare gains would accrue to developed countries, more specifically to consumers in protected and producers in unprotected markets.

A number of developing countries would also stand to gain from OECD policy reform. In general they are already net exporters of temperate or competing products. However, they are very few in number and belong largely to the group of the most advanced developing countries. The group of the least developed countries would in general be worse off. Very few of them are net exporters of temperate-zone or competing products.

Impacts on prices. In general, the results suggest that even a comprehensive policy reform package would have only a moderate impact on the level of world market prices (border prices). Supply for temperate-zone commodities in OECD countries is relatively responsive to price incentives, particularly in countries with substantial production potential and where farmers have traditionally been producing at world market price levels (in Oceania and, to a lesser extent, in North America). As prices increase, farmers in these countries would swiftly expand production. A significant impact on world prices occurs only for products where distortions are particularly high and the responsiveness of producers to higher prices is generally low, notably milk, for which prices are expected to increase by about 17 percent (Table 9.5).

Table 9.5	Impacts of partial and comprehensive policy reform on world commodity prices

	Partial policy reform (phase-out of market price support)	Complete policy reform (phase-out of all support)
	Changes in real prices relative to the baseline (baseline=100)	
Cereals	103	111
Wheat	104	119
Rice	104	111
Maize	99	106
Milk and dairy products	111	117
Beef	106	108
Sheep and goat meat	104	105
Pig meat	102	103
Poultry meat	103	104

Some developing country producers would also be responsive to higher prices, notably those in Brazil, Argentina, Malaysia or Thailand. There is even an additional production potential in those developed countries in which support would decline (e.g. in Europe). Many have put in place policy programmes that offset the output-enhancing effects of support and hold production below "normal" output levels (such as production quotas, extensification programmes and set-aside schemes). Policy reforms are assumed not only to remove subsidies, but also to lift these production constraints.

A further dampening effect on prices would occur because a removal of subsidies for all commodities would be likely to result in mutually offsetting effects for interlinked markets. The expected price changes for cereals are a case in point. While a removal of subsidies for cereals would put a brake on production and underpin international prices, the removal of support for livestock production would lower demand for feed-grains. This would offset much of the potential international price boost to cereals from lower subsidies given to cereal producers.

Although *producer prices* for the world as a whole would not be affected strongly, this small average impact masks more significant, but mutually offsetting, price effects in individual countries or regions. For example, the world average producer incentive prices for rice are expected to fall by only about 10 percent, but those in Japan would be as much as 85 percent below the levels assumed in the baseline scenario. At the same time, producer price changes could be very substantial for farmers in non-protected markets, such as dairy farmers in New Zealand.

Changes in *consumer prices* are also expected to be small, especially in OECD countries. For many commodities, the price of the primary product (e.g. cereals) accounts for only a small share of the total costs for the final consumer good (e.g. bread, noodles). The effect of liberalization would be significantly diluted by substantial processing and distribution margins, which can account for up to 90 percent of the value of the final product. In developing countries, the processing and distribution margins are smaller and thus the changes in the price of the primary product translate into more pronounced increases in consumer prices.

9.4.3 What makes it difficult for farmers in developing countries to reap the benefits from lower OECD distortions?

First, a look at protection and support levels by country and commodity suggests that agricultural trade distortions are concentrated in a few developed countries and the most distorted markets are those for temperate-zone commodities. Particularly high subsidies are provided to farmers in Japan, Norway, Switzerland and the EU, while other OECD countries such as Australia and New Zealand are producing at low costs without major subsidies (OECD, 2001e). In addition, these countries have the infrastructure in place that will allow them to capture the market shares that become available when subsidies are removed in the high support countries. Only a few developing countries have a comparative advantage in producing temperate-zone commodities such as milk and meat, wheat and coarse grains. As a result, a cut in OECD subsidies may primarily result in an exchange of market shares between OECD countries.

Second, even if and when trade liberalization results in higher and more stable international prices, it is unclear whether and to what extent these signals will be transmitted to farmers in developing countries. Inadequate infrastructure and inefficient marketing systems effectively insulate many farmers from world markets. In these cases, much of the price incentive that farmers would receive from world markets can be absorbed by the inefficiencies in their marketing and transportation systems.

Third, for products where the developing countries have a comparative advantage and even LDCs could benefit, e.g. coffee, cocoa, tea, spices and tropical fruit, developed countries' import tariffs have already been reduced and the effects of further liberalization are likely to be small. Tariff escalation remains a serious problem, but it is unclear how many LDCs could develop significant export-oriented processing industries even if tariff escalation were eliminated. The biggest distortions for these products are in developing countries themselves (USDA, 2001c). Their bound import tariffs for these products are higher than those in developed countries, both for the raw commodities as well as for the processed products (USDA,

2001c). Even for these countries, applied tariffs are generally much lower than bound rates, so further reductions in bound rates will have little effect unless they constrain the applied rates.

Finally, farmers in developing countries may not gain so long as their own domestic policies offset much of the price incentives from international markets. Most developing countries heavily taxed their agriculture throughout the 1970s and 1980s (through direct and indirect measures). Many have continued to do so over the 1990s. India, for instance, has submitted an AMS notification with an overall support level of about US$-24 billion, equivalent to a tax of 31 percent of the value of production (FAO, 2000e). Also farmers in Pakistan are producing under a net tax burden, although at a lower level. For China, PSE calculations suggest that its agricultural sector faced massive taxation for much of the 1980s and 1990s, at rates of 18 to 65 percent (OECD, 2001f). Particularly high rates were reported for rice (Webb, 1989) and pork (USDA, 1998b), farm products that are typically produced by China's smallholders.[22]

Over and above these direct burdens on agriculture, farmers in almost all developing countries have been handicapped by even higher effective rates of taxation caused by considerable non-agricultural tariffs. These tariffs make their inputs more expensive and bring them into a competitive disadvantage particularly *vis-à-vis* farmers in industrial countries who benefit from very low tariffs for manufactures (on average 4 percent). The non-agricultural tariffs make the effective burden for farmers even higher than the negative AMS or PSE calculations would suggest. The high effective burden from non-agricultural protection also explains why developing countries' agriculture stands to benefit the most from comprehensive trade liberalization. Out of the US$832 million of welfare gains from comprehensive trade liberalization as estimated by the World Bank (see above), agriculture would account for US$587 billion. Of this latter amount, US$390 billion would accrue to low- and middle-income countries, where much of this would come from a liberalization of the non-agricultural sectors in developing countries (World Bank, 2001c, p. 171).

9.5 Beyond the traditional trade agenda: emerging long-term trade policy issues

9.5.1 The new trade policy environment

The main focus of international trade policy has traditionally been on the conditions of access to markets. In the recent and emerging trade policy environment, the scope of the rules governing trade continues to expand. These include the health, safety and environmental rules that ensure quality and acceptability in discriminating markets, codes for the treatment of foreign direct investment, the regulation of conditions of competition and the codification of intellectual property rights.

Agricultural trade policy used to be dominated by farm-level issues, with the active participation of farm and commodity groups and those arguing for more protection. One major shift in the 1980s was the involvement of multinational food firms in the trade negotiations. This trend is likely to continue with the structural changes apparent in the food sector. First, the processing sector has a strong incentive to look for low-cost supplies. It therefore has the incentive to lobby governments for the ability to import those supplies from world markets, so as to remain competitive with firms located in countries where prices are lower. In many cases the low-cost food suppliers are in the Americas, as are the main competitors in the global marketplace. Hence one would expect continued pressure from the food industry to allow raw material prices to fall to roughly United States levels over a period of years. Given the disinclination of governments to support these prices indefinitely, the pressures from the food industry may well come to prevail in the end.

The tendency for international food companies to search for low-cost supplies will be reinforced by

---

[22] Negative support is not subject to reductions in multilateral trade negotiations. In fact, there are proposals to maintain negative support and receive credit for negative support in the calculations for the total AMS. In practice this could be implemented by adding up negative non-commodity-specific support with positive commodity-specific AMS and vice versa. The easiest way to remove the bias against agriculture, however, would be to remove taxation on agriculture. It may also be the cost-effective way for many developing countries. Taxation often works through procurement price systems that are easy to administer. Keeping both taxes and subsidies would also add to the administrative burden of policy implementation, an important advantage for many developing countries that often lack the necessary administrative system for more targeted policy measures.

pressure from those firms that are already operating in several countries. For these firms, including those in the distribution and retailing business, international trade is often intrafirm trade. Any restriction on the movement of food items within the firm will tend to cause problems for the firm, and hence will be resisted. But just as intrafirm movement of goods can be thwarted by government regulations, so too can the contractual obligations of firms that have come together in other forms of alliance. One would expect that those firms that have been pioneering supply chains, linking producers in one country to wholesale and retail outlets in another, would also find government restrictions on trade irksome. Thus one might expect these supply chains to add their voice to pressures for trade liberalization.

Many future changes in the global food system are in the direction of a more sophisticated agricultural industry, aware that future success depends on satisfying a variety of consumer tastes, and competing for a share of consumer spending with other goods and services (Moyer and Josling, 2002). As more actors become involved in the political process, the centre of gravity will shift perceptibly away from the primary producer. Policy will become less focused on unprocessed commodities, and the emphasis will switch to adding value to the raw material and marketing the final product. These changes will be crucial to the future of agricultural trade policy reform. In a world where farmers produce for the market, improvement in access to overseas markets can compensate in part for less domestic support. For those developing countries that can take advantage of the opportunities provided by the changing food habits of middle-class consumers, this could offer a way to use the food trade system as an engine of development.

9.5.2 The need for a consistent regulatory framework in global food trade

The process of globalization will increasingly require a consistent regulatory framework within which national regulations can be developed. At one level this is a technical task. National regulators need to iron out arbitrary differences that unnecessarily impede trade. But domestic bureaucratic and political pressures often complicate the technical task of avoiding incompatible regulations.

Some vested interests exploit regulatory differences for the sake of furthering protectionist interests. Others, in the name of reducing costs, push for a degree of harmonization that may be inappropriate. Complicating the task of devising technical regulations for global markets is the tendency to see such issues as touching upon national sovereignty. Politicians are not keen to cede authority to regulate their domestic food supplies to outside agencies or other governments. Domestic regulatory agencies also cling to accepted practices and standards, in part out of inertia and in part as a way of ensuring bureaucratic survival.

The debate about the regulation of food supplies has been given increased prominence in recent years by a wide range of public-interest groups. Some have focused on issues of consumer health, both the prevention of diseases and the promotion of better nutritional habits. Others focus on social and political implications of food production systems or identify food consumption with lifestyle choices. Food issues have thus become part of a broader social discourse among groups with different objectives, particularly in the debate over globalization.

Coherence in global food regulations can be achieved more easily if national food regulations are reformed in a way that reduces the scope for trade conflicts. Technical trade barriers will play an increasing role in this context. Agricultural exporters may be required to demonstrate that native species or human health are not endangered by their products, while simultaneously complying with standards that stipulate everything from ingredients to packaging materials. The regulatory environment for agricultural and agro-industrial producers is likely to become more complex in coming years, even though reform initiatives are currently under way in many countries to reduce the number and the rigidity of regulations faced by the private sector. With rising per capita incomes, demands for food safety, environmental amenities, product differentiation and product information increase among developed and developing countries alike. More and more, regulators are being asked to provide these services when markets fail to do so (Josling, Roberts and Orden, 2002).

Measures that regulate imports of new agricultural products, ranging from new animal genetics to new disease-resistant seeds, have also spawned disagreements between trading partners. New products, particularly GM commodities, have been at the centre of the most prominent recent debates over technical barriers to trade, as some importing countries consider genetic modifications to pose a risk to consumers or to biodiversity, or to violate ethical norms. Trade officials are drawn into the debate when exporters believe that lengthy regulatory review of new products is motivated by a desire to protect the commercial interests of domestic producers in importing countries, rather than by concerns about the safety of consumers or quality of the environment. There is reason to expect that in the near future the number of agricultural product and technology innovations, and the number of measures to regulate their entry into importing countries, will increase. Technical barriers will therefore remain an important topic of discussion in the international regulatory and trade policy arena well into the foreseeable future.

Food trade regulation is also becoming a major issue for developing countries and their role in the global economy. In particular, trade in processed food products will be of growing interest to developing countries. Exporters are finding the increase in value added a useful way of avoiding the "raw-material" trap, while importers need processed food products to meet the increasing demand for them. Decisions about the use of agricultural biotechnology to increase productivity, and the provision of technical assistance to permit developing countries to meet the high food standards in developed countries can have a major effect on their opportunities for trade and development.

9.5.3 Trade and international standards

As traditional market access barriers such as tariffs and quotas are reduced, the restrictions caused by safety and quality standards will become more apparent and important in determining an exporter's ability to gain access to markets. The WTO SPS Agreement governs the use of safety regulations in agricultural trade. Other quality attributes are covered under the Agreement on Technical Barriers to Trade (TBT) and the Agreement on Trade-Related Aspects of Intellectual Property Rights (TRIPS). Environmental protection measures, which were given the same status in the Doha Declaration as the protection of the life and health of animals, plants and humans, may be proposed during the negotiations as a basis for restricting trade. The SPS and TBT Agreements will not be renegotiated in the current talks (see Box 9.1 for an explanation of the SPS and TBT Agreements), but implementation issues related to these Agreements will be taken up, such as the need for longer transition periods and technical and financial assistance for developing countries.

These Agreements attempt to strike a balance between the concerns of importers and exporters. Importers wish to enforce quality and safety standards to protect the life and health of their people, plants and animals. Exporters have the right to expect such standards to be transparent, science-based and no more trade restrictive than necessary to meet the stated objective. In practice, balancing these concerns is very difficult. Science is not static, since knowledge evolves. Moreover, even where risk levels are known with scientific accuracy, their acceptability varies over time and between societies.

Since the WTO Agreements came into force, 18 separate cases involving safety or quality attributes of agricultural products have been filed under WTO dispute settlement procedures. Of these, five were settled bilaterally, three have been resolved through dispute settlement Panel and Appellate Body decisions, and the remaining ten are awaiting final resolution. Of the three "resolved" cases, the most contentious was the one launched by the United States and Canada over the EU's ban on the use of hormones in beef production (both for imported beef and beef produced within the EU). The Panel and Appellate Bodies found that the EU's ban on imports of beef produced using artificial growth-promoting hormones was not justified under the SPS Agreement because it was not based on a scientific risk assessment and there was not enough scientific evidence to support the ban. Since the EU has continued to ban such imports despite the results of the dispute settlement process, the complainants have the right to seek compensation by imposing higher tariffs on their imports from the EU equal to the value of the trade impairment suffered as a result of the ban. Thus, while this case has officially been resolved, it clearly has not resulted in a satisfactory

Box 9.1 The SPS and TBT Agreements

The two pillars on which the multilateral rules for food safety are built are the Sanitary and Phytosanitary (SPS) Measures Agreement and the Technical Barriers to Trade (TBT) Agreement. The SPS Agreement rests on two premises: that basing national standards on international norms would reduce conflicts and lower transaction costs; and that requiring scientific justification for standards that deviated from these international norms would make it more difficult for countries to shelter domestic industries behind unnecessarily restrictive health and safety regulations. In addition to setting out the rights and obligations of WTO Members, the SPS Agreement also establishes enforcement mechanisms. These mechanisms include notification procedures for informing other WTO Members of changes in SPS measures, the establishment of an SPS committee to discuss these issues on a continuing basis, and the use of WTO dispute resolution mechanisms for resolving conflicts between countries in a timely manner. These mechanisms include formal consultations between the parties to a dispute, followed by adjudication by a WTO panel and the WTO Appellate Body if required. Regulations that aim at protecting people, animals or plants from direct and definable health risks, such as the spread of disease, potentially allergic reactions or pest infestations, are covered by the SPS Agreement.

The TBT Agreement covers all other technical regulations. Like the SPS Agreement, the TBT Agreement aims to distinguish measures that are necessary for achieving some regulatory objective from disguised trade protection. Specifically, the TBT Agreement extends the GATT principles of national treatment and most favoured nation obligations. As under the SPS Agreement, the TBT Agreement also stipulates that countries avoid unnecessary trade impediments. Beyond these general trade-promoting provisions, the TBT Agreement is on the face of it more permissive than the SPS Agreement. The TBT Agreement does not limit a government's right to impose domestic trade restrictions when pursuing a legitimate goal in a non-protectionist way. The key provisions of the TBT Agreement define the basic concepts of "legitimate goals" and "non-protectionist actions" related to technical regulations promulgated by central government bodies. According to these provisions, an import regulation has to meet two conditions. First, the regulation should aim to fulfil a legitimate objective and, second, there should be no other less trade-restrictive measure available to fulfil the legitimate objective. The combination of "legitimate objective" and "least trade-restrictive" conditions form the core of the disciplines on domestic regulations imposed by the TBT Agreement. But whereas the SPS Agreement clearly requires a sufficient scientific basis for the measure in question, the TBT Agreement appears to set less strict standards and allows more discretion. However, technical regulations that refer in their objective to issues of a scientific nature are subject to evaluation based on the scientific knowledge available (Heumueller and Josling, 2001).

The TBT Agreement includes food regulations. Although it has played less of a role in food-related technical barrier trade disputes than the SPS Agreement, many of the current controversies have to do with the product and process attributes of food and not directly their safety. In the area of food quality, the TBT Agreement is most germane. Moreover, as countries find that their controversial food regulations are subject to successful challenge under the SPS Agreement, they may frame their objectives in a way that brings the measure under the TBT Agreement. The nature of the global food sector suggests there are rents to be attained from labelling and origin-identification regulations. Trade conflicts may well shift towards issues of the labelling of quality attributes and away from the more traditional health and safety issues.

solution either from the point of view of the parties to the dispute or from the institutional perspective of promoting a fair and transparent trading system.

Considerable time and cost are involved in pursuing a case through the WTO dispute settlement procedures and, as the beef hormone case illustrates, the final result of the process may be unsatisfactory. A number of countries have suggested that importers are increasingly using SPS and TBT measures as disguised protectionism.

This point was raised in several FAO country case studies (FAO, 2000e) and has been mentioned in proposals submitted to the ongoing agriculture negotiations (WTO, 2000f and 2001c). On the other hand, a number of countries have indicated that consumer safety and the protection of traditional food applications are increasingly important for them (WTO, 2000c and 2000e). This suggests that issues related to food safety and quality will become a source of increasing tensions in the agricultural trading system.

Health and safety standards. All governments accept responsibility for guarding the safety of the nation's food supply and the health of its plant and animal populations, and many also undertake to ensure food quality and to provide information to consumers as they make food-purchase choices. Yet countries face very different circumstances in their markets for food, and consumers can have vastly different concerns and susceptibilities. As a result, countries have developed quite diverse systems of regulations to safeguard plant, animal and human health, to ensure food product quality, and to provide consumer information (for a more comprehensive discussion, see Josling, Roberts and Orden, 2002).

As economies open up to trade there is increasing potential for conflicts to arise from the different ways of providing for food-related health and safety and from the differing levels of health protection afforded by food systems among countries. Trade conflicts can also arise from those aspects of food regulations that are not directly related to health. Firms selling processed farm products are motivated to seek protection for their trademarks and the image of their products, setting up potential conflicts with new entrants in the market. Many consumers expect basic nutritional or other information to be readily available, although this again can lead to charges of protection for domestic producers. Affluent consumers have also begun to take a greater interest in how their food is produced, i.e. whether the farms use environmentally sound practices, whether pesticides and other chemicals are used, and how animals are treated. Demands for regulations that impose standards in these areas add to the pressures on governments and increase the potential for trade conflicts. The emergence of new methods of production, such as the use of advances in biotechnology to "design" plants and animals, poses further regulatory challenges.

Trade conflicts over health, safety and quality standards are not new, but increased globalization of the food and agricultural sector has made these conflicts more visible. Governments need to handle these conflicts in a way that both upholds public confidence in food safety and product standards and preserves the framework for trade and the benefits of an open food system.

Environmental and labour standards. As noted in previous chapters, the production increases in prospect at the world level for the period to 2030 are significant. Thus, almost another billion tonnes of cereals must be produced annually by 2030, another 160 million tonnes of meat, and so on. This also means that pressure on resources and the environment will continue to mount. The challenge is how to produce the required increases of food in sustainable ways, while keeping adverse effects on the wider environment within acceptable limits.

The agro-ecological environments of individual countries differ in their ability to withstand adverse effects associated with increasing production, either because they are inherently more or less resilient or have more or less abundant resources or because such resources at present are more or less stretched from the past accumulation of stresses. Countries also differ as to their technological and policy capacity for finding solutions and responding to emerging problems.

Trade can help to minimize adverse effects on the global resource system, if it spreads pressures in accordance with the capabilities of the different countries to withstand and respond to them. Whether it will do so depends largely on how well the prices of each country reflect its "environmental" comparative advantage. This requires that, in addition to the absence of policy distortions that affect trade, the environmental "bads" generated by production be embodied in the costs and prices of the traded products. If all countries meet these conditions, then trade will contribute to minimizing the environmental "bads" globally as these are perceived and valued by the different societies, although not necessarily in terms of some objective physical measure, e.g. soil erosion, loss of biodiversity, etc. This latter qualification is important, because different societies can attach widely differing values to the same environmental resources relative to those of other things, such as export earnings, employment, etc. In the end, the values of environmental resources relative to those of other things are anthropocentric concepts and countries at different levels of development, of different cultural backgrounds and resource endowments are bound to have differing priorities and relative valuations.

Attempts to impose uniform environmental standards can stand in the way of countries profiting from trade based on their relative endowments of environmental resources. This may happen as agricultural policies are "greened", and pressure increases for multilateral rules that constrain countries with less restrictive environmental standards. This could act as a further hindrance to developing country exports, many of which might not face the same environmental pressures that agriculture in, say, northern Europe may have to contend.

Environmental regulations are likely to have the strongest impact on those agricultural activities that have the closest links with broad transnational environmental objectives, such as mitigating climate change and conserving oceanic resources. Thus fisheries and forestry may be more directly affected than agriculture. But more localized pollution issues such as pesticide runoff and water quality could still be important issues for crop and livestock agriculture in the future. Intensive livestock rearing may well have to absorb significant extra costs as countries strengthen their environmental regulations, and attempt to restrict trade from those countries without such regulations. And if the environmental impact of GMOs turns out to be more serious a problem than many scientists now assume, the next two decades could see a raft of new restrictions on the use of this technology that may prevent its spread and adoption.

At present it looks unlikely that the Doha Round of trade talks will establish any significant new rules on environment and trade. More likely is an agreement that clarifies the relationship between disciplines under WTO and those that might be undertaken in the context of a multilateral environmental agreement (MEA). The potential conflicts have been increased by the attraction of trade sanctions as ways of enforcing MEAs. Eventually one of these conflicts could have a damaging impact on the credibility of WTO and the trade system. Attempts to agree on the areas of overlap, such as allowable sanctions, are likely in the next few years.

The Doha Round has not included the contentious issue of labour rights on its agenda.

Countries have been reluctant to do so since the Seattle Ministerial meeting, at which labour standards played a role in the collapse of the talks. The likelihood is, however, that a closer relationship between the International Labour Office (ILO), with its "core labour standards", and WTO will be established. One would expect the debate on the extension of the ILO core standards away from human rights towards economic rights to be contentious, and it is possible that agriculture could be caught up in this debate. Currently there have been few conflicts that have emerged on the international agenda (as opposed to the domestic political landscape) on the issue of labour standards in agriculture.[23] But this situation may not last for the next three decades.

Trade and the conditions of competition. A new area on the Doha agenda is the issue of trade and competition. With increasing globalization it is clear that pressures will continue for some degree of harmonization in these policy areas. If the Doha Round were to establish some basic rules in this area, it could take another ten years thereafter to have them in place. Thus, within the timeframe of this study, the emergence of a set of rules governing the competitive behaviour of governments and firms is quite possible. Intellectual property protection was introduced in the Uruguay Round.

A global trade system may need global competition rules. But the way in which those rules will develop is not clear at present. While some are calling for full-scale negotiations on international competition policy, others maintain that the most that can be done is to make sure that each trading country has its own antitrust policy in place. But the minimalist approach is unlikely to be satisfactory for very long. The best policy for curbing misuse of market power in any one country is an open trade system. But the very openness of the trade system allows large firms to develop market power in the world market. Global competition policy should be more about market power in world markets than about enforcing competition policy in each national market.

An emerging competition issue is the concentration of market power in the agrofood distribu-

---

[23] One recent example is that of the use of child labour in the harvesting of cocoa in Africa. Firms are instituting voluntary schemes to avoid such practices, in large part to avoid the consumer reaction that took place in the footwear and clothing industries.

tion chain. This has two separate but related aspects. One is the use of market power by public agencies or by parastatals, given their ability to act in a restrictive way. This issue of "state trading" is coming to the fore in trade talks. It represents a concern among those countries that do not practise state trading that those that do can gain an "unfair" advantage through hidden export subsidies and import barriers. The issue of competition is also at the heart of another potential problem facing the agrofood system. Concentration of economic power is not only confined to public agencies that have monopoly rights in importing or exporting. Private firms can have significant market power to influence prices and the pattern of trade through restrictive business practices. Should there be any rules relating to the use of market power in international markets? What dangers should such rules try to prevent? Is the problem the withholding of supplies to raise the price of commodities? This seems relatively unlikely in the case of basic foods, but could happen with vital supply components. Or is the problem one of dumping and market disruption? The incorporation of antidumping rules in a set of more comprehensive competition regulations is the object of many trade economists. Whatever is agreed will have significant implications for global agriculture.

Trade-related aspects of intellectual property. Less central than the SPS and TBT Agreements but still important in the framework of multilateral food regulations is the TRIPS Agreement. The Agreement imposes on member countries an obligation to provide a minimum standard of protection to a range of intellectual property (IP) rights, including copyrights, patents and trademarks.[24] Two aspects are particularly important in the global food regulatory framework, namely the requirement to respect geographic indications, which are widely used in the wine and spirit sector as well as in cheese and other processed food industries, and the obligation to provide protection to new plant varieties (although not necessarily by patents) and to innovations in the area of microbiology.

The issue of the patentability of plant and animal varieties, as well as of GMOs, raises questions beyond the mere protection of IPR, such as questions concerning the rights of local communities and indigenous peoples, and the sovereign rights over natural genetic resources, biosafety and food security.[25]

The TRIPS Agreement is having a marked effect on the shape of domestic IP regulations, as it was designed to do. Although it is not harmonizing these regulations, it is establishing a template into which domestic regulations must fit. The objective is to avoid trade conflicts that inevitably arise if different countries have different coverage and use different instruments for IP protection. The provision of IP protection does not always facilitate trade since, by providing protection to existing rights holders, new entrants are discouraged. Overprotection can be a problem in this area of food regulation as well as in health and safety issues (see Box 9.2). In addition, IP protection holds significant implications for access to and transfer of technology, particularly to the developing countries. Access to most protected technologies and products, particularly in the seed and biotechnology area, is subject to the terms of licensing agreements dictated by a very small number of enterprises.

Multilateral disciplines for labelling regimes are set out in both the SPS and TBT Agreements of WTO. The TBT rules apply to all safety and quality labelling regimes, except those defined as an SPS measure, i.e. "labelling requirements that are directly related to food safety". Article IX of GATT establishes rules for marks of origin. The international rules for protection of IPR in the form of geographic indicators are also germane to understanding the WTO framework for the governance of information provision. The TRIPS Agreement sets out importers' obligations to protect geographic indicators, which are increasingly used to differentiate agricultural products in domestic and international markets. Resolution of conflicts under the TRIPS Agreement is subject to the WTO Dispute Settlement Mechanism.

---

[24] The three main intellectual property instruments are copyright, for artistic and literary works; patents and similar devices for inventions, industrial designs and trade secrets; and trademarks, signs and geographic indications for commercial identification. The rights holder is given exclusive ownership or user rights for a specified (in some cases indefinite) period of time. The TRIPS Agreement incorporates and extends previous intellectual property arrangements including the Berne Convention and the Paris Convention, as well as those administered by the World Intellectual Property Organization (WIPO).

[25] The Convention on Biological Diversity deals with most of these issues. In addition, the International Treaty on Plant Genetic Resources for Food and Agriculture provides for the conservation and sustainable use of genetic resources for food and agriculture as well as for the fair and equitable sharing of benefits arising from their use, in harmony with the Convention on Biological Diversity.

Box 9.2 Overprotection of intellectual property can present a threat to trade

The basic idea behind the Agreement on Trade-related Aspects of Intellectual Property Rights (TRIPS) is to avoid trade conflicts that can arise when countries rely on different instruments for the protection of intellectual property rights (IPR). To accomplish this goal, the TRIPS Agreement provides common guidelines for national legislators to design domestic rules and regulations for the protection of IPR, which are largely comparable and thus compatible with the legislation of trading partners.

In practice, however, there are a number of factors that impede rather than facilitate trade, particularly in the case of agricultural trade between developed and developing countries. First, obtaining IPR is a costly process, too costly for many developing countries, particularly for internationally recognized patents. Moreover, many developing countries fail to establish and protect their IPR simply because they are unaware as to what innovations are patentable. Similarly, as the number of patents and cross-patents held in countries abroad rises, many developing countries may simply not be aware of possible infringements of IPR that are held by trading partners abroad. Finally, the ability to obtain IPR and to earn royalties from these rights has created incentives to obtain patents for hitherto unprotected germplasm, including by private players from countries abroad. This practice is often referred to as biopiracy, particularly when patents are acquired after only marginal alterations of the original germplasm.

An interesting case in point is the trade conflict that has emerged out of a United States patent on a dry bean variety (the so-called Enola bean) that originated from Mexico. The conflict started when a United States plant breeder bought a small amount of dry bean seed in Mexico in 1994 and brought it back to the United States. He selected the yellow-coloured beans and multiplied them over several generations until he obtained a "uniform and stable" population. In 1999, the United States Patent and Trademark Office (PTO) granted a patent on the yellow beans, and a United States Plant Variety Protection Certificate followed shortly afterwards. Soon after the patent and Plant Variety Protection Certificate were issued, the plant breeder brought legal suits against United States trading companies that imported yellow beans from Mexico. He now asks for royalties to be paid on all yellow bean imports to the United States or otherwise to block yellow bean imports.

This controversy also illustrates that the current practice of IPR protection can have rather serious effects on trade flows. Similar controversies in other areas, notably over South Asian basmati rice, Bolivian quinoa and Amazonian ayahuasca (Indian chickpeas), suggest that agricultural exports from developing countries could be particularly strongly affected. The potential for future trade conflicts is significant. An increasing number of patents and cross-patents in the new transgenic plant varieties and the rapid growth in trade in transgenic organisms mean that IPR protection could have massive impacts on future agricultural trade flows. Given the fact that the IPR for most modern GM varieties are in the hands of developed-country companies, a large share of developing countries' agricultural exports could be subject to royalty surcharges or face import barriers in markets where these rights are protected.

9.6 Summary and conclusions

The agricultural trade of developing countries has seen a number of important changes. Agricultural exports grew much more moderately than exports of manufactures, resulting in a dramatic decline in the share of agricultural exports from about 50 percent of total exports in the early 1960s to less than 7 percent by 2000. Despite this overall decline in importance, some countries continued to rely heavily on agricultural exports whereby single commodities such as coffee, cocoa or sugar can account for more than half of total foreign exchange earnings.

The surplus in the overall agricultural trade balance of developing countries has virtually disappeared over the past 40 years, and the outlook to 2030 suggests that, as a group, they will increasingly become net importers of agricultural commodities. The group of LDCs already underwent this shift 15 years ago. Their agricultural imports are already twice as high as their agricultural exports. The trade balance of the LDCs will further deteriorate, and their current trade deficit will quadruple by 2030. Within agricultural trade, developing countries recorded a growing trade deficit for temperate-zone commodities, while their trade surplus for tropical products grew only moderately. The trade balance for competing products remained largely unchanged.

These shifts in trade flows have been brought about by market factors and policy influences. On the market side, income and population growth have fuelled robust growth in demand, which could not fully be matched by domestic supply. On the policy front, subsidies and protection in developed countries as well as taxation and industrial protection in developing countries augmented the effects that arose from the market side. Numerous studies suggest that the importance of policy effects is small relative to the effects of market forces and that policy reforms, at least if limited to the developed countries only, would not significantly alter the overall trade picture. These studies also show that the largest portion of welfare effects from freer trade and policy reform in developed countries would accrue to the developed countries themselves.

An analysis of the impacts of OECD policy reform confirms the rather limited impacts of policy factors, at least if policy changes are limited to the agricultural sector of developed countries. If developing countries also embarked on comprehensive reforms reaching beyond agriculture, the benefits could be more significant, with the major part of the additional gains going to developing countries. Farmers in developing countries could benefit the most from domestic reforms that encompassed a removal both of the direct bias against agriculture through taxation, and of indirect bias caused by macroeconomic distortions and industrial protection.

The reduction of OECD farm support may not be sufficient and may perhaps be of only limited benefit to developing countries. However, developing countries are likely to gain substantially from other reform measures, like a move towards a "de-escalation" of tariffs; abolition of consumer taxes; further reduction of the bias against agriculture in their own countries; more and deeper preferential access for the poorest of the poor (LDCs); and open borders for foreign direct investments (FDIs) to enable developing countries to compete more efficiently in international markets (see Chapter 10).

High priority should be given to investments in infrastructure to lower transaction costs for exports, and to investments that help enhance the quality of goods and allow developing countries to meet rising quality standards in international markets. Such investments could be most beneficial for products where developing countries have a comparative advantage, such as fruit and vegetables.

Future developments in the international trade policy agenda will be strongly affected by the speed and extent of farm policy reform in OECD countries. Most developed countries are currently modifying the method of giving protection to farmers in the direction of less trade distortion, although overall support levels remain high. But the crucial questions are to ascertain to what extent the reforms will be permanent, and whether they are the manifestation of a new paradigm, which takes government out of the game of supporting commodity prices and making farming decisions.

The next 30 years may also see a shift in focus within the overall trade agenda. As traditional market access barriers such as tariffs and quotas are reduced, the trade agenda is likely to shift away from traditional issues such as export competition or enhancing market access towards trade restrictions caused by safety, quality or environmental standards. Moreover, as a growing share of trade will be handled through ever larger and more transnationally active companies, there will be a growing need to establish global competition rules. A global marketplace will also augment pressures to work on global rules for the protection of IPR and of geographic indications. In parallel with the shift towards these new trade issues, the importance of the various WTO agreements is likely to change. Future trade negotiations will focus more on details in the SPS, TBT or TRIPS Agreements, and less on the rules and regulations set out by the Agreement on Agriculture.

This shift in the focus of the trade agenda will be accompanied by a change in the relative importance of countries within the multilateral trade negotiation process. Hitherto, the importance of new and emerging issues was largely confined to developed countries and thus the agenda has mainly reflected developed countries' trade concerns. Now developing countries are having an increasing influence on WTO and its deliberations. But at the same time the cohesion among developing countries is itself being weakened. Some see advantages in firm rules on intellectual property protection, while others fear that they will lose the ability to pursue traditional farming practices.

Splits have also arisen over traditional trade issues, notably between those countries that import agricultural goods and those whose main interest is in expanding export markets. Some developing countries wish to retain preferential access to developed country markets, while others see such arrangements as mainly harming other developing countries. Over the next 30 years there is a danger that these divisions may become more pronounced. Countries that are not integrating within the mainstream of the world economy may find themselves poorly served by the global trade system. By contrast, those that play a full role in the global economy will increasingly make use of trade facilitation in such areas as services, intellectual property and investment rules.

Globalization in food and agriculture

The preceding chapter focused on the role of trade and trade policies as driving factors for increasingly integrated markets. This chapter on globalization will expand the analysis by identifying the other main factors that drive global economic integration and analyse their main effects on food and agriculture. These are presented in three major sections. The first part provides a definition of the process of globalization, placed in its historical context. Emphasis is given to the importance of factors that reduce transaction costs, notably on the impacts of new transportation and communication technologies. The second section presents the main features of globalization in agriculture, and discusses why some countries have been successful in integrating their food and agricultural economies into the rapidly growing world markets, but also why others have largely failed to do so. This includes factors such as openness to trade and capital flows, ability to adopt external expertise and technologies, but also the importance of factors relating to a country's geographic location or its infrastructure endowment. The third part presents the options, the potential and the limits that developing countries are facing for future integration into global food and agriculture.

10.1 Globalization as an ongoing process

Globalization refers to the ongoing process of rapid global economic integration facilitated by lower transaction costs and lower barriers to movements in capital and goods. It has shown itself in a growing interdependence of the world's economies, rapidly rising trade flows, increases in capital movements and an increasing internationalization of production, often organized within and between multinational corporations. To a large extent, globalization has been brought about by a massive reduction in transaction costs, which in turn was made possible through more efficient transportation and communication facilities as well as innovations in organizing complex logistical processes. Trade and capital flows have also been boosted by a systematic reduction in trade and investment barriers. This process has brought about massive income gains for those who participated. Broadly, the integration into a larger and more competitive market has increased the returns to investment for producers and provided consumers with a greater variety of products at lower prices.

This process of growing integration of the world economy has also given rise to numerous concerns. Most prominent is the concern about the growing marginalization of entire countries or societal groups within countries. There are in fact many countries that have been left out of the overall economic integration and growth process, in some cases, despite sincere efforts to open up to foreign trade and capital flows. Over the 1990s, rapidly integrating economies recorded a per capita income growth rate of more than 4 percent p.a. while the income available per person in less integrated countries shrank by 1 percent annually (World Bank, 2001e). The rapid growth in agricultural trade has given rise to concerns that diseases and pests will be hard to control and contain at the local level. Moreover, there are sociocultural concerns that globalization could destroy the cultural heritage (including dietary habits) as well as traditional societal and social links that have evolved over centuries. Finally, there is widespread concern about a growing economic, social and cultural dependence on a few dominant countries or corporations, which are seen to have the potential of disempowering entire societies.

While the term "globalization" has been coined only recently, its driving forces and its principal impacts are of older date. Similar developments, albeit of a more limited scope, have characterized global economic development in the past. In particular, innovations that reduce transaction costs (better transportation and communication technologies) have always had a strong accelerating impact on global integration. A look back also suggests that the process of economic integration is a non-continuous one. It is often a process of waves that occur when new technologies are widely adopted around the world. Similarly, trade and investment liberalization is being negotiated and implemented in rounds. The two developments, technological change and liberalization, can be mutually reinforcing and create particularly noticeable waves of globalization. The current wave of globalization is driven by major technological breakthroughs in transportation and communication technologies (notably the Internet, mobile telephone technology and just-in-time systems) in tandem with various efforts to liberalize international trade and investment flows.

The first wave of globalization during the second half of the nineteenth century. The first wave of rapid global integration began in the second half of the nineteenth century and was brought about by a combination of breakthroughs in transportation and communication technologies. Trade between continents was boosted by the shift from sail to steamships, which resulted in a tremendous decline in transatlantic transportation costs as well as faster and more reliable connections. Trade in agricultural commodities, typically bulky, perishable or both, enjoyed a particular boost. Transatlantic trade in grains and oilseeds – previously circumscribed by high transaction costs – shot up sharply. This brought new land into production, most notably in the Midwest United States and some parts of Australia.

Agricultural trade was further accelerated by the advent of the railways, which resulted in a further and sharp reduction of transportation costs within continents. Lower transaction costs heightened competition and brought about not only a significant downward pressure on prices, but also a growing convergence of commodity prices across continents. For example, in 1870, a bushel of wheat sold for 60 cents in Chicago but for double that price in London. The difference was largely a result of high transportation costs from Chicago to London. With railroads and steamships,[4] transport rates between Chicago and London fell to 10 cents a bushel between 1865 and 1900 and the price differentials for wheat declined accordingly (Henderson, Handy and Neff, 1996).

The decline in transaction costs also had significant impacts on the overall volume of intercontinental trade, market shares and income. United States exports of grain and meat to Europe, for instance, increased from US$68 million in 1870 to US$226 million in 1880, which boosted farmers' incomes in the United States and the welfare of consumers in Europe. The new transportation facilities also reduced costs for internal shipments and, together with cheaper food supplies from abroad, increased food security at the local and regional level. For the first time in history, this brought about years of "lower agricultural production without famine" in Europe (Tilly, 1981).

---

[1] Between 1850 and 1913, global overseas transportation capacity increased by more than 500 percent. At the same time, tankers and vessels with cooling facilities vastly expanded the range of products exchanged within and across countries and continents.

Lower transportation costs also affected labour mobility and labour costs. Sixty million people migrated from Europe to North America and Australia to farm the newly available land. As land was abundant, it created an opportunity for many immigrants to earn an income that exceeded by far the wages they used to earn in Europe. In Europe itself, in turn, it created a relative labour shortage and an upward pressure on wages both in absolute terms and relative to land prices. Overall, immigration led to a narrowing of differences in wages in all globalizing regions. "Emigration is estimated to have raised Irish wages by 32 percent, Italian by 28 percent and Norwegian by 10 percent. Immigration is estimated to have lowered Argentine wages by 22 percent, Australian by 15 percent, Canadian by 16 percent and American by 8 percent" (Lindert and Williamson, 2001). Indeed, migration was probably more important than either trade or capital movements.

The backlash after 1914. With the end of the First World War, trade policy went into reverse and many nations stepped up border protection. The increase in tariffs was built on the notion that higher protection would help rebuild the domestic industries that had suffered or were destroyed during the war. The process started in Europe. France, Germany, Spain, Italy, Yugoslavia, Hungary, Czechoslovakia, Bulgaria, Romania, Belgium and the Netherlands all raised their import tariffs to levels comparable to those before the war. Even the United Kingdom, a free trade nation, declared that "new industries since 1915 would need careful nurturing and protection if foreign competition were not again to reduce Britain to a technological colony".

In June 1930, when the United States Congress passed the Hawley-Smoot Tariff Act, the United States joined in the new protectionist wave. Agricultural tariffs were increased particularly sharply, both in absolute terms and relative to industrial ones. In reaction to the sharp increase in United States tariffs, other countries enacted retaliatory trade laws. The spiralling tariff increases put a brake on global trade and reversed much of the liberalization that resulted from the first wave of globalization. Between 1929 and 1933, United States imports fell by 30 percent and, more significantly, exports fell by almost 40 percent. The sharp decline in trade aggravated the internal economic situation, and the depression in the United States intensified and engulfed much of the rest of the then economically integrated world.

The second wave of globalization, 1945-80. The experience gathered from the reversal to protectionist policies during the interwar period gave an impetus to a new wave of internationalism after the Second World War. The new wave of trade liberalization was, however, more selective both in terms of countries participating and products included. By 1980, developed countries' barriers to trade in manufactured goods had been substantially removed, but barriers for developing countries' agricultural products had been lowered only for those primary commodities that did not compete with agriculture in the developed countries. By contrast, most developing countries had erected trade barriers against imports from each other and from developed countries.

The resulting effect on trade flows was very uneven. For developed countries, the second wave of globalization was a spectacular success. Freer trade between them greatly expanded the exchange of goods. For the first time, international specialization within manufacturing became important, allowing scale economies to be realized. This helped to drive up the incomes of the developed countries relative to the rest of the world (World Bank, 2001e). For developing countries, it perpetuated the North-South pattern of trade, i.e. the exchange of manufactures for land-intensive primary commodities, and this impeded them in exploiting their comparative advantage in labour-intensive manufactures. In addition, as discussed below, many developing countries adopted a policy approach that was not conducive to a greater integration into the globalizing world economy.

The economic policy approach adopted in many developing countries during the 1950s and 1960s was strongly influenced by the work of Raul Prebish. Under the assumption of balanced trade and price stability, Prebish established the following relationship between the relative growth rates of an economy *vis-à-vis* its trade partners and the income elasticities for its exports and imports: $g_i / g_w = e_x / e_m$, where g_i and g_w are the trend growth rates of the economy and the rest of the world, and e_x and e_m the export and import income elasticities.

The policy message from this relationship was straightforward: if a country wants to grow more rapidly than the rest of the world, its export elasticity needs to be higher than its import elasticity. The actual situation in developing countries, however, was precisely the reverse. Typically, they exported primary goods with low income elasticities and imported manufactured products with high income elasticities. As a result, growth without a balance-of-payment constraint was assumed to be impossible without a continuous depreciation of the real exchange rate or the steady accumulation of foreign debt. This so-called "elasticity pessimism" was the main rationale behind the import substitution policies of this period.

For much of the 1950s and 1960s, import-substituting industrialization (ISI) was seen as a way out of this deadlock. ISI was based on the idea that domestic investment and technological capabilities can be spurred on by providing home producers with (temporary) protection against imports. Whether and to what extent ISI helped or hindered development remains controversial. On the one hand, the so-called "'consensus view" emphasizes that ISI policies were at the heart of the problems that many of their adopters encountered in the subsequent decades when they opened up their economies (for example, see OECD, 2001c). On the other hand, there are claims (Rodrik, 1997; Hausmann and Rodrik, 2002) that ISI worked reasonably well, notably in raising domestic investment and productivity. It has been stressed that numerous economies in Latin America and the Near East recorded robust growth under ISI policy regimes.

There is, however, a broad consensus that ISI was an ineffective response to weather the economic turbulence of the 1970s, which witnessed the abandonment of the Bretton Woods system of fixed exchange rates, two major oil shocks and other commodity boom-and-bust cycles. For agriculture, ISI strategies meant higher input costs and therefore negative effective protection, i.e. implicit taxation. In conjunction with explicit taxes on output and exports, ISI strategies created a considerable burden for agriculture in many developing countries, put a brake on agricultural export growth and slowed their integration into global agricultural markets. On average, for the period from 1960 to 1984, the bias against agriculture depressed the domestic terms of trade for agriculture by 30 percent. In the extreme cases of Côte d'Ivoire, Ghana and Zambia, the average bias against agriculture reached levels of 52, 49 and 60 percent, respectively (Schiff and Valdes, 1997).

The current wave of globalization. The last two decades of the twentieth century marked the beginning of a new wave of globalization. Like the first wave about a hundred years earlier, it was brought about by a combination of lower trade barriers and numerous technological innovations that strongly reduced transaction costs for movements not only of goods but also of people and capital. This is particularly apparent from the substantial increase in international migration and capital movements, which were of less importance during the second wave of globalization. Unlike its predecessors, this wave of globalization included many more developing countries, even though not all of them were able to harness globalization to their benefit. Particularly countries in sub-Saharan Africa failed to participate, resulting in a further widening of their income gap with both integrating Asian economies and, even more so, the fully globalized economies of the North.

Most countries in East Asia were able to reap substantial benefits by exploiting their comparative advantage of cheap and abundant labour. Some countries in Latin America and the Near East/North Africa region were also able to integrate fast. A common feature of successful integrators is an above-average shift towards exports of manufactures. Countries such as China, Bangladesh and Sri Lanka already have shares of manufactures in their exports that are above the world average of 81 percent. Others, such as India, Turkey, Morocco and Indonesia are swiftly approaching the world average. Another important change in the exports of successfully globalizing developing countries has been their substantial increase in exports of services. In the early 1980s, commercial services made up 17 percent of the exports of developed countries, but only 9 percent of the exports of developing countries. During the third wave of globalization, the export share of services in the former group increased slightly, to 20 percent, but for developing countries the share almost doubled to 17 percent (World Bank, 2001e).

10.2 The main features of globalization and the correlates of success

10.2.1 Freer trade and outward-oriented policies

Chapter 9 has already dealt with the main developments in global agricultural trade, its importance within overall trade and the structural changes that have taken place over the past 40 years. It also provided an overview of likely trade developments for the next 30 years and the trade policy issues that are expected to arise from the projected shifts in trade flows. In this section, emphasis is placed on the potential role of trade for development and poverty alleviation.

The links between trade, development and poverty have been subject to an extensive and heated debate. While proponents and opponents agree on the central importance of freer trade for increasing global welfare, there is considerable disagreement as to whether and to what extent freer trade can be harnessed by individual countries as a means to promote development and fight poverty. There is also considerable disagreement as to how the transition towards freer trade, i.e. the speed, timing and sequencing of liberalization measures, should evolve. Some of these issues will be addressed in the following section.

The consensus view. Economists have been asserting for a long time that trade liberalization is good for economic development, particularly in developing countries. The benefits from openness are assumed to arise from the efficiency gains that flow from superior resource-allocation decisions in more open markets (Bhagwati and Srinivasan, 1999). The result is an increase in economic growth. More recently there have also been numerous empirical studies that suggest that openness to trade and investment flows has had a positive effect not only on economic growth but also in helping to fight poverty. Among the most influential empirical studies are those by Edwards (1998) and by the World Bank (Dollar and Kraay, 2000, 2001). Wolf (2000) summarizes much of this literature.

In view of its importance for the ongoing policy debate, the main conclusions of the World Bank study are summarized below. The first concerns the link between growth and openness. Dollar and Kraay examine this relationship using an econometric study covering a sample of 72 developing countries. Avoiding some of the pitfalls of earlier studies by using a single indicator of openness (the ratio of trade to GDP), the authors arrive at a number of important conclusions:

- Weighted for population, the per capita income of the group of "globalizers" grew at 5 percent a year in the 1990s, compared with 1.4 percent for the "non-globalizer" group.
- Growth rates for the globalizers have been steadily increasing since the mid-1970s, while those for the non-globalizers fell sharply in the 1980s and recovered only marginally in the 1990s.
- Per capita income among the globalizers is rising more than twice as fast as in industrialized countries, while the non-globalizers are falling further behind. On a population-weighted basis, countries that are open are growing 3.6 percent a year faster than others. On this basis, average income in a globalizing economy would double every 14 years, compared with 50 years in a non-globalizing economy: a growth gap that would have profound implications for poverty reduction.

The second conclusion concerns the relationship between economic growth and poverty reduction. On the basis of an econometric exercise analysing economic growth in 80 countries over a period of four decades, it is argued that, on average, the income of the poor rises on a one-to-one basis with overall growth. In other words, poor people capture a share of any income increment that reflects their existing share of income distribution. As the authors say: "It is almost always the case that the income of the poor rises during periods of significant growth" (Dollar and Kraay, 2001).

On closer inspection, however, some of the numbers look less impressive. One reason for this is that averages have the effect of obscuring important differences between countries, especially when samples are weighted for population (since this means that large countries such as China have a disproportionate influence). Using an unweighted average, the per capita growth rate for the globalizers in the 1990s falls to 1.5 percent. Moreover, ten of the 24 countries in the group have growth rates of 1 percent or less. Further disaggregation reveals that one-third of the "globalizing" countries have lower average growth rates for the 1990s than the "non-globalizing" group.

The critique of the consensus view. The basic critique of the consensus view is that the link between openness and growth is one of correlation but not, or at least not necessarily, one of causation. Simply put, openness is essentially an economic outcome, captured (in the case of the World Bank study) by the ratio of trade to GDP, but not an input, i.e. a policy tool to arrive at higher growth.[2]

When focusing on the causal relationship between trade policy, growth and poverty reduction, the critics of the consensus view claim that it appears to be an upside-down version of reality (Rodrik, 2001 and Oxfam, 2002). In fact, they stress that some of the most successful globalizers are anything but radical liberalizers, while many of the most radical liberalizers have actually achieved very little in terms of economic growth and poverty reduction. They claim that no country has ever developed simply by opening itself up to foreign trade and investment and that practically all of today's developed countries embarked on their growth behind tariff barriers, and reduced protection only subsequently (Rodrik, 2001).[3]

There are also many examples in agriculture, where appropriate domestic policy settings and the timing and sequencing of liberalization steps have proved to be more important than a complete and immediate reduction of border protection. Some of today's most successful agricultural exporters (e.g. China and Viet Nam) established their international competitiveness under protection and import substitution regimes and embarked subsequently on "policy reforms".[4] In many cases, success was built on a promotion of export-led growth combined with a domestic investment and institution-building strategy to stimulate entrepreneurship and the willingness to assume risks. Another important factor has been that mechanisms are put in place to ensure

that excess capacities are cut back, and to create exit possibilities for non-performing sectors or actors, and that the opening-up process to international competition is phased in a determined manner (for examples, see below).

Notwithstanding the importance of temporary trade protection measures, the proponents of fast and full liberalization stress that no country has developed successfully by turning its back on international trade and long-term capital flows. Very few countries have grown over long periods of time without experiencing an increase in the share of foreign trade in their national product. In practice, it is hard to imagine that a country can create and sustain growth if it remains shut off from the forces of competition that help to innovate and upgrade its productivity. Moreover, it is equally hard to imagine that developing countries would not benefit from imported capital goods that are likely to be significantly cheaper than those manufactured at home. Policies that restrict imports of capital equipment raise the price of capital goods at home and thereby reduce real investment levels. Exports, in turn, are important since they permit the purchase of imported capital equipment.

The agricultural sector in many developing countries has been particularly adversely affected by the inward-oriented industrial development strategies of the 1950s and 1960s. In some countries the anti-agriculture bias remained a policy feature throughout the 1970s and 1980s (Schiff and Valdes, 1997). Import substitution policies for manufactures restricted capital good imports for agriculture, raised input costs and resulted in often significant negative effective rates of protection. This held back real investment levels in agriculture and slowed export performance in many developing countries. In some developing countries, industrial protection and restrictions on capital good imports for agriculture were accompanied by direct taxation of agricultural exports, placing agriculture at a disadvantage both relative to other sectors and vis-à-vis developed country competitors.

---

[2] Dollar and Kraay acknowledge this possibility, when declaring that "we use decade-over-decade changes in the volume of trade as an imperfect proxy for changes in trade policy" (Dollar and Kraay, 2001).

[3] Bussolo and Lecomte (1999) also stress that trade policy theory does not unambiguously suggest that protection has a negative impact on growth in developing countries. However, they emphasize that those countries that apply more open trade regimes, together with fiscal discipline and good governance, have enjoyed higher growth rates than those implementing restrictive policies. An open and simple trade policy can foster some external discipline, helping to reduce distortions on domestic markets, and to narrow the scope for wrong or unbalanced policies in other areas, as well as for rent-seeking and corruption that do not normally favour the poor.

[4] For much of the 1970s and 1980s, developing countries' agriculture was heavily discriminated against (see Chapter 9). Wherever and as long as agriculture was taxed, either directly or through macroeconomic measures, development slowed and international integration suffered.

Openness and development in agriculture – some country examples. *Viet Nam*'s rapid economic and agricultural development over the 1990s is now commonly regarded as one of the most successful development stories of the last decade. Annual GDP growth has been consistently high throughout the 1990s, averaging 7.6 percent. Over the same period, agricultural output has been growing at almost 5 percent per year, far outstripping demand in local markets (Government of Viet Nam, 2001). Poverty has declined substantially and the number of undernourished has dropped by 3 million people (FAO, 2001a).

Export markets provided an important source of demand to sustain growth. Over the 1990s, the value of agricultural exports shot up by a factor of 3.5 and, for a number of commodities such as coffee and rice, Viet Nam emerged as a leading exporter in world markets. By the end of the 1990s, rice and coffee exports combined generated about US$2 billion in foreign exchange earnings (1997/99 average), accounting for nearly 20 percent of the country's total merchandise exports.

The foundations for Viet Nam's rapid integration into the global market were laid in 1986 with the introduction of *Doi Moi*, Viet Nam's economic renovation programme. At the heart of the reform was a decollectivization process, through which farming families received most of the land. In tandem, farmers were allowed to increase sales to the market and agricultural taxes were reduced. Agriculture also benefited from other fiscal reforms, the creation of a Treasury system, and the reform of the banking system, which created a secure deposit base and allowed fiscal operations deep into the country's rural areas. These measures had a profound effect on society, encouraging entrepreneurship and willingness to take risk. Finally, *Doi Moi* offered "return" options to workers in the new factories, thereby reducing risk for internal migrants and further accelerating the fast development of rural areas.

There is no doubt that the success of the 1990s was also promoted by a growing openness in the global trading environment, in which Viet Nam's export performance benefited from declining tariffs and non-tariff barriers. As in many coun-tries, Viet Nam's economy also enjoyed all other benefits of globalization, such as cheaper and faster transportation and communications. However, while benefiting from improved market access abroad, Viet Nam was slow in removing its own border protection or its trade-distorting subsidies. Particularly, agricultural import tariffs have been raised repeatedly over the 1990s (see, for example, USDA, 1999c, 1999d and 2001b)[5] and subsidies have been provided with the aim of increasing agricultural production and exports. Fforde (2002) even maintains that the initial fast liberalization process in the early 1990s did not allow the country to build up enough expertise and competitiveness and put a brake on overall growth.

Policies also played an important role in managing the 2000/02 coffee crisis that severely affected large parts of Viet Nam's thriving agricultural sector. For example, a sizeable support programme was launched to help coffee growers regain international competitiveness. The programme includes subsidies to upgrade coffee quality and to reduce production costs. It promotes smaller, less centralized processing factories and warehouses suitable for the many different coffee-producing regions (USDA, 2001b) and supports the creation of overall and agricultural infrastructure and the shift towards improved coffee varieties. But the new policy package also initiated a rationalization process within Viet Nam's coffee economy. Changes in eligibility for the existing soft loan programme are probably the most important efforts in this context. Under the revised scheme, credit subsidies will not be offered to low-yield producers or inefficient operations, but only to potentially profitable farmers. In parallel, special preferences have been given to participating farmers to switch to arabica coffee or to improve their operation's effectiveness (USDA, 2001b).

Overall there are probably three important features that have contributed to the success of the coffee policy. First, policies play an active role in promoting production, particularly production for exports. Second, support is not an open-ended government commitment but is limited to kick-starting the process and helping the sector discover where its comparative advantage lies. And third,

---

[5] Viet Nam undertook, for instance, a far-reaching tariffication exercise in 1999 (effective December 1999), converting many non-tariff barriers into tariffs, in line with World Bank prescriptions. This process was in general accompanied by an increase in effective border protection, with tariff rates between 30 and 100 percent. Detailed tariff schedules are, for instance, available from USDA (1999c, 1999d).

once competitiveness is established, policies focus on the competitive producers and decline support to non-competitive ones. In doing so, competitive producers are helped through the price troughs while the non-competitive ones are encouraged to exit from the sector.[6]

A fairly similar set of reforms in China in the late 1970s set the stage for the impressive economic performance that has been the envy of poor countries since. Per capita GDP (at current prices) increased by a factor of nine and the value of exports by a factor of ten. Agricultural output tripled, as did agricultural exports, while the number of undernourished declined by 76 million people (from 1990/92 to 1997/99). In fact, China was the single largest contributor to the reduction of undernourishment during the 1990s, accounting for two-thirds of the progress made in fighting hunger (FAO, 2001a).

This rapid development process started with fairly simple initial reforms in the agricultural sector. The communal farming system was loosened and the so-called household responsibility system was introduced, allowing farmers to sell their crops on the free market once they had fulfilled their quota obligations to the state. The government remained actively involved in agricultural policy formulation and implementation. The overall process can be best described as one of active experimentation, in which production expanded rapidly under administrative pressure to fulfil production quotas, as well as under production incentives through input subsidies (water, fertilizer).[7] In tandem, policies were put in place that promoted the adaptation of new technologies from abroad to the domestic production environment (particularly the high-yielding varieties of the green revolution) which, over time, even enabled domestic researchers to take the lead in developing new applications (hybrid rice, etc.).[8] The importance of adopting external knowledge and technologies is discussed in Section 10.2.3 below. Finally, domestic policies also encouraged the exit from agriculture of unproductive farmers. These measures include the creation and promotion of township and village enterprises (TVEs) that helped absorb the excess labour of rural areas or, more recently, massive investments in rural infrastructure to reduce transaction costs and increase competitiveness of farmers and food processors in China's hinterland.

Unlike Viet Nam and China, sub-Saharan Africa largely failed to take advantage of the growing trade opportunities in global markets. Its share in global exports, for instance, dropped from 3.1 percent in the mid-1950s to 1.2 percent in 1990. This corresponded to an annual loss in export earnings of about US$65 billion. In trying to identify the contributing factors to this decline, a World Bank study (Amjadi, Reinke and Yeats, 1996) found that trade barriers abroad have not had a significant influence. On the contrary, once preferences were taken into account, tariffs conveyed significant competitive advantages over competing goods shipped from some other regions, and were even a positive factor for the location of commodity processing in Africa as opposed to some other foreign locations.[9] Similarly, non-tariff barriers (NTBs) of markets abroad did not account for Africa's poor export performance. In fact, the share of Africa's exports subject to NTBs (11 percent) is less than half the average for the group of developing countries.

To draw general lessons from a few success stories is difficult. Nonetheless, there are a few commonalities that characterize successful globalizers. To begin with, all of them have both outward-oriented policies and domestic produc-

---

[6] Viet Nam's active policy engagement in promoting and disciplining production may also be regarded as a special case of policies that have been pursued elsewhere in East Asia. "Where Korea differs from other developing countries in promoting big business, was the discipline the state exercised over these chaebols by penalizing poor performance and rewarding only good ones … The government as the controller of commercial banks was in a powerful position to punish poorly managed firms by freezing bank credit. As a result only three of the largest 10 chaebols in 1965 – Samsung, Lucky-Goldstar and Ssangyong – remained on the same list 10 years later. Similarly, seven of the largest 10 in 1975 remained on the same list in 1985" (Kim, in Nelson, 2000). The Korean Government was quick to shelve its plans for supporting particular firms or industries when new information suggested that productivity would lag (Westphal, 1981, p. 34).

[7] It should be noted that the system of incentives to increase or slow output, and even to leave or stay in agriculture, was accompanied by a rigid system of administrative measures that may not be at the disposal of policy-makers in market economies.

[8] Agricultural policies were accompanied by non-agricultural policy measures that aimed to facilitate structural change and gradually to liberalize the non-agricultural sector. The most important measures were the creation of township and village enterprises (TVEs), the extension of the "market track" into the urban and industrial sectors, and the creation of special economic zones to attract foreign investment.

[9] The authors warn that Africa may experience some losses through the Uruguay Round erosion of these preferences, although such losses should not be large.

tion incentives. Moreover, freer trade regimes are adopted after or in parallel with domestic policy reforms. The country examples also suggest that openness *per se* is unlikely to be a sufficient condition for a successful integration into the global economy. More important seems to be (i) that farmers can operate in the appropriate domestic incentive system; (ii) that the incentives are reduced where unproductive excess capacity is created and exit policies are in place; and (iii) that adjustment and reallocation costs are minimized, e.g. through appropriate timing, sequencing and pacing of policy measures.

10.2.2 Freer movement of capital and the emergence of transnational companies

Alongside the expansion of trade flows, another feature of globalization has been the rapid growth in international capital flows. Transnational corporations (TNCs) have been the driving force behind this rapid development and foreign direct investment (FDI) is the main instrument through which TNCs expand their reach beyond national boundaries. Through FDI, TNCs affect production levels and composition, production technologies, labour markets and standards, and eventually also trade and consumption patterns. Through their control over resources, access to markets and development

of new technologies, TNCs have the potential to integrate countries into global markets.

Foreign direct investment: level, flows, and distribution. Between 1989/94 and 2000, annual global FDI inflows increased more than sixfold, from US$200 billion to US$1 270 billion (UN, 2001c). The growth in FDI exceeded by far the growth in trade flows. Between 1991 and 1995 the average annual growth rate of FDI was 21 percent compared with 9 percent for exports of goods and non-factor services. Between 1996 and 1999, the difference increased, with FDI growing at an average rate of 41 percent and exports growing at 2 percent. In 2000, total sales of foreign affiliates amounted to US$16 trillion, compared with world exports of goods and non-factor services of US$7 trillion. Developed countries absorbed the major part (80 percent) of the FDI inflows but also accounted for a similar proportion of outflows.

As an increasing number of countries integrated into the global economy, FDI flows also became more evenly distributed and reached more countries in a substantial manner (UN, 2001c). By 2000, more than 50 countries (24 of which were developing) had accumulated an inward FDI stock of more than US$10 billion, compared with only 17 countries 15 years earlier (seven of them developing countries). The picture for outward FDI is similar: the

Table 10.1 Regional distribution of FDI inflows and outflows (billion US$)

	FDI inflows		FDI outflows	
	1989/1994	2000	1989/1994	2000
Developed countries	137.1	1 005.2	203.2	1 046.3
EU	76.6	617.3	105.2	772.9
Japan	1.0	15.8	9.0	32.9
United States	42.5	8.2	49	139.3
Other	17.0	363.9	40.0	101.2
Developing countries	59.6	240.2	24.9	99.5
Africa	4	8.2	0.9	0.7
Latin America and the Caribbean	17.5	86.2	3.7	13.4
Asia	37.9	143.8	20.3	85.3
Other	0.2	2.0	0.0	0.1
Central and Eastern Europe	3.4	25.4	0.1	4.0
World	200.1	1 270.8	228.3	1 149.9

Source: UN (2001c).

number of countries with stocks exceeding US$10 billion rose from ten to 33 (now including 12 developing countries, compared with eight in 1985) over the same period. In terms of flows, the number of countries receiving an annual average of more than US$1 billion rose from 17 (six of which were developing countries) in the mid-1980s to 51 (23 of which were developing countries) at the end of the 1990s. In the case of outflows, 33 countries (11 developing countries) invested more than US$1 billion at the end of the 1990s, compared with 13 countries (only one developing country) in the mid-1980s.

A closer look at the regional distribution, however, reveals that there is still a high concentration of FDI flows within developing Asia (Table 10.1). More than half of all FDI went to Asian economies, and within Asia, East and South Asia accounted for almost the entire inflow. At the other end of the scale, FDI inflows to Africa have remained minimal. While doubling in absolute terms, the continent's share in total inflows to developing countries fell by half, between 1989-94 and 2000, from 6.8 to 3.4 percent (Table 10.1).

TNCs and FDI in food and agriculture. The basis for the large TNCs that dominate today's global food economy was laid with the market concentration process in developed countries. In the United States, for instance, four meat-packing firms have traditionally controlled about two-thirds of the beef supply, and by the mid-1990s over 80 percent of the beef supply was controlled by four firms (OECD, 2001d). High levels of firm concentration also characterize the retail food distribution system in other OECD countries. For example, in Australia, over 75 percent of the retail food distribution system is controlled by three firms.

As the domestic markets for their products became increasingly limited, these large food processors extended their operations in two principal directions. First, they extended their reach "vertically" by taking over the principal operations along the food chain. The final result of this process is often a fully vertically integrated company with operations that cover the entire food chain from the "farmgate to the dinner plate". Second, they expanded horizontally, i.e. they extended their reach by branching into foreign markets. The combined process of horizontal expansion across countries and vertical integration within the company created the typical TNC in food and agriculture. These TNCs are frequently referred to as "food chain complexes" or "food chain clusters".

The three most advanced food chain clusters are Cargill/Monsanto, ConAgra and Novartis/ADM.[10] ConAgra, for example, one of the three largest flour millers in North America, ranks fourth in corn milling. It produces its own livestock feed and ranks third in cattle feeding, second in slaughtering, third in pork processing and fourth in broiler production. United AgriProducts is part of ConAgra and sells agrochemicals and biotechnology products (seeds) around the world. The conglomerate also owns its own grain trading company (Peavey). At the retail level it widely distributed processed foods through such brands as Armour, Swift and Hunt's, and is second only to Philip Morris as a leading food processor. The Novartis/ADM cluster also connects the different stages of food production from genes/seeds (Novartis and Land O'lakes) to grain collection (ADM) to processing across the globe from Mexico, the Netherlands, France, China and the United Kingdom. Alliances with IBP, the largest United States beef packer and second largest pork packer, extend its influence down the food chain (Heffernan, 1999).

A more recent feature within the process of vertical integration is that the food chain complexes have extended ownership and control from the agricultural downstream sector (food processing and marketing) into strategic parts of the upstream system. For instance, it is estimated that only three firms control over 80 percent of US maize exports and 65 percent of US soybean exports; only four firms control 60 percent of domestic grain handling and 25 percent of compound feed production (Hendrickson and Heffernan, 2002). While market concentration in certain parts of a country's food system is a well-established feature in many countries, these complexes have extended their influence across country borders and have created vertically integrated or coordinated production chains across the globe (OECD, 2001d).

---

[10] All of them are located in, and operate globally from, the United States. They control, through joint ventures and strategic alliances, important parts of the food industry that range from seeds to processed products such as meats, seafood and other foods (OECD, 2001d).

Table 10.2 shows the implantation of agrofood TNC subsidiaries from different home regions of the parent company arranged by host region of the subsidiary, i.e. how much and where TNCs have spread out their activities. It shows that most TNCs in the food industry operate from a western European or United States home base. Together they account for about 84 percent of all TNCs that have invested in markets abroad. Those from Asia are largely found in Asia, although there are significant numbers located in the EU and North America, and those from Latin America are predominantly in other Latin American countries. Europe and North America are both the home and the hosts to the vast majority of TNC subsidiaries, their stage of development acting both as push and pull forces. TNCs from the EU and the United States have, to a significant extent, also established foreign affiliates in developing countries. In both cases, Asia and Latin America are the most important destinations. TNCs from western Europe, for instance, have nearly as many foreign affiliates in Asia or Latin America as they have in North America. By contrast, Africa is home to very few subsidiaries, and those it has are almost entirely located within other African countries.

TNCs in food and agriculture: help or hurdle for rural development? The general view among experts – in developed and developing countries alike – is that FDI is a powerful catalyst for overall economic development. A number of recent publications (World Bank, 2001e and UN, 1999b) have documented the benefits that FDI can create for development. The 1999 issue of the *World Investment Report* (UN, 1999b) identifies five major advantages that are carried along to the host country alongside inflows of FDI: access to capital,[11] access to technology, access to markets, enhanced skills and management techniques and help to protect the environment

The UN report stresses that developing countries have vastly benefited from the rapid increase in FDI inflows during the 1990s, particularly through added productivity growth. Several other sources underline and quantify the potential that TNCs have in generating productivity gains (e.g. Sachs and Warner, 1995; Baily and Gersbach, 1995).[12] Baily and Gersbach (1995) stress that the potential for productivity gains is particularly large where TNCs reinvest profits in the host countries, create forward and backward linkages with local firms, upgrade the performance of a country's firms

Table 10.2   Number of subsidiaries of the 100 largest TNCs by region (1996)

Home region	Host region						Total
	Africa	Latin America and the Caribbean	North America	Asia	Eastern and central Europe	Western Europe	
Africa	58	0	0	1	3	2	64
Latin America	8	45	14	5	0	49	121
North America	52	390	1 295	234	114	818	2 903
Asia	9	37	103	587	1	90	827
Western Europe	84	233	312	268	104	1 948	2 949
Australasia	1	8	5	25	0	46	85
Total	212	713	1 729	1 120	222	2 953	6 949

Source: Agrodata (2000).

[11] The catalyst for the East Asian financial crisis in 1996 was a huge outflow of funds, as commercial banks and institutional investors called in loans. The resulting losses were equivalent to more than 10 percent of GDP for some countries (based on data in IMF, 1999). By contrast, FDI remained constant throughout this period.

[12] This is partly because of increased competition, partly a demonstration effect "…when companies based in one country set up operations in another, they carry with them the production processes and productivity levels of their home country" (Baily and Gersbach, 1995, p. 309).

through the provision of superior expertise and technologies and hence boost growth (Box 10.1). Also to be remembered is that TNCs are the world's chief repository of economically useful skills and knowledge and that technology flows are increasingly important components of FDI (UN, 1999b).

Despite the vast potential of FDI for rural development, there are a number of reasons to suggest that simply opening up a country's border to FDI may not be the best way to reap the benefits. There are substantial differences in the "quality" of FDI flows and governments may have to intervene in the process of channelling FDI. Furthermore, the complexity of the FDI package means that governments face trade-offs between different benefits and objectives. For instance, they may have to choose between investments that offer short- as opposed to long-term benefits; the former may lead to static gains, but not necessarily to dynamic ones. Moreover, the level of FDI inflows to developing countries can easily be overstated. TNCs can repatriate much of the profits that they produce from their investments in developing countries. In sub-Saharan Africa, for instance, an average of 75 percent of the profits have been repatriated annually between 1991 and 1997 (IMF, 1999). Also the costs of attracting FDI through tax revenues foregone in the host country can be substantial[13]– costs that need to be counted against the benefits these inflows bring.

There are also concerns that TNCs abuse their market power, add downward pressure on rural wages and disempower farmers through unfair contractual arrangements. These concerns often arise from the notion that the linkages between farmers and TNCs are based on contracts between unequal parties, one party consisting of a mass of unorganized small-scale farmers with little bargaining power and few of the resources needed to raise productivity and compete commercially, and the other party being a powerful agribusiness, offering production and supply contracts which – in exchange for inputs and technical advice – allow it to exploit cheap labour and transfer most risks to the primary producers. This imbalance in negotiating power has been described extensively for the international cocoa and coffee markets where smallholder farmers are at the starting-point of "buyer-driven supply chains" (e.g. Ponte, 2001; see also Box 10.2). Such an imbalance in negotiating power can affect the distribution of benefits along the food chain (Talbot, 1997; see also Ponte, 2001; Gibbon, 2000; and Gereffi, 1994). For example, the share of income retained by coffee producers dropped from 20 percent in the 1970s to 13 percent in the 1990s and is likely to have dropped further with the dramatic price decline for green coffee in 2001/02.

But it should also be noted that there have been important domestic factors that have squeezed the profit margins for producers. Some developing countries have creamed off farmers' profits through export taxes, export controls and mandatory sales. For instance, for much of the 1970s and early 1980s

---

Box 10.1 TNCs can be the source of major productivity gains

Baily and Gersbach (1995) carried out a comparison of labour productivity in Japan, Germany and the United States for a number of manufacturing sectors, including food and beer. The United States was most productive in both of these sectors, food productivity in Germany reaching 76 percent of the United States level while in Japan it was only 33 percent. For beer the comparable figures were Germany 44 percent, and Japan 69 percent. They relate relative productivity to a globalization index and find a significant positive relationship – high globalization leads to high relative productivity. The globalization index is a complex construct that takes into account the extent of the exposure of a country's firms in a particular industry to the productivity leader's firms, through trade, production by the productivity leader's subsidiaries in the country, or ownership. For food, the globalization index is very low in Japan, but at a medium level in Germany, and rising.

The authors conclude that the entrance of foreign firms is the most significant impetus to productivity upgrading in an industry and that the indirect effects (on local firms and the supply chain) may be more significant than the direct effects.

---

[13] For example, in the second half of the 1990s, the governments of Rio Grande do Sul and Bahia in Brazil gave General Motors and Ford financial packages worth US$3 billion to locate factories in their states (Hanson, 2001).

Box 10.2 The global coffee chain: changing market structures and power

The 1990s saw a number of major changes in the structure of the international coffee market. The main changes include a growing market concentration of trading and roasting companies; a growing product differentiation in high- and low-quality brands; and a redistribution of the value added along the marketing and processing chain.

Growing market concentration. The general deregulation of the international coffee market that followed the end of the International Coffee Agreement (ICA) in 1989 opened the way for a growing consolidation of the market. This process was particularly pronounced for roasting and trading, where a declining number of companies control an increasing part of the market. In 1998, the two largest coffee traders (Neumann and Volcafé) accounted for 29 percent of the total market, and the top six companies controlled 50 percent (Figure 10.1). Amid growing market concentration, some smaller and specialized companies have emerged, focusing on trade in the speciality coffee market (high-quality and specific origins).

Concentration within the group of coffee traders was promoted by higher international price volatility, which increased after the end of the ICA (Gilbert, 1995). Mid-sized traders with unhedged positions suffered considerable losses or found themselves too small to compete with larger traders. As a result, they either went bankrupt, merged with others, or were taken over by the majors. Within the exporting countries, the bureaucracy that was needed to monitor exports and ensure compliance with the quota restrictions of the ICA was no longer needed. Coffee boards and other parastatals that regulated export sales have been dismantled, the capability of producing countries to control exports disappeared and their ability to build up stocks decreased. Despite very low prices, current producer-held stocks are close to their lowest levels in 30 years.

Product differentiation. Despite the overall increase in market concentration there was a product differentiation into a system of first-line and second-line supply, subject to price premia and discounts. The differentiation was created by roasters who declined shipments from countries that could not guarantee a reliable minimum amount of supply. In the case of arabica, this minimum is around 60 000 tonnes a year (Raikes and Gibbon, 2000). Minimum supply requirements have created concerns that minor producers may become increasingly marginalized in the future. In addition, product segmentation will further encourage international traders to engage in major producing markets such as Uganda in order to satisfy their major roaster clients (Ponte, 2001).

Redistribution of the value added. Increased consolidation in the coffee industry has also affected the distribution of total income generated along the coffee chain. Talbot (1997) estimates that in the 1970s an average of 20 percent of total income was retained by producers, while the average proportion retained in consuming countries was almost 53 percent. Between 1980/81 and 1988/89, producers still controlled almost 20 percent of total income, while 55 percent was retained in consuming countries. In the 1990s, the situation changed dramatically. Between 1989/90 and 1994/95, the proportion of total income gained by producers dropped to 13 percent, and the proportion retained in consuming countries surged to 78 percent. The share of income retained by producers in the last three to four years is likely to have dropped further as a result of the current situation of oversupply and low prices for green coffee and the ability of roasters to maintain retail prices at relatively stable levels.

Figure 10.1 Market concentration in the coffee chain

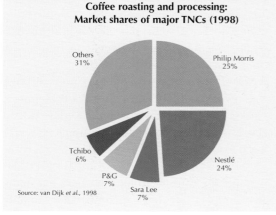

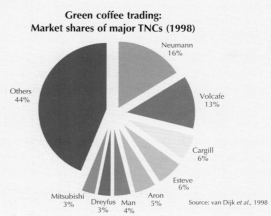

Source: van Dijk et al., 1998

cocoa farmers in Ghana were obliged to sell their crop to the government for as little as a twentieth of the world price (The Economist, 2002a, p. 44). Likewise, profit margins have also been squeezed for Ecuador's banana growers, largely in the absence of TNC activities. Ecuador's government has currently fixed the local price for bananas at US$2.90 per box, whereas the export price is as high as US$17 per box. The low farmgate price "has squeezed farmers' profits to almost nothing" (The Economist, 2002b, p. 54). Unlike Central America, where TNCs own almost all banana plantations, Ecuador's banana economy is dominated by some 6 000 small family farm producers.

Some experience with successful FDI in food and agriculture. The links created between TNCs and domestic firms are crucial factors that determine whether and to what extent a host country benefits from FDI. In the food industry, these linkages are forged between the TNC and the farmers or the local procurement company. The potential for linkage-intensive FDI is particularly substantial in food and agriculture. Linkage-intensive FDI is often the result of the need to process perishable inputs such as milk or fruit and vegetables (UN, 2001c). It can also be forced by logistical bottlenecks or by tariff barriers that make imported goods less competitive. Moreover, TNCs may face restrictions on landownership in many developing countries which can make it necessary for foreign affiliates to rely on domestic producers and to engage in efforts to develop new and upgrade existing suppliers.

These linkages from the foreign affiliate to the national farm sector can provide enormous benefits for farmers and their cooperatives and thus have considerable potential to stimulate rural development. Field research conducted in India (UN, 2001c) provides a number of interesting insights as to how these benefits are generated. It reveals that the four leading TNCs (i.e. Pepsi Foods Ltd, GlaxoSmithKline Beecham Ltd, Nestlé India Ltd and Cadbury India Ltd) on average sourced locally 93 percent of their raw material (tomatoes, potatoes, basmati rice, groundnuts, cocoa, fresh milk, sugar, wheat flour, etc.) and 74 percent of other inputs (such as plastic crates, glass bottles, refrigerators, ice chests, corrugated boxes, craft paper, etc.). Through these linkages the TNCs promoted overall development by means of the

following methods.

- *Collaboration in product development.* All four TNCs are engaged in product development with local research institutes or universities, to develop hybrid varieties of crops and vegetables and new agricultural implements, to alter cropping patterns and to raise productivity. For example, Pepsi Foods has evaluated more than 215 varieties/hybrids of chili, probably the largest scientific evaluation of chilies anywhere. Pepsi's technology in chili cultivation has raised its yield three times. In addition, Pepsi has developed 15 new agricultural implements to facilitate planting and harvesting in India.

- *Technology transfer and training.* New hybrid varieties, implements and practices have been transferred to suppliers (primarily farmers) through farmer training camps. Pepsi provides its contract farmers, free of charge, with various agricultural implements and hybrid seeds/plantlets, as well as process expertise. Cadbury India has a procurement and extension services team that imparts training to potential and existing suppliers in new techniques in planting, harvesting, quality control and post-transplantation care of cocoa crops through technical bulletins, video demonstrations, slides and charts and live demonstrations on the use of various agricultural implements.

- *Introduction of contract farming.* Farmers are contracted to plant the processors' crops on their lands and to deliver to the processors, at pre-agreed prices and quantities of output based upon anticipated yields and contracted area. Towards this end, a processor usually provides the farmers with selected inputs such as seeds/seedlings, information on agricultural practices, regular inspection of the crop and advisory services on crops. Farmers have the choice to leave some part of the output free from the contract arrangement to sell on the open market (see also Box 10.3).

- *Financial assistance* is provided to growers through the involvement of agricultural development banks. For example, GlaxoSmithKlineBeecham acts as a guarantor, enabling its suppliers to take bank loans.

Technology transfer to local farmers has had a positive impact on farm productivity. Tomato

Box 10.3 Formalizing the linkages in agriculture: the importance of contract farming

An extensive study by FAO (FAO, 2001i) brings together numerous examples from many developing countries that confirm the generally positive influence of TNCs on the agriculture of these countries. But the FAO study also shows that policies play an important role in promoting the benefits that TNCs or the local processors can provide for a country's agriculture. Most important, it shows that the underlying contracts between farmers and the company are crucial for success or failure. Numerous examples demonstrate how well-managed contract farming works as an effective tool to link the small-farm sector to sources of extension, mechanization, seeds, fertilizer and credit, and to guaranteed and profitable markets for produce. When efficiently organized and managed, contract farming reduces risk and uncertainty for both parties. The principal benefits laid out in the study are the following.

Increased productivity. In northern India, Hindustan Lever, a food processor, issued contracts to 400 farmers to grow hybrid tomatoes for processing. A study of the project confirmed that production yields and farmers' incomes increased as a result of the use of hybrid seeds and the availability of an assured market. An analysis of the yields and incomes of the contracted farmers compared with farmers who grew tomatoes for the open market showed that yields of the farmers under contract were 64 percent higher than those outside the project. In Sri Lanka, a flourishing export trade in gherkins has been built on contracts between companies and more than 15 000 growers with plots of around 0.5 ha each. On a much larger scale, more than 200 000 farmers in Thailand grow sugar cane for the country's 46 mills under a government-sponsored system that assigns growers 70 percent and millers 30 percent of total net revenue (FAO, 2001i).

Introduction of superior technologies. Small-scale farmers are frequently reluctant to adopt new technologies because of the possible risks and costs involved. In contract farming, private agribusiness will usually offer technology more effectively than government agricultural extension services, because it has a direct economic interest in improving farmers' production. Indeed, most of the larger corporations prefer to provide their own extension. In Kenya, for example, the South Nyanza Sugar Company (SONY) places strong emphasis on field extension services to its 1 800 contracted farmers, at a ratio of one field officer to 65 sugar-cane growers. The extension staff's prime responsibilities are focused on the managerial skills required when new techniques are introduced to SONY's farmers. These include transplanting, spacing, fertilizer application, cultivation and harvesting practices. Also, SONY promotes farmer training programmes and organizes field days to demonstrate the latest sugar-cane production methods to farmers.

Risks and problems. In addition, the FAO study emphasizes that contract farming can be a major tool for transferring skills and providing access to credit – features that are particularly important for smallholders. But the study also underlines that certain risks and problems can be associated with contract farming. Considerable problems can result if farmers perceive that the company is unwilling to share any of the risk, even if it is partly responsible for the losses. In Thailand, a company that contracted farmers to rear chickens charged a levy on farmers' incomes in order to offset the possibility of a high chicken mortality rate. This was much resented by the farmers, as they believed that the poor quality of the chicks supplied by the company was one cause of the problem. Inefficient management can lead to overproduction, and in some cases processors may be tempted to manipulate quality standards in order to reduce purchases. One of the biggest risks for farmers is debt caused by production problems, poor technical advice, significant changes in market conditions, or a processor's failure to honour contracts.

yields of local suppliers for Pepsi in Punjab, for example, rose from 16 tonnes/ha in 1989 to 52 tonnes/ha by 1999. In general, foreign affiliates may have contributed to better farming practices (e.g. hybrid seeds, transportation innovation) resulting in increased incomes and yields (McKinsey &Company, 1997).

10.2.3 International flows of knowledge and technological innovations

Probably the single most important transfer of external technology to developing countries' agriculture took place during the green revolution. The literature documents where the new technologies have been adopted, to what extent, and how swiftly this has been the case. This section will focus on why some countries managed to adopt, exploit and further enhance the new technologies, while others failed to do so. It will try to identify the policies that allowed some countries to embrace global technological innovations, but also describe why other countries are still struggling to adopt the technologies that developed country farmers have been using for many decades.

One of the most influential studies in this context (Griliches, 1957) underlines the importance of inventive adaptation. Griliches shows that farmers in Iowa and Illinois had long adopted high-yielding hybrid maize varieties suited to the Corn Belt states, while farmers outside the Corn Belt (e.g. in Alabama) continued to grow inferior traditional varieties. This had little to do with the farmers' capabilities. Instead, differences in the agro-ecological conditions between the Corn Belt and Alabama, together with the sensitivity of hybrid maize to these differences, resulted in the lack of adoption and the continuous technological distance between these maize-growing areas. As Griliches noted, "farmers outside the Corn Belt could not reap the benefits of the new technologies until the adaptive research had taken place to make the technologies available to the new environment".

In general, the same holds for the transfer of new technologies to developing countries. Farmers in the Philippines got no direct benefit from many decades of United States hybrid maize research that produced a tripling of United States maize yields. They indirectly benefited from previous hybrid research in the United States only after the research capacity was created to adapt the hybrid varieties to local conditions in the Philippines. In many African countries, farmers are still cut off from the benefits of hybrid maize varieties, not because they are unwilling to import the technology but simply because the technology has not been adapted to their local growing environments.

Perhaps even more important, much of the success of the green revolution was not or not primarily based on the fact that new technologies were made available to countries from outside.

While the new "foreign" technologies, i.e. high-yielding varieties, played an important role, there is ample evidence that the superiority of these new varieties was largely limited to the areas to which these new technologies were adapted. Evenson and Westphal (1994) documented how important the adaptation process to tropical environments was for the success of high-yielding rice varieties outside their subtropical homes.[14] "It was in the 1950s, after an Indica-Japonica crossing programme sponsored by the FAO and IRRI gave major impetus to rice improvement for tropical conditions, that the new technologies became available where they were needed. By 1965, many national rice breeding programs had been established in tropical conditions. India, for example had 23 programs in various locations. Around 200 rice breeding programs existed in some 40 countries by 1970. Most had, and have maintained, a close association with IRRI, which has served as a nodal point in the transfer of new germplasm."

Evenson and Gollin (1994) quantified the importance of adaptive research in the spread of high-yielding rice varieties during the green revolution. They show that national research centres played a crucial role in the adoption and spread of the new "technology". The International Rice Research Institute (IRRI), as the central and exogenous provider of the new technology, accounted for only 17 percent of all new varietal releases since 1965. IRRI played, however, a crucial role in generating the basic technology: it accounted for 65 percent of all new releases of parental varieties.

In the future, the very same factors will likely determine the extent to which the new agrobiotechnologies will be adapted and diffused to the economic and agro-ecological environments of developing countries, even though the issue is complicated by the fact that many of the new technologies are proprietary ones. Countries that put in place the basic infrastructure that promotes the adaptation to local environments are likely to gain the most. Again, Asian countries are likely to come first, followed by Latin America, while there is a danger that African countries will, once again, be left behind.

[14] Evenson and Westphal also provide an extensive documentation of the events that kick-started the spread of high-yielding rice varieties in the 1950s. *Inter alia*, they explain that: "The earliest rice improvement research activities were in Japan, where major gains were made in this century through improving Japonica landraces suited to subtropical regions. It was not until World War II that concerted efforts were made to improve the Indica landraces. As of that time, rice producers in Japan, Korea, Taiwan and parts of mainland China had achieved a 50-year technological lead over the tropical rice producing areas".

The factors that determine the success in reaping the benefits of new technologies have much in common with those that enable countries to reap the benefits of open markets. The experience of the green revolution suggests that the existence of new productivity- enhancing technologies alone is not a guarantor for a successful adaptation of these technologies. Likewise, opening up to international markets and reduction of border measures will not, on their own, ensure that the potential of freer trade can be fully exploited. Both openness to trade and to technological change are important, but what seem to be more important are the policies and institutions that allow countries to exploit the opportunities offered by openness. These factors can help to acquire the often tacit knowledge that enables countries to adapt new technologies to the domestic market environment, help them to exploit the demand potential of large international markets and employ trading rules to their advantage. Taken alone, access to markets is unlikely to create an exportable surplus. If not locally adapted, new technologies will not substantially increase productivity.

10.2.4 Greater mobility of people within and across national borders: migration and urbanization

International migration. Massive movements of labour have been a feature of all three waves of globalization. During the first great wave of modern globalization, from 1870 to 1910, about 10 percent of the world's population relocated permanently (World Bank, 2001e). International migration was much more modest and geographically limited during the second wave. The main reason was that only a limited number of countries were involved in the second wave of globalization, and where intense migration pressures occurred, strict immigration controls helped to put a brake on labour flows. These controls were somewhat relaxed during the third wave of globalization and had a powerful effect on transnational migration. By 1995, about 150 million people or 2.3 percent of the world's population lived in foreign countries (Taylor, 2000). Roughly half of this stock of migrants was in the industrial countries and half in the developing world. However, because the population of developing countries is about five times

greater than the population of the developed countries, migrants account for a much larger share of the population in developed countries (about 6 percent) than in poor countries (about 1 percent).

The freer movement of people has always had powerful effects on wages in poor and rich countries alike. Initially, different speeds of growth within and across countries promote inequality in wages and wealth, which in turn creates the economic pressure to migrate. Then migration itself, in addition to increased trade and capital flows, helps arrest or even reverse a growing wage inequality. The influx of low-wage labour puts downward pressure on wages in immigrant regions, while raising wages in the emigrant nations. Moreover, wealth is also redistributed when and to the extent that emigrants send back remittances to their countries of origin. As already mentioned in Section 10.1, Lindert and Williamson (2001) conclude that migration was overall a more important equalizing factor than either trade or capital movements.

Globalization also affects international migration in agriculture. During the first wave of globalization, migration was almost exclusively determined by different speeds of agricultural development. But even today many developed countries turn to foreign-born migrants as an important source of agricultural labour. Most rural migrants are attracted by higher wages in developed countries' fruit, vegetable and horticultural sectors. In the United States, for example, an estimated 69 percent of the 1996 seasonal agricultural workforce was foreign-born (Mines, Gabbard and Steirman, 1997). In California, by far the nation's largest agricultural producer, more than 90 percent of the seasonal agricultural workforce was foreign (Taylor, 2000). The majority (65 percent) of United States migrant farm workers originated from households in rural Mexico. Despite the high concentration of foreign-born workers in farm jobs, the vast majority of immigrants are employed outside agriculture, most in low-skill service and manufacturing jobs.

Agricultural migration is primarily a movement of low-skill labour from developing countries to the higher wage environments of developed countries. As such, it is unlikely to be associated with many of the typical concerns that emigration would lead to a "brain drain" in developing countries and deprive them of their most important capital for future development. In fact, empirical studies show that migrants seldom sever their ties with their source households after they migrate and family members who remain behind (often parents and siblings) reorganize both their consumption and production activities in response to the migrant's departure. Migrants (often children) typically share part of their earnings with their household of origin through remittances (Taylor, 2000). Remittances or savings accumulated abroad can even create the basis for future investments in the rural economies of their home countries.

The impacts in the countries of immigration are often more ambiguous. While an inflow of unskilled workers from developing countries benefits the highly skilled workers in host countries (their jobs are not threatened by these immigrants, and the presence of immigrants will lower prices for many things that the skilled workers consume, including food, restaurant and hotel services, and personal services), the same inflow will reduce the real wages of unskilled workers. Such competition in the low-wage sector has brought about political tensions within many host countries and has often resulted in increasingly restrictive immigration rules.

As immigration rules tightened and the economic incentives to immigrate remained unabated, illegal immigration and trafficking in human beings increased rapidly. The World Bank estimates that there is an annual inflow of about 300 000 illegal workers to the United States alone (World Bank, 2001e). Many more cross temporarily into the United States. In 1999 United States authorities apprehended 1.5 million illegal immigrants along the Mexican border. The great majority sent back to Mexico attempt to cross again within 24 hours. Illegal migration into the EU soared in the 1990s, from an estimated 50 000 p.a. in 1993 to half a million in 1999 (World Bank, 2001e).

Intranational migration and urbanization. As discussed in the preceding section, international migration has affected rural populations and agricultural labour forces in developed and developing countries alike. However, despite its importance for certain regions or countries (notably North America), international migration has become negligible compared with migration within national boundaries.

A look at the United Nations population projections by urban and rural areas reveals that a significant proportion of the world's population growth expected between 2000 and 2030 will be concentrated in urban areas. Urban population was estimated at about 2.9 billion in 2000, and is projected to be about 4.9 billion by 2030 (Figure 10.2). Most of the future increase will be in the cities of developing countries. The urban population in developing countries is projected to increase from 1.9 billion people in 2000 to about 3.9 billion people by 2030, thus accounting for almost the entire increment in developing countries' population growth. But only a part of it is caused by increased rural-urban migration. Also important will be the transformation of rural settlements into urban areas and, most important, natural urban population growth.

These shifts in population distribution are considerable. At the beginning of the 1960s, only about 20 percent of the developing countries' population lived in urban areas. By 2000 the share had risen to nearly 40 percent and is expected to rise to 56 percent by 2030. The rural-urban population ratio declined from about 3:1 in the 1960s to almost 3:2 in 2000 and will be close to 3:4 in 2030. Within the group of developing regions, urbanization will be most pronounced in developing Asia and sub-Saharan Africa. In other developing regions, notably Latin America, urbanization has already progressed to an extent that leaves little room for further growth in urban populations, at least relative to rural ones.

Numerous factors have promoted and will continue to promote urbanization in developing countries. These factors are often classified as push and pull factors, i.e. factors that either incite people to leave their rural homes or attract them to urban areas. Typical push factors include low and declining profitability of agricultural production, lack of non-agricultural employment opportunities and a general lack of services such as schools, medical treatment and entertainment. They have resulted from a general neglect of, or even an outright bias against, agriculture.

On the pull side, expectations for better services, housing, higher wages and more reliable sources of food are the main factors that attract migrants to urban areas. The typical drivers of globalization, notably better information facilities (television, etc.) have been instrumental in creating these expectations. There is, however, a large and widening gap between the expectations and the realities of urban areas in developing countries. Access to food, jobs and services is becoming more limited and other amenities that are often associated with "urbane" or "civilized" city life are entirely missing. This so-called premature urbanization is associated with numerous externalities. These include huge social costs caused by health and sanitation problems, urban poverty, crime, etc.

Despite these problems, it is widely accepted that urbanization is unstoppable, let alone reversible (The Economist, 2002c). Moreover, while it may not even be economically desirable to stop or reverse urbanization, it can be very profitable to slow the trend and mitigate or avoid the externalities associated with premature urbanization. The most important factor is a revival of rural areas in developing countries, which would amount to a reversal of the internal and external policy bias against agriculture in developing countries (as discussed in Chapter 8).

10.2.5 How important are geographic location and infrastructure?

Geographic location. Economists have long noted the crucial role of geographic location for economic development. Adam Smith, who is most remembered for his prescription of free market forces for economic development, emphasized that the physical geography of a region can influence crucially its economic performance (Smith, 1976). He contended that the economies of coastal regions, with their easy access to sea trade, usually outperform the economies of inland areas. Smith's rationale for the importance of geographic location is that productivity gains depend on specialization, and that specialization depends on the size of the market. The size of the market in turn depends on both the openness of markets and the costs of transport. Geography is a crucial factor in determining transport costs.

Empirical studies based on geographic information systems (GIS) have drawn renewed attention to the importance of the physical location for economic development. A frequent point of departure for the GIS-based analyses of location-based development questions is a map of income density,

Figure 10.2 Past and projected trends in urbanization of developing countries

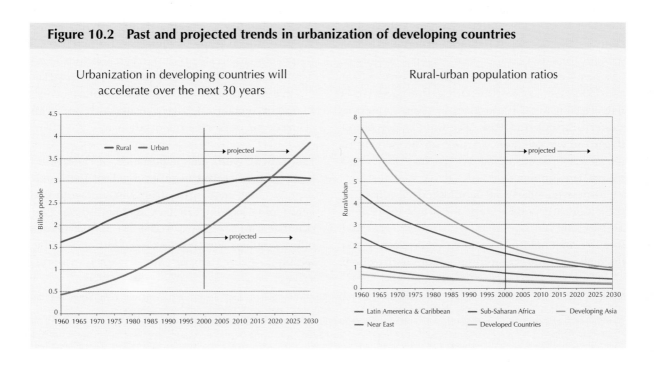

a measure of how much GDP is produced within a given area of land.[15]

Figure 10.3 provides such a map. The map underlines two principal geographic factors that affect economic well-being. First, almost all high-income countries are in the mid- and high latitudes, while nearly all countries in the geographic tropics are poor. Second, coastal economies have generally higher incomes than landlocked countries and coastal, temperate, northern hemisphere economies have the highest economic densities in the world. Indeed, outside Europe, there is not a single high-income landlocked country, although there are 29 non-European landlocked countries (Gallup, Sachs and Mellinger, 1999). Four areas – western Europe, Northeast Asia (coastal China, Japan and the Republic of Korea), and the eastern and western seaboards of the United States and Canada – are the core economic zones of the modern world. These regions are the overwhelming providers of capital goods in global trade, the world's financial centres, and the generators of a large proportion of global production.

A look at the regions within the United States, western Europe and temperate-zone East Asia that lie within 100 km of the coastline reveals that these areas account for a mere 3 percent of the world's inhabited land area, 13 percent of the world's population and at least 32 percent of the world's GDP measured at purchasing power parity. If coastal China is excluded from the calculations (since it lags far behind the other economies in this group), then the core coastal region has a mere 9 percent of the world's population but produces at least 30 percent of world GDP. According to WTO data (1995), just 11 countries in North America, western Europe and East Asia, with 14 percent of the world's population, account for 88 percent of global exports of capital goods (machinery and transport equipment).

Moreover, nearly all landlocked countries in the world are poor, except for a few in western and central Europe that are deeply integrated into the regional European market and connected by low-cost trade. Even mountainous Switzerland has the vast bulk of its population in the low-elevation cantons north of the Alps, and these population centres are easily accessible to the North Atlantic by land and river-based traffic. There are 35 landlocked countries in the world with a population greater than 1 million, of which 29 are outside western and central Europe. The difference in the average GDP per capita is striking: the landlocked countries outside western and central Europe have an average income of about US$1 771, compared with the non-European coastal countries, which have an average income of US$5 567. The difference in economic density is even greater, since the landlocked countries tend to be very sparsely populated[16] (Gallup, Sachs and Mellinger, 1999).

The most important points that arise from the inspection of GIS-based information can be summarized as follows:

- Coastal regions, and regions linked to coasts by ocean-navigable waterways, are strongly favoured in development relative to the hinterlands.

- Landlocked economies may be particularly disadvantaged by their lack of access to the sea, even when they are no farther than the interior parts of coastal economies.[17]

- Location advantages are particularly important for successful economic integration of agriculture and the food industry. Many agricultural commodities are either bulky, perishable or both, which leads to high transportation costs per unit value of product. High transportation costs mean that countries with poor market access conditions and inadequate infrastructure might remain effectively insulated, even if all trade barriers were removed.

Endowment with, and importance of, infrastructure. Infrastructure can offset much of the disadvantage that may arise from an unfavourable geographic location. In fact, in many developed countries and regions, access to infrastructure offsets possible disadvantages caused by unfavourable locations. The extensive and efficient transporta-

---

15 Ideally, such an income density map depicts information about population density and the spatial distribution of income. In the absence of the latter, Figure 10.3 simply combines population density with average per capita income levels (1997/99) for a given country.

16 Fifty-nine people per km^2 in landlocked countries compared with 207 people per km^2 in coastal countries.

17 Obvious reasons for this are: cross-border migration of labour is more difficult than internal migration; infrastructure development across national borders is much more difficult to arrange than similar investments within a country; and coastal economies may have military or economic incentives to impose costs on interior landlocked economies.

Figure 10.3 GDP density map of the world

Geographic distribution of Income: GDP density
(GDP: 1997/99 average, in 1995 constant USD)

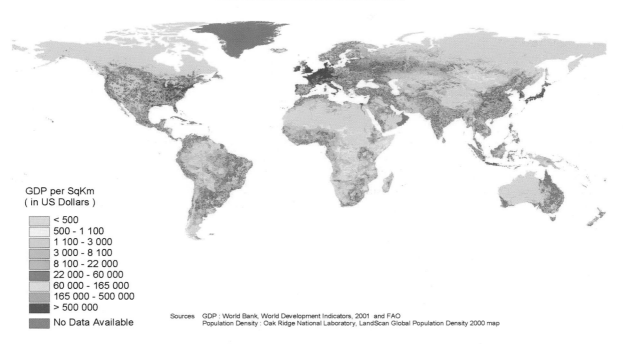

GDP per SqKm
(in US Dollars)

< 500
500 - 1 100
1 100 - 3 000
3 000 - 8 100
8 100 - 22 000
22 000 - 60 000
60 000 - 165 000
165 000 - 500 000
> 500 000
No Data Available

Sources GDP : World Bank, World Development Indicators, 2001 and FAO
Population Density : Oak Ridge National Laboratory, LandScan Global Population Density 2000 map

tion and communication systems in landlocked parts of Europe or North America effectively link these regions to one another and integrate them into world markets. A look across other regions of the world provides a more mixed picture. While parts of East Asia and Latin America hold relatively high stocks of infrastructure, Africa and many countries in sub-Saharan Africa in particular suffer from both unfavourable location and a lack of infrastructure.

The case of sub-Saharan Africa. Numerous studies (e.g. Finger and Yeats, 1976; Amjadi, Reinke and Yeats, 1996) have analysed and quantified the importance of infrastructure as a factor for the successful integration into international markets. Amjadi, Reinke and Yeats focus on sub-Saharan Africa and how the region's inadequate endowment with infrastructure weighs on its export performance. The study also documents the importance of insufficient infrastructure and related policies relative to other factors such as tariffs and non-tariff barriers.

The importance of the region's infrastructure relative to its competitors is also underlined by a comparison of the barriers that sub-Saharan African exporters face in markets abroad, relative to those its competitors faced when they embarked on export-oriented policies. For example, pre-Uruguay Round tariffs facing African exports to the EU, Japan and the United States averaged three-quarters of a percent (about 18 points lower than those the Asian newly industrializing countries [NICs] faced when they began their sustained export-oriented industrialization drive), and preferences give Africa an edge over some competitors (Amjadi, Reinke and Yeats, 1996). As long as transport is expensive, electricity unavailable or unreliable and access to phones restricted, the costs and risks of doing business remain high and the possibilities of reaping the benefits of globalization remain limited. But globalization also offers new opportunities to leapfrog traditional constraints. These new options will be discussed in Section 10.3 of this chapter.

10.2.6 What are the impacts on food consumption patterns?

The effects of freer trade in agriculture, the operations of TNCs in the global food sector as well as urbanization and migration become visible in changes in food consumption patterns. In general, these factors promote a convergence of food consumption patterns across different countries and regions. The channels through which these factors operate are either direct or indirect via the link of income growth in poorer countries. Rising incomes have an equalizing effect on food consumption patterns as they promote a shift of poorer consumers to higher-value food items, while higher-income segments are constrained by natural consumption limits even for higher-value food items.

A number of concerns revolve around the growing convergence in food consumption patterns. Some analysts see convergence as an indicator of a loss of cultural identity that reflects in part the growing market power of transnationally operating food enterprises ("McDonaldization"). Moreover, there are concerns that a fast convergence in food consumption patterns may have unexpected resource implications. A growing global convergence on, say, a typical United States diet is associated with rapid growth in feedgrain needs and thus with an extra burden on the available agricultural resource base. Another concern associated with a convergence in food consumption patterns is that food would have to travel over ever longer distances and that the externalities associated with these extra "food kilometres" are not, or not fully, reflected in the price of food. This section examines to what extent food consumption patterns have already converged and what the projections to 2015 and 2030 imply for future convergence.

Measuring convergence in consumption patterns. The comparison of food consumption patterns was undertaken on the basis of 29 primary product groups.[18] The need to compare diets of some 150 countries over a period of 70 years (1961 to 2030) strongly favoured the use of a single indicator, namely the consumption similarity index (CSI). This index measures the overlap in the diets of two countries by comparing how many calories consumed in two given countries originate in the same primary products. The CSI is expressed as:

$$CSI_{j,k} = 1 - \frac{1}{2}\sum_i \left| \frac{CAL_{ij}}{CAL_j} - \frac{CAL_{ik}}{CAL_k} \right|$$

A $CSI_{j,k}$ of 1 means that the diets of country j and country k are identical, i.e. that consumers draw the same number of calories from each of the 29 food categories distinguished in this study, while a $CSI_{j,k}$ of 0 means that the diets of the two countries are entirely different, i.e. that consumers in country j and k draw their calorie consumption from completely different food categories. In principle, the CSI allows the food consumption pattern of any country j to be compared with the one of any other country k.[19] It is important to stress that the CSI only captures the similarities in the structures of the diet in terms of primary products as defined in this study, but does not necessarily capture similarities in the final processed products that are actually consumed. This means that it measures to what extent consumers in two countries rely on a wheat-based or meat-based diet but not whether the wheat-based calories are consumed in the form of noodles or bread, or that meat is consumed in the form of hamburgers or traditional meat products. It is also important to note that the CSI measures similarities in diet structures, regardless of the absolute calorie intake levels. This can result in surprisingly high similarities in diets that are indeed very similar as far as the overall structures are concerned but very different regarding their respective levels of calorie intake (e.g. high shares of meat consumption in pastoral societies with low overall calorie consumption generate a high similarity with meat-intensive OECD diets).

The CSI has been used to compare the food consumption patterns across countries and over years. While CSI calculations have been undertaken for all combinations of countries, the results are only reported for the United States as a "comparator" country. All CSI coefficients are based on a comparison of any given country's diet with the one of the United States. Convergence over time is convergence towards the United States consumption patterns. The convergence in food

[18] The 28 food commodities given in Annex 1 of this study and a commodity, "other calories", i.e. all calories consumed and not covered by the 28 commodities of this study.

[19] Note that a CSI of 0.5 between country 1 and country 2 and a CSI of 0.5 between country 1 and country 3 do not mean that the overlap/similarity in the diets between country 2 and 3 has to be 0.5 as well.

consumption reported here may thus be regarded an "Americanization" rather than a globalization of food consumption patterns.

Evidence for convergence. A look at CSI developments suggests that diets have indeed become increasingly similar over time. The speed of convergence, however, differs markedly across countries. The traditional OECD countries form a cluster with consumption patterns that are very close to the United States diet. About 75 percent of the calories in many OECD countries originate from the same sources as in the United States. These countries are fully integrated into the global food economy and their food economies are tied to one another through effective and efficient transportation and communication infrastructures, similar food distribution systems, cold chains, etc. One of the striking features in the group of OECD countries is the high similarity in consumption patterns within the cluster of English-speaking countries (Australia, New Zealand and the United Kingdom) where 80 percent and more of all calories stem from the same foodstuffs.

Many of these countries share not only the same language, but have also a common food and cooking culture. The absence of language barriers and the similarity in food culture are important parameters for an effective and low-cost operation

of transnational food enterprises. These similarities allow them to employ the same or similar advertising and marketing strategies and thus reap economies of scale in their market penetration strategies. Finally, there is also a geographic element that plays a role in explaining similarities in diets. For instance, 85 percent of all calories consumed in Canada stem from the same primary commodities as in the United States. Similarly high values exist when diets of North African or western European countries are compared among each other (not with the United States).

Outside the group of the well-integrated western countries, the similarity with the United States consumption pattern is often considerably smaller. Again, there are a number of different groups of countries that exhibit different levels of overlap and different dynamics in moving towards United States food consumption patterns. Very dynamic change can be observed within the group of East and Southeast Asian countries. Japan's consumers are among the most dynamic adapters of a United States-type food consumption pattern. Starting from an overlap of only 45 percent in 1961, similarity had increased to about 70 percent in 1999 and is expected to reach a level of 75 percent in 2030 (Figure 10.4).

Outside the group of OECD countries, a number of different clusters can be identified across conti-

Figure 10.4 Food consumption convergence in OECD countries

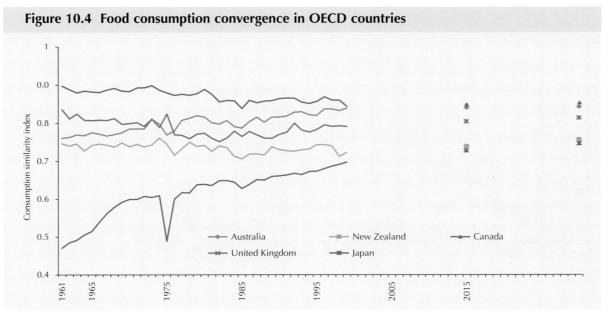

Note: The dip for Japan in 1975 reflects a change in diet caused by the drastic price increases in world markets during the world food crisis in the mid-1970s.

nents and regions. Within Africa, three major trends in consumption patterns emerge. First there is the group of North African countries, where consumption patterns are characterized by a grain-rich diet and where often more than 70 percent of the calorie intake stems from cereals, notably from wheat. Within this group, food consumption similarity reaches levels of more than 90 percent. Compared to a United States diet, however, similarity has reached a level of about 60 percent (Figure 10.5) and is, even by 2030, not expected to exceed about 65 percent. This could seem surprising at first sight, given the geographic vicinity to, and increasingly important economic integration with OECD markets. However, other factors override the integrating forces of globalization/Americanization on food consumption patterns. These are rooted in (i) the traditional food culture characterized by high consumption levels of wheat-based staples (bread, couscous); and (ii) the non-consumption of pork that limits the potential for shifts towards meat consumption.

In summary, the forces of globalization have had a significant impact on food consumption patterns and have resulted in a growing convergence of consumption patterns. Even though the relative importance of the various driving forces of globalization is difficult to gauge, openness to trade and investments, geographic location, income levels and growth and TNC activity are almost always associated with a rapid convergence in food consumption patterns. Many of these factors are interrelated. Well-integrated countries also often enjoy higher income growth that works, within the boundaries of income responsiveness and overall calorie intake levels, as a force for convergence. There are, however, factors of a longer-term nature that put a cap on the convergence of food consumption patterns. They include cultural and religious constraints as well as deeply rooted traditions in food consumption and preparation. As a result they limit convergence, even in the most integrated OECD markets, to a level of about 80 percent, a level that is not expected to be topped over the next 30 years. Outside the OECD area, convergence levels off at about 60 percent overlap.

10.3 Some options to integrate developing countries better

Multimodal transportation systems. The smart combination of various transportation modes (sea, air, rail and road) can help to overcome the financing constraints that many developing countries face in building up traditional infrastructure components. So-called multimodal transportation systems have gained a considerable momentum in integrating hitherto remote areas of Asia and Latin America (Box 10.4).

Figure 10.5   Food consumption convergence in Africa and Asia

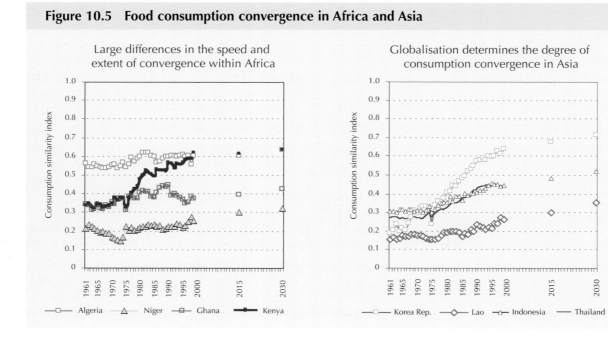

Box 10.4 Multimodal transport offers new opportunities for developing countries

The advent of multimodal transportation systems adds to numerous new opportunities for developing countries to integrate faster and more easily into the global economy. In China, for instance, a new service has been established connecting the country's hitherto isolated hinterland with Europe. Multimodal Logistics, a Rotterdam-based company, is now offering rail transport for containers from Rotterdam to northwest China. The Marco Polo Rail Express, as the service is commonly called, has two connecting points at Almaty and Druzhba. The transit time is between two and three weeks depending on its destination in China, which could be Alataw-Shankou, Jinghe, Wusu, Ürümchi, Turpan, Korla or Hami/Yumen (UNCTAD, 1999a).

There is also an effort to connect Singapore and Europe by rail. So far, India, the Islamic Republic of Iran, Pakistan and Turkey have been active in the project. Each participating country must bear the cost of completing the link within its national boundary. The aim is to reduce the transit time for deliveries from Singapore to Europe and vice versa by two weeks. There is also a proposal to construct a tunnel that will link Taiwan Province of China and mainland China in order to accommodate the increase in goods traded and transported between the two countries. The Taiwanese private sector has already responded by investing in new depots in Shanghai and Shenzen in China.

There are also plans to create a land bridge that would connect Latin America's Atlantic side with its Pacific coast. These plans have grown out of the anticipated congestion of the Panama Canal, which is projected to face serious capacity constraints within the next 15 years. The proposal is to integrate Bolivia's eastern and Andean railways to form a link to and from the Chilean port of Arica and the Brazilian port of Santos. The increase in traffic from such a link is expected to be between 2 and 20 million tonnes with a cut in transport costs of about US$16 per tonne (UNCTAD, 1999a).

The new transportation systems can be particularly efficient if combined with new communication tools. These new technologies offer considerable potential to overcome the location handicap that many remote areas in the developing world face; they could provide new trade opportunities for bulky or perishable goods that were previously excluded because of prohibitively high transaction costs. This in turn could help integrate agricultural producers and provide a stimulus to rural areas, perhaps comparable to the rapid expansion of agricultural production in the United States Midwest during the first wave of globalization.

Leapfrog traditional communication constraints: the Internet for trade facilitation. As during the first wave of globalization, the availability of more efficient transportation systems has been accompanied by the advent of more efficient communication systems. These new communication technologies enabled shippers to tailor volumes and delivery dates of goods to the precise needs of importers. The Internet now allows even smaller-sized companies to compete with their larger counterparts, who had gained a competitive advantage through the dedicated but more expensive electronic data interchange (EDI) systems. For the low initial cost of a personal computer, a modem and an Internet connection, anyone can now access the Internet. More and more shippers and carriers choose to do business through the Internet because of the lower administrative costs involved in conducting transactions. This means significant savings because carriers and shippers depend less on third-party value-added networks that are normally required to run EDI transactions.

There is also a growing expectation that the Internet will soon provide all the advantages that had previously been restricted to expensive EDI systems. Most important, Internet data transfer will become increasingly safe. Moreover, it is available 24 hours a day, allowing business deals to be made at the shipper's and carrier's convenience. It could provide niche markets for smaller carriers, enabling them to capture a greater market for small package deliveries.

Economic agglomeration and special economic zones. Despite the possibilities of exploiting further efficiencies in the existing infrastructure, investments in a uniform expansion of infrastructure in every location may be neither an efficient nor an affordable option for most developing countries, particularly in

Box 10.5 The benefits and limits of economic agglomeration

China's special economic zones are a well-known example of a successful formation of spatial agglomerations. These zones have attracted substantive but spatially limited public investments in infrastructure. They have offered free trade with otherwise protected countries and have thus been particularly successful in attracting foreign investment. Alongside foreign investment, skills, production techniques and management knowledge in private companies improved while domestic policy-makers gathered institutional knowledge and practical policy experience. This hands-on experience then helped to attract additional foreign capital and is now helping to develop China's hinterland.

Mauritius' export processing zone (EPZ) provides another example of a successful spatial agglomeration strategy. Operating under free-trade principles, it enabled an export boom in garments to European markets and an accompanying investment boom at home. The Mauritian EPZ was created as part of an overall development strategy in the 1970s. Given the small size of the home market, it was not surprising that Mauritius would benefit from an outward-oriented strategy. The challenge, however, was to smooth the adjustment process for the existing domestic garment sector that had been long protected under the country's ISI regimes. The EPZ scheme also provided a way around political difficulties. The EPZ generated new opportunities of trade and employment, without taking protection away from the import-substituting groups and from the male workers who dominated the established industries. The segmentation of labour markets early on between male and female workers, with the latter predominantly employed in the EPZ, was particularly crucial, as it prevented the expansion of the EPZ from driving wages up in the rest of the economy, thereby disadvantaging import-substituting industries. New employment and profit opportunities were created at the margin, while leaving old opportunities undisturbed. This in turn paved the way for the more substantial liberalization that took place in the mid-1980s and 1990s.

Where these linkages to the domestic sector are not developed, the success of the EPZ model is often less sustainable. For example, during the 1980s the Dominican Republic was able to diversify out of its dependence on agricultural commodity exports by expanding its production of garments for the United States market. However, the country's increasing share of the North American market owed less to domestic competitiveness than to the arrival of United States subsidiaries and their subcontractors in the country's EPZ. When wages increased, foreign investors relocated to lower-wage economies in Central America. Because the export industry never established domestic linkages or generated a national supply base, export growth did little to raise long-term capacity (Vincens, Martínez and Mortimore, 1998). Problems can also arise out of the extensive tax inducements granted in EPZs. EPZs typically offer tax concessions for five to ten years and, in some cases, as in Honduras, they are granted on a permanent basis (Agosin, Bloom and Gitli, 2000). The resulting revenue losses for the national governments can be substantial. For Bangladesh, the revenue losses associated with tax concessions in the EPZ amount to around US$84 million p.a.

areas where income and population densities are low. As an alternative, a number of manufacturing/ service agglomerations could be developed. Large areas with low population densities (sub-Saharan Africa and Central Asia) would still require several such locations and a considerable labour mobility to populate these places. Of particular interest from the perspective of globalization is the formation of dynamic economic regions and export processing zones (EPZs), which often thrive in the open trading environment (Scott, 1998). These zones can become centres of industry, producer services and urban amenities. Their growth derives from trade, the capacity to attract financing and skills, and the use of agglomeration effects to create networking relation-ships yielding the maximum of synergy. Recent research also suggests that advances in telecommunications, by increasing the frequency of contact between people, can motivate greater face-to-face interaction and make it more desirable to live in cities (Gaspar and Gleaser, 1999).

In most developing countries, a small number of cities – with the capital city in the lead – generate half or more of the country's GDP. These interlocking metropolitan areas comprising an economic region could become the principal "growth pole" for countries (Simmie and Sennet, 1999). These regions could be located in a single country as in Brazil, India and China or straddle two or three countries as they do in Southeast Asia.

Box 10.6 Why and when is two-way trade in agriculture important for developing countries?

There is growing recognition that developing countries can reap important benefits from two-way trade (TWT) and there is evidence that TWT can provide an avenue for successful globalization strategies.

First, a shift towards TWT is typically associated with a shift in trade to processed products and thus higher-value goods. Within a given infrastructure, a shift in trade towards higher-value goods reduces the share of transaction costs per unit of merchandise and thus helps overcome the geographic and infrastructure constraints faced by many developing countries. Second, TWT helps to cope with adjustments in factor markets arising from the large swings in international commodity prices witnessed over the last decades. As trade is largely the exchange of similar and processed products, shocks result in a reallocation of production factors within an industry, rather than between industries. The latter is typically a process that involves the discontinuation of activities in one sector and the loss of jobs, in order to move factors to other industries or sectors. This type of resilience against international shocks can be an important argument for policy-makers in developing countries to integrate their economies faster and more fully into international markets.

Third, TWT in food and agriculture offers economies of scale in the food industry. It enables domestic producers and processors to sell products that are homogeneous with respect to factor requirements but heterogeneous with respect to utilization and marketing to both domestic and foreign markets. Like increased natural protection, scale economies are particularly important when countries want or have to integrate their domestic economy into international markets and expose their domestic sectors to greater competition from abroad. This is a particular benefit for small developing countries, notably small islands, for which TWT could provide an interesting avenue to reap economies of scale, diversify trade and escape the volatility of price swings that they would face otherwise. Fourth, if intra-industry adjustments dominate, commitments towards freer trade are likely to be more comprehensive and to last longer. Empirical studies (Caves, 1981) suggest that increasing IIT following trade liberalization keeps pressures from import competition low. As a consequence, politicians are more likely to press ahead with the process of trade liberalization since the high political cost associated with resources shifting between industries is limited. Finally, IIT in food and agriculture enables those developing countries that are scarcely endowed with productive natural resources (land, water, climate, etc.), to create and foster trade opportunities independent of the ability to produce primary agricultural products. TWT in food and agriculture has boosted the food processing sector in many Asian economies and created a trade surplus for countries that lack the climatic and agronomic conditions for a flourishing agricultural export economy.

Most developed countries have vastly benefited from the productivity gains that economic agglomerations and clusters provide. Agglomeration gains were probably most pervasive during the second wave of globalization, when trade between developed countries became determined not so much by comparative advantage based on differences in factor endowments but by cost savings from agglomeration and scale. Because such cost savings are quite specific to each activity, although each individual industry became more and more concentrated geographically, the industry as a whole remained very widely dispersed to avoid costs of congestion.

However, while agglomeration economies are good news for those in the clusters, they are bad news for those left out. A region may be uncompetitive simply because not enough firms have chosen to locate there. As a result "a 'divided world' may emerge, in which a network of manufacturing firms is clustered in some 'high wage' region, while wages in the remaining regions stay low" (Yussuf, 2001).

Intra-industry trade in agriculture. Classical trade theory suggests that countries specialize in international trade according to differentials in production costs for different goods. These cost differentials can result either from efficiency differences in the use of production factors (Ricardian trade specialization) or from differentials in factor endowment (Heckscher/Ohlin). When developing countries specialized in producing and exporting agricultural products for which they seem to have a classical "comparative advantage", they were facing increasingly binding constraints for their potential to grow (relative to world markets). This was not necessarily as a result of limits in domestic factor endowments or as a result of trade barriers in markets abroad (see Chapter 9), but simply because the export markets they produced for exhibited low and declining demand elasticities, while their own import demand remained elastic.[20] This created the so-called "elasticity pessimism" and the economic rationale for the import-substituting industrialization (ISI) strategies pursued by many developing countries in the 1950s and 1960s (see Section 10.2.1 above).

Krugman (1986) showed how a country could overcome the elasticity constraint by embarking on intra-industry trade (IIT) or two-way trade (TWT) in differentiated products. The principle is as follows: if consumers have a certain taste for variety, each new differentiated product creates a niche and the corresponding demand. If the number of products produced in a given country is related to the size of the economy, then the countries with the fastest growth also tend to produce more products. Contrary to the traditional view, this mechanism does not need a price (exchange rate) or demand adjustment to equilibrate the trade balance. Instead, the mechanism works endogenously. The country with higher growth produces more product variety, which in turn generates its own export markets.

These microeconomic effects play a role in the growth process and interact with macroeconomic policies, notably with the exchange rate. For example, there is evidence from a comparison between the Republic of Korea and Taiwan Province of China that the latter was able to generate more product variety than the Republic of Korea and rely less on a continuous competitive devaluation to gain a market share (Oliveira Martins, 1992 and Feenstra, Yang and Hamilton, 1999). More recent work has also shown a positive and significant impact of product variety on relative export intensity and growth (Funke and Ruhwedel, 2001).

In practice, however, TWT in food and agriculture was largely limited to trade within developed countries. In developing countries, trade patterns in food and agriculture remained biased towards a traditional (inter-industry) trade specialization, which broadly reflected two major factors. First, a great number of developing countries experienced low GDP growth rates and failed to attain the GDP

Box 10.7 How has two-way trade been quantified?

To measure the order of magnitude and the development in IIT specialization, a modified Grubel-Lloyd (GL) index of two-way trade (TWT) has been computed. The modification of the GL index was necessary to account for the overall trade imbalance in food and agriculture that is characteristic of many developing countries. The TWT index for n products (i) and a given country j is computed as follows:

$$TWT_j = 1 - \left[0.5 \cdot \sum_{i=1}^{n} \left| \frac{X_{ij}}{\sum_{i=1}^{n} X_{ij}} - \frac{M_{ij}}{\sum_{i=1}^{n} M_{ij}} \right| \right]$$

where X_{ij}, M_{ij} are export and import values in current US$.

In general the TWT index measures the proportion of total trade (sum of values of imports and exports) that is composed of trade in "similar products". A value of the index close to one indicates that there is predominantly IIT (i.e. in differentiated products), while a TWT value close to zero suggests that trade is primarily inter-industry trade, i.e. in different products. The computations reported are made at the highest possible disaggregation level that is allowed by the data (the FAO trade database for food and agriculture comprises a maximum of 521 different products).

---

20 Even for the development of a competitive food processing sector, most developing countries are simply too small to reach the necessary economies of scale. Where developing countries have made significant inroads into food processing – for example, orange juice production in Brazil, canned pineapples in Thailand and soluble coffee production in Colombia and Brazil – the scale required for efficient production means that upstream access to raw materials and downstream access to markets must also be secured on a large scale. Many developing countries lack the raw materials, capital and market access to make processing viable (UNCTAD, 2000).

levels required to stimulate a greater diversification of demand and eventually rising IIT. Second, growing TWT goes hand in hand with the development of an internationally competitive food processing industry, a process for which most developing countries faced major constraints. Food processing industries are well established in the industrialized countries.

But indicators for trade diversification and specialization (Table 10.3) also show that the difference in the levels of TWT between developed and developing countries was not always as pronounced as in the 1990s. Throughout the 1960s and 1970s, IIT in agriculture has been low when compared to manufactures, in developed and developing countries alike. However, while TWT remained at low levels in developing countries, many developed countries recorded a rapid growth in TWT trade over the 1980s and 1990s. For the United States and the EU, for instance, the level of TWT in agriculture increased rapidly since the early 1970s, while the TWT coefficients remained largely unchanged throughout the developing world. In sub-Saharan Africa, TWT accounts for merely 16 percent of total agricultural trade and this share has remained unchanged since 1970. The level is higher in the Near East/North Africa and South Asia but there too IIT in food and agriculture stagnated.

A closer inspection of country-specific information suggests that the level of TWT in agriculture increases rapidly when trade barriers are reduced and even more so when countries integrate their economy into a common economic market (Figure 10.6). Such a change is accompanied not only by the increasing amount of IIT, i.e. by the volume of exports and imports of similar products, but also by a rapid increase in the number of products traded. For countries that are firmly integrated into a common market (the Netherlands, Belgium), TWT in food and agriculture has reached levels that are comparable to those attained in manufactures. Country-specific data also reveal that a rising TWT is often associated with a growing trade surplus in food and agriculture and vice versa. This suggests that the ability to generate a trade surplus in food and agriculture does not so much depend on the ability of a country to produce a certain amount of raw material as on the country's capacity to produce differentiated products and to cater for specific markets, including market niches.

Table 10.3 Two-way trade in food and agriculture, by region

Region	TWT (percentage)			Number of products traded		
	1969/71	1984/86	1997/99	1969/71	1984/86	1997/99
Developed countries	25.2	30.2	34.6	252	297	279
EU	28.4	33.4	45.3	278	329	395
North America	31.0	33.8	44.8	280	315	390
Other countries	18.5	21.3	26.8	199	236	332
Developing countries	17.4	17.6	20.4	98	119	194
East Asia	13.9	17.5	16.0	101	136	186
Latin America and the Caribbean	14.0	15.3	18.1	122	139	234
Near East/North Africa	29.2	28.2	29.5	120	146	200
South Asia	17.7	17.6	13.5	96	167	227
Sub-Saharan Africa	16.4	15.7	16.8	77	79	142
Transition economies	26.3	19.0	28.9	118	121	269
Eastern Europe	25.2	19.3	28.0	117	119	278

Source: own calculations.

Figure 10.6 Two-way trade in food and agriculture, the effects of economic integration

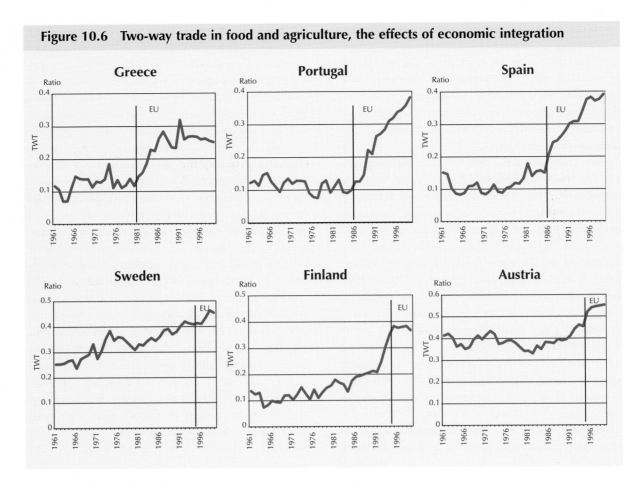

10.4 Concluding remarks

Globalization – the growing integration of economies and societies around the world – is a complex process that affects the world's food and agricultural economy in numerous ways. Cheaper and faster transportation, easier communication and the development of the Internet are important drivers. Also important are a growing number of international agreements that have codified and liberalized the flow of goods and capital. These factors have resulted in a rapid expansion of trade and FDI but also in the rise and growing influence of transnational companies. The impacts of these new factors have been very positive overall, even though the benefits are distributed unevenly. For example, globalization has helped to fight poverty and undernourishment in China, Viet Nam and Thailand, but has done little so far to integrate the poorest in sub-Saharan Africa, to improve their food security, or to enable the region's farmers to make significant inroads into markets abroad.

This raises the question as to what factors determine success or failure, integration or marginalization. Why have some countries been able to take advantage of the great development potential that globalization offers while others have failed to do so?

Some of the correlates of success or failure have been identified in this chapter. To begin with, openness to trade and capital flows, and the ability to adopt and to adapt technological innovations are undoubtedly among the important factors for success. Also geographic location and endowment with infrastructure can play a crucial role in determining whether a country thrives or falls further behind in an increasingly globalized economic environment. But probably most important are the domestic incentive system and the companion policies that facilitate the integration process.

A number of examples have been presented to document success and failure in the process of globalization. No claim is being made that these examples are representative or comprehensive. Nonetheless, the examples suggest that a number of common features are associated with success or failure in the process of global integration.

First, while openness to trade and investment flows is an important contributor to a successful global integration process, openness alone is not a guarantor for success. In many cases, openness has emerged gradually alongside overall economic and agricultural development. At the same time, no country has recorded high growth in the long term on the back of infant-industry protection and import substitution policies. Nowhere have insulation and protection spurred on agricultural growth and overall development in a sustainable manner.

Second, successful globalizers are masters in managing adjustment. They succeed in rationalizing excess capacities and create exit possibilities for farmers and new employment opportunities at minimal cost. The creation of township and village enterprises in China, the pruning of excess capacity in Viet Nam's coffee sector and credit restrictions imposed on unproductive chaebols in the Republic of Korea have been mentioned. Gradual adjustment is particularly important for agriculture, as a large part of the human and financial capital of the sector is fairly immobile in the short term. Managed transition provides an opportunity to reallocate resources or gradually depreciate them. Active adjustment management also helps to mitigate adverse impacts for the poor.

While the process of structural change often creates greater opportunities for the poor in the medium term, it also means that they have to bear short- to medium-term transitional costs that they are ill positioned to absorb. This is particularly true of trade reforms, where the adjustment costs often come upfront, while the benefits are seen only over a longer period of time. The policy measures to manage adjustment entail an appropriate mix, sequencing and phasing of trade reforms; they also include measures that prepare farmers and processors for international competition, e.g. through training and technical assistance. Even where and when border protection is largely removed, farmers can vastly benefit from measures that protect them from excessive price swings (e.g. China and Viet Nam).

Successful integration is also a process of learning and experimenting and there are measures that promote this learning process. The two-track system in China's agriculture (both in terms

of export orientation and in terms of free market plus government control) seems to work. It has allowed policy-makers to gather experience as to what system works and what does not, as to what farmers can do best and what not. Two-track systems have also proved useful in opening up to international competition without facing the costs of massive and rapid adjustments for a whole economy. Probably the most prominent examples for successful two-track systems are China's special economic zones and Mauritius' EPZ. There are various channels through which these two-track systems facilitate transition towards freer environments. They allow, for example, a country to provide foreign investors with special conditions that may be difficult to guarantee for the whole economy. They also help domestic companies to prepare for growing competition from abroad and allow policy-makers to adjust the domestic framework of competition policies to an environment of freer trade and capital flows.

Globalization has generated growth in FDI that exceeded growth in trade flows. FDI inflows can play a catalytic role for development. FDI provides not only an important source of finance. More important, it is a carrier of technology, skills and management techniques. But, as with trade, success rests not only on the degree of openness. As important as the quantity of inflows is the quality of FDI. High-quality FDI is characterized by low repatriation levels and intensive linkages to domestic farmers. Experience from FDI in India's food industry, in particular, has demonstrated its potential for promoting agriculture and overall rural development. TNCs provided farmers with better seeds, enhanced technologies and more stable prices and thus boosted crop yields and farm incomes. The contracts that forge these linkages are crucial for success. But there are also examples where FDI largely failed to create linkages with local farmers and even instances where TNCs have added to the marginalization of whole farm populations. Developments in the global coffee markets illustrate this point. There is evidence for a growing concentration in trading and processing of coffee and there is also evidence that TNCs managed to reap a growing share in the total value created in the coffee marketing chain. It is however less clear to what extent these developments reflect the abuse of market power and the absence of an appropriate competition policy framework that can address the new competition policies in a globalized market.

Globalization in agriculture has also been brought about by an internationalization of production technologies. The green revolution was the single most important vehicle in this process. Again, while some countries have been phenomenally successful in adopting these new technologies, others have largely failed to do so. As with trade and investment flows, the correlates of success are not merely openness to innovation. Numerous studies suggest that it is more important to create the appropriate domestic environment that allows local producers to employ the new technologies gainfully. In short, adoption has to be accompanied by adaptation to provide success. Similarly, success or failure in reaping the benefits of biotechnologies will depend less on availability than on the capacity to adapt the new technologies to the agronomic and economic environments that prevail in a specific location. Finally, geographic location and infrastructure endowment play a crucial role in successfully tapping potential world markets. There is ample evidence that the lack of infrastructure (not the existence of trade barriers) has been the crucial impediment that hindered sub-Saharan Africa's farmers from making significant inroads into OECD markets.

But there is also evidence that globalization can offer new opportunities to leapfrog old obstacles resulting from unfavourable locations or inade-quate infrastructure. In general, new technologies are cheaper and faster and can bring the most remote areas to the heart of the markets. These include Internet-based business communication systems as well as multimodal transportation systems. In sparsely populated regions, transaction costs can be reduced by promoting economic agglomeration.

In summary, globalization offers a great potential for farmers and the entire food sector of developing countries. Many developing countries are successfully tapping this potential, but not all of them are able to take full advantage of the new opportunities. The ability of a country to reap the benefits of globalization depends on factors such as openness to trade and capital flows, ability to adopt technological innovations, and also geographic location or infrastructure endowment. The various examples suggest that openness and outward-oriented policies characterize many successful globalizers but *per se* they are not guarantors for success. More important are the companion policies on the domestic front that facilitate integration into global markets. These are policies that provide appropriate transition periods towards freer trade; help adapt new, external technologies to the domestic environment; and provide competition policy settings and design contracts that also allow small-scale agriculture to thrive within the operations of TNCs.

CHAPTER

11

Selected issues
in agricultural technology

This chapter discusses selected issues in agricultural technology. First, it continues the evaluation started in Chapter 4 (Section 4.5.2) of the room for further yield improvements. It then discusses some technologies that could contribute to making agriculture more sustainable, such as integrated pest and nutrient management, conservation agriculture and organic agriculture. It continues with an assessment of prospects for agricultural biotechnology and concludes with some observations on the future agenda for agricultural research.

11.1 The scope for yield increases

As discussed in Chapter 4, world agriculture has derived more of its growth from an increased intensive use of land already under crops than from expansion of agricultural areas, even though area expansion has been and still is the main force in a number of countries, mainly in sub-Saharan Africa. Improved farming practices, irrigation, improved varieties, modern inputs, etc. all contributed to the growth of yields that underpinned many of the increases in agricultural production. This trend is expected to continue.

How far can this process go? Intensification and yield growth are subject to limits for reasons of plant physiology (see, for example, Sinclair, 1998)

and because of environmental stresses associated with intensification (see Murgai, Ali and Byerlee, 2001 and Chapter 12). Moreover, in many circumstances it is simply uneconomical to attempt to raise yields above a certain percentage of the maximum attainable. In considering the prospects and potentials for further growth in world agriculture, we address below the question: what are the gaps between the actual yields of any given crop in the different countries and those that are agronomically attainable given the countries' specific agro-ecological endowments for that crop? Naturally, what is agronomically attainable changes over time as agricultural research produces higher-yielding varieties and farming practices improve.

As discussed in Chapter 4, intercountry differences in average yields can be very large, but they do not always denote potential for growth in countries with low yields. As an example, Figure 11.1 shows the wheat yields (five-year averages 1996/2000) in the major wheat producers of the world. Yields vary from a high range of 6.0-7.8 tonnes/ha in four EU producers (the United Kingdom, Denmark, Germany and France) plus Egypt; through an upper-middle range of 3.0 to 4.0 tonnes/ha (China, Hungary, Poland and Italy) and a lower-middle range of 2.4-2.7 tonnes/ha of the United States, Spain, India, Romania, Ukraine,

Figure 11.1 Wheat yields (average 1996/2000)

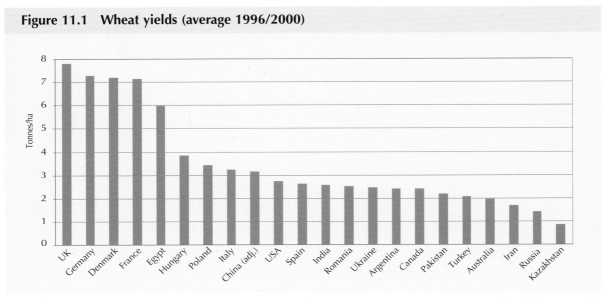

Note: Twenty-two countries with a production of over 4 million tonnes in 1996/2000 accounting for about 90 percent of world wheat output in 1996/2000

Argentina and Canada; down to the low-yield range of 1.0-2.2 tonnes/ha of Pakistan, Turkey, Australia, the Islamic Republic of Iran, Russia and Kazakhstan. Analogous wide yield differentials exist for all crops: those for maize in the major producers are shown in Figure 11.2.

The reasons why country average yields differ from one another are many. Some are agro-ecological, others socio-economic. Irrigation is important in the achievement of high yields in several countries, e.g. Egypt. In addition, agro-ecological and demand factors influence the mix of varieties of the same crop grown in each country, for example, low-yielding durum wheat versus common or soft wheat with higher yields. Given that we are interested in the physical/agronomic potential for yield growth, we need to separate out the part of these inter-country yield gaps that is caused by agro-ecological diversity from the part caused by other factors.

The results of the global agro-ecological zones (GAEZ) analysis (see Chapter 4) provide a way of controlling agro-ecological diversity in such inter-country comparisons. In a nutshell, GAEZ describe at a fairly detailed geographic grid the agro-ecological conditions prevailing in each country. GAEZ also have models defining the agro-ecological requirements for the growth of each crop. Based on this, GAEZ derive estimates for attainable yields for each crop and in each grid cell in the different countries under three technology (input use and

management) variants. A summary description of the procedure is given in Box 4.1. More detailed explanations are to be found in Fischer, van Velthuizen and Nachtergaele (2000).

The agro-ecologically attainable yields can be used to draw inferences about the scope for raising yields in countries where actual yields are "low" in relation to what is attainable for their agro-ecologies. Actual yield data in the agricultural statistics are normally available only as country national averages, not by agro-ecological environments. Therefore, for comparison purposes, the estimates of the agro-ecologically attainable yields for any given crop must also be cast in terms of national averages specific to each country's agro-ecological endowments in relation to that crop. Also, since we compare agro-ecologically attainable yields under rainfed conditions, the remainder of this section will focus on countries with predominantly rainfed agriculture to minimize the distortion caused by the unknown contribution of the normally higher irrigated yields.

For each crop, averaging out over the whole country, the yields for each grid cell give an estimate of "attainable" national average yield for that crop. These yields can be compared with actual national average yields to form an idea of the physical/agronomic scope for yield growth compatible with the country's agro-ecological endowments. In principle, countries with similar attainable averages

Figure 11.2 Maize yields (average 1996/2000)

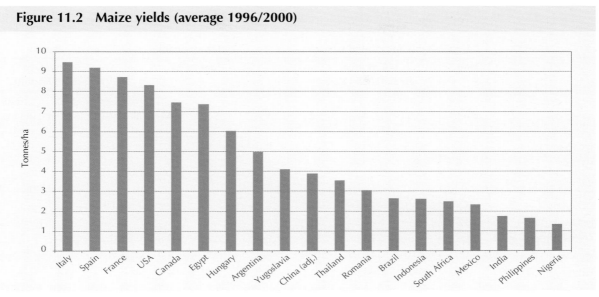

Note: Nineteen countries with a production of over 4 million tonnes in 1996/2000 accounting for about 90 percent of world maize output in 1996/2000

for any given crop and technology level may be considered to be agro-ecologically similar for that crop. Naturally, any two countries can have similar attainable yields but for very different reasons, e.g. in some countries the limiting factors may be temperature and radiation, in others soil and terrain characteristics or moisture availability. Nevertheless, the GAEZ average attainable yields for any crop can be taken as a rough index of agro-ecological similarity of countries for producing that crop under specified conditions.

For example, France and Finland have actual wheat yields of 7.1 tonnes/ha and 3.2 tonnes/ha, respectively (averages 1996/2000). This gap does not indicate that Finland has considerable scope for raising yields towards those achieved in France, because Finland's agro-ecology is much less suitable for growing wheat than France's. The GAEZ evaluation suggests that the agro-ecologically attainable yields in the two countries (i.e. controlling for agro-ecological differences) are 6.6 tonnes/ha and 3.7 tonnes/ha, respectively (rainfed wheat yields under high inputs). By contrast, France and Hungary are very similar as to their agro-ecological environments for wheat growing since both have agro-ecologically attainable yields of around 6.5 tonnes/ha. However, Hungary's actual yield is only 3.9 tonnes/ha, compared with France's 7.1 tonnes/ha. The gap indicates that there is considerable agronomic potential for yield growth in Hungary if a

host of other conditions (economic, marketing, etc.) were to become closer to those of France. However, this does not mean that it would be economically efficient for Hungary to emulate France's overall economic and policy environments in relation to wheat, e.g. the high support and protection afforded by the Common Agricultural Policy (CAP).

Table 11.1 shows the agro-ecologically attainable national average wheat yields for more countries and compares them with actual prevailing yields. These countries span a wide range of agro-ecological endowments for wheat production, with some countries having a high proportion of their "wheat land" in the very suitable category (Uruguay) and others having high proportions in the suitable and moderately suitable categories, e.g. Brazil, Paraguay and Sweden. Attainable average yields in these countries range from 7-7.5 tonnes/ha in Germany and Poland, through 5.0-5.8 tonnes/ha in the United States, Uruguay and Sweden and 4.0-4.8 tonnes/ha in Turkey, Russia, Canada, Australia, Argentina and Ethiopia, to 3.0-3.4 tonnes/ha in Paraguay, Brazil and the United Republic of Tanzania.

The divergence between economically efficient and agro-ecologically attainable yields can be very wide. For example, Uruguay and Sweden have nearly equal agro-ecologically attainable yields (5.0-5.3 tonnes/ha, although Uruguay has more land suitable for wheat growing than Sweden) but

Table 11.1   Agro-ecological similarity for rainfed wheat production, selected countries

| | Area suitable for rainfed wheat | | | | Yields attainable | | | | Actual | |
| | Total | % of area by suitability class | | | Tonnes/ha | | | | Average 1996/2000 | |
	mln ha	VS	S	M	VS	S	M	Average all classes	Area (mln ha)	Yield (tonnes/ha)
Germany	16.9	42.5	39.2	18.3	9.0	7.1	5.2	7.6	2.7	7.3
Poland	17.6	26.6	51.0	22.5	8.7	7.2	5.1	7.1	2.5	3.4
Japan	6.4	31.0	39.7	29.3	8.9	7.0	5.1	7.1	0.2	3.4
Lithuania	5.5	1.3	72.1	26.7	8.2	7.3	5.3	6.8	0.3	2.8
Belarus	16.5	1.2	64.8	34.0	8.2	7.4	5.4	6.7	0.3	2.5
United Kingdom	11.9	4.0	70.6	25.4	8.4	7.2	4.8	6.7	2.0	7.8
France	24.6	26.0	45.6	28.4	8.4	6.7	4.7	6.6	5.2	7.1
Italy	7.6	31.0	46.9	22.2	8.6	6.2	4.0	6.5	2.4	3.2
Hungary	6.1	11.6	51.5	36.9	8.5	6.8	5.2	6.4	1.1	3.9
Romania	8.4	14.6	50.8	34.5	9.1	6.8	4.5	6.3	2.0	2.5
Latvia	5.4	5.8	64.1	30.1	6.6	6.8	4.9	6.2	0.2	2.5
Ukraine	30.8	15.3	40.5	44.2	8.9	6.9	4.6	6.2	5.9	2.5
United States	230.4	18.8	54.1	27.1	6.5	6.1	4.6	5.8	23.7	2.7
Uruguay	13.8	66.7	28.8	4.5	5.8	4.5	3.2	5.3	0.2	2.3
Sweden	4.3	0.0	54.8	45.2	0.0	5.7	4.2	5.0	0.4	6.0
Turkey	7.6	8.2	31.3	60.4	5.7	5.9	4.0	4.8	9.1	2.1
Russia	167.4	7.5	36.5	56.0	6.2	5.5	3.5	4.4	24.8	1.4
Canada	42.2	10.7	35.0	54.3	6.3	5.6	3.1	4.3	10.9	2.4
Australia	24.3	17.5	38.0	44.5	6.2	4.5	3.2	4.2	11.1	2.0
Argentina	61.1	22.7	45.5	31.8	5.3	4.3	3.1	4.2	6.0	2.4
Ethiopia	10.5	26.3	43.0	30.7	5.1	4.1	3.0	4.0	0.9	1.2
Paraguay	6.9	0.0	39.8	60.3	0.0	4.2	2.9	3.4	0.2	1.4
Brazil	24.4	8.8	32.6	58.6	4.5	3.7	2.9	3.3	1.4	1.8
Tanzania, United Rep.	5.5	24.4	41.2	34.4	4.0	3.1	2.1	3.0	0.1	1.5
Myanmar	5.4	2.6	38.8	58.5	3.2	2.8	2.3	2.5	0.1	0.9

Note: Countries with predominantly rainfed wheat with over 5 million ha of land in the wheat suitability classes VS (very suitable), S (suitable) and MS (moderately suitable) under high input. See Box 4.1 for an explanation of classes. All data on potentials exclude marginally suitable land which in the GAEZ analysis is not considered appropriate for high-input farming.

actual yields are 6 tonnes/ha in Sweden (in practice exceeding what the GAEZ evaluation suggests as attainable on the average) and 2.3 tonnes/ha in Uruguay. In spite of Uruguay's yields being a fraction of those that are agro-ecologically attainable and of those prevailing in Sweden, it is not necessarily a less efficient wheat producer than Sweden in terms of production costs. Other examples of economically efficient wheat producers with low yields in relation to their agronomic potential

include Australia (2.0 tonnes/ha actual versus 4.2 tonnes/ha agro-ecologically attainable) and the United States (2.7 tonnes/ha versus 5.8 tonnes/ha).

The yield gap in relation to agronomic potential is an important element when discussing agronomic potentials for yield growth. For the countries in which we find large differences between actual and attainable, it seems probable that factors other than agro-ecology are responsible. Yields in these countries could grow some way towards bridging the gap between actual and attainable if some of these factors could be changed, e.g. if prices rose. We could then take the countries with a sizeable "bridgeable" gap and see their aggregate weight in world production of a particular crop. If the weight is significant, then the world almost certainly has significant potential for increasing production through yield growth, even on the basis of existing knowledge and technology (varieties, farming practices, etc.).

Among the major wheat producers, only the EU countries (the United Kingdom, Denmark, France and Germany) have actual yields close to, or even higher than, those attainable for their agro-ecological endowments under rainfed high-input farming. In all other major producers with predominantly rainfed wheat production (11 countries) the gaps between actual and attainable yields are significant. This is shown in Figure 11.3. These 11 countries account for 37 percent of world wheat production. If we assumed that half of their yield gap (attainable minus actual) were "bridgeable", their collective production could increase by some 60 percent without any increase in their area under wheat – an increment equal to 23 percent of current world output. Yield growth would also occur in the other countries accounting for the rest of world production, including the major producers with irrigated wheat not included in

those shown in Figure 11.3, such as China, India, Pakistan and Egypt. All this is without counting the potential yield gains that could come from further improvement in varieties, since the agro-ecologically attainable yields of the GAEZ reflect the yield potential of existing varieties.

Some states in India, such as the Punjab, are often quoted as examples of areas where wheat and rice yields have been slowing down or are even reaching a plateau. Fortunately, India is one of the few countries for which data at subnational level and distinguished by rainfed and irrigated area are available. Table 11.2 compares wheat and rice yields by major growing state with the agro-ecologically attainable yields, taking into account irrigation. It shows that, although yield growth has indeed been slowing down, in most cases actual yields are still far from agro-ecologically attainable yields (with a few exceptions such as wheat in Haryana). This suggests that there are still considerable bridgeable yield gaps in India.

The discussion above gives an idea of the scope for wheat production increases through the adoption of improved technologies and practices to bridge some of the gap that separates actual yields from obtainable yields. The broad lesson of experience seems to be that if scarcities develop and prices rise, farmers quickly respond by adopting such technologies and increasing production, at

Figure 11.3 Wheat: actual and agro-ecologically attainable yields (rainfed, high input)

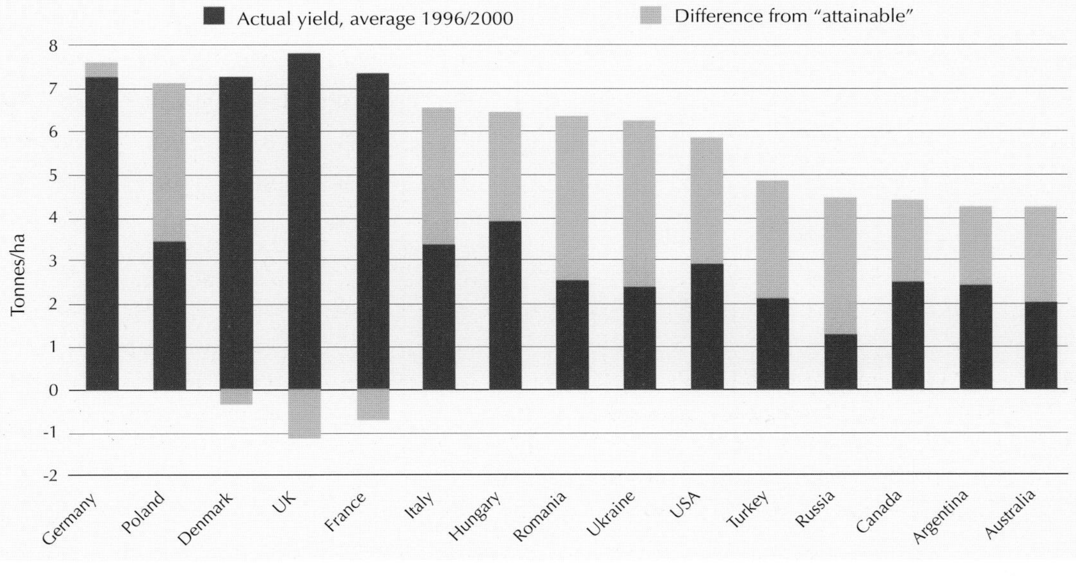

Note: Fifteen countries with a predominantly rainfed wheat production of over 4 million tonnes in 1996/2000

Table 11.2 Wheat and rice yields in India, by state[1]

Wheat	1972/74	1984/86	1993/95	1972-85	1986-95	Area 1997/98	Land under irrigation 1995	Prod. 1997/98	Actual yield 1997/98	Maximum attainable yield (AEZ) Rainfed	Irrigated	Weighted
	kg/ha			% p.a.		mln ha	%	mln tonnes		kg/ha		
India total	1 260	1 947	2 477	3.9	2.9	27.1	86	68.6	2 534	1 786	4 352	3 998
All states below						25.8	85	66.4	2 571			
Uttar Pradesh	1 123	1 933	2 423	4.8	2.6	9.4	93	23.0	2 503	2 932	5 143	4 990
Punjab	2 280	3 263	3 993	3.4	2.5	3.3	97	13.6	4 093	1 994	5 661	5 544
Haryana	1 683	2 840	3 663	4.3	3.1	2.1	49	8.1	3 788	2 634	5 481	4 020
Madhya Pradesh	773	1 147	1 660	3.6	3.9	4.6	71	7.8	1 684	1 008	3 824	2 996
Rajasthan	1 200	1 900	2 213	4.0	2.1	2.7	96	6.8	2 494	1 940	4 279	4 192
Bihar	875	1 593	2 060	6.6	3.5	2.1	88	4.5	2 165	1 401	4 113	3 798
Gujarat	1 673	2 000	2 280	2.2	1.9	0.7	81	1.7	2 400	877	3 173	2 737
Maharashtra	580	777	1 377	3.2	6.1	0.9	65	1.0	1 094	1 356	3 144	2 520

Rice (paddy)	1973/75	1984/86	1994/96	1973-85	1986-96	Area 1998/99	Land under irrigation 1997	Prod. 1998/99	Actual yield 1998/99	Maximum attainable yield (AEZ) Rainfed	Irrigated	Weighted
	kg/ha			% p.a.		mln ha	%	mln tonnes		kg/ha		
India total	1 630	2 215	2 830	2.2	2.6	44.0	51	126.4	2 871	2 516	8 161	5 395
All states below						40.5	50	116.1	2 867			
West Bengal	1 725	2 305	3 090	1.5	3.0	5.9	26	19.9	3 374	4 105	8 051	5 147
Uttar Pradesh	1 195	2 030	2 815	3.7	3.2	5.8	64	17.8	3 080	2 133	8 322	6 088
Andra Pradesh	2 350	3 150	3 875	3.2	2.1	3.8	96	15.0	3 909	1 638	8 182	7 894
Tamil Nadu	2 810	3 205	4 855	0.6	2.7	2.3	93	11.3	4 870	2 066	8 188	7 741
Punjab	3 185	4 655	5 011	3.4	0.6	2.4	99	11.9	4 963	1 463	8 914	8 847
Bihar	1 310	1 590	1 980	0.1	1.4	5.1	41	10.3	2 022	3 611	8 214	5 489
Orissa	1 265	1 670	2 125	1.2	3.0	4.5	37	8.7	1 944	2 180	7 457	4 132
Madhya Pradesh	1 000	1 405	1 725	1.2	2.5	5.4	24	7.4	1 385	1 450	7 905	2 973
Karnataka	2 635	2 850	3 570	1.2	3.2	1.4	68	5.1	3 678	1 916	8 131	6 136
Assam	1 500	1 650	2 015	0.7	2.5	2.5	21	5.0	2 028	6 426	7 733	6 700
Maharashtra	1 415	2 155	2 425	4.0	2.3	1.5	28	3.6	2 464	1 330	8 150	3 246

Note: [1] States in descending order of latest year production. Agro-ecological zone (AEZ) yields: rainfed under mixed inputs and irrigated under high inputs. The weighted AEZ yield (last column) was derived by applying the percentage of land under irrigation as a weighting factor.

Source for data: India Department of Agriculture Cooperation: Statistics at a glance, March 2001.

least those living in an environment of not too difficult access to improved technology, transport infrastructure and supportive policies. However, in countries with land expansion possibilities, the quickest response comes from increasing land under cultivation, including shifting land among crops towards the most profitable ones. Argentina's example is instructive: mostly from land expansion, it increased wheat production by 68 percent in 1996 and maize production by 48 percent in 1997 and another 25 percent in 1998, following price rises in the immediately preceding years.

Countries use only part of the land that is suitable for any given crop. This does not mean that land lies bare or fallow waiting to be used for increasing production of that particular crop. In most cases the land is also suitable for other crops and in practice is used for other crops (see Box 4.2). The point made here is that the gap existing between yields actually achieved and those obtainable under high-input technology packages affords significant scope for production increases through yield growth, given conducive socio-economic conditions, incentives and policies. The point is not that production increases can be obtained by expanding cultivation into land suitable for a particular crop, because such land may not be available if it is used for other crops.

Moreover, even if there probably is sufficient slack in world agriculture to support further increases in global production, this is small consolation to food-insecure people who depend for their nutrition on what they themselves produce. Such people often live in semi-arid agricultural environments where the slack for increasing production can be very limited or non-existent. The fact that the world as a whole may have ample potential to produce more food is of little help to them.

The preceding discussion may create the impression that all is well from the standpoint of potential for further production growth based on the use of existing varieties and technologies to increase yields. Nothing is further from the truth, for two main reasons:

■ Exploitation of the yield gaps as defined in the preceding discussion means further spread of the conventional high external input technologies, which is precisely what we should be trying to mitigate if we are to avoid aggravation of the related environmental problems.

■ Perhaps more important from the standpoint of meeting future demand, ready potential for yield growth does not necessarily exist in the countries where the additional demand will be, e.g. in the mature green revolution areas of India and other developing countries. When potential demand is in countries with limited import capacity, as is the case in many developing countries, such potential can be expressed as effective demand only if it can be predominantly matched by local production. As noted in Chapters 2 and 8, increases in local production in these countries, in addition to adding to food supplies, stimulate the demand for food because they create employment and incomes and stimulate the wider rural economy. In such circumstances, the existence of large exploitable yield gaps elsewhere (e.g. in Argentina or Ukraine) is less important than it appears for the evaluation of potential contributions of yield growth to meeting future demand.

It follows that continued and intensified efforts are needed on the part of the agricultural research community to raise yields (including through maintenance and adaptive research) in the often unfavourable agro-ecological and socio-economic environments of the countries where the additional demand will be. It is thought (see below) that biotechnology will play an important role here, as it has the potential to be a more efficient instrument than conventional plant breeding for overcoming constraints inherent in such environments (semi-aridity, susceptibility to pest infestations, etc.; see Lipton, 1999).

11.2 Technologies in support of sustainable agriculture

Various approaches have been developed in the past few decades to minimize the environmentally detrimental effects of agricultural production. Among the foremost of these are integrated pest management (IPM), Integrated Plant Nutrient Systems (IPNS) and no-till/conservation agriculture (NT/CA). Rather than as isolated technologies they should be seen as complementary elements of sustainable agriculture.

The conventional model of agricultural development stresses increased production and intensification through progressively specialized operations. By contrast, the approaches discussed in this section seek to meet the dual goals of increased productivity and reduced environmental impact. They do this through diversification and selection of inputs and management practices that foster positive ecological relationships and biological processes within the entire agro-ecosystem. With the help of participatory research and extension approaches, the principles of these technologies can be developed further into location-specific sustainable resource management systems. Even though each of these three approaches has some distinct features, many of the specific technologies used are, to various degrees, found in all of the approaches discussed in this section.

Sustainable agriculture is not a concretely defined set of technologies, nor is it a simple model or package that can be widely applied or is fixed over time. The lack of information on agro-ecology and the high demand for management skills are major barriers to the adoption of sustainable agriculture. For example, much less is known about these organic and resource-conserving technologies than about the use of external inputs in modernized systems.

11.2.1 Integrated pest management

Crop, forestry and livestock production systems throughout the world suffer losses caused by diseases, weeds, insects, mites, nematodes and other pests. The intensification of farming, forestry and livestock production favours pest buildup, and the high-yielding varieties and breeds utilized are often more susceptible to pests than traditional ones. The impact of many of these problems can be reduced with the help of pesticides but at a cost, including negative health and environmental effects. Because most chemical pesticides are hazardous to human health and toxic to many non-target organisms, there are potential hazards associated with their manufacture, distribution and application, particularly if pesticides are misused (GTZ, 1993). These hazards include exposure during handling or application, pesticide residues in or on foodstuffs, pollution of the environment (soil, groundwater, surface waters and air) and killing of non-target organisms. Because of the disruption of natural enemies, there has been a resurgence of existing pests and an

outbreak of new ones. Almost all economically significant pests are already resistant to at least one chemical pesticide.

The goal of IPM is to avoid or reduce yield losses by pests while minimizing the negative impacts of pest control. The term IPM was originally used to describe an approach to pest control with the primary aim of reducing the excessive use of pesticides while achieving zero pest incidence. The concept has broadened over time. Today IPM can best be described as a decision-making and action-oriented process that applies the most appropriate pest control methods and strategy to each situation. To ensure the success of this process, the presence and density of pests and their predators and the degree of pest damage are systematically monitored. No action is taken as long as the level of the pest population is expected to remain within specified limits.

IPM promotes primarily biological, cultural and physical pest management techniques, and uses chemical ones only when essential. Naturally occurring biological control is encouraged, for example through the use of alternate plant species or varieties that resist pests, as is the adoption of land management, fertilization and irrigation practices that reduce pest problems. If pesticides are to be used, those with the lowest toxicity to humans and non-target organisms should be the primary option. Precise timing and application of pesticides are essential. Broad spectrum pesticides are used only as a last resort when careful monitoring indicates they are needed according to pre-established guidelines. This broader focus, in which judicious fertilizer use is also receiving attention (see the next section), is also referred to as integrated production and pest management (IPPM).

The Centre for Research and Information on Low External Input for Sustainable Agriculture distinguishes three stages in the development of IPM (IPMEurope Web site, 2002). In the first stage, the concept of pest population thresholds and targeted pests was introduced. Later, diseases and weeds were added to address more comprehensively the many crop protection problems that farmers face. In the second stage, crop protection was integrated with farm and natural resource management. Indigenous knowledge and traditional cropping practices were studied and adapted, while proper natural resource management became important because of the role of biodiversity in biological control. A whole-farm approach was thus adopted and integrated crop management practised to solve the conflicting needs of agricultural production and the environment.

In the third stage came the integration of the natural and social sciences. Most IPM projects now develop around a dynamic extension model, the farmer field school (FFS), which emphasizes farmers' ability to experiment and draw conclusions, and enhances their ability to make decisions. The knowledge base has been expanded for a wide range of crops both in terms of new technologies and ecological aspects. Much of this IPM knowledge has still not reached the farm level and lacks site-specific adaptation.

Experience shows that IPM has economic and other benefits for farmers and farm households. However, national policy frameworks in many developing countries have tended to strongly favour pesticide use through subsidies that distorted prices. Because of this, alternative pest control measures, even where successful technically, are often not financially competitive and farmers are reluctant to adopt them. In addition, generally weak extension services lack the capacity for the intensive educational programmes needed to familiarize and train farmers in the use of IPM practices.

In spite of these problems, IPM has been introduced successfully in many countries and for many different crops such as rice, cotton and vegetables. In Cuba, IPM has been integrated successfully into organic farming. Where farmers have had no previous access to chemical pesticides, the introduction of plant protection based on IPM is the preferred option to avoid financially and environmentally costly overdependence on pesticides.

IPM applied to rice has shown good to dramatic improvements in production, in some cases simultaneously reducing costs. Human capacities and networks developed for rice will continue to provide support for new initiatives. Combined with the proven successes, they will promote the introduction of IPM in other crops or cropping systems, particularly vegetables and cotton. Unfortunately, a quantitative evaluation of the uptake in terms of hectares covered and reduced pesticide use is only available for a few projects, making a global or regional estimate of its present and future use impossible.

11.2.2 Integrated Plant Nutrient Systems

Any agricultural crop production – extensive or intensive, conventional or organic – removes plant nutrients from the soil. Nutrient uptake varies according to soil types and the intensity of production. An increase in biomass production results in a higher plant nutrient uptake. Imbalance in the availability of nutrients can lead to mining of soil reserves of nutrients in short supply and to losses of plant nutrients supplied in excess. Insufficiency of one plant nutrient can limit the efficiency with which other plant nutrients are taken up, reducing crop yields. For a farming system to be sustainable, plant nutrients have to be replenished. The nutrient mining that is occurring in many developing countries is a major but often hidden form of land degradation, making agricultural production unsustainable.

IPNS aim to maximize plant nutrient use efficiency by recycling all plant nutrient sources within the farm and by using nitrogen fixation by legumes to the extent possible. This is complemented by the use of external plant nutrient sources, including manufactured fertilizers, to enhance soil productivity through a balanced use of local and external sources of plant nutrients in a way that maintains or improves soil fertility (FAO, 1998e). At the same time IPNS aim at minimizing plant nutrient losses to avoid pollution of soils and water and financial losses to the farmer.

At the plot level, IPNS are designed to optimize the uptake of plant nutrients by the crop and increase the productivity of that uptake. At the farm level, IPNS aim to optimize the productivity of the flows of nutrients passing through the farming system during a crop rotation. The decision to apply external plant nutrients is generally based on financial considerations but is also conditioned by availability and perceived production risks.

Advice on quantities of nutrients to be applied may be based on empirical results from experiments in farmers' fields, which provide information on the impact of combined nutrient applications, timing of nutrient supply and sources of nutrients on crop yields. In the absence of such detailed information, knowledge of the quantities of nutrients removed by crops at the desired yield level provides a starting-point for estimating nutrient requirements.

Improved plant nutrition management will be important for environmentally and economically sustainable crop production, be it conventional or organic. However, the rate of spread of IPNS and their implications for the use of mineral fertilizers in agricultural production cannot be predicted in isolation. Precise management of fertilizer use can raise efficiency by 10 to 30 percent and should therefore be included in all production systems aiming for sustainability, even if they do not emphasize IPNS.

11.2.3 No-till/conservation agriculture

By far the largest extent of agricultural land continues to be ploughed, harrowed or hoed before every crop. These conventional tillage practices aim to destroy weeds and loosen the topsoil to facilitate water infiltration and crop establishment. This recurring disturbance of the topsoil buries any soil cover and may destabilize the soil structure so that rainfall can cause soil dispersion, sealing and crusting of the surface. An additional problem of conventional tillage is that it often results in compacted soils, which negatively affect productivity.

This negative impact of soil tillage on farm productivity and sustainability, as well as on environmental processes, has been increasingly recognized. In response to the problem, no-till/conservation agriculture (NT/CA) has been developed. NT/CA maintains and improves crop yields and resilience against drought and other hazards, while at the same time protecting and stimulating the biological functioning of the soil. Various terms are used for variants of NT/CA in different countries, depending on the perceived importance of one or another aspect of the approach: zero tillage; minimum or low tillage; *plantio directo na palha* (direct planting in straw); *siembra directo permanente* (permanent direct seeding); and conservation tillage.

The essential features of NT/CA are: minimal soil disturbance restricted to planting and drilling; maintenance of a permanent cover of live or dead vegetal material on the soil surface; direct sowing; crop rotation combining different plant families (e.g. cereals and legumes); adequate biomass generation; and continuous cropland use. In some countries the above-mentioned systems might lack some essential features of NT/CA and will therefore not have the same beneficial effects.

Soil cover is needed to protect the soil from the impact of rainfall, which would destroy the porosity of the soil surface, leading to runoff and erosion. Crops are seeded or planted through this cover with special equipment or in narrow cleared strips. Direct planting or seeding is linked with NT/CA, since any more general tillage would bury most or all of the vegetal cover. Crop sequences are planned over several seasons to minimize the buildup of pests or diseases and to optimize plant nutrient use by synergy among different crop types and by alternating shallow-rooting crops with deep-rooting ones. When the same crop or cover crops are repeated on the same piece of land each year, NT/CA is an imperfect and incomplete system, because diseases, weeds and pests tend to increase and profits tend to decrease (Derpsch, 2000). The cropland is being used continuously and no burning of residues is allowed.

Besides protecting the soil against erosion and water loss by runoff or evaporation, the soil cover also inhibits the germination of many weed seeds, minimizing weed competition with the crop. After the first couple of years of NT/CA on a field, the stock of viable weed seeds near the soil surface usually declines, often to the point where weed incidence becomes minor, with remnant populations at scattered spots in the field. In the first few years, however, herbicides may still need to be applied. Systems without continuous soil cover or crop rotation may not even reduce the incidence of weed in the long term (e.g. wheat monoculture in the United States).

After a number of years, yields have often risen to some 20 to 50 percent higher than what they were before under conventional procedures. The yields also become less variable from year to year. Labour costs can be significantly lower, and labour demand is distributed much more evenly over the year. Input costs are lower as well, particularly for machinery once the initial investments have been made. In mechanized farming less fuel is needed and smaller tractors can be used or fewer draft animals are needed for a given area; in areas without these power sources, the heavy manual work preparatory to crop establishment is drastically reduced (see also Section 4.6.2).

There are several reasons, however, for the continued dominance of conventional tillage-based agriculture. There is a natural reluctance to change approaches that have been working in past years or for decades. Conventional wisdom on the benefits of ploughing and a lack of knowledge on the resulting damage to the soil system tend to maintain plough-based agriculture. Also, the transition to NT/CA is not free of cost, nor particularly simple. During the transition years, there are extra costs for tools and equipment. Higher weed incidence may increase herbicide costs initially and the yields and resilience against drought will improve only gradually.

A more important impediment to the successful introduction of NT/CA is probably the required

---

Box 11.1 No-till development support strategy: the Brazil experience

Large-scale expansion in Brazil to the current more than 10 million ha started in about 1980, after small and local initiatives during the 1960s. Large farmers used methods and equipment first from the United States and later from local manufacturers. Small farmers, with animal or small mechanical draught power, followed more than a decade later. During this period, small manufacturers together with innovative farmers designed smaller prototypes and started producing and marketing equipment adapted to small farms, including knife rollers to manage crop residues and combined direct seeders/fertilizer applicators.

The success of NT/CA in Brazil cannot be attributed to technical parameters alone. In conjunction with technical innovation, an effective participatory approach to adaptive research and technology transfer was adopted that tied farmers into a development strategy suited to their specific requirements. Institutional support was demand driven and concentrated on training and education that equipped participating farmers with the skills to adapt and refine NT/CA on their own farms. The cornerstones of the development support strategy were:

- *close collaboration between researchers, extensionists, the private sector and farmers* for the development, adoption and improvement of NT systems;

- *onfarm trials and participatory technology development;*

- *strengthening of farmers' organizations;* creation of local "Friends of the Land Clubs" where farmers exchange information and experiences and improve their access to extension and other advisory services as well as input and output marketing;

- *close cooperation with existing and new cooperatives* concentrating primarily on marketing and training for vertical diversification into livestock and processing;

- *aggressive dissemination strategy of technical, economic and environmental information* through the media, written documents, meetings and conferences – controlled and managed by producers' organizations (Friends of the Land Clubs) with emphasis on farmer-to-farmer exchange of experiences;

- *the national NT farmers' organization FEBRAPDP* played a significant role in advocating and supporting the promotion of NT/CA on large and small farms. As NT systems are complex to manage and require efficient farm management, training in record-keeping and a holistic understanding of farming systems' dynamics have been an integral aspect of support to small farmers;

- *private-public partnerships;* agro-input companies (Zeneca and Monsanto) supported demonstration projects in large and small farms through the provision of inputs and extension services;

- *targeted subsidies;* short-term subsidies played a significant part in supporting small farmer adoption of NT practices. In Paraná much of the hand-held or animal-drawn equipment was acquired with financial support from the state in the context of development programmes (mainly World Bank). Subsidized or free equipment is still made available to groups of farmers. Apart from economic constraints to adoption, the rationale for public subsidies has been the generation of offsite benefits from NT adoption. In some instances, private companies provided equipment for small farmers;

- *integration of crops and livestock;* special attention has been paid to the incorporation of crops and livestock (including poultry, hog and fish farming). A particular challenge is the development of rotational grazing patterns on cover crops, which do not jeopardize the sustainability of NT systems;

- *incorporation of environmental considerations;* correcting watershed degradation (e.g. soil erosion, pollution of streams and lakes and road damage) was a key reason for the adoption of NT farming practices. Environmental awareness raising among farmers also resulted in central facilities for the disposal of pesticide containers, household sanitation and recovery of gallery forests.

Source: Evers and Agostini (2001).

complex management skills. Any production system that includes crop rotation (see also Section 11.3 on organic agriculture) is more complex as it calls for coherent management over more than one or two crop seasons. Farmers will need to understand the new system and the reasons for the various procedures, and adapt them to their specific needs and conditions to balance crop rotation with market requirements. In mixed agriculture-livestock systems, practices such as stall-feeding or controlled grazing will need to replace free grazing on harvested fields.

NT/CA farmers appear to be keen to learn and embrace new developments (Derpsch, 2000). Being acquainted with more holistic management approaches to farming, many NT/CA farmers have introduced aspects of organic agriculture or converted entirely to organic agriculture where a market for organic products exists. On the other hand, some organic farmers have successfully adopted NT farming. Moreover, NT/CA farmers have also been faster adopters of IPM approaches than conventional ones (Pieri *et al.*, 2002).

The initial introduction of NT/CA in a new area, its adaptation to the environmental, social and economic conditions and its validation and demonstration in representative farms depend partly on the people involved. They require the determined and sustained efforts of competent, innovative governmental or non-governmental organizations and an active learning attitude of some of the most change-minded farmers and farmers' groups as well as the extension staff. Once NT/CA has been shown to work well on several farms in a given environment, the practices tend to spread spontaneously over large areas. Farmers need professional contacts with each other and local manufacturers need to be in a position to supply the necessary tools and equipment. During the initial phase many farmers will need some financial support in the form of loans or grants.

Some or all elements of NT/CA have been applied by farmers so far on between 50 and 60 million ha worldwide. Almost half of this is in the United States, where the area under zero tillage tripled over the last decade to about 23 million ha (USDA, 2001e), responding to government conservation requirements and to reduce fuel costs. But a considerable share of this is under monoculture and misses two essential features, namely full soil cover and adequate crop rotation, and cannot therefore be classified as NT/CA. In Paraguay about half of all the cropped land is under elements of NT/CA, mainly zero tillage. The area increased from about 20 000 to almost 800 000 ha between 1992 and 1999 because the government assisted by sharing part of the initial costs of conversion.

The spread of NT/CA approaches in the next three decades is expected to be considerable but, in addition to the constraints mentioned above, expansion will for several reasons vary widely across countries. Investment is needed to restore nutrient-depleted soils before crop residues can be produced in adequate amounts to satisfy the needs of livestock and maintain a soil cover. In arid areas without irrigation, the amounts of crop residues generally will not be sufficient for effective NT/CA systems. In some countries, established extension services or staff have been actively discouraging farmers from converting to NT/CA, while in others the scientific or extension institutes are not able to initiate the onfarm experiments needed to adapt and validate NT/CA systems locally. Even under favourable circumstances, it can take years before the new production system is widely known, understood and appreciated. A further ten years might be needed for its practical application over a large part of the country or a major farming system area (for example, the South Asian rice-wheat area, or the Brazilian *cerrados*).

11.3 Organic agriculture

Organic agriculture is a production management system that aims to promote and enhance ecosystem health, including biological cycles and soil biological activity (Box 11.2). It is based on minimizing the use of external inputs, and represents a deliberate attempt to make the best use of local natural resources. Methods are used to minimize pollution of air, soil and water (FAO/WHO, 1999), although they cannot ensure that products are completely free of residues, because of general environmental pollution. Organic agriculture comprises a range of land, crop and animal management procedures. Unlike food labelled as "environmentally friendly", "natural" or "free-range", organic agriculture is circumscribed by a set of rules and limits, usually enforced by inspection and certification mechanisms. Other terms used, depending on the language, are "biological" or "ecological". "Biodynamic" refers to commodities that are produced according to organic and other additional requirements.

Synthetic pesticides, mineral fertilizers, synthetic preservatives, pharmaceuticals, GMOs, sewage sludge and irradiation are prohibited in all organic standards. Plant nutrient or pesticide inputs derived directly from natural sources are generally allowed, as is a minimum of pretreatment before use (water extraction, grinding, etc.). Industrially produced pesticides, for example, may not be applied in organic agriculture, but an extract of neem (*Azadirachta indica*) leaves, which have biocidal properties, is currently allowed.

Most industrial countries, but few developing countries, have national organic standards, regulations and inspection and certification systems that govern the production and sale of foods labelled as "organic". At the international level, the general principles and requirements applying to organic agriculture are defined in the Codex guidelines (FAO/WHO, 1999) adopted in 1999. The growing interest in organic crop, livestock and fish products is mainly driven by health and food quality concerns. However, organic agriculture is not a product claim that organic food is healthier or safer, but rather a process claim intending to make food production and processing methods respectful of the environment.

Organic agriculture, broadly defined, is not limited to certified organic farms and products only. It also includes non-certified ones, as long as they fully meet the requirements of organic agriculture. This is the case for many non-certified organic agricultural systems in both developing and industrial countries where produce is consumed locally or sold directly on the farm or without labels. The extent of these systems is difficult to estimate since they operate outside the certification and market systems (El-Hage Scialabba and Hattam, 2002).

Organic practices that encourage soil biological activity and nutrient cycling include: manipulation of crop rotations and strip cropping; green manuring and organic fertilization (animal manure, compost, crop residues); minimum tillage or zero tillage; and avoidance of pesticide and herbicide use. Research indicates that organic agriculture significantly increases the density of beneficial invertebrates, earthworms, root symbionts and other micro-organisms (fungi, bacteria) (FiBL, 2000). Properly managed organic agriculture reduces or eliminates water pollution and helps conserve water and soil on the farm. Some countries (e.g. France and Germany) compel or subsidize farmers to use organic techniques as a solution to nitrate contamination in groundwater.

Box 11.2 What is an organic production system designed to do?

- Enhance biological diversity within the whole system.

- Increase soil biological activity.

- Maintain long-term soil fertility.

- Recycle wastes of plant and animal origin in order to return nutrients to the land, thus minimizing the use of non-renewable resources.

- Rely on renewable resources in locally organized agricultural systems.

- Promote the healthy use of soil, water and air as well as minimize all forms of pollution that may result from agricultural practices.

- Handle agricultural products with emphasis on careful processing methods in order to maintain the organic integrity and vital qualities of the product at all stages.

- Become established on any existing farm through a period of conversion, the appropriate length of which is determined by site-specific factors such as the history of the land and the type of crops and livestock to be produced.

Source: FAO/WHO (1999).

11.3.1 Land under organic management

Growth rates of land under organic management are impressive in western Europe, Latin America and the United States. Between 1995 and 2000, the total area of organic land tripled in western Europe and the United States. In the United States, land under certified organic agriculture has been growing by 20 percent p.a. since 1989, while in western Europe average annual growth rates have been around 26 percent since 1985 (with greatest increases since 1993). In 1999 alone, the United Kingdom experienced a 125 percent growth of its organic land area. However, these dramatic increases must be viewed against the small starting base levels. In some cases they may reflect a reclassification of land rather than an actual switch in farming systems.

Policy measures were instrumental in persuading small farmers to convert to organic farming by providing financial compensation for any losses incurred during conversion. The guidelines established by the organic agriculture community in the 1970s were formalized by national and supranational legislation and control systems (e.g. first in Denmark in 1987, followed in 1992 by Australia and the EU: Reg. no. 2092/91). The role of organic agriculture in reaching environmental policy objectives, including sustainable use of land set-aside (Lampkin and Padel, 1994), led to the adoption of agri-environmental measures that encourage organic agriculture (e.g. the 1992 reform of the Common Agricultural Policy and accompanying EU Reg. no. 2078/92).

The estimates given below are derived from a compilation of available information by the Foundation Ecology and Agriculture (SÖL) in Germany. In the absence of a statistical database on organic agriculture in FAO, this is the most comprehensive source available (Willer and Yussefi, 2002). SÖL reports a global area of land under certified organic agriculture of 15.8 million ha of which 48.5 percent are located in Oceania (Australia 7.7 million ha); 23.5 percent in Europe (with Italy having the highest area, nearly 1 million ha); 20 percent in Latin America (with Argentina having 3 million ha); 7.4 percent in North America (United States nearly 1 million ha); 0.3 percent in Asia; and 0.1 percent in Africa.

With 3 million ha, Argentina accounts for more than 90 percent of the certified organic land in Latin American countries and has the second largest area of organically managed land in the world after Australia. In both countries, however, most of this is grassland. Because of the large size of organic ranches in the pampas, the average size of organic farms is 3000 ha in Argentina. Some 85 percent of Argentina's organic production is exported.

In 2000, agricultural land under certified organic management averaged 2.4 percent of total agricultural land in western Europe, 1.7 percent in Australia, 0.25 percent in Canada and 0.22 percent in the United States. In most developing countries, agricultural land reported under certified organic production is minimal (less than 0.5 percent of agricultural lands). However, some traditional farms in developing countries have adopted modern organic management to improve their productivity, especially in areas where pesticides and fertilizers are inaccessible. The extent of such non-market organic agriculture is difficult to quantify but some attempts have been made. The Ghanaian Organic Agriculture Network, for example, estimates that there are around 250 000 families in South and East Africa farming around 60 million ha on an organic basis. Anobah (2000) estimates that over one-third of West African agricultural produce is produced organically.

A number of industrial countries have action plans for the development of organic agriculture. Targets are set for the sector's growth and resources are allocated to compensate farmers during, and sometimes after, the conversion period, and to support research and extension in organic agriculture. For example, the United Kingdom increased the budget of the Organic Farming Scheme to support conversion to organic agriculture by 50 percent for 2001-02 (£20 million per year) and allocations in the United States for the organic sector include US$5 million for research in 2001. India and Thailand have established their own organic standards to facilitate exports and to satisfy domestic demand. China, Malaysia and the Philippines are at present working towards establishing national standards.

11.3.2 Yields and profitability

Typically, farmers experience some loss in yields after discarding synthetic inputs and converting their operations from conventional, intensive systems to organic production. Before restoration of full biological activity (e.g. growth of soil biota, improved nitrogen fixation and establishment of natural pest predators), pest suppression and fertility problems are common. The degree of yield loss varies and depends on inherent biological attributes of the farm, farmer expertise, the extent to which synthetic inputs were used under previous management and the state of natural resources (FAO, 1999h). It may take years to restore the ecosystem to the point where organic production is economically viable.

Transition to organic management is difficult for farmers to survive without financial compensation, especially in high intensive input agriculture and in degraded environments. After the conversion period, organic agriculture achieves lower yields than high external input systems. Depending on the previous management level of specialization, yields can be 10 to 30 percent lower in organic systems, with a few exceptions where yields are comparable in both systems. In the medium term, and depending on new knowledge, yields improve and the systems' stability increases. In the longer term, performance of organic agriculture increases in parallel with improvements in ecosystem functions and management skills.

Yields do not usually fall, however, when conversion to organic agriculture starts from low-input systems (often traditional) that do not apply soil-building practices. A study from Kenya indicates that, contrary to general belief, organic agriculture in the tropics is not constrained by insufficient organic material (to compensate for the non-use or reduced use of external inputs), but instead shows a good performance (ETC/KIOF, 1998).

As discussed in Chapter 4, average fertilizer consumption will rise in developing countries. The average figure masks, however, that for many (especially small) farmers the purchase of manufactured fertilizers and pesticides is and will continue to be constrained by their high costs relative to output prices or simply by unavailability. Organic agriculture emphasizes understanding and management of naturally occurring production inputs (such as farmyard manure, indirect plant protection and own seed production) as an alternative to enhance yields. It will not be possible for organic agriculture to attain the high yields achieved with the use of synthetic inputs in high-potential areas. But organic management offers good prospects for raising yields and the sustainability of farming in resource-poor and marginal areas, and can raise the productivity of traditional systems while relying on local resources (Pretty and Hine, 2000).

For example, India is collecting nationwide information regarding the experiments being carried out in organic agriculture, with a view to reintroducing it as part of its traditional "*rishi*

Figure 11.4 Farm area under certified organic management

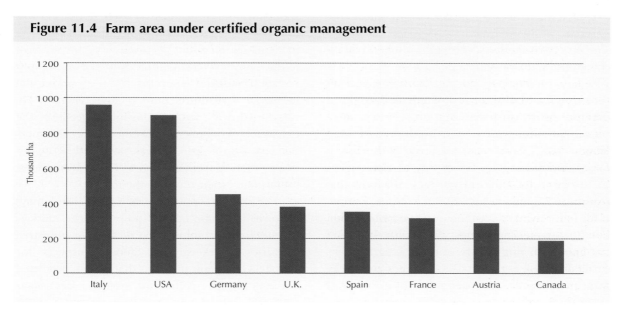

Source: Willer and Yussefi (2002).
Note: Australia (7.7 million ha) and Argentina (3 million ha) are not included in this graph because most of the organic land is extensive grassland

agriculture". In Latin America, hundreds of thousands of indigenous farmers along the Andes have turned to the organic movement to reinstate sophisticated agricultural practices developed by the Incas. Individual small family vegetable plots and groups/associations managing organic produce for domestic urban markets and small informal fairs are widespread. Cuba adopted organic agriculture as part of its official agricultural policy, with investments in research and extension, to compensate for shortages in external inputs. In 1999 (non-certified) organic urban agriculture (in self-provision gardens, raised container beds and intensive gardens) produced 65 percent of the country's rice, 46 percent of fresh vegetables, 38 percent of non-citrus fruit, 13 percent of roots, tubers and plantains and 6 percent of eggs (Murphy, 2000).

In organic systems, external inputs such as fertilizers, herbicides and machinery are replaced by labour, most often increasing women's work. Labour can either be a major constraint to organic conversion, or an employment provider to rural communities. Often the introduction of organic agriculture shifts gender distribution of labour as men prefer to be involved with mechanized agriculture. Women rarely own land and are dependent on access to common property. Since access to credit frequently requires land as collateral, women (and landless people) are largely excluded from the formal credit market. As a result, women seek methods that require little external inputs. Organic agriculture facilitates women's participation as it does not rely on financial inputs and access to credit.

The economic performance of organic agriculture in industrial countries (mainly in Europe) is determined by financial support from governments, premium prices for produce and high labour costs. An extensive analysis of European farm economics in terms of labour use, yields, prices, costs and support payments, concludes that profits on organic farms are, on average, comparable to those on conventional farms (Hohenheim, 2000). However, only a few studies have assessed the long-term profitability of organic agriculture. Profitability of organic systems relates to whole-farm production (total production of a variety of species and not single crop yields) over the entire rotation period. This includes both marketed products and non-food production (to feed animals and soils). Incomes achieved over a given season may appear high because of price premiums when excluding the low profits over rotational seasons. Supply constraints faced by organic farmers, which are expected to increase if the sector expands, include the provision of adequate organic inputs such as organic seeds (e.g. GMO-free), natural pest enemies and mineral rocks (e.g. rock phosphate).

An increased organic food supply above a certain level would lead to a decline in premium prices. A study for Denmark (SJFI, 1998) concludes that the primary agricultural sector income may not fall if fewer than 25 percent of Danish farmers were to adopt organic methods. Most countries are far below such a threshold.

11.3.3 Demand for organic products

On the demand side, promotion and marketing strategies of retailers and supermarkets, in particular of major food-retailing chains, have created new market opportunities for organic agriculture in industrial countries. Food-retailing chains, which also stock and promote organic foods as a tool to improve their public image, account for a major share of the retail markets for fresh as well as processed organic foods. Concerns about growth-stimulating substances, GM food, dioxin-contaminated food and livestock epidemics (such as bovine spongiform encephalopathy) have given further impetus to organic food demand as consumers increasingly question the safety of conventional foods. The most recent outbreak of foot-and-mouth disease has added to concerns over the soundness of industrial agriculture. Several governments have responded with declarations of targets for the expansion of organic production. Many consumers perceive organic products as safer and of higher quality than conventional ones. These perceptions, rather than "science", drive the market.

The market opportunities arising from these concerns have also opened possible niche markets for developing country exporters. Major industrial countries' markets offer good prospects for suppliers of organic products that are not produced domestically (e.g. coffee, tea, cocoa, spices, sugar cane and tropical fruit) as well as off-season products (such as fruit and vegetables) and processed foods. Liberalization and privatization policies in developing countries open the way for a greater role for organic entrepreneurs and producers' organizations. Markets for value-added products such as organic commodities can help counterbalance falling commodity prices and withdrawal of government support for agricultural inputs and other services. Price premiums range from 10 to 50 percent over prices for non-organic products. There is also government support for organic exports. Examples include the organic coffee programme of the Coffee Development Authority in Uganda; the promotion of organic exports by India's Ministry of Commerce; and support by the Argentinean government for the export of over 80 percent of the country's organic produce.

The size of domestic organic production is not necessarily related to the importance of organic markets. Australia, which has the world's largest area of organic land, most of which is grassland, has a market of US$170 million of organic food retails. Japan, on the other hand, which has only 5 000 ha of organic lands, is the second largest world organic market (US$2.5 billion of retail sales in 2000). The largest markets of organic foods are in western European countries (Germany being the most important market at present), Japan and the United States. The UNCTAD/WTO International Trade Centre (ITC, 1999) estimated retail sales of organic foods in the largest markets at US$20 billion in 2000, of which US$8 billion in Europe and the United States each, and US$2.5 billion in Japan.

In spite of dramatic growth rates, sales of organic agricultural products in industrial countries in 2000 represented less than 2 percent of total food sales at the retail level. However, in particular countries and for particular products, the market share of organic agricultural products can be appreciably larger. Organic food sales in Germany are 3 to 4 percent of total sales, while individual commodities such as organic milk products have over a 10 percent market share, with the figure for organic ingredients in baby foods in the range of 80 to 90 percent. Organic coffee, which accounts for 0.2 percent of world coffee consumption, accounts for 5 percent of the United States coffee market (Vieira, 2001). Some 100 developing countries produce organic commodities in commercial quantities, most of which are exported to industrial countries. Where they exist, devel-

oping countries' organic markets are still very limited and food is sold mainly in specialized stores in large cities. ITC (1999) estimates an annual sales growth of organic food between 10 and 40 percent over the medium term, depending on the market. Thus, organic food retail sales could grow from an average of 2 percent of total sales in 2000 to a share of 10 percent in major markets in a few years.

11.3.4 Long-term prospects for organic agriculture

The future growth of organic agriculture will depend more on supply constraints than on developments in demand, at least over the medium term. The tendency so far has been for the rate of demand growth to outstrip the rate of growth in available supplies. Developing countries are just starting to benefit from organic market opportunities but present conditions benefit primarily large producers and operators.

The supply and quality of organic raw material and rules governing organic production and processing might limit the extent to which developing countries could satisfy the demand for organic food in industrial countries. Organic food trade might be discouraged by difficulties in complying with foreign standards and costly control systems, especially if international equivalency is not established. Access to inspection and certification, as well as the need to develop new methods of processing organic food, are major challenges that are likely to be taken up by large and established food companies (Kristensen and Nielsen, 1997). Multinational food companies are expected to contract for and certify organic foods. In particular, the growth of processed organic foods will be facilitated by these companies' capacities to assemble ingredients from different parts of the world and to guide production to meet their specific needs. At the same time there are numerous opportunities for developing country producers and exporters to enter the markets for value-added organic products using simple technology.

Further long-term impetus towards adoption of environmentally friendly farming systems, including organic agriculture, will stem from moves towards decoupling agricultural support from purely production-oriented goals. There will be increasing emphasis on support to agriculture's role in providing public goods. Agricultural and environmental policies, including those responding to food safety concerns, will play a large role in facilitating or hindering the adoption of organic agriculture.

Besides financial support for conversion and regulations to protect the claim of organic producers, public investment in research and training is fundamental for such a knowledge- and management-intensive production system. It is still difficult for farmers and extension services to draw on a wide selection of well-researched methods and approaches. This often limits adoption to the most innovative farmers. Organic agricultural research receives only a fraction of the funds going to biotechnology research.

In developing countries, non-market organic agriculture and domestic certified organic agriculture are expected to increase in the long term. In particular, areas where economic growth is lagging (e.g. sub-Saharan Africa) and external inputs are unavailable or unaffordable, non-market organic agriculture could contribute to achieving local food security.

By about 2015, some organically produced tropical commodities (e.g. coffee, cocoa, cotton and tea) should have a moderate market share. The current tendency towards organic convenience food in industrial countries will increase, in particular for tropical beverages, baby food and frozen vegetables.

The oilcrops trade (especially soybeans and rape) is subject to major changes as oilcrops are the focus of biotechnology development. Future evidence on the safety of GM oilcrops might either increase their production or create new markets (and exports) for organic oilcrops. Cases are emerging where, because of the advent of GM crops, organic production will be constrained or no longer feasible; for example, organic farmers in Canada can no longer grow organic canola because of GM canola contamination in west Canada.

European governments' year 2010 targets for conversion of agricultural land to organic agriculture are ambitious: some countries (the Netherlands and Norway) aim to have 10 percent of agricultural land converted to organic agriculture. Germany has set a target of 20 percent. The United Kingdom Organic Food and Farming Targets Bill aims to increase total organic area to 30 percent (and domestic organic food retails to 20 percent). Denmark and Italy each aim at 10 percent and Sweden at 20 percent, as early

as 2005. In view of the present levels and these targets, the EU, on average, might possibly have a quarter of its total agricultural land under organic management by 2030.

It is hard to make estimates on future expansion of area under organic management in developing countries. Expansion will depend on technological innovations and unforeseen factors that challenge agricultural development as a whole, similar to the development of organic agriculture in Europe. Here it took 30 years for organic agriculture to occupy 1 percent of agricultural land and food markets, but food safety concerns resulted in its recent spectacular and unforeseen increase.

11.4 Agricultural biotechnology

This section focuses on the potential, risks and likely benefits of agricultural biotechnology to 2030. The benefits of agricultural biotechnology arise from its potentially large contribution to productivity gains and quality improvements. Productivity gains encompass essentially all factors of agricultural production: higher returns on land and livestock, labour and capital or simply lower input requirements per unit of outputs. This may mean higher crop and livestock yields, lower pesticide and fertilizer applications, less demanding production techniques, higher product quality, better storage and easier processing, or enhanced methods to monitor the health of plants and animals. Ultimately, higher productivity will result in lower prices for food and fibre, a benefit for all consumers but particularly important for the poor who spend a relatively large share of their incomes on food and fibre.

Higher productivity also holds the key in the fight against rural poverty. The underlying mechanisms of the productivity-poverty nexus have been discussed in Chapter 8. Biotechnology holds the promise of boosting productivity and thus raising rural incomes, in much the same way as the green revolution did in large parts of Asia during the 1960s to 1980s. It could kick-start a new virtuous cycle of productivity growth, increased output and revenues.

But there are also numerous risks and uncertainties associated with these new technologies that have given rise to a host of concerns and questions.

The most important of these is whether and how developing countries can actually harness the potential of biotechnology to promote production and the productivity of the poor. This in turn raises other questions. Whether and to what extent are the needs of developing countries being taken into account in current research efforts? How fast and to what extent have GM crops been adopted by developing countries? Which crops took the lead? Are the products developed by and for developed countries suited to the economic and ecological environments of developing countries and to what extent will developing countries develop their own biotechnology applications? More specifically, will "orphan" crops such as millet or bananas, which often play a vital role in the livelihoods of the poor, receive sufficient attention by new research? Will farmers in developing countries be trained and equipped to reap the benefits of the new technologies? Will the proliferation of GM-based crops and livestock further weigh on biological diversity? How can consumer concerns about environmental safety and potential human health hazards be taken into account, at low costs and without unduly distorting international trade? The parameters that determine the answers to these questions are changing quickly and it is therefore impossible to provide definite answers, particularly in view of the long-term perspective of this study. Instead, the following section will discuss some of the factors that are likely to affect the development and adoption of these new technologies in the future.

11.4.1 What is agricultural biotechnology?

Many traditional forms of biotechnology continue to be used and adapted. Some biotechnologies, such as manipulating micro-organisms in fermentation to make bread, wine or fish paste, or applying rennin to make cheese, have been documented for millennia.

Modern biotechnology takes various forms. These include: (i) tissue culture, in which new plants are grown from individual cells or clusters of cells, often bypassing traditional cross-fertilization

and seed production; (ii) marker-assisted selection (MAS), in which DNA segments are used to mark the presence of useful genes, which can then be transferred to future generations through traditional breeding using the markers to follow inheritance; (iii) genomics, which aims to describe and decipher the location and function of all genes of an organism; and (iv) genetic engineering, in which one or more genes are eliminated or transferred from one organism to another without sexual crossing. A GMO, also referred to as a living modified organism (LMO) or transgenic organism, means any living organism that possesses a novel combination of genetic material obtained through the use of modern biotechnology.[1]

Marker-assisted selection. Traditionally, plant breeders have selected plants based on their visible or measurable traits (phenotype). But this process was often difficult, slow and thus financially costly. MAS helps shorten this process by directly identifying DNA segments (genes) that influence the expression of a particular trait. The markers are a string or sequence of nucleic acid that makes up a segment of DNA.[2] As more and more markers become known on a chromosome, it is possible to create a detailed map of the markers and corresponding genes that codify certain traits. Using detailed genetic maps and better knowledge of the molecular structure of a plant, it is possible to analyse even small bits of tissue from a newly germinated seedling. Once the tissue is analysed, it is possible to check whether the new seedling contains the specific trait.

These new techniques are also important because they are not stigmatized by the negative attributes associated with GMOs, which have resulted in growing concerns about the safety of these new products for consumers and the environment (see below). They have revolutionized conventional breeding and help accomplish significant genetic improvements across almost all crops and livestock. And, should consumers' concerns *vis-à-vis* GMOs become more important, they could become the most crucial biotechnological application in the future.

---

1 This definition of LMO is taken from the Cartagena Protocol on Biosafety, Article 3(g). In Article 3(i), "modern biotechnology" is defined as "the application of: a. *In vitro* nucleic acid techniques, including recombinant deoxyribonucleic acid (DNA) and direct injection of nucleic acid into cells or organelles, or b. Fusion of cells beyond the taxonomic family, that overcome natural physiological reproductive or recombination barriers and that are not techniques used in traditional breeding and selection".

2 Of particular importance are the so-called quantitative trait loci that represent economically significant expressions of traits such as higher yields, improved quality or better resistance to diseases or various forms of abiotic stress.

Genomics. Genomics is the science of deciphering the sequence structure, the variation and the function of DNA in totality. More important than merely discovering and describing all genes of an organism is to describe the functions of the genes and the interactions between them. So-called functional genomics will help to discover the functionality of all genes, their functional diversity and the interactions between them. Functional genomics is expected to accelerate genetic improvement, the discovery of traits and to help solve intractable problems in crop production.

Recent progress in the mapping of the entire rice genome sequence, with the complete sequence expected to be delivered in 2004, represents a first important step towards understanding the overall architecture of the crop and provides valuable information for other techniques such as MAS or genetic transformation. But this would not yet include a full description of the biological functions of the various DNA sequences and their interactions, which would be a much more important step towards improved varieties. Many more years are likely to pass before all functions of all rice genes will be fully understood.

Genetically modified organisms. *Current trends and applications.* The first GMOs became commercially available in the mid-1990s. Since then, their importance has grown at an astounding pace. The number of GM varieties and species has increased rapidly and the area sown to GM crops has multiplied (some illustrative examples are given in Table 11.3). But the adoption across countries has been very uneven, with almost the entire expansion taking place in developed countries. Similarly, despite the rising variety of GM products available, commercial success has been concentrated on a few varieties or traits, notably herbicide-tolerant (Ht) maize and soybeans as well as *Bacillus thuringiensis* (Bt) cotton and maize. In 2001, Ht soybeans accounted for 63 percent of the area under transgenic crops, followed by Bt maize with a share of 19 percent (ISAAA, 2001).

Insect resistant traits. "Pest-protected" varieties were among the first GM crops to be developed, for the purpose of reducing production costs for farmers. Insect-resistant GMOs have been promoted both as a way of killing certain pests and of reducing the application of conventional synthetic insecticides. For more than 50 years, formulations of the toxin-producing bacteria *Bacillus thuringiensis* (Bt) have been applied by spraying, in the same way as conventional agricultural insecticides, to kill leaf-feeding insects. Studies on the safety of Bt for humans have not revealed any adverse effects on health.

In the late 1980s, scientists began to transfer the genes that produce the insect-killing toxins in bacteria *Bacillus thuringiensis* into crop plants. The intention was to ensure that all cells in these GMOs produced the toxin. Although no efforts were made to increase the growth rates or yield potential of the GM crops with these innovations, farmers have welcomed Bt crops because of the promise of better insect control and reduced costs. However, in the United States, the impact of Bt GMOs on crop yields and the number of conventional insecticide applications have varied widely by location and year. This is partly because of differences between the intended potential impact of the GM crops on target pests and their actual field performance. Some of these differences were a result of the uneven distribution of the toxin within the plants as they grew, some resulted from variations in target and non-target pest populations, and others were the result of toxins accumulating in plant-feeding insect pests, causing mortality of predators and parasites that ate those pests.

Herbicide-tolerant traits. The insertion of a herbicide-tolerant gene into a plant enables farmers to spray wide-spectrum herbicides on their fields killing all plants but GM ones. For that reason, the new GM seeds opened new markets for themselves and for herbicides. In fact, these crops contain a slightly modified growth-regulating enzyme that is immune to the effects of the active ingredient and allows it to be applied directly on the crops and kill all plants not possessing this gene.

Virus resistance. Virus-resistant genes have been introduced in tobacco, potatoes and tomatoes. The insertion of a resistance gene against potato leaf roll virus protects the potatoes from a virus usually transmitted through aphids. For that reason, it is expected that there will be a significant decrease in the amount of insecticide used. The introduction of a virus resistance gene in tobacco may offer similar benefits

Stacked traits. The so-called stacked traits embody a combination of properties introduced through

GM technologies. The most important applications at present are combinations of herbicide tolerance (Ht) and insect resistance (Bt). A number of other combinations have already become commercially available, such as herbicide-resistant maize varieties with higher oil contents. In the future, the addition of more traits with specific value will be added with combinations of stacked traits that provide insect tolerance, herbicide resistance and various quality improvements such as high lysine and/or low phytate content, possibly even in conjunction with higher oil content.

GM farm animals and fish. While there was considerable growth in the development and commercialization of GM crops, GM livestock have largely remained outside commercial food production systems. At the experimental level, more than 50 different genes have been inserted into farm animals, but these efforts still require substantial skill and are not as routine as those for plants. Early research in the development of transgenic farm animals has also been accompanied by manifestations of perturbed physiology, including impaired reproductive performance. These experiences raised ethical problems of animal welfare.

So far, the prospect of foods from transgenic farm animals has not been well received by consumers. Surveys consistently show that the public accepts transgenic plants more easily than transgenic animals. Experimenting with and altering animals are less acceptable prospects and have broader implications. Various cultures and religions restrict or prohibit the consumption of certain foods derived from animals. The use of certain pharma-ceutical products from transgenic animals, however, seems more acceptable to the public.

Highly successful research has been carried out on GM fish, but no GM fish have entered the market as yet (see Chapter 7). Most GM fish are aquaculture species that have received genes controlling the production of growth hormones, which raises the growth rate and yield of farmed fish. Ethical questions on the welfare and environmental impact of these GM fish have been raised, but it is also argued that GM fish share many attributes of conventionally selected alien fish species and genotypes, both of which are proven and accepted means of increasing production from the aquatic environment (FAO, 2000f).

How important are GMOs for agriculture? *Current trends*. The importance of GM crops has risen dramatically following the first endeavours of larger-scale commercialization in the mid-1990s. The six-year period from 1996 to 2001 witnessed a 30-fold increase in the global area grown with GM crops. With more than 52 million ha in the year 2001 (Table 11.4), the area planted with GM crops has reached a level that is twice the surface of the United Kingdom. At the same time, the number of countries growing GM crops has more than doubled.

This impressive growth notwithstanding, the annual increments in GM crop area have been levelling off both in absolute terms and in terms of percentage growth. This reflects to a large degree a saturation effect, as certain GM traits (soybeans) account already for a considerable share of the overall area. In addition, there was an actual decline

Table 11.3   A selection of commercially available and important GMOs

GMO	Genetic modification	Source of gene	Purpose of genetic modification	Primary beneficiaries
Maize	Insect resistance	*Bacillus thuringiensis*	Reduced insect damage	Farmers
Soybean	Herbicide tolerance	*Streptomyces spp.*	Greater weed control	Farmers
Cotton	Insect resistance	*Bacillus thuringiensis*	Reduced insect damage	Farmers
Escherichia coli K 12	Production of chymosin or rennin	Cows	Use in cheese-making	Processors and consumers
Carnations	Alteration of colour	Freesia	Produce different varieties of flowers	Retailers and consumers

Source: FAO (2000f).

Table 11.4 Area under GM crops, globally, from 1996 to 2001

| | Number of countries | Million ha | Change in area over previous year | |
			%	Million ha
1996	6	1.7		
1997		11.0	550	9.3
1998	9	27.8	153	16.8
1999	12	39.9	44	12.1
2000	13	44.2	11	4.3
2001	13	52.6	19	8.4

Source: ISAAA (2001).

in the area planted to GM canola, which is attributed to lower canola prices and the introduction of non-GM herbicide-tolerant varieties in Canada. But this slowdown also coincided with growing consumer concerns in developed countries that the new crops could jeopardize biosafety and pose a serious risk to human health. These fears have led to a growing pressure for legislation to label GM food, to increase the stringency of requirements for their approval and release or even to outright active resistance (Graff, Zilberman and Yarkin, 2000). These concerns were particularly forcefully voiced by consumers in developed countries. As a result, nearly all the additional area grown with GM crops came from developing countries, while area used for GM crops virtually stagnated in developed countries (Table 11.5). Canada's GM area even declined, leaving the United States and Argentina as the principal growers of GM crops with an overall share of 91 percent.

Which GM crops are important? Soybeans, maize, canola and cotton represent almost 100 percent of area grown with GM crops globally in 2001 (Figure 11.5). Ht soybeans alone account for 58 percent of all GM crops. Ht soybeans are not only the most important transgenic crop but, after Bt cotton, also the most rapidly growing one. The rapid market penetration of these first GM crops is impressive, particularly when compared with the introduction of similar technologies, such as hybridized varieties of maize and sorghum. In 2001, GM soybeans accounted for 63 percent of all area under GM crops. GM varieties of maize, cotton and canola accounted for 19, 13 and 5 percent, respectively.

What is in the pipeline? *Input-oriented technologies.* The next important improvement is likely to result from a further market penetration of so-called stacked traits, combining the benefits from two or more genetic modifications. The first stacked traits of cotton and maize (Bt/Ht cotton and maize) have already been released, offering both herbicide tolerance and insect resistance. In parallel, herbicide tolerance and insect resistance are planned to be extended to other varieties, notably sugar beet, rice, potatoes and wheat, while new releases of virus-resistant varieties are expected for fruit, vegetables and wheat. Fungus-resistant crops are in the pipeline for fruit, vegetables, potatoes and wheat. In addition, efforts are being made to create new traits with greater tolerance to drought, moisture, soil acidity or extreme temperatures. Chinese researchers claim to have developed salt-tolerant varieties of rice, which could help mitigate water scarcity and allow land lost to salinization problems to be recovered. The potential to cultivate marginal land appears to be particularly interesting for poorer farmers who are often more dependent on these environments. However, the ability of poor farmers to pay for these new technologies may be much more limited. This suggests that both speed of research and speed of introduction in the field are likely to be less impressive than for the first generation of GM crops.

Output-oriented technologies. A shift in focus is expected with the transition from the first to the second generation of GM crops. The new generation of GM crops is expected to offer higher output and better quality of the produce. Many of these new traits have already been developed but have not yet been released on the market. They include a great variety of different crops, notably soybeans with

Table 11.5   Area under GM crops by country, 1999 and 2001

	1999		2001		1999-2001	
	Area	Share in global area	Area	Share in global area	Change in area	
	Million ha	Percentage	Million ha	Percentage	Million ha	Percentage
Developed countries	32.8	82	39.1	75	6.3	19.2
United States	28.7	72	35.7	68	7	24.4
Canada	4	10	3.2	7	-0.8	-20.0
Australia	0.1	<1	0.2	<1	0.1	100.0
Others	<0.1	<1	<0.1	<1	n.a.	n.a.
Developing countries	7.1	18	13.5	24	3.6	90.1
Argentina	6.7	17	11.8	23	3.3	76.1
China	0.3	1	1.5	1	0.2	400.0
South Africa	0.1	<1	0.2	<1	0.1	100.0
Others	<0.1	<1	<0.1	<1	n.a.	n.a.
Total	39.9	100	52.6	100	12.7	31.8

Source: ISAAA (2001) and own calculations.

higher and better protein content or crops with modified oils, fats and starches to improve processing and digestibility, such as high-stearate canola, low phytate or low phytic acid maize. Another promising application is cotton with built-in colours that would spare the need for chemical dyes.

First efforts have been made to develop crops that produce nutraceuticals or "functional foods": medicines or food supplements directly within the plants. As these applications can provide immunity to disease or improve the health characteristics of traditional foods, they could become of critical importance for improving the nutritional status of the poor.

In the pipeline are also a number of non-food applications of GM technology. These include speciality oils (e.g. jet engine lubricants), biodegradable thermoplastics, hormones, "plantibodies" (e.g. human antibodies for treatment of infectious and auto-immune disease), vaccines or pharmaceuticals (e.g. anticancer drugs such as taxol) (Thomashow, 1999). Non-food applications that have already reached practical importance include a transgenic variety of *Cynara cardunculus* thistle, which is grown in Spain for electricity generation and GM poplars grown in France for paper production which demand less energy and produce less waste during processing.

The success of the second generation of GM crops will ultimately depend on their profitability at the farm level and their acceptance by consumers. Unlike the first-generation products, quality-focused products such as functional foods provide a higher perceived benefit for the consumer. This may increase the risk that consumers are willing to assume, suggesting a high market potential for the second generation of GM crops.

Specific products in the pipeline

Soybeans

■ High oleic soybeans contain less saturated fat than that in conventional soybean oil. The oil produced is more stable and requires no hydrogenation for use in frying or spraying. For that reason, this variety has a "health" image.

■ Soybeans with improved nutritional traits for animal feeding contain higher levels of two amino-acids (lysine and methionine) which will reduce the need for higher-cost protein meals in the preparation of feed mixes.

■ High-sucrose soybeans have better taste and greater digestibility.

Rapeseed and canola

■ High-lauric variety produces an oil containing 40 percent of lauric acid for chemical and cosmetic purposes.

Figure 11.5 GMO crops by country and crop

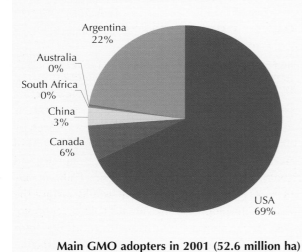

Main GMO adopters in 2001 (52.6 million ha)

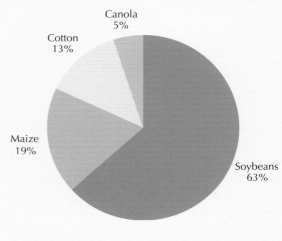

Main GMO crops in 2001 (52.6 million ha)

Source: ISAAA (2001).

■ High-stearate variety produces oil high in stearic acid, solid at room temperature without hydrogenation. This oil could be used for baking, margarine and confectionery foods that cannot use liquid oils.

Maize

■ Several researches, both conventional and biotechnological, aim to produce value-enhanced maize that will offer improved nutritional traits for livestock. Since grain is fed primarily as a source of energy, many of the new value-enhanced varieties aim to increase the content or availability of energy. But some new varieties will also include more protein and better amino-acid balances, which would reduce the need to buy supplemental feed ingredients.

Cotton

■ Coloured cotton is already available on a niche market basis. This trait would reduce the need for chemical dyes. Fibre quality improvement, such as polyester-type traits, would make sturdier fabrics.

■ Chinese researchers are breeding a new strain of cotton that includes rabbit keratin. Fibres of this cotton are longer and more resilient and they have an increased ability to maintain warmth. Research is also being carried out to develop wrinkle-resistant cotton and even fire-retardant qualities.

Prospects for the nearer term. Given the enormous speed of progress in generating and adopting new biotechnology applications, any longer-term outlook is necessarily speculative. Somewhat greater confidence, however, can be attached to forecasts of possible developments over the nearer term. The following short-term developments are discernible.

First, adoption rates for GM crops are likely to increase in developing countries. With the rapid adoption of Bt cotton in China, GM crops have made an important inroad into a potentially important market. China's GM potential rests not only on the sheer size of its agriculture but also on the particular importance for China of soybeans, maize, and tobacco – crops in which GM traits have been introduced successfully elsewhere. Moreover, an approval of the respective GM traits in China may have important knock-on effects in other developing countries. The significant catch-up potential in some developing countries, however, masks limiting factors and constraints that prevail elsewhere. To the extent that the new GM products favour capital-intensive and labour-extensive environments, the incentives to adopt these technologies are limited in other developing countries.

Second, growth of the area under traditional GM crops such as Ht soybeans and Bt maize is likely to slow down. This is in part a reflection of the impressive growth in the past, which limits the remaining potential. GM soybeans, for instance, already account for two-thirds of soybean area worldwide and for an even larger share of developed countries' soybean area. An expansion of GM soybeans must therefore come from an overall growth in soybean area rather than from a shift out of non-GM soybean production. Growth may also be curbed because of food safety and environmental concerns that have received particular attention in Europe.

Third, there is a considerable growth potential for new GM applications in developed countries. Examples include GM fish varieties or GM crops for renewable energy. Other possibilities are GM-based nutraceuticals or GM applications for health and cosmetic applications. As these new applications are likely to produce much wider benefits than just cheaper food and feedstuffs, consumers in developed countries are also more likely to accept greater risks and thus to adopt these non-traditional applications at faster rates.

Prospects for the medium and longer term. Substantial progress has been made over the last five years, both in terms of theoretical advances and practical importance of biotechnology. These advances over such a short timespan make it impossible to identify specific products that are likely to dominate developments in biotechnology over the next 30 years. It is, however, easier to identify the overall parameters that are likely to affect future trends.

The overall direction of research and development is likely to be determined by economic incentives. Developments in prices of production factors and products are critical in the context of a 30-year outlook. These, in turn, will be crucially affected by future changes in the relative abundance/scarcity of production factors, notably land, capital and labour. In developed countries, costs of labour may increase relative to land, which would favour the further development of labour-saving technologies. In developing countries, by contrast, factor proportions may change in the opposite direction with increasingly abundant labour and increasingly scarce land. This would favour labour-intensive and land-saving technologies. The critical question in this regard is whether and to what extent the private sector will cater for diverging needs and to what extent investments from the public sector are needed to reconcile these needs.

Many of the currently available technologies have catered for land-intensive and labour-extensive environments. This is particularly noticeable

Box 11.3 Golden rice: a polarized debate

There is still considerable uncertainty as to how much of the potential of GM crops can be harnessed for the benefit of the poor. The most prominent example in this context is the so-called "golden rice", a betacarotene-enriched variety that was developed with the help of free-of-charge licences from a number of life science companies. The proponents of the technology claim that "golden rice" provides a low-cost means to alleviate one of the gravest health problems of the developing world. The main goal of this development was the creation of a tool to help combat vitamin A deficiency (VAD), a public health problem that affects 118 countries and more than 400 million people worldwide, especially in Africa and Southeast Asia, and that most affects young children and pregnant women. As betacarotene is provided through rice which is the main staple food in many developing countries, the distribution is largely self-targeting.

Opponents, however, underline that these new varieties include too little betacarotene and that it would be impossible to cover needs through golden rice alone. They claim that the returns on the investment of US$300 million are relatively small and that the same effects could be achieved by a combination of existing tools. Critics also argue that VAD is not best characterized as a problem, but rather as a symptom of broader dietary inadequacies associated with both poverty and agricultural transition from diverse cropping systems to rice monoculture. It would therefore be more important to have a more varied diet rather than relying too heavily on "a magic-bullet solution" while leaving poverty, poor diets and extensive monoculture intact. Moreover, opponents suggest that golden rice could have counterproductive impacts on nutritional problems by curtailing the progress made in educating people to diversify their diet and increase the diversification of agriculture production. Finally, if only a limited number of varieties were genetically modified and widely cultivated, this would have a negative impact on crop biodiversity.

for GM crops, where productivity gains are based on savings in input needs (labour, capital) even when output (yields) is stagnant or declining. This was one of the main factors that contributed to the high adoption rates for GM crops in developed countries. If today's relative factor proportions are a guide to the future, the incentives to adopt these new technologies in developing countries are likely to be subdued. Moreover, private investors have little incentive to provide proprietary technologies where the chances of recouping investments in research and development are small. This suggests that the public sector will have to play a significant role in providing the technologies to cater for the specific needs of developing countries.

11.4.2 Why agricultural biotechnology matters to developing countries

The principal benefits. *Productivity gains.* Biotechnology has the potential to increase crop and livestock yields. The first generation of GM crops was largely input-oriented and provided the same or only marginally increased yield potential. The fact that some GM crops rendered higher yields in practice largely resulted from the effect that "built-in" inputs such as pesticides have reduced output losses that are typically caused by inappropriate or inadequate input applications. Moreover, the fact that GM technology embodies this expertise directly into the seeds is particularly important for environments where sophisticated production techniques are difficult to implement or where farmers do not command the management skills to apply inputs at the right time, sequence and amount. This suggests a much larger potential for GM crops (stacked traits of Bt/Ht cotton or Bt/Ht maize) in developing countries even for the first generation. The second generation of GM crops is expected to raise both the volume of output and the quality of the produce. These technologies are currently being tested but only a few traits are available in practice.

More, cheaper and better food. A second factor arises from the prospects for lower prices for better food. Higher productivity lowers production costs and will ultimately result in lower food prices. While this is not in itself a guarantor of improved food security or reduced poverty, more and better food at lower prices is particularly important for poor consumers. They would particularly benefit where GM products offer less expensive and nutrient-enriched food staples, which account for a large share of their food expenditure.

A higher capacity to feed a more populous world. The capacity that GM crops offer to produce more and better food is even more important when the future food needs of growing populations are considered. Much of the incremental food production in the future has to come from higher yields, yet the potential to raise actual yields through more traditional agronomic improvements such as earlier ploughing, scotch carts, higher fertilizer and pesticide applications is declining. A slowdown in yield growth has already been observed in some high-intensity systems in Asia, where the gap between yields attained by farmers and the economic maximum yield has narrowed noticeably.

The potential to save and improve resources or recoup marginalized land. A fourth factor is the potential to save resources or recoup marginalized land. Empirical studies suggest that the poor are cultivating the most marginal agronomic environments, and that they are more often dependent on these marginal growing environments than other groups.[3] These marginal production environments are often characterized by drought or moisture stress, extreme temperatures, soil salinity or acidity. The potential to grow food in such environments is therefore doubly important in the fight against hunger and poverty: the potential to produce food where food is needed most helps ease the food problems of the poor directly. Moreover, to recoup land that was lost through environmental stress (e.g. soil salinity or acidity) could help contain further encroachments on areas with high environmental value or high sensitivity.

More and better non-food products. GM crops could also offer a more attractive way to produce non-food products. Plants with higher energy conversion and storage capacities could be bred and make a more meaningful contribution to alternative energy use in the future. If successfully implemented on a large scale, this could boost agriculture's role as a carbon sink. Transgenic plants and animals could significantly extend the possibilities of various areas of technology and overcome some

---

[3] In fact, these environments are available to the poor because only the poor are willing to accept the low factor returns (wages) for the inputs they can provide (low-skill labour).

Box 11.4 GURTs: technical aspects and possible impacts

What are GURTS?

The acronym GURTs stands for genetic use restriction technologies and refers to biotechnology-based switch mechanisms to restrict the unauthorized use of genetic material. Two types of GURTs can be distinguished: variety use restriction (V-GURTs), rendering the subsequent generation sterile (the so-called "terminator" technologies), and use restriction of a specific trait (T-GURTs), requiring the external application of inducers to activate the trait's expression.

... and what are their principal impacts?

Agricultural biodiversity. Impacts on agricultural biodiversity will vary across different farming systems. In low- and medium-intensity farming systems a change from local to GURT varieties may imply a loss of agricultural biodiversity while in high-intensity farming systems the impact may be minor.

The environment. While the environmental containment aspect of GURTs may reduce potential risk associated with eventual outcrossing, there remains a possibility of pollination of neighbours with GURTs pollen, leading to yield drops in cultivated areas, as well as to alteration of wild ecosystems.

Research and development. By stimulating further investment, GURTS may increase agricultural productivity in certain farming systems. However, restricted introgression of genes from GURTs into local gene pools may reduce incentives for farm-level breeding, if desirable traits in introduced GURTs varieties cannot be accessed, widening the technological and income gap between resource-poor and better-off farmers.

Market structure. While strengthened control over the use of GURTs products may likely increase investment in further breeding, GURTs may well reinforce the concentration and integration trends in the breeding sector in such a way as to lead to possibilities for misuse of monopoly power, rendering farmers fully dependent on formal seed supply systems.

Food security. GURTs could also increase the seed insecurity of resource-poor farmers who cannot afford to purchase seed and who depend on the local grain market for their seed needs. This may generate a low level of acceptance by low-income farmers in developing countries.

Source: FAO (2001k)

of the traditional constraints that medical research and applications face today.

The principal concerns. Notwithstanding the potentially large benefits of GM technology for developing countries, there are growing concerns that these new technologies are associated with significant costs, risks and problems.

Market concentration in the seed industry. Some concerns have arisen out of the significant market concentration in the seed industry. In 1998, 60 percent of the world market for seeds was controlled by just 35 companies. One company alone controlled over 80 percent of the market for GM cotton, 33 percent of the market for GM soybeans and 15 percent of the GM maize market (Then, 2000). This growing horizontal concentration is accompanied by an increasing vertical concentration between seed producers and agrochemical companies whereby larger agrochemical companies have been absorbing the few large seed companies that resulted from the horizontal

consolidation process within the seed industry. The trend towards larger and more integrated operations (the so-called life science companies) was largely driven by the chemical industry. Chemical firms were looking for partners in the seed industry to protect the value of their intellectual property rights (IPR) in patented herbicides (Just and Hueth, 1999). The consolidation process between the agrochemical and the seed industry is currently being extended to a third stage, as the life science companies broaden their reach through strategic alliances with major trading companies such as Cargill or ADM. While this concentration process has offered new possibilities to reap scale effects and to overcome barriers in creating and commercializing GM products, it has also given rise to concerns that these non-competitive market structures may impose significant social and private costs (Phillips and Stovin, 2000). These are only now being considered.

Intellectual Property Rights (IPR). The impact of the application of IPR, the mechanisms for their enforcement and the excludability that is associated with them is another source of concern. In general, the excludability is of critical importance in encouraging private research in all sectors. Without it, innovators would not be able to recoup their investments, private research would languish, productivity gains would slow down and social welfare would suffer. Recognition of the importance of IPR has brought about a strengthening of legal protection for biotechnology processes and products and spurred on significant private investment in biotechnology. But the strengthening of IPR has also given rise to concern. First, the scope of intellectual property protection may be too wide, thereby choking off spillovers, follow-on innovations and diffusion. Second, IPR afford private companies the possibility of protecting the alteration of a single gene derived from freely accessible germplasm that has been generated by farmers and public research efforts over centuries. Developing countries in particular believe that they should be compensated for their contributions to existing genetic resources. The International Treaty on Plant Genetic Resources for Food and Agriculture (PGRFA), adopted in November 2001, addresses these concerns (Box 11.5). It could assume a pivotal role in facilitating access to plant genetic resources in the future and in safeguarding traditional indigenous contributions to the breeding process (farmers' rights).

Biosecurity. A third area of public concern revolves around the risk that biotechnology applications in food and agriculture pose to human health and the environment. Consumers in all countries would like assurances that GM products reaching the market have been adequately tested, and that these products are being monitored to ensure safety and to identify problems as soon as they emerge. Because of the complexity of food products, research on the safety of GM foods is thought to be more difficult than carrying out studies on components such as pesticides, pharmaceuticals, industrial chemicals and food additives. Through the Codex Alimentarius Commission and other fora, countries discuss standards for GMOs and ways to ensure their safety. One

approach, which is being used in assessing the risks of GMOs, derives from the concept of substantial equivalence.[4] If the GMO-derived food is judged to be substantially equivalent to its traditional counterpart, then it is considered as safe as its conventional counterpart. If it is not, further tests are conducted.

Critics claim that only 1 percent of public research funds has been allocated to assess the risks associated with the introduction of GM technologies. It is suggested that the experience with traditional counterparts cannot be applied to products based on GM technology, as the substantial equivalence approach implies, and that the new technologies require a new risk assessment approach. Underestimating or ignoring the risks means that external costs associated with the technology are not fully accounted for and that the welfare gains of the new technology may be overstated. The recent accidental use for human food consumption of GM maize that contains a potentially allergen protein has reinforced such concerns.

Genes can end up in unexpected places. The artificially inserted genes might be passed on to other members of the same species, and perhaps to other species. Antibiotic-resistance genes are often inserted into GMOs as markers so that researchers can tell whether gene transfer has succeeded or not. These genes may be transferred to bacteria within the human body with yet unclear impacts. While this technique is now being replaced, other problems may remain. There is even a possibility that the gene for herbicide resistance may transfer to weeds, with potentially disastrous impacts for agriculture and food security.

Genes can mutate. It is still unclear what impact the artificially inserted gene has on the stability of the genome. There are claims that it may cause more unexpected mutations. While mutations could be neither new nor necessarily bad, GMOs may cause unexpected and undesirable instability.

"Sleeper" genes could accidentally be switched on. Organisms can contain genes that are not activated except under certain conditions, for example under the influence of pathogens or as a result of certain

---

[4] Substantial equivalence acknowledges that the goal of the assessment is not to establish absolute safety but to consider whether the GM food is as safe as its traditional counterpart, where such a counterpart exists. It is generally agreed that such an assessment requires an integrated and stepwise, case-by-case approach. Factors taken into account when comparing a GM food with its conventional counterpart include: (i) identity, source and composition; (ii) effects of processing and cooking; (iii) the transformation process, the DNA itself and protein expression products of the introduced DNA, and effects on function; and (iv) potential toxicity, potential allergenicity and possible secondary effects; potential intake and dietary impact of the introduction of the GM food.

Box 11.5 The International Treaty on Plant Genetic Resources for Food and Agriculture

A new International Treaty on Plant Genetic Resources for Food and Agriculture (PGRFA) was adopted by the FAO Conference in November 2001. The main areas covered by the treaty include: (i) a multilateral system of access and benefit sharing of plant genetic resources for major food crops; (ii) an agreement on access to *ex situ* genetic resources not covered by the Convention on Biological Diversity (CBD); and (iii) a recognition of the contributions of local and indigenous communities and farmers to PGRFA (farmers' rights). The PGRFA covered by the treaty include most major food crops (cereals such as rice, wheat, maize, sorghum and millet; grain legumes such as beans, peas, lentils, chickpeas and cowpeas; roots and tubers such as potatoes, sweet potatoes, cassava and yams), plus a list of forages (32 genera). The treaty will enter into force once it has been ratified by 40 or more countries. This is expected to be in 2003 or 2004.

Provision of access and benefit sharing. The treaty provides for facilitated access to material in the multilateral system for the purposes of food and agriculture research, breeding and training in this area. It obliges signatories to provide access to PGRFA listed in the multilateral system for the purposes listed above. The treaty also provides that benefits arising from the use, including commercial use, of PGRFA under the multilateral system shall be shared fairly and equitably through exchange of information, access to and transfer of technology, capacity building and the sharing of the benefits arising from commercialization. It includes special provisions for monetary benefit sharing in the case of commercialization of products that are PGRFA and that incorporate material accessed from the multilateral system.

Conservation of PGRFA. The treaty also calls for an integrated approach to the exploration, conservation and sustainable use of PGRFA and includes specific provisions on surveying, inventorying and collecting PGRFA, as well as on *in situ* and *ex situ* conservation. Explicit reference is given to "onfarm" conservation by farmers, as distinct from in situ conservation of wild PGRFA. It requires parties to develop and maintain appropriate policy and legal measures that promote the sustainable use of PGRFA.

Farmers' rights. The treaty also addresses the need to "recognize the enormous contribution that the local and indigenous communities and farmers of all regions of the world, particularly those in the centres of origin and crop diversity, have made and will continue to make for the conservation and development of plant genetic resources which constitute the basis of food and agriculture production throughout the world". Three substantive elements of farmers' rights are included: (i) protection of traditional knowledge relevant to PGRFA; (ii) the right to participate equitably in sharing benefits arising from the utilization of PGRFA; and (iii) the right to participate in making decisions, at the national level, on matters related to the conservation and sustainable use of PGRFA.

Source: Cooper and Anishetty (2002)

weather conditions. The "promoter" gene that is used to insert the new gene could activate "sleeper" genes, potentially in inappropriate circumstances.

Allergens can be transferred. Genes that cause allergies could be transferred into another species. The problem is twofold: it extends the range of potentially allergen products and creates uncertainty as to what products are potentially allergens. For example, an allergenic Brazil-nut gene was transferred into a transgenic soybean variety. It was found in testing, and the soybean was not released.

Sterility could be transferred. There is the theoretical risk that a dominant gene from a GURT plant could be passed on through cross-pollination to non-GURT plants, thereby reducing their fertility rate. However, this risk would be relatively small and, even if it were to happen, the inherited dominant non-germination gene would anyway be self-eliminating.

Controls over GM releases are inadequate. In 2000, a maize variety intended only for animal feed was accidentally used in products for human consumption. There is no evidence that this variety was dangerous to humans, but it could have been.

Animal welfare is at risk. There is evidence of abnormal physiology in some transgenic animals. Some effects on animals are unpredictable and could range from benign to distressful to dangerous.

Unintended effects on the resource base. Unusual traits in a plant could have unintended effects on the farming system. For example, a wheat variety capable of extracting more nutrients from the soil may exhaust the soil. Plants bred for land that has been made saline by unsuitable irrigation may enable a farmer to use even more brackish water, destroying the land completely.

Loss of biodiversity. GM plants could compete with traditional farmers' varieties, causing loss of crops that have been bred for millennia to cope with local stresses. For example, the existence of traditional potato varieties in Latin America permitted a recovery from the catastrophic potato blight in Ireland in the 1840s. Today, traits from farmers' varieties and wild relatives are often used to improve climate tolerance and disease resistance. GM crop varieties might also cross, and thus compete, with wild relatives of crops such as wheat and barley. This is especially risky in the developing world, where wild relatives may be found growing close to farmed crops.

11.4.3 Who benefits and who bears risks and costs?

The principal problem: disjoint risks and benefits. As with any new technology, there are winners and losers associated with the use of GMOs and biotechnology. At country level, the costs and benefits accrue to different stakeholders and cause concerns about, or even the outright rejection of, the new technology. Addressing and reconciling these problems are part of the policy response of the respective country. A second, less common source of concern emerges when risks and benefits accrue to stakeholders in different countries. This either requires international policy coordination or leaves externalities unaddressed. The introduction of GM technology is associated with both dimensions of the problem, i.e. there are disjoint risk and benefits within and across countries. An analysis of these disjoint costs and benefits may help identify appropriate policies. It may also provide insights as to what directions the new technologies will take.

Rich versus poor countries. The risks and benefits associated with an innovation are the principal determinants for the degree of adoption by a country. The willingness to assume risks may therefore be disconnected from the extent and possibility of capturing the benefits of GM technologies. The disconnection of risks and benefits affects numerous stakeholders: consumers versus producers, developed versus developing countries and private companies versus public research institutions. For example, the benefits of GM maize – and thus the willingness of rich societies to assume the associated risks – may be too small to pursue the technology. The benefits of the same technology for poor societies may be large, but their ability to pay for it is too

small to develop it. If left unaddressed, such disjoint interests can result in a situation that neglects the interests of the poor. Bridging these gaps calls for appropriate policy action and for international policy coordination.

High-value versus low-value goods. The willingness of a society to bear the risks of a new technology is positively related to the benefits drawn and expected from it. The benefit from cheaper food staples such as rice, maize or soybeans is likely to be small for rich consumers in the north and high for poor consumers in the south. Rich consumers are therefore unlikely to assume the same risks as poor consumers and, if the benefits are sufficiently small, they are rational to reject the new technology altogether.

Consumers in the developed countries accept the higher risks for functional food, medical applications or cosmetics as the (perceived) benefits from these applications exceed the risks that they carry. But given the small benefits from less expensive food staples for the same consumers, staples such as rice, wheat or coarse grains are likely to be most affected by a decline in research expenditure. Even more so are tropical staples such as some roots and tubers. But such food staples are of critical importance for consumers in developing countries, particularly for the rural poor.

What does it mean for the direction of research? The different perception and importance of risks and benefits in developed and developing countries could reduce the speed of progress and change the direction of GM research and development. In the north, which has the funds to afford research, consumers perceive the risks associated with these technologies to be high and their benefits to be small. Yet the greatest need and the greatest benefits are in the south, where the ability to pay for the development of the technologies is small. The allocation of risks and benefits could mean that the north may reduce investments in these technologies, as the consumers in the north are unwilling to accept their products, while the consumers in the south, in need of the products, are unable to pay for the technology.

Where the south can afford to develop and import the technology, this may adversely affect trade in the final products. GM-based food and fibre may be faced with growing non-tariff barriers, which reflect the lower willingness to accept risks in the north or an added form of protecting domestic agriculture. A smaller market volume for GM crops in turn may adversely affect the profitability of developing the new technology. A lack of market access may render research and development non-viable for the south.

What policies could help to reconcile risks and benefits? Appropriate policy action could help reconcile some of the conflicting interests of the various stakeholders.

Addressing the concerns of consumers in developed countries: more transparency. Part of the risk that makes consumers in developed countries reluctant to accept GM products is not actual risk but perceived risk. In part, this risk perception reflects a lack of transparency and calls for measures that help to maximize transparency for the consumer. Appropriate labelling is an important step towards higher transparency and thus towards lower risk. It would also help facilitate trade since labels help products to comply with international standards. However, excessive labelling requirements may result in "regulatory capture", which in itself is detrimental to trade. Moreover, labelling of GM food requires segregation of GM from non-GM products and can generate substantial additional costs and incentives for fraud. The higher costs of segregation are a particular problem for poorer countries that often lack the necessary regulatory capacity.

The second part of the risk that consumers in developed countries face is real. It is a reflection of insufficient testing and premature releases of GM crops for field applications. This calls for better risk assessment procedures and commensurate rules and regulations to minimize the risks associated with applying GM technology.

Addressing the concerns of developing countries: lowering seed costs. A crucial factor that limits access to the new technologies in developing countries is the high cost of GM seeds. To the extent that high seed costs reflect the monopoly rents from a lack of competition in the life science industry, antitrust measures could provide a remedy. In addition, national and international research institutions themselves could charge royalties for the germplasm that has traditionally been provided free of charge to the private sector. These payments could then be used to promote targeted GM research for developing countries. This, of course, may reduce the overall incentive for the private sector to invest in crop research and could result in an overall loss in productivity.

Private-public partnerships in GM research. The public sector could explore a number of routes to collaborate with the private sector to target developing countries' specific needs, such as poverty reduction, public goods provision and the capture of spillover effects. National and international research institutions could define specific breeding tasks, or the development of field-proven varieties with certain characteristics and put them out for competitive tender. Companies could use their patented germplasm as an input into work under such tenders, but the final product would have to be made available free of IPR charges or genetic technical use restrictions (Lipton, 1999).

11.5 Directions for agricultural research

Agricultural research has been crucial in meeting the challenge of increasing food production faster than population growth over the past 50 years. A main characteristic of research efforts was the focus on increasing productivity through a set of technologies in what has come to be known as the green revolution. The impressive global achievements mask considerable regional differences. Asia received most attention while sub-Saharan Africa was largely bypassed, as were, within many countries, the remote and poorest communities that did not have the resources, physical and financial, to capitalize on the potential of the new high-input demanding technologies (Conway, 1997). Also, the impacts of the new technologies on the environment (see Chapter 12) were largely ignored in the early years of the green revolution.

What has emerged most strongly in various reviews of the green revolution is that development of technologies by themselves is not enough. Relevant and sustainable technology innovation must be planned, developed, tested and delivered within a broad-based agricultural and rural development framework. Nor can productivity be the sole criterion to guide technology development; the potential implications of technologies for agro-ecological stability and for sustainability and equitability must be fully addressed at the "drawing board" stage. This has fundamental implications for the planning of future agricultural research strategies.

To meet the food security needs of an expanding global population in the decades ahead and to reduce poverty, there is a need to maintain and increase significantly agricultural productivity on land at present available across the developing world and at the same time to conserve the natural resource base. This will require (i) increasing productivity of the most important food crops both on the more fertile soils and on marginal lands; (ii) exploring possibilities for limiting the use of chemical inputs and substituting these inputs with biologically based inputs; (iii) more precise use of soil, water and nutrients in optimized integrated management systems; and (iv) increasing production efficiency and disease tolerance in livestock.

These challenges call for a comprehensive and complex research agenda that must integrate current advances in the molecular sciences, biotechnology and plant and pest ecology with a more fundamental understanding of plant and animal production in the context of optimizing soil, water and nutrient-use efficiencies and synergies. Effective exploitation of advances in information and communication technology will be necessary not only to facilitate the necessary interactions across this broad spectrum of scientific disciplines but also to document and integrate traditional wisdom and knowledge in the planning of the research agenda and to disseminate the research results more widely. This agenda calls for a three-dimensional research paradigm that integrates scientific investigation across genetics and biotechnology, ecology and natural resources and not least socio-economics to keep in focus the development environment that characterizes the livelihoods and food security of the poor.

The first dimension (genetics and biotechnology) has been discussed in the preceding section. The following section will focus on the other two dimensions.

Ecology and natural resources management research.
A comprehensive understanding of the ecology of all life forms within the farming system is a prerequisite to the development of sustainable agriculture in the context of sound knowledge-driven use and management of the natural resources with which the farming system interacts. Recent advances in ecology have been described alongside molecular genetics as the second great revolution in modern biology. The past decade or so has seen the use of mathematical modelling, the articulation of comprehensive hypotheses and advances in experimental design in support of more precise laboratory and field experiments. These trends have transformed the study of ecology into a rapidly developing science (Begon, Harper and Townsend, 1990), which should lead to a better understanding of the complex dynamics that are at work within agricultural systems. Increasingly, natural resource management calls for closer collaboration between ecologists and agricultural scientists whether they are addressing technical (biotic or abiotic), social or economic dimensions of agricultural development. By definition, agro-ecosystem research embraces the "ecological and socio-economic system, comprising domesticated plants and/or animals and the people who husband them, intended for the purpose of producing food, fibre or other agricultural products" (Conway, 1987).

Within the biophysical boundaries of the ecosystem, ecological research has much to contribute at three levels: (i) at the level of the plant, its pests and predators; (ii) at the level of the plant and its competition from weeds; and (iii) at the level of the plant rooting system and the roles of beneficial and competing micro-organisms in the capture and utilization of soil nutrients. Research at the plant-pest-predator level is opening up new insights in pest control, in one measure through the genetic development of pest-resistant plant varieties and, at the other end, in the refinement of IPM.

At the level of plant-weed competition, the geneticist's approach has been to develop herbicide-resistant crops, while the ecology-based approach has been to develop crop rotations and intercropping and high-density systems that minimize weed damage. Both approaches require still further research, particularly with greater focus on marginal lands and food crops that are mostly grown by the poor. More fundamentally, a better understanding of plant genetics and of plants' relation to other competing plants may offer new insights to stable cropping systems. This would provide a more informed basis on which to design cropping systems and rotations that can sustain higher and more stable yields.

At the plant rooting levels, there is a need for much more fundamental research at the physical, biochemical/physiological and genetic bases of plant-micro-organism interactions and symbiosis (Cocking, 2001). Nutrient utilization and biological nitrogen fixation lie at the heart of this research, which can have enormous benefits for low-input agriculture and the poor. The benefits to sustainable agriculture and the environment are obvious in the reduction of dependence on chemically produced nitrogenous fertilizer and associated greenhouse gas emissions. Future ecosystem research must also address soil nutrient availability and utilization more comprehensively within alternative cropping systems in the context of NT/CA, integrated cropping and crop-livestock systems.

At the community and watershed levels, agro-ecology, ecosystem and natural resources management research must address the biophysical and socio-economic dimensions of resources use and their potential enhancement and/or depletion. In this context, CGIAR has recently articulated a comprehensive agenda of integrated natural resources management research that embraces the more important topics that need to be addressed (CGIAR, 2000):

- *Water:* model system flows (river-basin level) allocated across multiple users, with particular attention to onsite and offsite effects, develop recharge balance models for aquifers at risk of excessive drawdown.
- *Forests:* characterize the complexity of forest systems and the range of stakeholders who interact with them, and develop strategies to influence the global policy agenda.
- *Fisheries:* identify the types of farming systems and agro-ecologies into which integrated aquaculture-agriculture can be sustainably incorporated.
- *Livestock:* develop databases, models and methods for analysing livestock-based systems to help identify priorities for research and development interventions.
- *Soils:* develop soil erosion models for various multifunctional land use systems, and nutrient balance and flow models.
- *Carbon stocks:* document and model alternatives for above and below ground carbon stocks and relate changes in those stocks to global climate change impact.

- *Pest and disease incidence:* describe, define and track key insect pests using GIS and develop models that relate incidence to agroclimatic conditions.
- *Biodiversity:* project alternative scenarios of functional biodiversity under different land management systems.

Research priorities for the poor. Defining research priorities (and the development of stakeholder-specific research agendas) is becoming a complex and increasingly sophisticated process. It demands the interactive interpretation of information flows between the scientist and the end-user of research outputs. The criteria for making strategic choices among alternative research programmes vary depending on the stakeholder. In private-sector-funded research, the ultimate criterion is profit in one form or another. In academic research the goal is often loosely defined around scientific advancement and knowledge. And in national and international research institutions the broad objective is primarily the production of information and products for the "public", usually termed "public goods".

Establishing research priorities that specifically address the needs of the poor or multidimensional goals such as food security, sustainability, conservation of biodiversity and natural resources, becomes increasingly difficult as the distance (socio-economic as well as spatial) between the research planners and the target beneficiaries widens. In essence, bridging this gap is the challenge of the new research agenda, not only at the research planning stage but also at the interpretation and field-testing stage of the technologies and other research outputs.

In determining research priorities, it is also vitally important to understand how new technologies may influence the lives and livelihoods of the poor. New technologies can lead to increased productivity on the family farm, resulting in increased food consumption and family incomes. Such technologies can result in lower food prices, benefiting a wide range of the urban and rural poor. Growth in agricultural output generally leads to increased employment opportunities for the landless and the poor in both the rural and urban non-farm economy. Perhaps not emphasized adequately in research programmes is that new technologies can lead to the production of food crops rich in specific micronutrients that are often deficient in the diets of the poor.

Research to underpin a new technology revolution with greater focus on the poor must put special emphasis on those crop varieties and livestock breeds that are specifically adapted to local ecosystems and that were largely ignored throughout the green revolution. These include crops such as cassava and the minor root crops, bananas, groundnuts, millet, some oilcrops, sorghum and sweet potatoes. Indigenous breeds of cattle, sheep, goats, pigs and poultry and locally adapted fish species must also receive much greater priority. A particular focus in the new research agenda should be on plant tolerance to drought, salinity and low soil fertility as nearly half of the world's poor live in dryland regions with fragile soils and irregular rainfall (Lipton, 1999).

Research modalities and dialogue. Research that addresses only one component in the development chain, for example crop yield potential, will not result in an equitable or sustainable increase in food production. New research efforts should address a minimum of four critical questions at key salient points along the research continuum, from the conception/planning stage to the stage of application of the outputs by the targeted beneficiary. The key questions are: (i) whether the technology will lead to higher productivity across all farms, soil types and regions; (ii) how the technology will affect the seasonal stability of production; (iii) how the technology will impinge on the sustainability of the targeted farming system; and (iv) what are the sectors that will benefit most (or lose out) as a result of the widespread adoption of the technology. It is comparatively easy to tailor these questions to specific research programmes depending on the nature of the research to be undertaken, but it is much more difficult to arrive at well-supported answers, in particular at the research planning stage. This research challenge needs scientists from a range of disciplines and from different agencies, both public and private, to engage in close collaboration, not only among themselves but also with the intended beneficiaries – the farmers – either directly or through the extension services.

Effective dialogue among all scientists and extension workers in this research development continuum also calls for a new information-sharing mechanism that embraces transparent interactive dialogue and easily accessible information. Modern information and communication technology can provide the vehicle for this information sharing and dialogue, opening up the possibility of a global knowledge system through which the sharing of global knowledge on all emerging technologies relating to food and agriculture can be effectively realized (Alberts, 1999). This in turn will lead to the strengthening of the research process at all stages.

National government and international donor support for research has declined significantly over the past decade, despite compelling evidence on very high rates of return to investment in agricultural research and in particular in genetic improvement programmes for crops and livestock. This is particularly worrying at a time when there is a widely shared consensus on the absolute need and importance of strengthening agricultural research. While more and more funds go into biotechnology research, the other areas mentioned above are trailing behind. This is especially true for research focusing on marginal areas and crops. The private sector can and must contribute more than just funding. As outlined above, its expertise, technologies and products are essential to the development and growth of tropical agriculture based on rapidly advancing biotechnologies and genetically engineered products. It is argued by some that incentives (e.g. in the form of tax concessions) should be offered to induce private sector participation. It is also argued that collaborative partnerships with private sector companies or foundations in well-articulated and mutually beneficial agreements can mobilize the required cooperation and make significant contributions to agricultural research in the developing countries.

To conclude, the scientific community bears a responsibility to address ethical concerns. On the one hand, it must ensure that the technologies and products of research do not adversely affect food safety or risk damage to the environment. In this context, timely and transparent communication of relevant research findings and their interpretation (risk analysis) to all pertinent audiences including the general public is required. On the other hand, scientists, together with public servants, politicians and private sector leaders, bear a more profound humanitarian responsibility to ensure that all people can realize their most fundamental right – the right to food.

Agriculture and the environment: changing pressures, solutions and trade-offs

12.1 Introduction

Agriculture places a serious burden on the environment in the process of providing humanity with food and fibres. It is the largest consumer of water and the main source of nitrate pollution of groundwater and surface water, as well as the principal source of ammonia pollution. It is a major contributor to the phosphate pollution of waterways (OECD, 2001a) and to the release of the powerful greenhouse gases (GHGs) methane and nitrous oxide into the atmosphere (IPCC, 2001a). Increasingly, however, it is recognized that agriculture and forestry can also have positive externalities such as the provision of environmental services and amenities, for example through water storage and purification, carbon sequestration and the maintenance of rural landscapes. Moreover, research-driven intensification is saving vast areas of natural forest and grassland, which would have been developed in the absence of higher crop, meat and milk yields. But conversely, intensification has contributed to the air and water pollution mentioned above (Nelson and Mareida, 2001; Mareida and Pingali, 2001), and in some instances reduced productivity growth because of soil and water degradation (Murgai, Ali and Byerlee, 2001).

Quantification of the agro-environmental impacts is not an exact science. First, there is considerable debate on their spatial extent, and on the magnitude of the current and long-term biophysical effects and economic consequences of the impact of agriculture. Much of the literature is concerned with land degradation, especially water erosion. Moreover, most of the assessments are of physical damage, although a few attempts have been made to estimate the economic costs of degradation as a proportion of agricultural GDP. Scherr (1999) quotes estimates of annual losses in agricultural GDP caused by soil erosion for a number of African countries, which can be considerable. Unfortunately these aggregate estimates can be misleading, and policy priorities for limiting impacts based on physical damage may not truly reflect the costs to the economy at large. Second, the relative importance of different impacts may change with time, as point sources of pollution are increasingly brought under control and non-point sources become the major problem. Lastly, offsite costs can be considerably greater than onsite costs.

These important analytical limitations are apparent from recent estimates of the external environmental costs of agriculture in various developed countries given in Pretty et al. (2001). These estimates suggest that in developing countries over the next 30 years greater consideration should be given to air pollution and offsite damage because their costs may exceed those of land and water pollution, loss of biodiversity and onsite damage. It should be noted that a large proportion of these environmental costs stems from climate change and its impacts, which are still very uncertain (see Chapter 13).

It is generally accepted that most developing countries will increasingly face the type of agro-environmental impacts that have become so serious in developed countries over the past 30 years or more. The commodity production and input use projections presented in Chapters 3 and 4 provide an overall framework for assessing the likely impacts of agricultural activities on the environment over the next 30 years in developing countries. Several large developing countries already have average fertilizer and pesticide application rates exceeding those causing major environmental problems in developed countries. Similarly, some developing countries have intensive livestock units as large as those in Europe and North America that are regarded as serious threats to waterbodies (OECD, 2001a).

Moreover, the experience of agro-environmental impacts in developed countries can give advance warning to developing countries where agro-ecological conditions are similar to those in OECD countries. Developing countries are likely to face similar problems when adopting similar patterns of intensification. They can use the experience of developed countries to identify some of the policy and technological solutions to limit or avoid negative agro-environmental impacts, and to identify the trade-offs. They can also estimate the economic costs (externalities) of the agro-environmental impacts of intensive agriculture that are not currently reflected in agricultural commodity prices, and these costs can provide a basis for policy and technology priority setting.

It will be argued that higher priority than is currently the case should be given to lowering agriculture's impact on air and water. The remainder of this chapter assesses the changing pressures on the environment from agriculture, using the projections for land, water, agrochemical input and technological change given in earlier chapters. It examines the main technology and policy options for limiting agriculture's negative impacts on the environment and widening its positive ones. Finally, it considers the range of situations and trade-offs that may influence the uptake of these options. The important issue of climate change is examined both here and in Chapter 13. This chapter examines the role of agriculture as a driving force for climate change, while Chapter 13 examines the impact of climate change on agricultural production and food security.

12.2 Major trends and forces

It is clear from the crop production projections presented in Chapter 4 that the key issue for the future is the environmental pressure from intensification of land use, rather than land cover or land use changes alone. Some 80 percent of the incremental crop production in developing countries will come from intensification and the remainder from arable land expansion (Table 4.2). Thus the dominant agro-environmental costs and benefits over the projection period will continue to be those stemming from the use of improved cultivars and higher inputs of plant nutrients and livestock feeds, together with better nutrient management and tillage practices, pest management and irrigation. Nonetheless, extensification of agriculture in environmentally fragile "hot spots" or areas high in biodiversity will also remain of continuing concern.

The positive benefits of these changes will include a slowdown of soil erosion and at least a slower increase in pollution from fertilizers and pesticides. Likely outcomes on the negative side are a continuing rise in groundwater nitrate levels from poor fertilizer management, further land and yield losses through salinization, and growing air and water pollution from livestock.

The main agro-environmental problems fall into two groups. First, there are those that are global in scale such as, for example, the increase in atmospheric concentrations of the GHGs carbon dioxide (CO_2) through deforestation, and nitrous oxide (N_2O) arising from crop production (Houghton *et al.*, 1995; Mosier and Kroeze, 1998). The second group of problems is found in discrete locations of the major continents and most countries, but at present has no substantive impact at the global level. Examples are the salinization of irrigated lands and the buildup of nitrate fertilizer residues in groundwater and surface water. These problems first emerged in the developed countries in the 1970s as a consequence of agricultural intensification. However, they have become of increasing importance in some developing countries during the past decade or so, and are destined to become more widespread and more intense unless there is a break from current policy and technological trends.

Most of the negative impacts from agriculture on the environment can be reduced or prevented by an appropriate mix of policies and technological changes (see, for example, UN, 1993; Alexandratos, 1995; Pretty, 1995; and Conway, 1997). There is growing public pressure for a more environmentally benign agriculture. Countries also have to comply with the WTO Agreement on Agriculture and the UNCED Conventions (particularly the Framework Convention on Climate Change). This forces countries to reduce commodity price distortions and input subsidies, and encourages them to remove other policy interventions that tend to worsen agro-environmental impacts, and to integrate environmental considerations explicitly into agricultural policies.

At the national level, there is now a range of policy options available to correct past agro-environmental mistakes and to prevent or limit future ones. The main problems were first recognized in those developed countries that embarked on agricultural intensification in the 1940s and 1950s, e.g. France, the United Kingdom and the United States. These countries started to formulate corrective measures soon afterwards. Their experience can help other countries embarking on intensification to avoid or moderate some of the problems. Some developing countries, for example, have introduced institutional mechanisms to promote environmentally benign technologies more rapidly than the developed countries at a comparable point of economic development. Moreover, the responses have not been only at the public sector level. Farmers in both developed and developing countries have also made a significant contribution by spontaneously creating or adopting environmentally benign technologies or management practices.

At the international level, there is now wide endorsement of the precautionary principle, under which countries accept the need to introduce corrective actions at an early stage and possibly before all of the scientific justification is in place (UN, 1993).

International action has also been taken to strengthen research on the biophysical changes that agriculture is causing (Walker and Steffen, 1999), and to monitor the key indicators of agro-ecosystem health (ICSU/UNEP/FAO/UNESCO/WMO, 1998; OECD, 1991, 2001b) so as to understand and give advance warning of any threats to agricultural sustainability.

12.3 Changing pressures on the environment

12.3.1 Agriculture's contribution to air pollution and climate change

Public attention tends to focus on the more visible signs of agriculture's impact on the environment, whereas it seems likely that the non-visible or less obvious impacts of air pollution cause the greatest economic costs (Pretty *et al.*, 2001). Agriculture affects air quality and the atmosphere in four main ways: particulate matter and GHGs from land clearance by fire (mainly rangeland and forest) and the burning of rice residues; methane from rice and livestock production; nitrous oxide from fertilizers and manure; and ammonia from manure and urine.

Pollution from biomass burning. Soot, dust and trace gases are released by biomass burning during forest, bush or rangeland clearance for agriculture. Burning is traditionally practised in "slash and burn" tropical farming, in firing of savannah regions by pastoralists to stimulate forage growth and in clearing of fallow land and disposing of crop residues, particularly rice. This burning has had major global impacts and has caused air pollution in tropical regions far away from the source of the fires.

Two developments should result in an appreciable fall in air pollution from biomass burning. Deforestation is often achieved by burning, or fire is used after timber extraction to remove the remaining vegetation (Chapter 6). The projected reduction in the rate of deforestation will slow down the growth in air pollution. The shift from extensive to intensive livestock production systems (Chapter 5) will reduce the practice of rangeland burning under extensive grazing systems, although the latter systems seem likely to remain dominant

Table 12.1   Agriculture's contribution to global greenhouse gas and other emissions

Gas	Carbon dioxide	Methane	Nitrous oxide	Nitric oxides	Ammonia
Main effects	Climate change	Climate change	Climate change	Acidification	Acidification Eutrophication
Agricultural source (estimated % contribution to total global emissions)	Land use change, especially deforestation	Ruminants (15)	Livestock (including manure applied to farmland) (17)	Biomass burning (13)	Livestock (including manure applied to farmland (44)
		Rice production (11)	Mineral fertilizers (8)	Manure and mineral fertilizers (2)	Mineral fertilizers (17)
		Biomass burning (7)	Biomass burning (3)		Biomass burning (11)
Agricultural emissions as % of total anthropogenic sources	15	49	66	27	93
Expected changes in agricultural emissions to 2030	Stable or declining	From rice: stable or declining From livestock: rising by 60%	35-60% increase		From livestock: rising by 60%

Main sources: Column 2: IPCC (2001a); column 3: Lassey, Lowe and Manning (2000); columns 4, 5 and 6: Bouwman (2001); FAO estimates.

in parts of sub-Saharan Africa. The growth in the contribution of crop residues may also slow down because of the projected very slow growth in rice production (Chapters 3 and 4). Climate change itself, however, may cause temperatures to rise in the dry season, increasing fire risks and thus increasing pollution from biomass burning in some areas (Lavorel *et al.*, 2001).

Greenhouse gas emissions. For some countries, the contribution from agriculture to GHG emissions is a substantial share of the national total emissions, although it is seldom the dominant source. Its share may increase in importance as energy and industrial emissions grow less rapidly than in the past while some agricultural emissions continue to grow (Table 12.1). There is increasing concern not just with carbon dioxide but also with the growth of agricultural emissions of other gases such as methane, nitrous oxide and ammonia arising from crop and livestock production. In some countries

these can account for more than 80 percent of GHG emissions from agriculture.

The conversion of tropical forests to agricultural land, the expansion of rice and livestock production and the increased use of nitrogen fertilizers have all been significant contributors to GHG emissions. Agriculture now contributes about 30 percent of total global anthropogenic emissions of GHGs, although large seasonal and annual variations make a precise assessment difficult (Bouwman, 2001). Tropical forest clearance and land use change were major factors in the past for carbon dioxide emissions, but are likely to play a smaller role in the future (see Chapters 4 and 6). More attention is now being given to methane (CH_4) and nitrous oxide (N_2O), since agriculture is responsible for half or more of total global anthropogenic emissions of these GHGs (Table 12.1).

Methane from ruminant and rice production. Methane is a principal GHG driving climate change. Its warming potential is about 20 times more powerful than carbon dioxide.[1] Global methane emissions amount at present to about 540 million tonnes p.a., increasing at an annual rate of 20-30 million tonnes. Rice production currently contributes about 11 percent of global methane emissions. Around 15 percent comes from livestock (from enteric fermentation by cattle, sheep and goats and from animal excreta). The livestock contribution can be higher or lower at the national level depending on the extent and level of intensification. In the United Kingdom and Canada the share is over 35 percent.

The production structure for ruminants in developing countries is expected to increasingly shift towards that prevalent in the industrial countries. The major share of cattle and dairy production will come from feedlot, stall-fed or other restricted grazing systems and by 2030 nearly all pig and poultry production will also be concentrated in appropriate housings. Much of it will be on an industrial scale with potentially severe local impacts on air and water pollution.

The livestock projections in this report (Chapter 5) entail both positive and negative implications for methane emissions. The projected increase in livestock productivity, in part related to improvements in feed intake and feed digestibility, should reduce emissions per animal. Factors tending to increase emissions are the projected increase in cattle, sheep and goat numbers and the projected shift in production systems from grazing to stall-feeding. The latter is important because storage of manure in a liquid or waterlogged state is the principal source of methane emissions from manure, and these conditions are typical of the lagoons, pits and storage tanks used by intensive stall-feeding systems. When appropriate technologies are introduced to use methane in local power production, as has been done in some South and East Asian countries, the changes can be beneficial. If emissions grow in direct proportion to the projected increase in livestock numbers and in carcass weight or milk output (see Chapter 5), global methane emissions could be 60 percent higher by 2030. Growth in the developed regions will be slow, but in East and South Asia emissions could more than double, largely because of the rapid growth of pig and poultry production in these regions.

Rice cultivation is the other major agricultural source of methane. The harvested area of rice is projected to expand by only about 4.5 percent by 2030 (Table 4.11) depending on yield growth rates, and possibly on the ability of technological improvements to compensate for climate-change-induced productivity loss if this becomes serious (Wassmann, Moya and Lantin, 1998). Total methane emissions from rice production will probably not increase much in the longer term and could even decrease, for two reasons. First, about half of rice is grown using almost continuous flooding, which maintains anaerobic conditions in the soil and hence results in high methane emissions. However, because of water scarcity, labour shortages and better water pricing, an increasing proportion of rice is expected to be grown under controlled irrigation and better nutrient management, causing methane emissions to fall. Second, up to 90 percent of the methane from rice fields is emitted through the rice plant. New high-yielding varieties exist that emit considerably less methane than some of the widely used traditional and modern cultivars, and this property could be widely exploited over the next ten to 20 years (Wang, Neue and Samonte, 1997).

Nitrous oxide. Nitrous oxide (N_2O) is the other powerful GHG for which agriculture is the dominant anthropogenic source (Table 12.2). Mineral fertilizer use and cattle production are the main culprits. N_2O is generated by natural biogenic processes, but output is enhanced by agriculture through nitrogen fertilizers, the creation of crop residues, animal urine and faeces, and nitrogen leaching and runoff. N_2O formation is sensitive to climate, soil type, tillage practices and type and placement of fertilizer. It is also linked to the release of nitric oxide and ammonia, which contribute to acid rain and the acidification of soils and drainage systems (Mosier and Kroeze, 1998). The current agricultural contribution to total global nitrogen emissions is estimated at 4.7 million tonnes p.a., but there is great uncer-

---

[1] Power is measured in terms of the global warming potential (GWP) of a gas, taking account of the ability of a gas to absorb infrared radiation and its lifetime in the atmosphere.

Table 12.2   Global N₂O emissions

Million tonnes N p.a.	Mean value		Range	
Natural sources				
Oceans	3.0		1-5	
Soils total, of which:	6.0		3.3-9.7	
Tropical soils Wet forest		3.0		2.2-3.7
Dry savannahs		1.0		0.5-2.0
Temperate soils Forests		1.0		0.1-2.0
Grasslands		1.0		0.5-2.0
Subtotal natural sources	**9.0**		**4.3-14.7**	
Anthropogenic sources				
Agriculture total, of which:	4.7		1.2-7.9	
Agricultural soils, manure, fertilizer		2.1		0.4-3.8
Cattle and feedlots		2.1		0.6-3.1
Biomass burning		0.5		0.2-1.0
Industry	1.3		0.7-1.8	
Subtotal anthropogenic sources	**6.0**		**1.9-9.6**	
Total all sources	**15.0**		**6.2-24.3**	

Source: Mosier and Kroeze (1998), modified using Mosier *et al.* (1996).

tainty about the magnitude because of the wide range in estimates of different agricultural sources (Table 12.2).

Nitrogen fertilizer is one major source of nitrous oxide emissions. The crop projections to 2030 imply slower growth of nitrogen fertilizer use compared with the past (Table 4.15 and Daberkov *et al.*, 1999). Depending on progress in raising fertilizer-use efficiency, the increase between 1997/99 and 2030 in total fertilizer use could be as low as 37 percent. This would entail similar or even smaller increases in the direct and indirect N₂O emissions from fertilizer and from nitrogen leaching and runoff. Current nitrogen fertilizer use in many developing countries is very inefficient. In China, for example, which is the world's largest consumer of nitrogen fertilizer, it is not uncommon for half to be lost by volatilization and 5 to 10 percent by leaching. Better onfarm fertilizer management, wider regulatory measures and economic incentives for balanced fertilizer use and reduced GHG emissions, together with technological improvements such as more cost-effective slow-release formulations should reduce these losses in the future.

Livestock are the other major source of anthropogenic nitrous oxide emissions (Mosier *et al.*, 1996; Bouwman, 2001). These emissions arise in three ways.

First, from the breakdown of manure applied as fertilizer, primarily to crops but also to pastures. The proportion of manure thus used is difficult to estimate, but it is probably less than 50 percent. Moreover, there are opposite trends in its use. In the developed countries, growing demand for organic foods, better soil nutrition management and greater recycling is favouring the increased use of manure. In the developing countries, with a strong growth in industrial-scale livestock production separate from crop production, and with decreasing labour availability, there is a trend to rely more on mineral fertilizers to maintain or raise crop yields. The second source is dung and urine deposited by grazing animals. The emissions from this source are higher for intensively managed grasslands than for extensive systems (Mosier *et al.*, 1996). Similarly, emissions from animals receiving low-quality feeds are likely to be less than with higher-quality feeds. Since shifts are expected from extensive to intensive production systems and from low- to higher-quality feeds, it can be assumed that there will be an increase in N_2O emissions from deposited dung and urine. The third source is from the storage of excreta produced in stall-feeding or in intensive production units. This may produce a reduction in emissions since, on average, stored excreta produce about half as much N_2O as excreta deposited on pastures (Mosier *et al.*, 1996).

Changes in manure production over time have been estimated using the projected growth in livestock populations (allowing for differences between cattle, dairy, sheep and goats, pigs and poultry). The amounts per head have been adjusted on a regional basis to allow for projected changes in carcass weight and milk output. Emission rates have been adjusted for the assumed shifts from extensive to stall-fed systems. Based on these assumptions and estimates, the total production of manure is projected to rise by about 60 percent between 1997/99 and 2030. However, N_2O emissions are projected to rise slightly more slowly (i.e. by about 50 percent) because of the switch from extensive to stall-fed systems. This relative environmental gain from intensification has to be seen against the rise in ammonia and methane emissions and the probable growth in point-source pollution that the intensive livestock units will generate. This latter cannot be quantified but is a very serious problem in a number of developed and developing countries (de Haan, Steinfeld and Blackburn, 1998).

Ammonia. Agriculture is the dominant source of anthropogenic ammonia emissions, which are around four times greater than natural emissions. Livestock production, particularly cattle, accounts for about 44 percent, mineral fertilizers for 17 percent and biomass burning and crop residues for about 11 percent of the global total (Bouwman *et al.*, 1997; Bouwman, 2001). Volatilization rates from mineral fertilizers in developing countries are about four times greater than in developed countries because of higher temperatures and lower-quality fertilizers. Losses are even higher from manure (about 22 percent of the nitrogen applied).

Ammonia emissions are potentially even more acidifying than emissions of sulphur dioxide and nitrogen oxides (Galloway, 1995). Moreover, future emissions of sulphur dioxide are likely to be lower as efforts continue to reduce industrial and domestic emissions and improve energy-use efficiency, whereas there is little action on reducing agriculture-related emissions. The ammonia released from intensive livestock systems contributes to both local (Pitcairn *et al.*, 1998) and longer-distance deposition of nitrogen (Asman, 1994). This causes damage to trees and acidification and eutrophication of terrestrial and aquatic ecosystems, leading to decreased nutrient availability, disruption of nitrogen fixation and other microbiological processes, and declining species richness (UNEP/RIVM, 1999).

The livestock projections of this study are based on changes in both animal numbers and in productivity, as determined by changes in carcass weight or milk output per animal. It is assumed that the volume of excreta per animal, which is the main source of the ammonia, increases over time in proportion to carcass weight, which in turn is a reflection of the increase in the use of feed concentrates. Table 12.3 gives estimates of ammonia emissions in 1997/99 and 2030 using these assumptions. The projected increase for the developing countries (80 percent) is significantly higher than the increase (50 percent) given in Bouwman *et al.* (1997).

These projections have three main environmental implications. First, all the developing regions potentially face ammonia emission levels that have caused serious ecosystem damage in the developed countries. Second, emissions may continue to rise in the developed countries, adding to the already serious damage in some areas. And third, in East Asia and Latin America, a high proportion of the emissions will come from intensive pig production systems, in which emission reduction is more difficult. Since many of these intensive production units are located where there is a large demand, the downwind and downstream effects are likely to be concentrated near large urban populations, often on river plains and coastal plains that are already subject to a high chemical and particulate pollution load.

12.3.2 Agriculture's impact on land

In recent decades the most important environmental issues concerning land have been land cover change, particularly deforestation, and land use intensification, especially its impact on land degradation. The future picture concerning land

Table 12.3 Ammonia emissions implied by the livestock projections

	Number of animals 1997/99	2030 (in millions)	Emissions NH$_3$ 1997/99	2030 ('000 tonnes N p.a.)
World				
Total			**30.34**	**48.60**
Cattle and buffalo	1 497	1 858	13.09	19.51
Dairy	278	391	5.35	9.98
Sheep and goats	1 749	2 309	2.02	3.50
Pigs	871	1 062	6.62	9.25
Poultry	15 119	24 804	3.27	6.35
Developing countries				
Total			**21.35**	**38.55**
Cattle and buffalo	1 156	1 522	9.33	15.34
Dairy	198	312	3.63	8.08
Sheep and goats	1 323	1 856	1.59	3.02
Pigs	579	760	4.52	7.02
Poultry	10 587	19 193	2.29	5.09
Industrial countries				
Total			**6.67**	**7.24**
Cattle and buffalo	254	243	3.03	3.18
Dairy	41	44	1.02	1.21
Sheep and goats	341	358	0.33	0.34
Pigs	210	220	1.51	1.58
Poultry	3 612	4 325	0.78	0.93
Transition countries				
Total			**2.32**	**2.80**
Cattle and buffalo	87	94	0.74	1.00
Dairy	39	35	0.69	0.69
Sheep and goats	85	95	0.10	0.14
Pigs	81	82	0.59	0.65
Poultry	920	1 287	0.20	0.32

Note: Figures for the base year are from Mosier and Kroeze (1998).

cover change is a continuing slowing down of the conversion of forests to areas for crop or livestock production; no appreciable change in grazing area; and continued growth of protected areas. However, in the case of land degradation, there are still widely differing opinions about future trends, and the empirical basis for making firm projections remains weak.

Land cover change. In the past much of the pressure for land cover change came from deforestation, but this is likely to slow down in future. The process of deforestation will continue in the tropics but at a decelerating rate, and in a number of major developing countries the extent of forest will actually increase. In Latin America, for example, governments have removed some of the policy distortions that encouraged large farmers and companies to create pastures on deforested land. However, deforestation by smallholders has not been reduced. In West Africa, almost no primary rain forest is left, and over 80 percent of the population is still rural and growing at 2.5-3 percent p.a. – a situation that is likely to continue until non-agricultural employment opportunities are found.

The total extent of grassland is likely to decrease. In most developing regions the general trend is away from extensive grazing towards mixed farming and improved pastures or intensive feedlot and stall-fed systems. It is assumed that there will be no substantial development of new grassland. Some of the more marginal rangelands and pastures are likely to be abandoned as herders and other livestock producers leave the land for better-paid jobs outside agriculture, as happened in parts of Europe after about 1950 (CEC, 1980). In the absence of grazing pressure these areas will revert to forest or scrub. Some of the better grazing land will be converted to cropland or urban land, with the loss being compensated by improving productivity on the remaining land rather than by clearing new land. At the national level some countries will diverge from this global picture. Countries such as China, some Commonwealth of Independent States countries and parts of South America still have the potential for major increases in the use of natural grasslands (Fan Jiangwen, 1998).

Net arable land expansion in developing countries is projected to fall from about 5 million ha p. a. over the past four decades to less than 3.8 million ha p.a. over the period to 2030 (Table 4.7). This net increase takes into account the area of cropland going out of use because of degradation, which some argue is in the order of 5 to 6 million ha p.a. (UNEP, 1997). It also takes into account land abandoned because it is no longer economically attractive to farm or as a result of changes in government policy, and land taken up by urbanization. It does not, however, take account of the area that is restored to use for reasons other than crop production, e.g. as part of agricultural carbon sequestration policies. In the future such areas could be substantial.

Even though an overall slowdown in the expansion of agricultural land is expected, there are three areas of concern. First, the frequency or intensity of cultivation of formerly forested slope lands will probably increase. Second, further

drainage of wetlands will result in loss of biodiversity and of fish spawning grounds, with increased carbon dioxide but lower methane emissions. Third, high-quality cropland will continue to be lost to urban and industrial development.

Cultivation of slope lands. Uplands are particularly prone to water erosion where cultivated slopes are steeper than 10 to 30 percent, lack appropriate soil conservation measures and rainfall is heavy. Substantial areas of land are at risk (Bot, Nachtergaele and Young, 2000), although it is not possible to make global or even regional estimates of how much of this is currently cropped land. In Southeast Asia, land pressure caused by increasing population has extended the use of steep hill slopes particularly for maize production. This has led to a very significant increase in erosion on lands with slopes of over 20 percent (Huizing and Bronsveld, 1991).

There are two main environmental concerns for the future. First, more forest may be cleared for cultivation, resulting in the loss of biodiversity and increased soil erosion. Second, existing cultivated slopes may be cropped more intensively, leading to greater soil erosion and other forms of land degradation (Shaxson, 1998).

In countries such as Bhutan and Nepal with limited flat land left to develop, almost all of the additional land brought into cultivation in the future will be steep lands prone to erosion unless well terraced, or protected by grass strips, conservation tillage, etc. In Nepal, for example, soil erosion rates in the hills and mountains are in the range of 20 to 50 tonnes/ha/p.a. in agriculture fields and 200 tonnes/ha/p.a. in some highly degraded watersheds (Carson, 1992). Crop yields in these areas declined by 8 to 21 percent over the period 1970-1995. Such losses seem likely to continue, unless there are substantial changes in farmer incentives for soil conservation and wider knowledge at all levels of the economic and environmental benefits that conservation provides. In countries with less acute land pressure there may be a slowing down in the expansion of cropped slope lands, but there will still be pressures for the intensification of cultivation and hence the risk of greater erosion.

Deforestation of slope lands was a serious concern in the past but may slow down over the projection period. On the one hand, a good part of the 3.8 million ha annual net new cropland over the period to 2030 will probably come from forest conversion. A high proportion will have steep slopes and will be in zones with high rainfall, so the water erosion risk will be high unless suitable management techniques are adopted (Fischer, van Velthuizen and Nachtergaele, 2000). On the other hand, large areas of existing crop and grazing land – much of which is likely to be slope land – could revert to forest and scrub because of land abandonment and outmigration.

Reclamation of wetlands. Historically, the reclamation of wetlands has made a major contribution to agricultural growth and food security. Significant parts of the rich croplands of the Mississippi basin in the United States, the Po Valley in Italy and the Nile Delta in Egypt are reclaimed wetlands. The developing countries have over 300 million ha of natural wetlands that are potentially suitable for crop production. Some of the wetlands will inevitably be drained for crop production. Part will be in countries with relatively large land areas per capita but limited areas with adequate rainfall or irrigation potential, e.g. Senegal. Part will be in countries where much of the potential arable land is not well suited for sustainable agriculture because of steep slopes or thin, fragile soils, so the development of wetlands is therefore a more attractive option, e.g. Indonesia.

Wetlands are flat, and by definition well watered. In the case of some Sahelian countries they are potentially important contributors to food security (Juo and Lowe, 1986). Past experience in the inland valleys of the Sahelian belt suggest, however, that reclamation for agriculture has been of doubtful benefit despite huge international investments. Many irrigation schemes have failed through mismanagement and inadequate infrastructure maintenance, civil unrest and weak market development. The soils in this region are potentially productive once certain constraints are overcome, e.g. acid sulphate, aluminium and iron toxicity and waterlogging.

Assuming that the present rate of drainage of wetlands declines, the total conversion over the projection period will amount to a relatively small part of the 300 million ha they currently cover, but even this would carry some environmental risks. There may be damage to the hydrological functions of wetlands, such as groundwater recharge and natural flood relief, and disruption of migration routes and overwintering grounds of certain birds.

In central China, for example, around half a million ha of wetlands have been reclaimed for crop production since about 1950, contributing to a reduction of floodwater storage capacity of approximately 50 billion m^3 (Cai, Zhao and Du, 1999). There is strong evidence that wetland reclamation is responsible for about two-thirds of this loss in storage capacity, and thus for about two-thirds of the US$20 billion flood damage in 1998 (Norse *et al.*, 2001). Similar links have been established for the severe 1993 floods in the United States (IFMRC, 1994). Therefore it is important to introduce appropriate planning and regulatory mechanisms to ensure that any future wetland development is undertaken with the necessary safeguards, as is the case in the United States and a number of other developed countries (Wiebe, Tegene and Kuhn, 1995).

Loss of and competition for good arable land. As populations grow, much good cropland is lost to urban and industrial development, roads and reservoirs. For sound historic and strategic reasons, most urban areas are sited on flat coastal plains or river valleys with fertile soils. Given that much future urban expansion will be centred on such areas, the loss of good-quality cropland seems likely to continue. In fact the losses seem inevitable, given the low economic returns to farm capital and labour compared with non-agricultural uses. Such losses are essentially irreversible and in land-scarce countries the implications for food security could be serious.

Estimates of non-agricultural use of land per thousand persons range from 22 ha in India (Katyal *et al.*, 1997), to 15-28 ha in China (Ash and Edmunds, 1998) and to 60 ha in the United States (Waggoner, 1994). The magnitude of future conversions of land for urban uses is not certain, nor how much of it will be good arable land. There is no doubt, however, that losses could be substantial. In China, for example, the losses between 1985 and 1995 were over 2 million ha, and the rate of loss to industrial construction has increased since 1980 (Ash and Edmunds, 1998).

The projected increase in world population between 2000 and 2030 is some 2.2 billion.

Table 12.4   Global Assessment of Human-induced Soil Degradation (GLASOD)[2]

Region	Total land affected (million ha)	Percentage of region degraded	
		Moderate	Strong and extreme
Africa	494	39	26
Asia	747	46	15
Australasia	103	4	2
South America	243	47	10
Central America	63	56	41
Europe	219	66	6
North America	96	81	1
Total	1 964	46	16

Source: Oldeman, Hakkeling and Sombroek (1991).

Assuming that the conversion of land for non-agricultural purposes is an average of 40 ha per thousand persons, the projected loss on this account would be almost 90 million ha. Even assuming that all of this would be land with crop production potential, this would be only a fraction of the global balance of potential cropland that is as yet unused (see Chapter 4). However, in heavily populated countries such as China and India which have very limited potential for cropland expansion, even small losses could be serious. In China, this issue has been of growing concern for a number of years. Land loss to urban and industrial development in central and southern China has been partially compensated by the conversion of grasslands in the northeast to crop production (Ash and Edmunds, 1998). But although this new land is very fertile, the growing season is short and allows only one crop per year, compared with the two to three crops per year possible on the land lost in the central and southern areas. However, it seems likely that the global trading system will be able to meet China's potential food import needs to 2030 and beyond (Alexandratos, Bruinsma and Schmidhuber, 2000).

Land degradation.[3] The assessment of land degradation is greatly hindered by serious weaknesses in our knowledge of the current situation (Pagiola, 1999; Branca, 2001). According to some analysts, land degradation is a major threat to food security, it has negated many of the productivity improvements of the past, and it is getting worse (Pimentel et al., 1995; UNEP, 1999; Bremen, Groot and van Keulen, 2001). Others believe that the seriousness of the situation has been overestimated at the global and local level (Crosson, 1997; Scherr, 1999; Lindert, 2000; Mazzucato and Niemeijer, 2001).

Area of degraded land. The most comprehensive global assessment is still the Global Assessment of Human-induced Soil Degradation (GLASOD) mapping exercise (Oldeman, Hakkeling and Sombroek, 1991; Table 12.4).

The results are subject to a number of uncertainties, particularly regarding the impact on productivity and the rates of change in the area and severity of degradation. A follow-up study for South Asia addressed some of the weaknesses of GLASOD. It introduced more information from national studies and greater detail on the different forms of degradation (FAO/UNDP/UNEP, 1994). The broad picture for South Asia remains similar in the two studies: 30-40 percent of the agricultural land is degraded to some degree, and water erosion is the most widespread problem (Table 12.5).

[2] GLASOD defines four levels of degradation: light = somewhat reduced agricultural productivity; moderate = greatly reduced agricultural productivity; strong = unreclaimable at the farm level; extreme = unreclaimable and unrestorable with current technology.

[3] Defined as a process that lowers the current or potential capabilities of the soil to produce goods or services, through chemical, physical or biological changes that lower productivity (Branca, 2001).

Table 12.5   Shares of agricultural land in South Asia affected by different forms of degradation

Type of land degradation	Percentage of agricultural land affected
Water erosion	25
Wind erosion	18
Soil fertility decline	13
Salinization	9
Lowering of the water table	6
Waterlogging	2

Source: FAO/UNDP/UNEP (1994), p. 50-51.

Despite these improvements in techniques of assessment, a number of serious difficulties remain in using them for perspective analysis. They are still heavily based on expert judgement, for entirely justified reasons. There is no clear consensus as to the area of degraded land, even at the national level. In India, for example, estimates by different public authorities vary from 53 to 239 million ha (Katyal *et al.*, 1997). Land degradation is very variable over small areas, e.g. as a consequence of differences in soil type, topography, crop type and management practice, so impacts are highly site specific. They can also be time specific: soil erosion impacts can vary in the short term because of interannual differences in rainfall, with no yield reductions in high rainfall years but appreciable losses in dry years (Moyo, 1998). Some forms of degradation are not readily visible, for example, soil compaction, acidification and reduced biological activity. Lack of data and analytical tools for measuring such differences prevents or limits estimation of their impact on productivity, and makes scaling up to the national or regional level problematic. There are no internationally agreed criteria or procedures for estimating the severity of degradation and most surveys do not make reliable assessments. Few if any countries make systematic assessments at regular intervals that permit estimation of rates of change. Finally, major changes in socio-economic conditions, improved market opportunities, infrastructure and technology over the medium to long term can induce farmers to overcome degradation (Tiffen, Mortimore and Gichuki, 1994).

Impact of degradation on productivity. Does degradation have a serious impact on onfarm produc-

tivity, and on offsite environments through wind and water dispersal of soil? Because degradation is normally a slow and almost invisible process, rising yields caused by higher inputs can mask the impact of degradation until yields are close to their ceiling. They thus hide the costs to farmers of falling input efficiency (Walker and Young, 1986; Bremen, Groot and van Keulen, 2001). In Pakistan's Punjab, for example, Murgai, Ali and Byerlee (2001) question whether technological gains can be sustained because of the severe degradation of land and water resources.

Moreover, the experimental methods commonly used to determine impacts on yields have a number of weaknesses (Stocking and Tengberg, 1999). Estimation of the economic costs can be equally complex. First, there may be impacts not just on yield levels, but on grain quality (e.g. drop in protein quality), yield stability or production costs or any combination of these (Lipper, 2000). Second, it is necessary to separate out the effect of the different factors involved in total factor productivity growth (Murgai, Ali and Byerlee, 2001). There are also offsite impacts such as siltation of streams and reservoirs, loss of fish productivity, raising water storage costs and the incidence of flood damage, and these are normally more serious than the onfarm impacts.

Given the above estimation difficulties, it is not surprising that there is little correspondence between global assessments of productivity loss and national or local realities. According to GLASOD, most areas in six states of the United States were classified as moderately degraded. According to the definition, this should result in reduced agricultural productivity. In reality, however, crop yields in these areas have been rising steadily for the past 40 years.

Crosson (1997) suggests that recent rates of land degradation and particularly soil erosion have had only a small impact on productivity, and argues that the annual average loss for cropland productivity since the mid-1950s was lower than 0.3 percent. Similarly, according to GLASOD there is almost no cultivated land in China that is not degraded in one way or another. Almost all the areas of rice cultivation in south and southeast China are classified as being affected by high or very high water erosion, and large wheat-growing areas southeast of Beijing are classified as suffering from medium levels of chemical deterioration (salinization). Yet, in spite of this, China was able to increase its wheat production between the early 1960s and mid-1990s from about 16 to 110 million tonnes, and its paddy production from 63 to 194 million tonnes. Moreover, more detailed assessments suggest that there has been little deterioration in China's (and Indonesia's) soils since the 1950s and they may have actually improved up to the 1980s (Lindert, 2000.) Water erosion from most rice fields is very low (Norse *et al.*, 2001). Nonetheless, rising yields may have masked productivity losses.

Oldeman (1998) estimates the global cumulative loss of cropland productivity at about 13 percent, but there are large regional differences. Africa and Central America may have suffered declines of 25 and 38 percent respectively since 1945. Asia and South America, on the other hand, may have lost only about 13 percent, while Europe and North America have lost only 8 percent. UNEP (1999) does not accept Crosson's assessment and argues

that land degradation is so bad that it has negated many of the gains in land productivity of recent decades. Support for this view comes from detailed analysis of resource degradation under intensive crop production systems in the Pakistan and Indian Punjab (Murgai, Ali and Byerlee, 2001).

Three issues arise here. First, these estimates are global and regional averages, whereas the implications of degradation for agricultural production and food security are primarily local and national. On the one hand, there are complex trade-offs and compensatory mechanisms involved in some forms of degradation. A high proportion of eroded soil is redeposited elsewhere in the catchment area, where it tends to boost productivity, but it may also silt up reservoirs and irrigation canals. For example, in the United States some 45 percent is redeposited locally, and some 46 percent in lakes, reservoirs and other impoundments (Smith *et al.*, 2001). On the other hand, there are also a number of so-called hot spots where the degradation is already serious and could get worse (Scherr and Yadav, 1996). These hot spots include some of the developing countries' most fertile river basins (Table 12.6), which play a vital role in food security.

Second, the most visible degradation tends to be on marginal lands, whereas the bulk of food production occurs on more favourable lands, particularly irrigated areas (Norse, 1988). On such favoured lands relatively low rates of degradation could have large impacts on productivity and yields, although there are some grounds for concluding that this will not be the case. There is

Table 12.6 Regional hot spots of land degradation

Region	Salinization	Erosion
South and West Asia	Indus, Tigris and Euphrates river basins	Foothills of the Himalayas
East and Southeast Asia	Northeast Thailand and North China Plain	Unterraced slopes of China and Southeast Asia
Africa	Nile Delta	Southeast Nigeria, the Sahel, mechanized farming areas of North and West Africa
Latin America and the Caribbean	Northern Mexico, Andean highlands	Slopes of Central America, the semi-arid Andean Valley and the *cerrados* of Brazil

Source: Scherr and Yadav (1996).

growing evidence that farmers are able to adapt to environmental stress in ways that limit degradation (Mortimore and Adams, 2001; Mazzucato and Niemeijer, 2001).

Third, there are factors at work that may reduce degradation in the coming decades. These include the wider use of direct measures to prevent or reverse degradation (Branca, 2001), and indirect measures such as improved irrigation techniques and water pricing to reduce salinization. The spread of NT/CA will limit the damage caused by conventional tillage. Nonetheless the results from the Punjab show no grounds for complacency regarding the sustainability of some high-input crop production systems, and point to the need for increased fertilizer-use efficiency and reduced salinization.

Future areas of degraded land and loss of productivity. Will the area of degraded arable and pastoral land expand in the future, or deteriorate further, e.g. because of population pressure or the projected intensification of production?

Although the projections presented in Chapter 4 do not directly address the issue of land degradation, they do contain a number of features that can be used to assess how some forms of degradation may decline or become more serious in the future.

First, about one-third of the harvested area in developing countries in 2030 is projected to be *irrigated* land (Table 4.8), up from 29 percent in 1997/99. This is generally flat or well-terraced land with little erosion. However, parts may be at risk from salinization, particularly in more arid zones (Norse *et al.*, 2001). In addition, a quarter of the harvested *rainfed* land is estimated to have slopes of less than 5 percent, which are generally not prone to heavy water erosion. Their annual soil losses of around 10 tonnes per ha should be reduced where it is economically feasible to do so, but such rates could be tolerated for several hundred years before they have an appreciable impact on crop production. In all, around half of the cropland will not be markedly prone to soil erosion, although it may be subject to other forms of land degradation including salinization, nutrient mining, soil acidification and compaction.

Second, the global area of rainfed land under NT/CA could grow considerably, bringing benefits

such as reduced soil erosion, reduced loss of plant nutrients, higher rainfall infiltration and better soil moisture-holding capacity (see Chapter 11 and Section 12.4.3 below). NT/CA will have positive effects on the physical, chemical and biological status of soils. Organic matter levels of soils, for example, are likely to rise. Organic matter is a major source of plant nutrients and is the glue that holds soil particles together and stabilizes the pore structure. Organic matter makes soils less vulnerable to wind erosion and functions as a sponge for holding water and slowing down its loss from the crop root zone by drainage or evaporation. Moreover, the nutrients added to the soil as organic residues are released more gradually than those from mineral fertilizers and are therefore less prone to leaching, volatilization or fixation (Avnimelech, 1986). In addition, higher soil organic matter levels are commonly associated with greater levels of humic acid, which increases phosphate availability and thus can be very beneficial in those areas with strongly phosphate-fixing soils found in sub-Saharan Africa and Latin America.

Third, fertilizer consumption and fertilizer-use efficiency are projected to rise. This will bring benefits in terms of higher soil fertility and soil organic matter levels. Soil erosion will diminish because of the positive impact on root proliferation, plant growth and ground cover of increased phosphate and potassium (associated with more balanced fertilizer inputs).

These conclusions are broadly consistent with projections made by the International Food Policy Research Institute (IFPRI) (Agcaoili, Perez and Rosegrant, 1995; Scherr and Yadav, 1996). That is, global losses to degradation are likely to be small, but losses could be significant in some localities and regions. However, soil productivity loss from land degradation could be much more serious if the above gains from NT/CA, greater fertilizer use and fertilizer-use efficiency, and other forms of soil and water conservation are less than we estimate, and crop yield growth slows appreciably, as projected in Chapter 4.

On the other hand, there are sound reasons to believe that some of the fragile lands most prone to degradation will be abandoned. This will not necessarily result in additional pressures for deforestation and cropland development, because high rates of urbanization and rural-urban migration are projected for the future. This outmigration could, for example, reduce degradation stemming from the cultivation of slope lands, and lower some of the pressure on grazing land in the Sahel and other semi-arid and arid areas.

This is likely to have similar effects to those experienced in western European countries from the 1950s and 1960s onwards (CEC, 1980; Baldock *et al.*, 1996), and in Eastern European countries since the 1990s. Here rural outmigration and the restructuring of agriculture led to the abandonment of steep slopes and other marginal land and reduced pressure to develop any more land. Substantial areas of marginal land were abandoned and reverted to forest or scrub. In France this amounted to around 3 percent p.a. in the 1960s and early 1970s (Faudry and Tauveron, 1975). In Italy around 1.5 million ha were abandoned in the 1960s, of which some 70 percent was slope land, with decreases of 20 percent in some provinces (CEC, 1980). The decline was very rapid, and closely related to sharp falls in agricultural employment.

Rural outmigration and agricultural restructuring have also been occurring in many developing countries and are projected to continue. This trend is most noticeable in countries such as China, where urbanization is accompanied by a shift to alternative income sources. It follows, therefore, that a significant percentage of the slope land cultivated at present could be abandoned over the next three decades in many developing countries, with a substantial proportion reverting to forest. However, in densely populated rural economies, such as Indonesia, where the rate of population growth is still over 2 percent p.a., population drift to cities has not significantly reduced the density of rural settlement, or improved the economic livelihoods of the majority of the rural population.

Desertification is a serious form of dryland degradation, given priority by the international community in the form of the United Nations Convention to Combat Desertification (UNCCD). In the 1970s and 1980s it was argued that the Sahara was spreading rapidly southwards as part of an irreversible expansion of the world's deserts. Since then, counter-arguments have been growing in force, backed up by strong empirical evidence from remote-sensing activities (Nicholson and Tucker, 1998; Prince, Brown de Colstoun and Kravitz,

1998). This has shown that the desert margins are quite dynamic because of natural climate variation. The problem is more one of localized dryland degradation because of overgrazing, excessive fuel collection, bad tillage practices and inappropriate cropping systems.

Nonetheless, there has been some expansion of the deserts and dryland degradation (Dregne and Chou, 1992). Degradation of vegetation and native habitat is the major reason for species extinction in many semi-arid and subhumid environments. Rapid rates of species loss, particularly of beneficial insects, birds and other predators may reduce the capacity for natural suppression of the pests, diseases and weeds that are among the greatest threats to current levels of agricultural production. However, quantification is not precise (Dregne and Chou, 1992). The most extreme estimates suggest that about 70 percent of the 3.6 billion ha of drylands are degraded, although this is likely to be an overestimate. More probing analysis is highlighting the resilience and adaptability of crop and livestock systems in vulnerable areas such as the Sahel (Behnke, Scoones and Kerven, 1993; Mortimore and Adams, 2001).

Looking to the future, there seem to be several ongoing positive forces that could have a significant effect. First, the contraction in extensive livestock production should take the pressure off some drylands and reduce dryland degradation. Second, gains in productivity on favourable lands should allow some of the marginal drylands to revert to range or scrub. Third, the spread of irrigation, water-harvesting techniques and measures to avoid or overcome salinization should improve the sustainability of dryland agriculture. Fourth, the continued adoption of NT/CA permits greater rainfall infiltration and improves soil moisture-holding capacity. Fifth, better drought- and grazing-tolerant crops and grasses will be created through gene transfer, although this is unlikely to have much impact before 2015. Sixth, countries such as China and India are making major efforts to restore degraded land in arid areas (Sinha, 1997). Finally, there are widespread efforts to restore saline and other degraded soils that have gone out of production. In the view of some analysts, restored lands could total 200-300 million ha by 2025 (GCSI, 1999). On the negative side, at least in the medium term, are the expansion or intensification of cropping in semi-arid areas and further losses from salinization.

Overgrazing has been one of the central environmental concerns related to livestock activities. It can lead to the degradation of grasslands and desertification in semi-arid areas, while on steep slopes it can cause serious soil erosion. Scherr and Yadav (1996) highlight overgrazing in parts of the Caribbean and North Africa, and the grazing of slopes in mid-altitude areas of Asia as areas of concern. UNEP's *Global Environment Outlook* (UNEP, 1997) also lays emphasis on devegetation and land degradation from overgrazing. However, gaseous emissions and water pollution from livestock systems could be of greater global and more widespread national concern than overgrazing, although the latter will remain a serious threat in some areas. This is consistent with the detailed analysis of the impact of livestock on the environment given in de Haan, Steinfeld and Blackburn (1998).

There is a growing consensus that the importance of overgrazing has been misjudged in the past, particularly in sub-Saharan Africa. This was in part caused by poor understanding of rangeland ecology, and in part by the lack of appreciation of traditional range management practices in arid and semi-arid areas (Behnke, Scoones and Kerven, 1993). The alleged overgrazing in the Sahel, for example, is mainly a consequence of natural climate variability, i.e. low rainfall in some years, and poor stock management rather than overstocking *per se* (Fleischhauer, Bayer and Lossau, 1998). There is very little lasting impact on the vegetation, which is more robust than once thought. In recent decades, however, the establishment of permanent watering-points and the restrictions placed on traditional migratory routes through border checks have led to greater impact in drought periods than probably occurred in the past. In Australia, for example, poor planning and management of water-holes have led to serious overgrazing and reductions in pasture species. The periodic overstocking that does occur is often the consequence of institutional and infrastructural problems constraining the marketing of livestock.

Looking ahead to 2030, it is reasonable to assume that overgrazing will not cause major increases in land degradation globally. Pastoralists will continue to move their livestock around to exploit spatial differences in growing conditions if they are allowed to. The lack of opportunities to raise the productivity of extensive livestock systems, and changing income sources and aspirations among livestock producers will continue to cause shifts to more intensive systems on the better lands closer to urban markets, or out of agriculture altogether. Finally, some of the institutional and infrastructural constraints that have encouraged overgrazing in the past will be reduced as growing urbanization and income growth stimulate market improvements.

12.3.3 Environmental dimensions of water use and water pollution by agriculture

Many water management and pollution issues have grown in importance in recent decades, such as growing competition with the urban and industrial sectors for the available water supply; poor irrigation water-use efficiency; overextraction of groundwater; reduced infiltration of rainwater into soils and reduced water recharge because of deforestation and land degradation; declining crop yields and water quality related to waterlogging and salinization; contamination of groundwater and surface water from fertilizers, pesticides and animal wastes; and the risk of greater aridity and soil moisture deficits towards the end of the projection period in some areas of sub-Saharan Africa and South Asia because of climate change.

Overextraction of groundwater. Overextraction of groundwater is widespread in both developed and developing countries. It arises when industrial and domestic and agricultural withdrawals of water exceed the rate of natural recharge. In some areas, particularly in the Near East/North Africa region, irrigation draws on fossil aquifers that receive little or no recharge at a level that is not sustainable (Gleick, 1994). In substantial areas of China and India groundwater levels are falling by 1-3 metres p.a. The economic and environmental consequences are serious and will get worse in the absence of appropriate responses. Irreversible land subsidence, especially in urban and peri-urban areas, causes serious structural damage to buildings, drainage systems, etc. Overextraction in coastal areas causes saltwater to intrude into freshwater aquifers, making them unfit for irrigation or drinking-water without costly treatment. Lowering of the water table increases pumping costs. It will take many years to achieve the investments and other changes required to limit overextraction, so several million ha of irrigated land may either go out of production or be faced with unsustainable operating costs.

Waterlogging and salinization. Both these problems are commonly related to irrigation mismanagement. Waterlogging restricts plant growth. It arises from overirrigation and inadequate drainage, and in many cases precedes salinization. Over 10 million ha are estimated to be affected by waterlogging (Oldeman, Hakkeling and Sombroek, 1991).

Salinization results from the buildup of dissolved solids in soil and soil water, and can occur in rainfed areas with inherently susceptible soils (as in parts of Australia) as well as in irrigated areas. UNEP considers that salinization is the second largest cause of land loss, but there are wide differences in the estimates of the area affected and of the area going out of production. Oldeman, Hakkeling and Sombroek (1991) estimate the total affected area to be over 76 million ha but do not differentiate between irrigated and rainfed areas. It seems possible that some 20 percent of the total irrigated area is affected and some 12 million ha of irrigated land may have gone out of production (Nelson and Mareida, 2001). The problem can be very serious at the subregional and national level. India, for example, has lost about 7 million ha (Umali, 1993; FAO/UNDP/UNEP, 1994; FAO, 1999b). In some semi-arid countries, 10 to 50 percent of the irrigated area is affected to a greater or lesser degree (Umali, 1993; FAO, 1997b, 1997e), with average yield decreases of 10 to 25 percent for many crops (FAO, 1993a; Umali, 1993). Unfortunately there are little or no time series data to allow reliable estimates of the rates of change in the salinized area. It could be 1-1.5 million ha p.a. and increasing (Umali, 1993) but it is difficult to quantify. Of particular concern are those irrigated areas in semi-arid regions that support large rural populations, such as the western Punjab and Indus valley where large areas of waterlogged saline land are spreading through the intensively irrigated plains.

Pollution by fertilizers. Since the 1970s extensive leaching of nitrate from soils into surface water and groundwater has become an issue in almost all industrial countries (OECD, 2001a). In large areas of the EU, for example, concentrations are near to or exceed the maximum permitted concentration of 50 mg per litre or 50 ppm (parts per million). This nitrate poses a risk to human health and contributes to eutrophication of rivers, lakes and coastal waters. The bulk comes from diffuse sources arising from mineral fertilizer and manure use on both crops and grasslands. The problem is now serious in parts of China and India and a number of other developing countries, and will get worse (Zhang *et al.*, 1996).

The problem occurs primarily when N application rates exceed crop nutrient uptake. The risk depends on crop type and yield, soil type and underlying rocks (Goulding, 2000). The risk of high nitrogen (and some phosphate) losses through leaching and runoff can become serious unless there is good fertilizer management (Hydro, 1995; MAFF, 1999). There are large regional and crop differences in fertilizer application rates per hectare (Daberkov *et al.*, 1999), and large spatial and temporal differences in nutrient levels and fertilizer-use efficiency on similar soil types. Hence the projected changes in average application rates given in Table 4.15 are not a good indicator of the risk of nitrate losses. In the United Kingdom, the present average application rate for all arable crops is about 150 kg N/ha, but the range is 25-275 kg N/ha with application rates of more than 150 kg on over 35 percent of arable land. In parts of China the situation is even more extreme with some rice farmers applying over 870 kg N/ha, almost four times the national average. As stated above, high application rates as such should not be a problem as long as crop yields are commensurate, but they become problematic when application rates exceed crop nutrient uptake. In contrast, most crop production in sub-Saharan Africa takes place without the benefit of mineral fertilizers and soil fertility remains very low or declining (Chapter 4).

Chapter 4 assumes substantial gains in fertilizer-use efficiency and hence a relatively modest aggregate growth rate in N fertilizer demand, projected to decline to less than 1 percent p.a. by 2030. However, extensive areas in both developed and developing countries already receive large nitrogen fertilizer applications that are not commensurate with the availability of adequate soil moisture, other nutrients and management practices needed to attain high yields. Even modest increases in nitrogen fertilizer application could cause problems when yield growth stagnates, leading to nutrient-use inefficiencies and severe pollution. Four difficulties arise here. First, these losses can occur at N application rates below the economic optimum, since current fertilizer prices do not include the cost of environmental externalities (Pretty *et al.*, 2001). Second, maximizing the efficiency of N use is complex and difficult (Goulding, 2000). Third, it may take many years for improvements in fertilizer-use efficiency to result in reductions in nitrate losses and decades for groundwaters to recover from nitrate contamination. Fourth, the situation could be particularly serious for the production of vegetables, because they are often grown in or close to urban areas so that there is fairly direct contamination of the drinking-water sources for large numbers of people.

Water pollution arising from agriculture has other dimensions. Nitrogen and phosphate enrichment of lakes, reservoirs and ponds can lead to eutrophication, resulting in high fish mortality and algae blooms. This is important because of the growing importance of aquaculture (Chapter 7). Algae blooms release toxins that are poisonous to fish and humans. The human risks are a growing problem in some developed countries and potentially even more serious in warmer developing countries with more intense sunshine (Gross, 1998).

Further intensification of fertilizer use may also add to the widespread problem of soil acidification (Scherr, 1999). A combination of improved nutrient management and liming could limit this but ammonia emissions from livestock and nitrogen fertilizers, discussed above, also add to soil acidification through acid rain. The conjunction is causing serious ecosystem damage in some developed countries, and this could also occur in developing countries, particularly in East Asia with the rise of industrial-scale pig and poultry production.

Pesticide pollution. Pesticide use has increased considerably over the past 35 years. Recent regional growth rates have ranged between 4.0 and 5.4 percent (Yudelman, Ratta and Nygaard, 1998). This has led to serious water pollution in OECD countries (OECD, 2001a). Pesticide pollution is now appearing in developing countries as well, exacerbated in some instances by the availability of cheap, out-of-patent, locally produced chemicals.

Future pesticide consumption is likely to grow more rapidly in developing countries than in developed ones (Morrod, 1995), although the introduction and spread of new pesticides may occur more rapidly in the latter. The environmental implications of this growth are difficult to assess. For example, application rates per hectare have gone down, but the new pesticides are biologically more active. Improved screening methods for pesticide safety and environmental health legislation have helped to reduce the mammalian toxicity of pesticides and to assess other potential environmental damage. On the other hand, the adoption of improved application techniques has not progressed sufficiently in the past decade, particularly in the case of sprayers, so that a high proportion of pesticide still fails to reach the target plant or organism (Backmann, 1997). This situation is unlikely to change in the near future.

Over the period to 2030 several factors could create significant breaks from recent trends in pesticide use, and could reduce pesticide contamination of groundwater and surface water, soil and food products. Developed countries are increasingly using taxes and regulatory measures to reduce pesticide use (DME, 1999; DETR, 1999). The rapid growth in demand for organically grown food will continue to reduce the use of pesticides. Research in "smart" pesticides using advances in biotechnology, knowledge of insect hormones and insight into the ecological basis of pest control, etc. is likely to result in safer control methods within the next decade or so (Thomas, 1998). There will be further development of IPM for crops other than rice, which should help to reduce the use of insecticides and, to a lesser extent, fungicides and herbicides (Yudelman, Ratta and Nygaard, 1998). However, shortages of farm labour, the reduced use of flood irrigation for rice and the spread of minimal tillage systems could lead to major increases in the use of herbicides and herbicide-resistant crops.

Pollution from livestock wastes. Water pollution also arises from intensive dairying and from the landless rearing of pigs and poultry, particularly in East Asia. Here peri-urban industrial-scale pig and poultry production has caused serious environmental damage, especially water pollution, unparalleled in the industrialized world (de Haan, Steinfeld and Blackburn, 1998). The problem arises from discharges or runoff of nitrogen and other nutrients into surface waters because of bad waste management and from environmental impacts of feed and fodder production (Hendy, Nolan and Leng, 1995; de Haan, Steinfeld and Blackburn, 1998). In the medium term these problems seem bound to increase, although the technological means of overcoming them are neither complex nor very costly.

12.3.4 Loss of biological and ecological diversity

In the main, recent land cover changes have reduced the spatial distribution of species rather than causing their extinction, although this has happened and will continue to happen on a more limited scale. The loss of wild relatives of crops and of native crop varieties that are better adapted to unfavourable or changing environmental conditions could be particularly serious for crop introduction or breeding programmes to adapt to climate change.

The projections of land cover and land use change do not explicitly examine changes in biodiversity,[4] but they do provide some proxy indicators. These can help in assessing how the impacts of agriculture on biodiversity might evolve over the projection period. The focus of this chapter is on the environmental and ecosystem impacts of the changes, rather than on plant genetic resource issues. Pressures on non-agricultural biodiversity from land clearance and the inappropriate use of agrochemicals may in future grow more slowly because of the increase in protected areas and land restoration.

However, there will be increasing pressures on biodiversity within agricultural production systems. This will stem primarily from the intensification of production. Together with economic forces, intensification will lead to farm and field consolidation; reduction of field margins; clearance and levelling of adjacent wastelands so that they can be cultivated; further expansion in the use of modern varieties; greater use of pesticides; and higher stocking rates for grazing animals. These trends can lead to the destruction of the habitats of beneficial insects and birds that help to keep crop pest populations under control, and to other losses in biodiversity. However, assessment of these losses in developing countries is severely limited by lack of data on quantitative and qualitative changes in pesticide use, livestock densities and wildlife populations.

The effects of agriculture on non-agricultural biodiversity can be positive as well as negative, depending upon the situation. In the United Kingdom, for example, the intensification of pastoral systems has been an important factor behind the decline in bird populations. On the other hand, in Norway around half of the threatened species depend on agricultural landscapes and therefore the conservation of biodiversity is closely related to the protection of such landscapes (Dånmark, 1998), including grazing systems that prevent pastures from reverting to scrub or woodland.

Nonetheless, intensification will have a major positive impact by reducing the need to convert new land to agriculture. CGIAR has estimated that land saved through yield gains over the past 30 years from CGIAR research on seven major crops is equivalent to 230-340 million ha of forest and grassland that would have been converted to cropland in the absence of these gains (Nelson and Mareida, 2001). Their estimate excludes the land savings that stemmed from research on other crops, from national and private research systems and from farmers' own research and development. Some estimates of land savings resulting from all past research efforts and agricultural intensification amount to more than 400 million ha (Goklany, 1999).

Agriculture's main impacts on wild biodiversity fall into four groups. First, there is the loss of natural wildlife habitat caused by the expansion of agriculture. This has been a major force in the past, and will continue in the future, although much more slowly. The projections of Chapter 4 suggest that an additional 120 million ha of arable land will be required over the next 30 years. Inevitably these will involve a reduction in the area of natural forests, wetlands and so on, with attendant loss of species.

Second, there is the general decline in species richness in managed forests, pastures and field margins, and the reduction of wild genetic

---

4   Biodiversity includes genetic differences within each species, diversity of species and variety of ecosystems.

resources related to domesticated crops and live-stock. There are comprehensive and well-maintained *ex situ* germplasm stocks for the major crops, and gene transfer and other advanced plant breeding tools have opened up new possibilities for genetic improvement. Nevertheless, these losses in the wild could be serious for future crop and livestock breeding. They cannot be quantified at present, although advances in molecular biology may provide the tools needed for more robust monitoring.

Third, there is the reduction of wild species, including micro-organisms, which help to sustain food and agricultural production, for example through soil nutrient recycling, pest control and pollination of flowering crops. This can be regarded as damage to the life support system for agriculture, given the vital role some of these species play in soil fertility maintenance through nitrogen and carbon cycling. Such losses are of increasing importance with the shift to integrated farming and the growing emphasis on IPM. The intensive use of mineral fertilizers is known to change soil microbe populations (Paoletti, 1997), but does not appear to disrupt nutrient recycling. Intensive grazing lowers plant species richness in pastures but the long-term consequences of this are not known. In developed countries, loss of insect-eating bird species, as a result of reduction or removal of field margins or pesticide use, has been firmly linked with increases in crop pest damage. This problem may arise increasingly in developing countries.

Lastly, there is the reduction in wild species that depend for habitat, food, etc. on agriculture and the landscapes it maintains – the habitats, flora and fauna that would not exist without agriculture. Richly diverse chalk grasslands, for example, would revert to scrub or woodland without grazing pressures, with the loss of ground-nesting bird species, butterflies and herbaceous plants. The reduction of wild species is most apparent in those EU countries that have lost large areas of hedges, ditches, shrubs and trees through field and farm consolidation. Losses have also arisen from extensive use of insecticide and herbicide sprays with consequent spray drift on to field margins and other adjacent ecological niches. Increased stocking rates on extensive pastoral systems have led to a decline in birds that either nest on such land or are predators of rodents, etc. living on these lands.

The impact of livestock production on biodiversity takes two main forms: high grazing pressures and reseeding of pastures. Grazing pressures are likely to rise with time in some areas, particularly where marketing infrastructure is weak and there are few alternative livelihoods, even though there is a continuing shift to limited or zero grazing such as feedlots and stall-feeding systems. In other areas, however, the area of grazing land and pastures will probably decline as the more marginal areas are abandoned. Some pastures will be converted into cropland and urban and industrial land. Such land use changes can be appreciable. In western Europe, for example, the area of meadows and pastures declined by 10 percent between 1970 and 1988 (OECD, 1991). This decline was associated with the rise in stocking rates. On rough pastures these commonly increased by 50 to 100 percent between 1970 and 1990 (Pain, Hill and McCracken, 1997) and they have risen even more on improved pastures. Such increases in stocking rates have been linked to the loss of certain bird species from large areas of the United Kingdom and Europe and lower populations elsewhere (Pain and Pienkowski, 1996). Similar stocking rate increases are projected for parts of sub-Saharan Africa and Asia, so they are also likely to suffer losses in bird and other wildlife populations. It is also likely that there will be a shift to more intensive pasture systems. This will most probably involve some reseeding of natural meadows, and hence loss of native grassland plants. Intensification of pastures normally also involves the application of high levels of organic or mineral fertilizers, leading to nitrate or phosphate loss to water systems. The experience of the developed countries indicates that these impacts can be substantial.

12.3.5 Perturbation of global biogeochemical cycles

Agriculture plays a significant role in the anthropogenic perturbation of several biogeochemical cycles, notably the nitrogen, phosphate and sulphur cycles. In the nitrogen (N) cycle, ammonia and nitrous oxide emissions from agriculture are significant, but there is also perturbation of N fixation. The manufacture of nitrogenous fertilizers, burning of fossil fuels and cultivation of leguminous crops have resulted in anthropogenic N fixation exceeding natural N fixation since about 1980, by an increasing margin. Some analysts suggest that "over the next few decades this alteration (of the N fixation cycle) will undoubtedly become even more severe" (Walker and Steffen, 1999), for example, if the use of nitrogen fertilizer more than doubles between 1990 and 2050, thereby causing a pro rata increase in direct and indirect N_2O emissions (Smith, 1999). However, the projections of N fertilizer use and leguminous crop cultivation given in Chapter 4 and of manure production discussed earlier in this chapter all point to a slowing down in the growth of agriculture's contribution to this perturbation. Moreover, there are a number of other changes, e.g. in land use management, which should also reduce perturbations of the N cycle. For example, the expected increase in the area under NT/CA and other measures to counter soil erosion should reduce the loss of nitrogen in eroded organic matter from arable land. But in contrast, other land management practices such as burning may result in increased nitrogen losses, particularly if climate change results in increased summer maximum temperatures and greater fire risk in savannah and other fire-prone ecosystems (Lavorel *et al.*, 2001).

12.4 Current and emerging solutions

It is clear from Section 12.3 that some of the most serious environmental pressures stem from agricultural intensification. This process started in the developed countries over 50 years ago and some of the environmental problems became clear in the 1960s (Alexandratos, 1988). A number of them have been overcome while others remain and seem likely to grow in severity for the next decade or so (OECD, 2001a). The policy and technological successes and failures of developed countries can be of great help to those developing countries that are now suffering the environmental damages of intensification. They may even help those countries and farmers who have yet to intensify production to avoid some of the environmental problems that could arise. Many of the required policy, regulatory and technological actions are known. If pursued, such actions could result in a more favourable agro-environmental future than that outlined in the preceding section.

Reduction of pollution by fertilizer. The EU and North America have used a number of research and regulatory measures to limit pollution from fertilizers, such as research on slow release and other less polluting formulations; tighter emission and discharge standards for fertilizer factories and higher fines; public and private advisory (extension) services; physical limits on the use of manure and mineral fertilizers; and application of the nutrient budget approach.

These actions have not been enough to prevent serious buildup of nitrate in drinking-water sources and the eutrophication of rivers, lakes and estuaries (OECD, 2001a, 2001b; EEA, 2001). Since the early 1990s an increasing number of countries have been introducing economic measures in the form of pollution taxes on mineral fertilizers.

All of these actions are or will be relevant to developing countries. They can be formulated in the framework of a strategy for integrated plant nutrition (see Chapter 11). Some countries will have to remove a number of other distortions, such as direct and indirect subsidies to mineral fertilizer production or sales (e.g. energy subsidies to nitrogen fertilizer). They will also need to phase out low-efficiency fertilizers such as ammonium carbonate, and provide adequate funding to extension services so that they are not even partially dependent on the sale of fertilizer.

The consumer-led drive towards organic agriculture will limit fertilizer pollution in some areas, as will the adoption of NT/CA over a much larger area.

Reducing pesticide pollution. A number of important lessons can be drawn from past experience. First, rigorous testing procedures must be in place to determine the safety of pesticides before they are allowed on the market. The developed countries have suffered in the past from weaknesses in this regard, and have had to tighten their procedures. It is important that pesticide safety information is shared with developing countries through the International Plant Protection Convention (IPPC) and other mechanisms. In addition, as more developing countries become pesticide producers and develop their own products, it is essential that they implement their own testing, licensing and control procedures.

Second, even where the above measures are in place, environmental problems can arise from the accumulation of pesticide residues along the food chain, in soils and in water, e.g. the buildup of atrazine in water supplies in Europe and the United States. There must be comprehensive and precise monitoring systems to give early warning of residue buildup. The international sharing of information, e.g. through the Codex Alimentarius, provides valuable support to developing countries that lack adequate monitoring and testing facilities. Moreover, in the context of consumer safety and the WTO agreements, rigorous procedures must be in place to ensure food safety and enable agricultural exports.

Third, pest control measures should be implemented in a strategic framework for IPM (Chapter 11), which aims to avoid or minimize the use of pesticides. In recent years a number of developed countries have concluded that even with the above measures some farmers are still applying too much pesticide, or pesticide accumulation in the environment has not been reduced. They have decided, therefore, to use economic as well as regulatory measures, and to impose pollution taxes on pesticides so as to create economic incentives to reduce their use. Such taxes appear to be a valid option for a wider range of countries and situations, although it may be some time before many developing countries have the institutional capacity to implement them.

Development and expansion of no-till/conservation agriculture. The biological, environmental and economic advantages of NT/CA have been described in Chapter 11. The wider adoption of NT/CA depends on raising awareness among politicians and farmers of the benefits of conservation agriculture. Government policies need to be directed towards creating the appropriate conditions for its uptake. Farmers need to see how it meets their specific needs. The lessons learned from the farmer-to-farmer training approach used successfully for IPM in Asia could be of help. For example, Brazilian farmers who have benefited greatly from NT/CA could share their experience with farmers in Africa and help them to adapt the technique to their own conditions. Greater national research and development efforts and international assistance will be needed to develop the technique for other agricultural environments and production systems.

Improving water management. Water scarcity and intersectoral competition for water are major problems. Reduced groundwater recharge because of deforestation and soil degradation is also an important issue. The most serious direct environmental problem is salinization. Three main actions could be used to limit salinization: (i) greater investment in better drainage and distribution canals, even though planners have been slow to act on this option in the past; (ii) better water management, for example through the increasing involvement of farmers in water users' associations and similar bodies; and (iii) stronger economic incentives for water conservation, which are growing as governments increasingly implement water-pricing policies and as competition from other sectors drives up the price.

Promotion of organic farming. While the technological approaches described above are measures to reduce the negative effects of conventional agriculture on the environment, organic agriculture (for environmental reasons) does not use any industrial fertilizer or pesticide inputs at all. The use of such inputs commonly has negative effects on the environment; however, their non-use does not necessarily make agricultural production sustainable. Soil mining and erosion, for example, can be problems in organic agriculture. Organic farming can also cause serious air and water pollution – for example, the overuse of manure or badly managed applications can increase ammonia in the air and nitrate in groundwater.

The rapid expansion of organic production during the past decade has already made an appreciable contribution to pollution reduction and agricultural sustainability in Europe (FAO/COAG, 1999). Three aspects need to be clarified. First, badly managed organic agriculture can result in some of the same pollution problems that arise from conventional agriculture, but not in others such as those associated with the use of industrial inputs and production systems described in the preceding sections. Second, although the rate of expansion has been fast, the proportion of agricultural land involved is small. Current policies in many EU countries aim at a considerable increase in the agricultural land under organic farming (see Chapter 11). Third, most of the pressure for the switch to organic farming is in the developed countries (FAO/COAG, 1999).

The environmental and economic benefits of organic farming can be increased in a number of ways. These include introduction of policies that bring prices of industrial inputs in line with their full economic costs, including externalities; improvements in product standards, certification and labelling to give consumers confidence that they are buying genuine organic foods (FAO, 1999f); establishment of an internationally agreed accreditation mechanism, particularly procedures to gain international equivalence of organic product standards; greater government assistance to farmers wishing to switch to organic farming; regulations to enforce or encourage the use of organic farming as a means of overcoming or reducing problems such as the buildup of nitrate in groundwater; increased research to widen the range of organic agricultural techniques; improvements in the availability of or access to organic inputs, e.g. GMO-free seed, rock phosphate and manure; and capacity building in extension systems, farmers' cooperatives and national accreditation bodies to remove barriers to the expansion of organic farming, particularly in extension services that still promote approaches centred on the intensive use of mineral fertilizers and pesticides.

Improving livestock waste management. The main actions required here are the following. Development of national strategies for livestock waste management to provide the general framework for local action; improved policy and regulatory framework, with clearly defined and enforceable discharge and emission standards and effective waste disposal charges; meaningful penalties for breaches in regulations, and an expanded range of economic instruments to discourage poor livestock waste management, e.g. pollution taxes to limit waste discharges from livestock farms and fisheries; well-targeted programmes to disseminate best practices; strengthened guidelines to optimize the location of livestock production units and to prohibit the development of certain types of intensive livestock units in unsuitable areas; increased support for the adoption and dissemination of appropriate technological measures, with emphasis on introducing better techniques for livestock waste management already available in developed countries; drawing lessons from North America and the EU on the failure of regulatory approaches

alone to achieve adequate livestock waste management; and improved donor coordination and environmental impact assessment (EIA) to ensure that international projects have adequate provision for sound livestock waste management.

12.5 Physical and economic trade-offs

Earlier sections have shown that agriculture is an industry with substantial environmental consequences upstream and downstream as well as onfarm. It is evident that crop production and food security cannot be achieved at zero environmental cost. The issue therefore is whether environmental costs can be minimized so that future food security is not at risk.

The trade-offs involved are multidimensional. They vary over time and space, between different environmental goods and services, and between different developmental goals. It is for society to decide which trade-offs are acceptable and which ones can be mimimized, but this raises the question of whose society, and who in society should decide. A few examples of the various types of trade-offs are mentioned below.

Between countries and regions. The increasing switch in developing countries to intensive, grain-fed livestock production systems close to urban markets has reduced or prevented overgrazing of rangeland. But it has transferred part of the environmental burden to developed country exporters of feedgrains, because row crops such as maize are more susceptible to soil erosion and nitrate leaching (although improvements in farming practices have been reducing this damage in recent years). The past high rates of growth for dairy, pigs and poultry in some major developing countries were only possible through the use of large quantities of feed concentrates imported from North America and other developed countries.

Spatially within countries. This expansion of intensive livestock production has also led to a number of environmental trade-offs within countries. First, it tends to shift the environmental burden from non-point to point pollution, with greater discharges of concentrated liquid and solid wastes, causing serious water pollution. Second, it reduces the grazing pressures on vulnerable semi-arid pastures and steep slopes, but it separates arable and live-stock production. This limits or prevents the return of livestock manure to cropland, which is often vital in raising crop yields and maintaining soil fertility. Hence some countries, notably the Netherlands, have introduced regulations on stocking rates or manure recycling. Moreover, the production of large quantities of feedgrain and fodder crops can lead to serious soil erosion, and to environmental problems stemming from fertilizer and pesticide use. The higher stocking rates of intensive systems lead to loss of biodiversity from trampling and reseeding (Pain, Hill and McCracken, 1997), and more concentrated emissions from manure and urine, causing ecosystem acidification locally and GHG accumulation globally (Bouwman et al., 1997).

Over time within countries. The European experience shows how forest clearance for food production can buy time while technology and international trade catch up with population growth, allowing marginal cropland eventually to be reforested (Norse, 1988). Most European countries converted forests to cropland prior to the application of mineral fertilizers and modern crop breeding techniques. However, the loss of biodiversity from such deforestation may be permanent, and it may be impossible to re-establish the original forest ecosystem.

Over space and time. Erosion from steep slopes that are difficult to cultivate and inherently unstable is commonly followed by the redeposition of sediments in reservoirs, in river valleys and estuaries up to 1 000 km or more away. The result with time is the creation of flat lands that are easy to cultivate. In South and East Asia such lands have been able to sustain crop production for thousands of years. On the other hand, by reducing the storage capacity of drainage systems this erosion can contribute to severe flooding, loss of human life and serious economic losses (FAO, 1999g), and it may be impossible to restore eroded slopes to their original vegetation.

Between food security and the environment. Poor farmers in various parts of the world are mining soil nutrients because they lack access to sufficient organic manure or mineral fertilizer (Bremen, Groot and van Keulen, 2001). They know that their land use practices cause environmental damage that may ultimately endanger future food

security but immediate food needs take priority (Mortimore and Adams, 2001).

Greater intensification of cropland use versus greater loss of biodiversity and GHG emissions from deforestation. The introduction of new technologies that lead to higher yields or returns on existing land reduces the need for further land development. This can save a considerable amount of forest and rangeland (Nelson and Mareida, 2001) and eventually allow marginal cropland to be taken out of production and used for more sustainable systems, e.g. agroforestry, forestry, pastures and recreation. However, even under well-managed sustainable systems such as IPNS and IPM, intensification can lead to more fertilizer and pesticide pollution (Goulding, 2000), greater GHG emissions from nitrogen fertilizer and loss of biodiversity on intensively grazed pastures.

Reduction of soil erosion and water pollution versus greater pesticide use. NT/CA, minimum tillage and related approaches to land management have multiple environmental and farm income benefits, yet may require greater use of herbicides. However, initial fears that NT/CA would lead to greater use of herbicides have not been fully confirmed, as herbicide use can be reduced or eliminated in systems following all the principles of NT/CA, once a new agro-ecosystem equilibrium has been established. Using green cover crops to reduce nitrate leaching during the autumn and winter may increase carry-over of weeds, pests and diseases and lead to greater pesticide use.

The potential environmental benefits versus risks of GM crops. As discussed in Chapter 11, GM crops can have a number of environmental benefits such as (i) reduced need for pesticides, particularly insecticides (e.g. Bt maize and cotton) and herbicides (e.g. Ht soybeans), although these gains are not necessarily permanent as pests can overcome the resistance of GM crops; (ii) lower pressures for cropland development and deforestation because of higher yields from existing land; and (iii) increased opportunities to take marginal land out of production for set-aside or to cultivate some crops less intensively. The technologies involved can produce cultivars that can tolerate saline soils and thereby help to reclaim degraded land. On the other hand, there are a number of possible environmental impacts and risks, such as the overuse of herbicides with herbicide-tolerant varieties and accumulation of herbicides in drinking-water sources; herbicide drift from cropped areas, killing plants in field margins, and hence leading to the death of insects and birds in or dependent on field margins; death of beneficial insects feeding on GM crops; and crossing of GM crops with wild relatives and particularly with related weed species, e.g. red rice, possibly leading to the development of herbicide-resistant weeds.

The most vocal concerns about agricultural pollution and ecosystem damage tend to come from environmentalists in developed countries. But in a number of respects an improved environment is a luxury good that these countries can now afford. In earlier times they had different priorities (Alexandratos, 1995). Until the 1960s, when most people were concerned with improving their incomes, diversifying their diets and general welfare, protection of the environment was a low priority for all but a small minority (Reich, 1970; Nicholson, 1976), and some serious environmental problems arose. Since then income growth, education and better understanding of the environmental consequences of different agricultural practices and lifestyles have led to a growing consensus that governments should do more to protect the environment and that the public should pay more for environmental protection and food safety.

Industrial countries have the economic and technical capacity to introduce additional measures to protect the environment, and can afford the higher food costs that may follow as a consequence of these actions. In short, they are more able to pay for trade-offs between environment and development, although current actions seem unlikely to prevent some growth in agricultural pollution over the next 20 years or so (OECD, 2001a).

Environmentalists and government officials in developing countries are no less aware of the negative environmental consequences of agricultural growth. However, their responses are constrained by inadequate finance for the necessary research, particularly in sub-Saharan Africa; lack of institutions and support services that could raise awareness of potential ways to minimize or eliminate trade-offs; and the need to avoid measures that raise food prices because a high proportion of people are unable to buy adequate food even at current prices.

12.6 Concluding remarks

It has been argued that during the next decades environmental trade-offs will be more difficult than in the last few decades, with fewer win-win situations and more obvious losers than winners (OECD, 2001a). This is not necessarily the case for agriculture in many developing countries if market signals are corrected so that they include the value of environmental goods, services and costs; give farmers everywhere the incentive to produce in a sustainable way; overcome the negative impacts of intensive production technologies; give resource-poor farmers the support they need to react to environmental and market signals (Mortimore and Adams, 2001); and North and South work together to remove production and trade distortions (McCalla, 2001).

There are many opportunities for placing agriculture on a more sustainable path over the next decades, with benefits for both farmers and consumers. For example, measures resulting in higher nitrogen fertilizer-use efficiency and IPM reduce production costs for the farmer and provide safer food, and at the same time they are cheaper than drinking-water treatment in reducing nitrate and pesticide residues.

Future agro-environmental impacts will be shaped primarily by two countervailing forces. Environmental pressures will tend to rise as a result of the continuing increase in demand for food and agricultural products, mainly caused by population and income growth. They will tend to be reduced by technological change and institutional responses to environmental degradation caused by agriculture. The early implementation of available policy and technological responses could reduce negative agro-environmental impacts or slow their growth, and speed up the growth of positive impacts.

Agricultural intensification is required for food security and for the conservation of tropical forests and wetlands. The main priority is to decouple intensification from the environmental degradation caused by some current approaches to intensification, by reshaping institutional structures and market signals. Research and farming practices must also be redirected towards greater use of biological and ecological approaches to nutrient recycling, pest management and land husbandry (including soil and water conservation).

This decoupling has already started in some countries, but it will take time before it has appreciable effects. Hence agro-environmental impacts in the nearer future will be largely a continuation or acceleration of present trends. In particular, there will be a further slowdown in deforestation and rangeland clearance for crop production. Thus the main quantitative impacts on the environment will stem from the intensification of production on existing cropland, rather than from expansion of cropland. There will be increasing pressure on some marginal lands, but progress in research and better off-farm employment opportunities seem likely to lead to the abandonment and natural recovery of some marginal lands in Asia and Latin America. There will also be moderate increases in the area under irrigation. Drainage development and better irrigation water management will help to limit or reduce soil damage from waterlogging and salinization. Lastly, increased intensification of production on existing arable land will have two main characteristics. There will be enhanced use of precision farming and other advanced technologies, for example sophisticated plant breeding and controlled release of mineral fertilizers. And the growth of fertilizer and pesticide use will slow down because of regulatory measures and consumer demand for organically grown produce.

The focus of concern is likely to shift from the onfarm impacts of physical land degradation towards chemical and biological impacts, and from onsite towards offsite and downstream impacts of air and water pollution. Soil erosion may be reduced in important crop production areas by the projected shifts in technology. However, air and water pollution from mineral fertilizers and intensive livestock production will increase, with more widespread nitrate contamination of water resources, eutrophication of surface waters and ammonia damage to ecosystems.

The slowing down of deforestation will reduce the rate of loss of biodiversity, but the intensification of cropland and pasture use seems likely to increase such losses. The general picture for desertification is less certain, but the abandonment or reduced use of extensive semi-arid grazing lands should lower the risk of desertification.

The overall pattern of future agro-environmental impacts is one of trade-offs between increased agricultural production and reduced pressures on the environment. Intensification of crop production on existing cropland reduces the pressure to deforest, but tends to increase water pollution by fertilizers and pesticides. Similarly, the switch from extensive to intensive livestock lowers grazing damage to rangelands but, for example, may increase water pollution from poorly managed manure storage.

On the other hand, intensification of production on the better lands allows the abandonment of erosion-prone marginal lands, and improvements in fertilizer-use efficiency and IPM together with the expansion of organic farming are projected to slow down the growth in use of mineral fertilizers and pesticides. Similarly, the concentration of livestock into feedlots or stalls makes it more feasible to collect and recycle manure and to use advanced systems for water purification and biogas production.

In an increasing number of situations the trade-offs are becoming less serious. Thus, for example, NT/CA may reduce overall pesticide use; reduce soil erosion, fossil energy inputs and drought vulnerability; and raise carbon sequestration, natural soil nutrient recycling and farm incomes. Factors such as these lead to the overall conclusion that agro-environmental impacts need not be a barrier to the projected production path because they can be reduced considerably through the adoption of proven policies and technologies.

It is one thing to project the potential for a reversal or slowdown in the growth of agriculture's negative impacts on the environment. It is quite another matter to make such a future a reality. This will need a multidimensional approach and the integration of environmental concerns into all aspects of agricultural policy. Such actions were first proposed more than a decade ago (FAO, 1988) but are only now being pursued in a partial manner by some developed countries. Governments need to exploit the complementary roles of regulatory, economic and technological measures. Actions are needed at the global, regional, national and local level. None of these actions will be easy, but the real achievements of some countries and local communities over the past 30 years in promoting sustainable agricultural development show what could be achieved over the next 30 years, given more coherent efforts.

Climate change and agriculture: physical and human dimensions

13.1 Introduction

There is a wide scientific consensus that global climate is changing in part as a result of human activities (IPCC, 2001b), and that the social and economic costs of slowing it down and of responding to its impacts will be large (OECD, 2001a). In the past there has been a steady rise in average global temperature: the 1990s were around 0.6°C warmer than the late 1890s. The 1990s were the warmest decade since the beginning of instrumental record-keeping in 1860, and the warmest in the past thousand years on the basis of tree rings and other proxy measurements. Moreover, there has been an increase in the number of heat waves and a reduction in the frequency and duration of frosts in many parts of the world. It is now generally accepted that this climate change is the result of increasing concentrations of carbon dioxide, methane, nitrous oxide and other greenhouse gases (GHGs) in the atmosphere (IPCC, 2001a).

However, there are large uncertainties as to when and where climate change will impact on agriculture production and food security. Climate models can now simulate part of the natural climate variability well enough to give confidence in their projections of future changes outside the natural range (IPCC, 2001b). The latest predictions for the year 2100 are slightly higher than earlier ones and suggest that global average temperatures could progressively rise by up to 6°C under the business-as-usual scenario of the Intergovernmental Panel on Climate Change (IPCC). These predictions are less clear as to the magnitude and timing of the changes and impacts at the regional, subregional and national level, and this uncertainty will remain for some time to come (IPCC, 2001c). Continuing development of global circulation models (GCMs) since the IPCC's 1995 report has resulted in some markedly different conclusions regarding spatial and temporal shifts in climate. In some regions at least, e.g. in Europe, climate change impacts until 2050 seem likely to be less than those arising from natural variability (Hulme et al., 1999), but there will be substantial differences within Europe (Parry, 2000).

It is generally agreed that agricultural impacts will be more adverse in tropical areas than in temperate areas. Developed countries will largely be beneficiaries: cereal productivity is projected to be higher in Canada, northern Europe and parts of the former Soviet Union compared with what it would have been in the absence of climate change. By contrast a number of today's poorest developing countries are likely to be negatively affected

(IPCC, 2001c). Here the next 50-100 years will see widespread declines in the extent and potential productivity of cropland (Fischer *et al.*, 2001) particularly in sub-Saharan Africa and southern Europe (Parry, 2000; Parry *et al.*, 1999). Some of the severest impacts seem likely to be in the currently food-insecure areas of sub-Saharan Africa with the least ability to adapt to climate change or to compensate for it through greater food imports.

Around the rising trend in average temperature and rainfall, interannual and seasonal variation will increase. This will result in more frequent and more intense extreme events, and in greater crop and livestock production losses. Climate variation is already the major cause of year-to-year fluctuations in production in both developed and developing countries, and of food insecurity in developing countries (FAO, 2001f). For the period up to 2030, alterations in the patterns of extreme events will have much more serious consequences for chronic and transitory food insecurity than shifts in the patterns of average temperature and precipitation. There is evidence that extreme events were already becoming worse towards the end of the 1990s, and there is rising confidence in projections that they will increase in frequency and severity well before 2030 (Easterling *et al.*, 2000; IPCC, 2001b, 2001c). These extreme events have a disproportionately large impact on the poor because their crops, livestock, homes, food stores and livelihoods are at risk from floods and droughts and they have few or no savings to carry them through bad periods. Such impacts can be missed by GCMs operating at broad spatial and temporal scales, since extreme events commonly result in short-term and relatively localized food shortages that are masked by shifts in national production stemming from normal climate variability.

This chapter is devoted primarily to a review of climate change and food security issues and interactions. It examines how climate change may alter the agriculture and food security outcomes expected in the absence of climate change. The chapter's assessment of the possible impacts of climate change on food security should be considered in the context of the following limitations and assumptions. The time horizon of this study is 2030. This chapter therefore does not cover the 2050-2080 period during which the IPCC and others project increasingly serious climate-change-induced shifts in food production potential in currently food-insecure areas (IPCC, 2001a, 2001b; Fischer *et al.*, 2001). Thus, the modest impacts on aggregate food production proposed here are a reflection of the shorter time frame, rather than of any undue optimism about the longer-term situation.

The emission of GHGs by agriculture is discussed in Chapter 12. In this chapter, four other dimensions of the interaction between agriculture and climate change will be considered. First, agriculture's role as an important moderator of climate change through the sequestration of carbon in the soil and in long-lived products, and through the growing of biofuels to replace fossil fuels. Second, the positive and negative impacts of climate change on agricultural production and on natural ecosystems. Third, the implications for food security. Household and national incomes will generally be rising, allowing people to be less reliant on subsistence agriculture and more able to buy their food needs, and allowing countries to compensate for domestic food deficits through greater imports. However, a significant number of countries and communities may continue to be bypassed by development. Fourth, the clear need for changes to agricultural policies and technologies which in the short term could combat climate variability and natural resource degradation, but would also reduce or avoid possible food security impacts of future climate change, for example, NT/CA (see Chapter 11). Such measures have gained in importance now that carbon sink projects will qualify for credits under the clean development mechanism (CDM) of the Kyoto Protocol.

13.2 Agriculture as a moderator of climate change

Chapter 12 (Section 12.3.1) discussed the important role of agricultural activities as a driving force for climate change through the emission of GHGs. At the same time there is a growing appreciation of agriculture's positive contribution to climate change mitigation through carbon sequestration and the substitution of biofuels for fossil fuels. These contributions are likely to be of growing economic and environmental importance in the context of the Kyoto Protocol.

The benefits of carbon sequestration. In the past, attention has been focused on the role of forestry in carbon sequestration. This role will remain important in the future (see Chapter 6). In addition, however, crop and livestock production can also play a significant role through the sequestration of substantial amounts of carbon as soil organic matter (SOM) derived from crop residues, manure and better-managed grasslands. The additional benefits of this sequestration will diminish with time.

Global estimates of the potential contribution of cropland to carbon sequestration are in the range of 450-610 million tonnes of carbon p.a. (equivalent to some 1 640-2 220 million tonnes of carbon dioxide) for the next 20-30 years (GCSI, 1999). There is, however, considerable uncertainty about the potential gains from improved crop and livestock management practices (Lal and Bruce, 1999). In the United States changes in cropping practices (particularly conservation tillage and crop residue management, improved cropping systems and land restoration) could sequester about 140 million tonnes of carbon p.a. – nearly 10 percent of total United States emissions of all GHGs (Lal et al., 1999). United States and United Kingdom studies show that permanent set-aside could sequester large amounts of carbon if it is forested or unmanaged (Cole et al., 1996; Cannell et al., 1999). Thus improved land management can enhance the role of agricultural soils as a major sink for carbon dioxide (CO_2) and as a compensating mechanism for agriculture's contribution to GHG emissions (Lal, Kimble and Follet, 1998) although it may be a decade or more before cultivated land is transformed from a net source to a net sink of carbon. Improved land management can therefore help countries to meet their obligations under international agreements to reduce net emissions of GHGs. Moreover, under the provisions of the CDM of the Kyoto Protocol, international support for improved land management to sequester carbon could also further sustainable agriculture and rural development (SARD) by providing other environmental and economic benefits (FAO, 2000c). The latter include reduced soil erosion and nitrate leaching, greater rainfall infiltration, higher soil moisture levels and lower energy costs.

Many of the required technological and land management changes could take place over the period to 2030. These could include shifts in land use, for example reversion of cropland in industrial countries to managed forests and pastures or to natural ecosystems as part of permanent set-aside; changes in cropping patterns, e.g. biomass cropping; adoption of NT/CA with improvements in tillage practices and residue management (see Chapter 11); better soil fertility and water management; and erosion control. All of these land management changes are based on known technologies and husbandry practices that have other benefits, including improved soil moisture availability to crops and higher yields, or reduced fossil fuel use as in the case of NT systems.

The crop production projections of this study, together with earlier FAO work on the biomass yield of different crops (FAO, 1978), give an estimate of the biomass of crop residues left in the field. Total global non-harvested residues (primarily crop stalks and roots) for 15 of the most important crops were around 4.7 billion tonnes p.a. in 1997/99 and are projected to rise to 7.4 billion tonnes by 2030. Depending on the region, these residues amount to between 2.4 and 6.2 tonnes per harvested hectare. These values are higher than those used for other global estimations (Lal and Bruce, 1999), but similar to those found in studies for Australia, Canada and the United States (Dalal and Mayer, 1986; Douglas et al., 1980; Voroney, van Veen and Paul, 1981). Under tropical conditions, residues can be much higher. Cowpeas, for example, produce up to 24 tonnes of residues per ha (Diels et al., 1999).

There are significant crop and regional differences in the proportion of crop residues that are left on the soil surface or incorporated in it. For most crops it is assumed that 25-50 percent of residues are returned to the soil as organic matter, and that half of this biomass is carbon. With these assumptions, gross carbon sequestration by the 15 crops could rise from 620-1 240 million tonnes p.a. to 960-1 910 million tonnes p.a. by 2030 (Table 13.1). If this is scaled up to include the harvested area for the remaining crops, the global estimate for 2030 rises to 1 170-2 330 million tonnes of carbon. Taking into account that these estimates refer to gross carbon sequestration, they are fairly close to other recent estimates (GCSI, 1999; Lal and Bruce, 1999; Batjes, 1999).

These estimates do not take account of the potential gains from NT/CA or from improved soil

Table 13.1 Estimated gross carbon sequestration per year by cropland soils

	Total carbon (million tonnes)		Carbon (tonnes/ha)	
	1997/99	2030	1997/99	2030
Sub-Saharan Africa	34-67	74-147	0.30-0.60	0.47-0.95
Latin America and the Caribbean	62-124	110-220	0.66-1.33	0.83-1.65
Near East/North Africa	27-54	46-91	0.52-1.04	0.75-1.50
South Asia	97-194	168-337	0.53-1.07	0.87-1.73
East Asia	182-363	267-534	0.84-1.69	1.17-2.34
Industrial countries	168-336	227-455	0.90-1.80	1.16-2.32
Transition countries	49-97	64-128	0.45-0.90	0.53-1.05
World	618-1 236	956-1 912	0.65-1.30	0.88-1.76

Source: FAO calculations.

erosion control. However, the switch to NT/CA systems, which started in the late 1960s in developed countries and in various developing countries in the 1970s, could add to the amount of carbon sequestered (Friedrich, 1996; Derpsch, 1998). The gains vary according to agroclimatic conditions: 0.5-1.0 tonnes of carbon/ha/p.a. in the humid temperate areas, 0.2-0.5 in the humid tropics and 0.1-0.2 in the semi-arid zones (Lal and Bruce, 1999). The NT/CA area grew very rapidly over the last few years, but compared to total arable land the area is still small. There are large areas of South and East Asia where NT/CA could be applied but as yet it is hardly used. On the Loess Plateau of China, for example, its use is barely out of the experimental stage, yet it could help to sequester some 4 million tonnes of carbon p.a. (CCICED, 1999).

Assuming that another 150 million ha of rainfed cropland will be using NT/CA by 2030, sequestering 200-400 kg carbon/ha/p.a. (Lal *et al.*, 1999), this would represent a further 30-60 million tonnes carbon/p.a. There would also be other environmental benefits in the form of reduced soil erosion and better water retention, plus savings in fossil fuel use (Frye, 1984). Even greater gains of 500-800 kg carbon/ha/p.a. could be achieved where marginal cultivated land is taken out of crop production and replaced by grass or legume forages (Lindwall and Norse, 2000). Moreover, degraded land that has gone out of production or contributes little to food security, e.g. saline soils, could be restored to sequester carbon at the rate of 100 kg/ha/year in temperate areas and 200-300 kg/ha/p.a. in tropical and subtropical areas (Lal, Kimble and Follet, 1998; GCSI, 1999). The total area of saline soils that could be restored to boost carbon sequestration is over 126 million ha (GCSI, 1999). Assuming that 2 million ha of saline lands are restored each year over the next 30 years, the total carbon sequestered each year could be about 12 million tonnes by 2030.

The rates of carbon sequestration presented above are only order of magnitudes. Potential rates of carbon sequestration in response to improved management vary widely as a function of land use, climate, soil and many other factors. The rate of sequestration gradually declines before reaching a limit and can be especially high during the first few years. As a result, short-term studies tend to overestimate the rate of carbon sequestration. For some of the activities sequestering carbon it may take 40 to 100 years before saturation levels are reached. However, the IPCC considers the estimates for most of these to be less reliable than estimates about many of the activities sequestering carbon over shorter periods. For NT/CA sequestration is particularly high during the early years. Should conventional tillage be reintroduced though, the sequestered carbon will be rapidly released (FAO, 2001h). Nonetheless, agricultural carbon sinks are needed to "buy time" in coping with CO_2 emissions.

Potential contribution of biofuels. Biofuels used for cooking and heating already make a significant contribution to GHG reduction. However, there is considerable uncertainty about the future rate of substitution of renewable biofuels for fossil fuels. The technological potential and environmental benefits are clear, and some modelling exercises have projected large increases in the area under biofuel crops. The uncertainty stems from economic and political factors. For the foreseeable future, the energy from biofuel crops will continue to be more expensive than that from fossil fuels. If, however, carbon taxes were imposed on fossil fuels so that their cost to consumers included the external costs of their use, including the costs of climate change, then biofuels would be much more competitive. In addition, if more governments introduced positive discrimination for biofuels, then their production could expand rapidly and make a significant contribution by 2030.

13.3 Climate change impacts on agriculture

13.3.1 Climate change to 2030

Global average temperatures are projected to rise by about 1°C by 2030 (i.e. well outside the natural range). Higher latitudes will warm more rapidly than lower ones, land areas will warm more rapidly than the oceans, and polar sea ice will decrease more in the Arctic than in the Antarctic. Consequently, average temperatures in the higher latitudes may rise by 2°C, possibly double the increase in the tropics. Projected changes in precipitation show even greater regional differences, with major grain-producing areas of South America showing increases and parts of Central America and South Asia suffering from decreased precipitation and higher soil moisture deficits (IPCC, 2001b).

Broadly speaking, climate change is projected to increase global mean precipitation and runoff by about 1.5 to 3 percent by 2030 (IPCC, 2001b). There will be greater gains in the higher latitudes and the equatorial region but potentially serious reductions in the middle latitudes. Parts of Central America, South Asia, northern and southern Africa and Europe could suffer appreciable falls in available water resources. Moreover, there could be significant subregional differences, e.g. northern and southern Europe are projected to undergo significant shifts in climate-change-induced runoff

but not western and central Europe (Hulme *et al.*, 1999). Estimation of the impact of changes in precipitation is further complicated by the interplay of two effects: changes in precipitation and rises in water-use efficiency associated with the CO_2 fertilization effect.

13.3.2 Impacts on agriculture

Climate change will have a range of positive and negative impacts on agriculture. Up to 2030 the greatest impacts could come from increased frequency and intensity of extreme events. Climate variability is currently the dominant cause of short-term fluctuation in rainfed agricultural production of sub-Saharan Africa and South Asia, and substantial areas of other developing regions. The most serious form is drought, when rainfall drops substantially below the long-term mean or fails at critical points in crop development. In semi-arid and subhumid areas, these rainfall deficits can dramatically reduce crop yields and livestock numbers and productivity. Such fluctuations can be countered by investment in irrigation or by food imports, but these options are not always open to low-income countries or remote regions. Indeed, the availability of water for irrigation may be reduced by the increased frequency and intensity of droughts together with long-term changes in surface water runoff or evapotranspiration, and this may reduce irrigated food production.

Although semi-arid and subhumid areas are generally the ones given the most attention in climate impact studies, humid areas are also vulnerable to climate variability. They can suffer from changes in the length of the growing season (Wilkie *et al.*, 1999) and from extreme events, notably tropical cyclones causing damage from high winds and floods. Such disasters are shorter-lived and more localized than those associated with droughts and other forms of climate variability and so fewer people may be affected. However, the consequences for their food security can be equally severe. Not only do they lose current crops and livestock, but in cases where perennial trees are lost or spawning grounds seriously damaged, they also lose future crops and fish catches. They may lose their stored food, homes and possessions, including irrigation infrastructure, livestock and tools, so that the negative consequences on food security may be felt for several years after the event. On the other hand, since these extreme events are relatively localized, other crop-producing areas within the same country can often provide the food needed in affected areas.

Recent research has suggested that some impacts of climate change are occurring more rapidly than previously anticipated (IPCC, 2001c). The impacts will stem primarily from:

- regional temperature rises at high northern latitudes and in the centre of some continents;
- increased heat stress to crops and livestock, e.g. higher night-time temperatures, which could adversely affect grain formation and other aspects of crop development;
- possible decline in precipitation in some food-insecure areas, such as southern Africa and the northern region of Latin America, although the main impacts will occur after 2030;
- increased evapotranspiration rates caused by higher temperatures, with lowering of soil moisture levels;
- concentration of rainfall into a smaller number of rainy events with increases in the number of days with heavy rain, increasing erosion and flood risks – a trend that is already apparent (Easterling et al., 2000);
- changes in seasonal distribution of rainfall, with less falling in the main crop growing season;
- sea level rise, aggravated by subsidence in parts of some densely populated flood-prone countries;
- food production and supply disruption through more frequent and severe extreme events.

These impacts fall into three main groups, i.e. direct and indirect impacts of climate change *per se*, and impacts from enhanced climate variation (extreme events), though with a degree of overlap.

13.3.3 Direct impacts – changes in temperature and precipitation

Crops. Changes in temperature and precipitation will bring changes in land suitability and in crop growth. The projected net effect will be an increase in the area of land in higher latitudes suitable for crop production, because of milder and shorter winters, but a decrease in land suitability in arid and semi-arid areas. The changes will be qualitative as well as quantitative. In the East African highlands, higher temperatures may result in land becoming unsuitable for wheat but more suitable for other grains. The effects on potential yields will follow the same pattern as land suitability, with yield gains in middle to higher latitudes and losses in the lower latitudes. There may be some gains in tropical highlands where at present there are cold temperature constraints.

The overall effects of climate-induced changes in land and crop suitability and yields are small compared with those stemming from economic and technological growth. By 2020 world cereal production might be only about 0.5 percent less than what it would have been in the absence of climate change (IPCC, 2001c; Parry *et al.*, 1999), although this decline might be much greater by 2050 or later. The largest regional reduction would be in Africa where cereal production is projected to decline by 2-3 percent. This potential fall could be compensated by a relatively small increase in yields or imports. But this regional picture hides important subregional differences. Parts of central and northern Africa may experience small increases in cereal yields.

The rise in atmospheric concentrations of carbon dioxide not only drives global warming but can also be a positive factor in tree and crop growth and biomass production. It stimulates photosynthesis (the so-called CO_2 fertilizer effect) and improves water-use efficiency (Bazzaz and Sombroek, 1996). Up to 2030 this effect could compensate for much or all of the yield reduction coming from temperature and rainfall changes. Recent work for the United States suggests that the benefits from CO_2-induced gains in water-use efficiency could continue until 2095 (Rosenberg *et al.*, 2001).

Forestry. As with crop production, CO_2 fertilization effects will combine with those of climate change. This will make it difficult to determine net impacts on forestry, but these effects are likely to be small before 2030. The developed countries seem likely to be the major beneficiaries. Given the higher temperatures at high latitudes and the CO_2 fertilization effect, boreal and north temperate forests in North America, northern Asia and Europe and parts of China are likely to grow more rapidly before 2030. Tropical forests may decline in area and productivity, because of decreased rainfall and higher temperatures, with some loss of biodiversity. However, dieback of tropical forests, i.e. progressive death from environmental or pest causes, could be a concern in parts of northern South America and central southern Africa (Hadley Centre, 1999).

Livestock. Some grasslands in developing countries are projected to deteriorate progressively as a result of increased temperature and reduced rainfall but this is unlikely to occur until after 2030 (DETR, 1997). Much of this grassland is of moderate or low productivity and is in any case projected to decline in importance with the continued shift to intensive livestock production systems in more humid areas (see Chapter 5). Of more significance to livestock production is the rise in temperature over the period to 2030, and the CO_2 fertilization effect. These will favour more temperate areas (i.e. primarily the developed countries but also Argentina and China) through reduced need for winter housing and for feed concentrates (because of higher pasture growth). Many developing countries, by contrast, are likely to suffer production losses through greater heat stress to livestock. Fodder and forage yields may be lower because of reduced precipitation but this may be compensated by the CO_2 fertilization effect.

Fisheries. Some of the earliest negative impacts may be on fisheries rather than on crops. There are three impacts of concern: higher sea temperatures, changes in ocean currents and sea level rise (discussed below). Most of the effects will occur after 2030 or even 2050, but may intensify greatly thereafter (IPCC, 2001c).

Average sea temperatures in northern latitudes are already rising rapidly (in particular in the North Sea). Sea temperature rise can disrupt ocean currents and fish breeding patterns. It can reduce surface plankton growth or change its distribution, thereby lowering the food supply for fish, and cause the migration of mid-latitude species to northern waters (Reid *et al.*, 2000).

The net effect may not be serious at the global level but could severely disrupt national and regional fishing industries and food supplies. It is already a serious issue in Europe, where climate-change-induced impacts on cod populations could compound the effect of current overfishing in the North Sea, causing permanent damage to fish stocks

if no action is taken (O'Brien *et al.*, 2000). In middle and southern latitudes coral bleaching and destruction through higher water temperatures could damage important fish breeding grounds.

13.3.4 Direct impacts – sea level rise

Sea level rise induced by global warming could lead to loss of land through flooding and saltwater intrusion, and damage to mangrove swamps and spawning grounds. Sea levels are rising at about half a centimetre p.a., and are likely to continue at this rate for several decades even if there is rapid implementation of international agreements to limit climate change. Thus sea levels could be 15-20 cm higher by 2030 and 50 cm by 2100 (IPCC, 2001b), increasing the flood risk in large parts of South and East Asia and placing populations and agriculture at risk (Gommes *et al.*, 1998). Three valuable production systems will be most affected: vegetable production that tends to be irrigated and heavily concentrated around urban areas threatened by saltwater intrusions; aquaculture systems sited in areas at or below sea level; and coastal fisheries dependent on spawning grounds in mangrove swamps and other coastal wetlands threatened by sea level rise, although some adjustment might take place through sediment deposition and the accumulation of organic matter.

Because tropical cyclones will increase in frequency and intensity, there will be more extreme high-water events and more severe storm surges penetrating further inland (IPCC, 2001a; Nicholls, Hoozemans and Marchand, 1999). Although most impact assessments have been on gradual sea level rise, these sea surges may pose the greatest risk to food security. Nicholls, Hoozemans and Marchand (1999) conclude that by 2080 the number of people vulnerable to flooding from sea surges in a typical year will be five times greater than those vulnerable to sea level rise. Earlier work suggests that 90 percent of these vulnerable people would experience flooding on an annual basis (Baarse, 1995). Migration to coastal zones because of the better employment opportunities associated with urbanization and industrialization and the overextraction of groundwater in urban areas will compound the problem. In Bangkok, for example, these trends have led to marked subsidence (up to several metres in the last century).

Even without climate change, population growth and urbanization will increase the number of people at risk from coastal flooding, possibly from about 200 million in 1990 to nearly 500 million by 2030 (Nicholls, Hoozemans and Marchand, 1999). Sea level rise alone will not raise this number substantially by 2030, but other expected developments, involving serious interactions between river flooding and sea level rise, could do so. These include greater river runoff because of increased precipitation inland, reduction of river width through siltation and urban and industrial development, and an increase in storm surges penetrating further inland (Arnell, 1999).

13.3.5 Indirect impacts

Indirect impacts operate primarily through effects on resource availability, notably water resources, and on ecosystems as they respond to shifts in temperature and precipitation; and through the loss of biodiversity, although the latter will have little impact by 2030.

Large changes are predicted in the availability of water resources because of reductions in runoff and groundwater recharge. Substantial decreases are projected for Australia, India, southern Africa, the Near East/North Africa, much of Latin America and parts of Europe (Hadley Centre, 1999). The main decrease will be after 2030 but there could be negative effects on irrigation in the shorter term. Moreover, the greater frequency of summer droughts in the interior of mid-latitude continents could raise the incidence of wildfires.

There will be changes in the distribution and dynamics of major pests. Although only small average temperature changes are projected to 2030, they are nonetheless large enough to bring about substantial shifts. In addition, fewer cold waves and frost days could extend the range of some pests and disease vectors, and favour the more rapid buildup of their populations to damaging levels.

Much of central and northern Europe could become more vulnerable to important pests and diseases such as Colorado beetle of potatoes and Karnal bunt of wheat (Baker *et al.*, 1999) as they expand their range north. Although control measures are known for these diseases there will still be some yield loss and associated production input and environmental costs. However, this is not just an issue for temperate areas. In subtropical Australia temperature rises up to 2 °C could favour the spread of the Queensland fruit fly and force production to shift substantially southwards (Sutherst, Collyer and Yonow, 1999).

The important changes in pest dynamics are increases in pest carryover (particularly overwintering in temperate regions) and population dynamics, since the life cycles of some major pests are extremely dependent upon temperature (Gommes and Fresco, 1998). Higher temperatures may foster larger pest populations, and may extend the reach of insect carriers of plant viruses, as in the case of aphids carrying cereal viruses, which are currently held in check by low winter or night temperatures. No attempt has been made to quantify these losses but they could be appreciable in terms of lower yields and higher production costs.

Finally, greater temperature extremes seem likely to give rise to higher wind speeds, and there may be increases in the occurrence of hurricanes. This will result in greater mechanical damage to soil, plants and animals; impacts on plant growth from greater wind erosion and sandblast damage; and drowning of livestock. Natural resource management decisions, both on the farm and at national level, could reduce or intensify the impacts of these factors on food security. For example, concerted efforts to promote IPM could lessen the impact of pest and disease outbreaks. Conversely, poor land management practices and inadequate protection for the diversity and stability of ecosystems could aggravate soil erosion and other damage.[1]

13.4 Implications of climate change for food security

13.4.1 Introduction

Up to 2030 the impact of climate change on global food production may be small, within the range

---

[1] In the case of hurricane Mitch in Honduras in 1998, where over 50 percent of agricultural infrastructure and production was reportedly severely affected or completely destroyed, human factors such as large-scale deforestation, the cultivation of marginal lands without soil conservation practices and a lack of adequate watershed management were largely blamed for severely compounding the effects of the hurricane. The Lempira Sur region, utilizing a diverse agroforestry farming system, suffered less damage than the rest of Honduras and was able to provide food aid for other parts of the country.

that normal carryover stocks, food aid and international trade can accommodate. During the 1992/93 drought in southern Africa, for example, crop production in some countries was reduced by as much as 50 percent. Yet there was no famine (Chen and Kates, 1996), and the negative food security impacts were relatively short-lived, although serious for some communities. National and international action was able to limit the increase in the numbers of undernourished. Nonetheless, Parry *et al.* (1999) consider that the 2-3 percent reduction in African cereal production they project for 2020 is sufficient to raise the numbers at risk from hunger by some 10 million people.

However, food security depends more on socio-economic conditions than on agroclimatic ones, and on access to food rather than the production or physical availability of food (FAO, 2001c; Smith, El Obeid and Jensen, 2000). Therefore, the implications of climate change for food security are more complex than the relations used by most of the current impact assessments. Future food security will be determined largely by the interplay of a number of factors such as political and socio-economic stability, technological progress, agricultural policies, growth of per capita and national incomes, poverty reduction, women's education, drinking-water quality (Smith and Haddad, 2001), and increased climate variation.

It is important to be clear about the respective roles and relative contributions to food security of these factors, and how they interact. For example, poverty is a major factor in food insecurity (FAO, 2001a), and urbanization can play an important role in improving physical access to food during serious droughts, although there are a number of positive and negative factors involved (FAO, 2000d). Urban wages are generally above rural wages, but urban food and housing costs can be higher, so actual food purchasing power in urban areas might in some cases be lower. Up to 2030 or even 2050, projected growth in incomes, urbanization and crop production for developing countries are likely to have a much greater impact on food security than the effect of climate change in reducing average cereal yields or the area suitable for grain production (Fischer *et al.*, 2001).

However, there will be problems arising from increased climate variability. Climate change may affect, for example, the physical availability of food

from production by shifts in temperature and rainfall; people's access to food by lowering their incomes from coastal fishing because of damage to fish spawning areas from sea level; or lowering a country's foreign exchange earnings by the destruction of its export crops because of the rising frequency and intensity of tropical cyclones.

In food-insecure countries, there is often a large seasonal as well as interannual variation in the ability of people to grow or purchase food. In parts of Africa, there is the so-called "hungry season" prior to the new harvest, when grain prices tend to rise substantially as stocks fall and lead to temporary food insecurity. Such features are lost in the annual or seasonal averages of most analyses of long-term food production and climate change impacts on agriculture, but they are important in determining people's ability to purchase food.

There is also the question of spatial variation of climate impacts, and the level of countries' ability or inability to exploit this to overcome local food production deficits. Inability generally stems from weaknesses in infrastructure or institutions, although it is reasonable to project improvements in these respects over the next 30 years. These features are not captured in climate impact assessment models, yet they are very important since quite large negative impacts on production from climate change will not necessarily result in diminished food security. Large countries such as India and China contain a range of agroclimatic situations, and droughts and floods in one area can be compensated by production from unaffected areas and carryover stocks. Thus, when parts of northeast and central China were seriously flooded in 1998, local food production losses were readily replaced by food from elsewhere. In countries in which agriculture is a small proportion of GDP, any food deficits from extreme events can normally be covered by imports, and by 2030 it is expected that more countries will be in a position to compensate for climate change impacts on domestic food production by imports from elsewhere.

13.4.2 Socio-economic developments and vulnerability to climate change

Given the above, it is necessary to examine food security in the context of the future agricultural and wider economic situation, which is likely to be quite different from today's in a number of respects. One only has to look back 30 years to see the need for this. For example, in the 1970s Bangladesh was being classified as incapable of functioning properly and with little hope of survival, and South Asia, particularly India, was considered to be the most food-insecure region, whereas sub-Saharan Africa was thought to have better food prospects (IFPRI, 1977). In reality, during the last decades sound agricultural policies, investment in irrigation, etc. have enabled Bangladesh and India to overcome their large food production deficits, whereas sub-Saharan Africa suffered from poor agricultural performance and prolonged food shortages for much of the same period.

Looking ahead 30 years, a number of today's food-insecure countries seem likely to have overcome their food production or food access problems, with much of the remaining food security problem concentrated in sub-Saharan Africa (see Chapter 2). Given the relatively high economic growth projected for most Latin American and Asian countries, they should be able to overcome any negative impacts of climate change on food production by increasing food imports. This demonstrates that it is not enough to assess the impacts of climate change on domestic production in food-insecure countries. One also needs to (i) assess climate change impacts on foreign exchange earnings; (ii) determine the ability of food-surplus countries to increase their commercial exports or food aid; and (iii) analyse how the incomes of the poor will be affected by climate change.

No matter how the climate changes, any impacts will be on a food security situation very different from the present. The structure of most developing economies will have shifted closer to that of today's developed countries. Food production will have changed in response to new technologies and changes in comparative advantage. Food consumption and food security will have changed because of shifts in consumer preferences and higher per capita incomes.

Economic growth in non-agricultural sectors and an increase in urbanization and non-agricultural employment will make people's incomes less dependent on agriculture. People may have easier and more reliable access to food during extreme events and thus become less vulnerable to climate change. The increasing role of home remittances in raising the food purchasing power of the rural poor has reduced seasonal and long-term food insecurity. This has been particularly the case in sub-Saharan Africa, where 20-50 percent of rural incomes now commonly come from off-farm sources, and increasing amounts of food are purchased rather than home produced (FAO, 1998d; Reardon, Matlon and Delgado, 1994; UNSO, 1994; Turner, 2000). This situation is likely to continue for the next 30 years. Provided government policies and infrastructural improvements allow food imports to flow readily to drought-affected and other natural disaster areas, their food security situation will become less dependent on local production. Fewer people will be vulnerable, as long as prices do not go up (although this is unlikely, as discussed in Section 13.4.5).

13.4.3 Climate change and crop production

Cereal yields play a key role in the food security of the poor. Recent estimates suggest that, relative to the no climate change situation, yields could change by -5 to +2.5 percent depending on the region (Table 13.2). In many but not all countries it may be possible to overcome this by expanding cultivated land, because there are still substantial suitable areas that could be brought into cultivation (Chapter 4). Furthermore, very small and quite feasible annual improvements in yields could compensate for a potential 5 percent yield reduction from climate change (Chapters 4 and 11), although in the regions facing the most negative potential impacts, yield increases were hard to realize in the past.

The regions and countries where food security is most at risk from sea level rise include South Asia, parts of West and East Africa, and the island states of the Caribbean and Indian and Pacific Oceans. They include deltaic areas that are difficult and costly to protect, yet play an important role in food production, e.g. in Bangladesh, Myanmar, Egypt, India, Thailand and Viet Nam. The concerns for food security are particularly great where farm sizes are already too small to provide adequate subsistence and where conversion of uplands to food production cannot compensate for the loss of coastal land. A number of the areas at

Table 13.2   Potential changes in cereal yields (percentage range, by region)

	2020	2050
Sub-Saharan Africa		
Sahel and southern Africa	-2.5 to 0	-5 to +5
Central and East Africa	0 to +2.5	-5 to +2.5
Latin America and the Caribbean		
Tropics and subtropics	-2.5 to 0	-5 to -2.5
Temperate	0 to +2.5	0 to +2.5
Near East/North Africa	-2.5 to +2.5	-5 to +2.5
South Asia	-2.5 to 0	0 to -5
East Asia	-2.5 to +2.5	-2.5 to +2.5
Canada and the United States	-5 to +2.5	-10 to 0

Source: Parry *et al.* (1999).

risk are in low-income countries that may not undergo appreciable economic development over the projection period, and so might find it difficult to undertake the necessary protective investments (Nicholls, Hoozemans and Marchand, 1999).

Three factors will affect food security: the loss of cropland and nursery areas for fisheries by inundation and coastal erosion; saltwater intrusion; and flood damage to crops and food stores. Each of these will eliminate livelihoods and lower agricultural production and incomes.

The loss of cropland could be substantial. India, for example, has more than 6 500 km^2 of low-lying coastal land, much of which is cultivated. Asthana (1993) estimated that a 1-metre sea level rise in India would result in the loss of some 5 500 km^2. Rises of this magnitude are not foreseen before 2100 according to the IPCC's latest estimates. Losses by 2030 could be from 1 000 to 2 000 km^2. Assuming an average farm size of 1.5 ha this could represent the loss of some 70 000 to 150 000 livelihoods. In the case of Bangladesh a similar rate of sea level rise by 2030 could result in the loss of 0.8-2.9 million tonnes of rice p.a., offsetting yield gains arising from changes in temperature, precipitation and atmospheric CO_2 concentration (Asian Development Bank, 1994).

The inland movement of saltwater into the aquifers used for irrigation, with negative impacts on crop yields, is already of significant proportions in some North African countries because of exces-

sive extraction of groundwater, and it will be intensified by sea level rise. Yet over the next 30 years much could be done to overcome this problem, e.g. by the introduction of new GM varieties of wheat, rice, oilcrops and green vegetables that are tolerant to saline conditions (see Section 13.5). However, yield losses may also occur through physical-chemical damage to the soil by salinization, so other measures will be required. Nonetheless, the food security impacts of saltwater intrusions could be quite small if appropriate policy and technology changes are made.

Flood damage to crops and food stores could be important at the national, local and household level: at the national level, in cases where agriculture is the main source of export revenues to pay for the imports of the development goods essential for economic growth, and of food to cover shortfalls in domestic production; locally, if public or private food stores are destroyed, shortages and higher prices can be expected; and at the household level, where season-to-season storage of food is essential to insulate families from pre-harvest price rises.

13.4.4 Implications for livelihoods and incomes

Food insecurity is in most cases caused by poverty. Food purchasing power depends on a person's income level and on food prices. For farmers, incomes depend mainly on the quantity and quality of the natural resources they have to produce food. Consequently, any impact of climate change on land and water resources, on agricultural and non-agricultural livelihoods, and on the prices of food or of other agricultural commodities sold to purchase food could have an important impact on food security. With the possible exception of sub-Saharan Africa, it seems doubtful that climate change will have an appreciable impact on agricultural livelihoods and incomes over the period to 2030. The wide range of domestic and international factors governing national economic performance could swamp any small effects resulting from climate change.

However, climate change will have some adverse effects on incomes and income distribution. A number of groups are particularly vulnerable, namely: low-income groups in drought-prone areas with poor food distribution infrastructure; low- to medium-income groups in flood-prone areas who may lose stored food and possessions; farmers whose land is submerged or damaged by sea level rise or saltwater intrusions; fishers who suffer falling catches from shifts in ocean currents, or flooding of spawning areas or fish ponds; and food or export crop producers at risk from high winds.

On the other hand, some of the short- to medium-term negative impacts on food security may lead to positive outcomes in the longer term. For example, increasing aridity may accelerate the migration of low-wage agricultural workers to urban centres where wages are higher and there is more secure access to food markets.

Increased frequency of extreme events could have substantial impacts on the economic performance of some countries and regions, and on transitory food insecurity. The Mozambique floods of 2000, for example, have been estimated by the World Bank to have reduced economic growth by 2 to 3 percentage points, and caused damage in excess of total export earnings. The 1998 floods in China caused over US$100 billion damage, and for the main provinces affected, the damage amounted to the equivalent of 3 to 4 percent of their GDP. Cambodia suffered similar economic losses from floods in 2000. In each case the number of people considered to be transitory food insecure increased ten to 100-fold or more. However, their recovery normally took place within months (FAO/GIEWS, 2000a) and the overall impact on national food production was quite small because of good harvests in other areas or seasons.

Finally, it is important to consider how policy responses to climate change could affect livelihoods and incomes. This aspect could become of increasing importance through the CDM and efforts to substitute fossil fuels by renewable ones, opening up new opportunities for job creation and income improvements. First, carbon sequestration and trading in carbon emission permits could improve the overall sustainability of agriculture (see Section 13.2). They could raise farm incomes and create new agricultural livelihoods. There could be growing competition for land and labour resources in some areas between biofuel production, carbon mitigation activities and food production, but such impacts are likely to be small over the next 20-30 years. Second, new non-fossil energy systems, particularly wind power, could provide marginal areas such as the slope lands of southwest India with new livelihoods and lower energy prices for rural electrification.

13.4.5 Implications for food prices

The analysis in Chapters 3 and 9 suggests that, independently of climate change, real world market agricultural prices will remain more or less constant or decline slightly over the projection period.

Climate change to 2030 may reduce the costs of crop and livestock production in some temperate areas, according to IPCC projections (IPCC, 2001b), for example, from milder winters, longer growing seasons and the reduced need for winter concentrate feeds for livestock. In contrast, some humid tropical and semi-arid areas of developing regions may face rising production costs, e.g. because of rice yield declines from higher night temperatures, higher irrigation costs and salinization induced by sea level rise.

The net effect of these regional differences could be downward price pressures in developed countries and upward pressures on prices in developing countries, but in both cases the movements in real prices would be relatively small to 2030. Parry et al. (1999) conclude that (other factors remaining equal) climate-induced cereal yield declines could push up global prices in 2020 by about 5 percent (and by implication much more in parts of Africa), with substantially greater rises by

2050-2080. Cereals tend to be more sensitive to climate change than other food crops, and many developing countries are growing net importers of cereals. Therefore they could become more vulnerable to climate-induced increases in grain prices.

However, most studies suggest that in the short term the net impact of climate change on current cereal areas is likely to be positive and in the longer term the area suitable for cereal production could expand considerably (IPCC, 2001c; Fischer *et al.*, 2001). Hence, even in the context of climate change, world market prices for cereals are likely to remain relatively stable. In addition, price developments may be partly offset through the implementation of present and future WTO Agreements on Agriculture. The gap between international and national prices should narrow, so that movements in national prices should follow movements in world market prices more closely. National and local prices, however, will still be perturbed by extreme events and more direct international to domestic price links will moderate these fluctuations but not eliminate them.

Technological change and infrastructural improvements allowing better flows of food from surplus to deficit areas could also offset some of the pressure on national and local prices. Given the slow progress of the last decades, however, there is great uncertainty whether all of the required national and regional infrastructural improvements will take place over the next 30 years. In south Mozambique, for example, maize prices in the spring of 2000 increased rapidly because of food shortages following the floods. At the same time, however, maize prices in north Mozambique were about half those in the south and declining. Yet the high transport costs from the north to the south made it cheaper to import maize from South Africa (FAO/GIEWS, 2000b).

Extreme events affect food prices in characteristic ways: price increases can be very rapid and large, particularly where both household and commercial stocks are lost, and transport is disrupted; price changes can be very localized, with appreciable differences between urban and rural areas with restricted access to outside supplies; and price increases can be short-lived, i.e. weeks rather than months. These points show how critical general economic development will be in reducing the vulnerability of countries to climate change and to increased frequency and intensity of extreme events.

It is important to bear in mind that changes in international commodity prices estimated by the models used for climate change impact studies do not necessarily relate closely to the food prices actually paid by consumers and hence to the ability of low-income groups to purchase their food needs. For example, bread is increasingly a purchased good rather than a home-baked food even for the rural poor, and the cost of the cereal may be less than 25 percent of the purchase price, with the rest coming from processing, distribution and marketing costs (Norse, 1976). Hence, even if climate change increases farmgate or international food prices over the next 30 years, this increase may have a much smaller impact on consumer prices, and limited effects on the food security of those low-income groups that purchase most of their food from the retail sector.

13.5 Technology and policy options

Many of the actions required to mitigate or to adapt to climate change can also be justified in terms of present needs. Many do not require large capital investments, and can be appropriate for poor smallholders as well as large farmers. They do not have to be justified on the basis of the uncertain economic benefits of lowering some climate change impacts. For example, improved water conservation would help to overcome current aridity as well as reduce the impact of any future deterioration in rainfall. Most of the actions would also contribute to the wider objective of alleviating poverty and improving access to food rather than just safeguarding the production of food.

13.5.1 Greenhouse gas reduction and abatement

The priority actions to lower agriculture's role as a driving force for climate change are clear from Section 12.3.1 in Chapter 12: reduction of methane and nitrous oxide emissions from mineral fertilizers, manure, livestock wastes and rice production. The wider benefits are also clear, e.g. lower production costs through greater fertilizer-use efficiency and better waste recycling, and reduced air and water pollution. The policy response options in the agrochemical sector include removing any subsidies on energy inputs and introducing carbon taxes to promote energy-use efficiency in fertilizer and pesticide production.

Other policy options include general actions to promote sustainability through conservation agriculture, together with specific measures such as environmental taxes on nitrogen fertilizers; promotion of precision placement and better timing of fertilizer and manure applications; development of rice cultivars emitting less methane; adoption of direct seeding and better water management for rice to reduce methane emissions; better feed quality for livestock; improved livestock waste management; promotion of biofuel crops to replace fossil fuels; and expansion of agroforestry.

13.5.2 Climate change impact mitigation and adaptation

Several actions need to be taken to mitigate and adapt to climate change. First, comprehensive support mechanisms must be formulated to help farmers adapt to climate change and to increase production under more variable conditions. Such mechanisms could include approaches to crop production which improve the resilience of farming systems.

Second, given the probability of higher incidence of drought, aridity, salinity and extreme events, greater priority will need to be given to the following measures:

- maintenance, both onsite and offsite, of a broad genetic base for crops and development and distribution of more drought-tolerant crop varieties and livestock breeds;
- breeding for greater tolerance of crops, livestock and fish to higher temperatures;
- development of salt-tolerant varieties of wheat, rice and oilcrops;
- improving the resilience of agricultural ecosystems by promoting NT/CA and practices such as agroforestry that utilize and maintain biological diversity;
- raising the efficiency of rainwater use and groundwater recharge by conservation agriculture, etc. and that of irrigation water by appropriate pricing policies, management systems and technologies;

- supporting pastoral and other livestock production systems, many of which are already food insecure. Activities should be centred on maintaining livestock mobility and providing location-specific investment in supplementary feed production, veterinary services and water supply (Sandford, 1995), and on improving the marketing of livestock during droughts and making it easier to restock after droughts or floods; and
- developing improved sea defence and flood management systems in sea level rise and storm surge situations, where these are economically viable.

All these actions have the benefit of helping to ameliorate the impact of current climate variation as well as countering future adverse effects of climate change.

13.5.3 Reducing or avoiding food security impacts

The IPCC now expresses high confidence in the projected increase in the frequency and intensity of relatively localized extreme events including those associated with El Niño, notably droughts, floods, tropical cyclones and hailstorms (IPCC, 2001b). The impacts of these increases will fall disproportionately on the poor (see Box 13.1).

All developing regions are considered by the IPCC to be vulnerable to increased droughts and floods. These extreme events could pose significant threats to food security, requiring policy action and investment both outside and within the agricultural sector.

For many countries the key to reducing food insecurity will be better disaster preparedness planning, although actions to lower the sensitivity of food and agricultural production to climate change will clearly be important to cope with the longer-term impacts of climate change. Many of the actions in response to drought and sea level rise should be conceived on the pattern of disaster management strategies being developed to reduce agricultural vulnerability to tropical storms (FAO, 2001g). The objectives of such strategies include avoiding or minimizing death, injury, lack of shelter and food shortages, loss of property or livelihoods of poor households, and preparing funding and procedures for large-scale relief and rehabilitation. Such strategies may be implemented through:

- the development of early warning and drought, flood- and storm-forecasting systems;
- preparedness plans for relief and rehabilitation;
- introducing more storm-resistant, drought-tolerant and salt-tolerant crops;
- land use systems that stabilize slopes and reduce the risk of soil erosion and mudslides;
- constructing livestock shelters and food stores above likely flood levels;
- equipping fishers with communication systems and safety devices so that they can benefit from early storm warnings, and credit systems so they can quickly replace any lost boats or equipment.

13.6 Conclusions

The projections of this study point to the likelihood of an appreciable increase in carbon sequestration by agricultural soils. Although the gains will eventually level off, they will extend the time available to introduce other measures with longer-term benefits. Thus, agriculture's role as a driving force for climate change could still increase, but its contribution to climate change mitigation will rise through greater carbon sequestration and increased resilience to climate variation.

The main impacts of climate change on global food production are not projected to occur until after 2030, but thereafter they could become increasingly serious. Up to 2030 the impact may be broadly neutral or even positive at the global level. Food production in higher latitudes will generally benefit from climate change, whereas it may suffer in large areas of the tropics. However, there could be large intraregional disparities in the medium term, e.g. western, central and eastern Africa could experience a reduction in cereal production and southern Africa an increase.

Up to 2030 these potential decreases in food production are relatively small and most countries should be able to compensate for climate change impacts by improving agricultural practices.

Box 13.1 Food-insecure regions and countries at risk

The IPCC and the other assessments considered in this chapter conclude that the main regions and countries at risk from climate change are the following:

Climate change
- Countries of arid, semi-arid and subtropical Asia, sub-Saharan Africa, Near East/North Africa and Latin America where temperatures are above the optimum for crop growth or even close to their maximum temperature tolerance, and already result in heat stress in livestock and fish.
- Water-scarce countries of arid, semi-arid and subtropical Asia, sub-Saharan Africa, Near East/North Africa and Latin America (northeast Brazil) where reduced stream flow and water recharge, higher transmission losses from irrigation systems and greater evapotranspiration from crops may lower irrigation and increase water stress in crops and livestock.

Sea level rise
- West Africa (Gulf of Guinea, Senegal and the Gambia), southern Mediterranean (Egypt), East Africa (Mozambique), South and Southeast Asia (Bay of Bengal), the Caribbean and island states of the Indian and Pacific Oceans.

Extreme events
- Droughts for much of semi-arid and subhumid Africa (particularly the Sahel, Horn of Africa and southern Africa), South Asia and northeast Brazil.
- Floods in deltas and their immediate hinterland during storm surges.
- Floods in river valleys and lake basins of all regions, including temperate ones, during abnormally long or intense rainfall events.
- High winds associated with tropical cyclones in Central America and the Caribbean, South and East Asia.

Priority should be given to raising the resilience of agricultural ecosystems, increasing the cropped area, and raising and diversifying yields through improved access to genetic resources and technologies. Moreover, with the exception of sub-Saharan Africa, the growing income of developing countries should make it possible for many of them to choose between greater food imports and greater mitigation and adaptation by their agricultural sector to overcome climate change impacts. The world's traditional cereal exporters should be able to meet any increase in demand, either because their production potential will be boosted by climate change, or because they will have the capacity to adapt to climate change and overcome any negative impacts.

Up to 2030, the most serious and widespread agricultural and food security problems related to climate change are likely to arise from the impact on climate variation, and not from progressive climate change, although the latter will be important where it compounds existing agroclimatic constraints. However, the more frequent extreme events will not necessarily increase food insecurity in all situations, given the other economic and social changes taking place. Given the likely structural change in the sectoral composition of the economy and of employment in developing countries, access to food will increasingly be determined by urbanization and non-agricultural incomes. As a result, food security in some countries will improve and they will become less vulnerable to climate change. Developed countries will also experience more frequent extreme events but it seems possible that these will not have a sustained impact on their food export potential.

Nonetheless, low-income groups in many countries will remain vulnerable to short- to medium-term supply constraints arising from climate change. The basic food security issue will remain that of poverty and the lack of food purchasing power.

Although the impacts of climate change on food production and food security up to 2030 may be relatively small and uncertain, those projected for the remainder of the century are larger and more widespread. By 2100 climate change could pose a serious threat to global and local food security. It is therefore vital that action be taken now to counter this threat. Actions should include measures to reduce agriculture's role as a driving force for climate change, through the reduction of GHG emissions, as well as measures to mitigate and adapt to climate change.

Institutional changes are going to be as important as or more important than technological ones. Institutional actions will be needed to raise national preparedness and reduce rural and urban poverty to enable vulnerable low-income groups to purchase all of their basic food requirements. Policies for agricultural development will need to emphasize the importance of improving not just the production capacity of agricultural ecosystems but also their diversity and resilience. It is vitally important to initiate the institutional and technological changes now, because of the long lead times for the development of new technologies and for the improvement of road and rail links between food-deficit and surplus areas, and between ports or railheads and isolated rural areas.

Appendixes

Countries and commodities

Developing countries

Africa, sub-Saharan	Latin America and the Caribbean	Near East/North Africa	South Asia
Angola		Afghanistan	Bangladesh
Benin	Argentina	Algeria	India
Botswana	Bolivia	Egypt	Maldives
Burkina Faso	Brazil	Iran, Islamic Rep.	Nepal
Burundi	Chile	Iraq	Pakistan
Cameroon	Colombia	Jordan	Sri Lanka
Central African Republic	Costa Rica	Lebanon	
Chad	Cuba	Libyan Arab Yamahiriya	
Congo	Dominican Republic	Morocco	
Congo, Dem. Rep.	Ecuador	Saudi Arabia	
Côte d'Ivoire	El Salvador	Syrian Arab Republic	
Eritrea	Guatemala	Tunisia	
Ethiopia	Guyana	Turkey	
Gabon	Haiti	Yemen	
Gambia	Honduras	Near East, other[3]	
Ghana	Jamaica		
Guinea	Mexico		
Kenya	Nicaragua		**East Asia**
Lesotho	Panama		Cambodia
Liberia	Paraguay		China
Madagascar	Peru		Indonesia
Malawi	Suriname		Korea, Dem. Rep.
Mali	Trinidad and Tobago		Korea, Rep.
Mauritania	Uruguay		Lao PDR
Mauritius	Venezuela		Malaysia
Mozambique	Latin America, other[2]		Mongolia
Namibia			Myanmar
Niger			Philippines
Nigeria			Thailand
Rwanda			Viet Nam
Senegal			East Asia, other[4]
Sierra Leone			
Somalia			
Sudan			
Swaziland			
Tanzania, United Rep.			
Togo			
Uganda			
Zambia			
Zimbabwe			
Sub-Saharan Africa, other[1]			

Note: Data on land with rainfed production potential as well as estimates of land in use by agro-ecological class are available for all countries except Namibia, Maldives, Mongolia and the groups "other" in sub-Saharan Africa, Latin America, Near East/North Africa and East Asia.

[1] Cape Verde, the Comoros, Djibouti, Guinea-Bissau, Sao Tome and Principe, Seychelles.
[2] Antigua, Bahamas, Barbados, Belize, Dominica, Grenada, Netherland Antilles, Saint Kitts and Nevis, Saint Lucia, Saint Vincent and the Grenadines, Bermuda.
[3] Cyprus, Kuwait, United Arab Emirates.
[4] Brunei, Macau, Solomon Islands, Fiji, French Polynesia, New Caledonia, Vanuatu, Papua New Guinea, Kiribati.

Industrial countries

European Union*	**Other western Europe**	**North America**	**Oceania**	**Other developed countries**
Austria		Canada	Australia	
Belgium	Iceland	United States	New Zealand	Israel
Denmark	Malta			Japan
Finland	Norway			South Africa
France	Switzerland			
Germany				
Greece				
Ireland				
Italy				
Luxembourg				
Netherlands				
Portugal				
Spain				
Sweden				
United Kingdom				

* In the analysis the European Union was treated as one country group (EU-15).

Transition countries

Eastern Europe and the former Yugoslavia SFR	**Commonwealth of Independent States**	**Baltic states**
Albania		Estonia
Bosnia and Herzegovina	Armenia	Latvia
Bulgaria	Azerbaijan	Lithuania
Croatia	Belarus	
Czech Republic	Georgia	
Hungary	Kazakhstan	
Poland	Kyrgyzstan	
Romania	Moldova, Republic	
Slovakia	Russian Federation	
Slovenia	Tajikistan	
The Former Yugoslav Republic of Macedonia	Turkmenistan	
Yugoslavia	Ukraine	
	Uzbekistan	

Commodities covered

Crops	Livestock
Wheat	Beef, veal and buffalo meat
Rice, paddy	Mutton, lamb and goat meat
Maize	Pig meat
Barley	Poultry meat
Millet	Milk and dairy products (in whole milk equivalent)
Sorghum	Eggs
Other cereals	
Potatoes	
Sweet potatoes and yams	
Cassava	
Other roots	
Plantains	
Sugar, raw[1]	
Pulses	
Vegetables	
Bananas	
Citrus fruit	
Other fruit	
Vegetable oil and oilseeds (in vegetable oil equivalent)[2]	
Cocoa beans	
Coffee	
Tea	
Tobacco	
Cotton lint	
Jute and hard fibres	
Rubber	

[1] Sugar production in the developing countries analysed separately for sugar cane and sugar beet.

[2] Vegetable oil production in the developing countries analysed separately for soybeans, groundnuts, sesame seed, coconuts, sunflower seed, palm oil/palm-kernel oil, rapeseed and all other oilseeds.

Note on commodities

All commodity data and projections in this report are expressed in terms of primary product equivalent unless stated otherwise. Historical commodity balances (supply utilization accounts – SUAs) are available for about 160 primary and 170 processed crop and livestock commodities. To reduce this amount of information to manageable proportions, all the SUA data were converted to the commodity specification given above in the list of commodities, applying appropriate conversion factors (and ignoring joint products to avoid double counting, e.g. wheat flour is converted back into wheat while wheat bran is ignored). In this way, one supply utilization account in homogeneous units is derived for each of the commodities of the study. Meat production refers to indigenous meat production, i.e. production from slaughtered animals plus the meat equivalent of live animal exports minus the meat equivalent of all live animal imports. Cereal demand and trade data include the grain equivalent of beer consumption and trade.

The commodities for which SUAs were constructed are the 26 crops and six livestock products given in the list above. The production analysis for the developing countries was, however, carried out for 34 crops because sugar and vegetable oils are analysed separately (for production analysis only) for the ten crops shown in the footnote to the list.

Summary methodology of the quantitative analysis and projections

This appendix gives a very brief account of the approach followed in this study. For a more extensive treatment, the reader is referred to Appendix 2 in Alexandratos (1995). The final part of this appendix discusses briefly some of the problems with the data and the assumptions, and explains why only one scenario was designed for this study.

Summary methodology

In projecting the likely evolution of the key food and agricultural variables, a "positive" approach has been followed, aiming at describing the future as it is likely to be (to the best of our knowledge at the time of carrying out this study), and not as it ought to be from a normative point of view. The study therefore does not attempt to spell out actions that need to be taken to reach a certain target (for example the World Food Summit target of halving the number of chronically undernourished persons by no later than 2015) or some other desirable outcome sometime in the future. The second overriding principle of the approach followed in this study was to draw to the maximum extent possible on FAO's in-house knowledge available in the various disciplines present in FAO, so as to make the study results represent FAO's "collective wisdom" concerning the future of food, nutrition and agriculture.

The quantitative analysis and projections were carried out in considerable detail in order to provide a basis for making statements about the future concerning individual commodities and groups of commodities as well as agriculture as a whole, and for any desired group of countries. For this reason the analysis was carried out for as large a number of individual commodities and countries as practicable (see Appendix 1). Another reason for the high degree of detail has to do with the interdisciplinary nature of the study and its heavy dependence on contributions provided by FAO specialists in the different disciplines. Such contributions can find expression only if the relevant questions are formulated at a meaningful level of detail. For example, a useful contribution can only be obtained from crop production experts, if questions of yield growth potential are addressed separately for maize, barley, millet and sorghum, not for coarse grains as a group, and preferably disaggregated in terms of agro-ecological conditions because, say, irrigated barley and rainfed semi-arid barley are practically different commodities for assessing yield growth prospects. Moreover, statements on, and projections of, land and water use cannot be made unless all the major crops are accounted for. For example, cereals account for about 55 percent of the total harvested area of the developing countries. An analysis limited to cereals would not provide a sufficient basis for exploring issues of land scarcities and possibilities of expansion in the future.

The variables projected in the study are (i) the demand (different final and intermediate uses), production and net trade balances for each commodity and country; and (ii) key agro-economic variables, i.e. for crops: area, yield and production by country and, for the developing countries only, by agro-ecological zone (irrigated and rainfed with the latter subdivided into dry semi-arid, moist semi-arid, subhumid, humid land and fluvisols/gleysols); and for livestock products: animal numbers (total stock and offtake rates) and yields per animal.

A significant part of the total effort is devoted to the work needed to create a consistent set of historical and base year data. For the demand-supply analysis, the overall quantitative framework for the projections is based on supply utilization accounts (SUAs). The SUA is an accounting identity, showing for any year the sources and uses of agricultural commodities in homogeneous physical units, as follows:

Food (direct consumption) + Industrial non-food uses + Feed + Seed + Waste = Total domestic use = Production + (Imports - Exports) + (Opening stocks - Closing stocks)

The database has one such SUA for each commodity, country and year (1961 to 1999). The data preparation work for the demand-supply analysis consists of the conversion of the approximately 330 commodities for which the primary production, utilization and trade data are available into the 32 commodities covered in this study, while respecting SUA identities (see the Note on commodities in Appendix 1). The different commodities are aggregated into commodity groups and into "total agriculture" using as weights world average producer prices of 1989/91 expressed in "international dollars" derived from the Geary-Khamis formula as explained in Rao (1993). The growth rates for heterogeneous commodity groups or total agriculture shown in this study are computed from the value aggregates thus obtained.

A major part of the data preparation work, undertaken only for the developing countries, is the unfolding of the SUA element "production" (for the base year only, in this case the three-year average 1997/99) into its constituent components of area, yield and production that are required for projecting production. For crops, the standard data in the SUAs contain, for most crops, also the areas (harvested) and average yields for each crop and country. These national averages are not considered by agronomists to provide a good enough basis for projections because of the widely differing agro-ecological conditions in which any single crop is grown, even within the same country. An attempt was therefore made to break down the base year production data from total area under a crop and an average yield into areas and yields for rainfed and irrigated categories. The problem is that such detailed data are not generally available in any standard database. It became necessary to piece them together from fragmentary information, from both published and unpublished documents giving, for example, areas and yields by irrigated and rainfed land at the national level or by administrative districts, supplemented by a good deal of guesstimates.

No data exist on total harvested land, but this can be obtained by summing up the harvested areas reported for the different crops. Data are available for total arable land in agricultural use (physical area, called in the statistics "arable land and land in permanent crops"). It is not known whether these two sets of data are compatible with each other, but this can be evaluated indirectly by computing the cropping intensity, i.e. the ratio of harvested area to arable land. This is an important parameter that can signal defects in the land use data. Indeed, for several countries the implicit values of the cropping intensities did not seem to make sense. In such cases the harvested area data resulting from the crop statistics were accepted as being the more robust (or the less questionable) ones and those for arable area were adjusted in consultation with the country and land use specialists (see Alexandratos [1995] for a discussion of these problems).

The bulk of the projection work concerns the drawing up of SUAs (by commodity and country) for the years 2015 and 2030, and the unfolding of the projected SUA item "production" into area and yield combinations for rainfed and irrigated land and, likewise, for livestock commodities into the underlying parameters (number of animals, offtake rates and yields).

The overall approach is to start with projections of demand, using Engel demand functions and exogenous assumptions on population and GDP growth.[1] Subsequently, the entry point for the projections of production is to start with provisional projections for production for each commodity and country derived from simple assumptions about future self-sufficiency and trade levels. There follow several rounds of iterations and adjustments in consultation with specialists on

---

[1] Population data and projections are from the United Nations (UN, 2001a; medium variant) and GDP projections are largely based on the World Bank (2001c), extended to 2030.

the different countries and disciplines, with particular reference to what are considered to be "acceptable" or "feasible" levels of calorie intakes, diet composition, land use, (crop and livestock) yields and trade. Accounting consistency controls at the commodity, land resources (developing countries only), country and world levels have to be respected throughout. In addition, but only for the cereal, livestock and oilseeds sectors, a formal flex-price model was used (FAO World Food Model; FAO, 1993b) to provide starting levels for the iterations and to keep track of the implications for all variables of the changes in any one variable introduced in the successive rounds of inspection and adjustment. The model is a partial equilibrium model, composed of single commodity modules and world market feedbacks leading to national and world market clearing through price adjustments. It is emphasized that the results of the model projections (whether the single Engel demand functions or the flex-price model) were subjected to many rounds of iterative adjustments by specialists on countries and of many disciplines, particularly during the phase of analysing the scope for production growth and trade. The end product may be described as a set of projections that meet conditions of accounting consistency and to a large extent respect constraints and views expressed by the specialists in the different disciplines and countries.

It should be emphasized here that the projections presented in this study are definitely not "trend extrapolations", whether the term is used to denote the derivation of a future value of any variable by simple application of its historical growth rate to the base year value (exponential trend) or the less crude notion of using time as the single explanatory variable in functional forms other than exponential, e.g. linear, semi-log, sigmoid, etc. For one thing, projecting all interlinked variables on the basis of estimated functions of time is a practical impossibility; for another, projecting any single variable at its historical growth rate (which could be negative, zero or very high) often leads to absurd results. Therefore, the term "trend" or "trend extrapolation" is not appropriate for describing these projections.

Summary statements of the methodologies of supporting and complementing analyses are given in the main body of this report: for example, the approaches followed for estimating the number of chronically undernourished people (Box 2.1), for deriving estimates of land with potential for rainfed agriculture (Box 4.1), for estimating water requirements in irrigated agriculture (Box 4.3), and for deriving projections of fertilizer consumption (Section 4.6).

Data problems

The significant commodity and country detail underlying the analysis requires the handling of huge quantities of data. Inevitably, data problems that would remain hidden and go unnoticed in work conducted at the level of large country and commodity aggregates come to the fore all the time. Examples of typical data problems are given below.

Data reliability. When revised numbers become available in the successive rounds of updating and revision of the historical data, it is not uncommon to discover that some of the data were off the mark, sometimes by a very large margin. It may happen therefore that changes projected to occur in the future have already occurred in the past. A typical case is presented in Chapter 2, Box 2.2. There the point is made that the revisions of the population (downwards) and food production data (upwards) in Nigeria implied that in the previous (1995) edition of this study, the projections were based on a food security situation in Nigeria that was worse than the one actually prevailing, assuming the new revised data are nearer reality. Another example: at the time of writing of this report (mid-2002), the latest available revised trade data for Namibia show significant rice imports that were not present in the previous data. They lead to an increase by 50 kg of the per capita food consumption of cereals for the base year 1997/99 of this study. The revised total food consumption is 2 600 kcal/person/day, up from the 2 090 kcal/person/day before the revisions. As a result, the revised estimate of undernourishment for 1997/99 is 9.5 percent of the population, down from the 33 percent before the data revisions.

Obviously, there is not much that can be done about this problem as errors in historical data become apparent after completing the projections. Such changes in the historical data also bedevil attempts to compare in any degree of detail the projections of earlier editions of this study with the actual outcomes for the latest year for which data are available. The comparisons occasionally shown in Chapters 3 and 4 are for the developing countries as a whole: net cereal imports (Figure 3.7) and production and yields of wheat, rice, maize and other coarse grains (Figure 3.12 and Box 4.4). As such, they are not greatly influenced by significant revisions in the historical data of individual countries.

Unbalanced world trade. A second data problem relates to the large discrepancies often encountered in the trade statistics, i.e. world imports are not equal to world exports. Small discrepancies are inevitable and can be ignored but large ones pose serious problems since in the projections exporting countries must produce export surpluses equal to the net imports of other countries. For example, the sugar exporters had net exports of 32.3 million tonnes in 1997/99 while importers had net imports of only 28.5 million tonnes, leaving a world imbalance of 3.7 million tonnes. In the projections, the importers are estimated to need net imports of 35.2 million tonnes in 2015, an increase of 23 percent. If the discrepancy in the base year were to be ignored, the export surplus of the exporters should also be 35.2 million tonnes, i.e. only 9 percent above the 32.5 million they exported in 1997/99, thus greatly distorting the analysis of their export prospects. By necessity, the unsatisfactory solution of assuming that a discrepancy of roughly equal magnitude to that of the base year will also prevail in the future had to be adopted (see Table 3.23, Chapter 3).

There are good reasons why discrepancies arise in the trade statistics, e.g. differences in the timing of recording of movement of goods in the exporting and importing country, although this can hardly explain some very large discrepancies, e.g. world exports of refined sugar are 20 percent higher than world imports, while for concentrated orange juice world exports (85 percent of them from Brazil) are double world imports. At the same time, world imports of single strength orange juice exceed world exports by an almost equal amount.

Some problems with the exogenous assumptions. As an example, the impossibility of foreseeing which countries may face extraordinary events leading to their being worse off in the future than at present is mentioned. In Chapter 2 it was noted that several countries suffered declining levels of food consumption, some of them in the form of collapses within the span of a few years, e.g. the Democratic People's Republic of Korea, Iraq, Cuba, Afghanistan, the Democratic Republic of the Congo and many transition economies. In most cases such collapses result from the occurrence of difficult to predict systemic changes or crises, or from outright unpredictable events, such as war or civil strife. It is impossible to predict which countries may be in that class in the future. Therefore, in the projections each and every country is shown with a higher food consumption per person than at present, some significantly better, others less so and several remaining with critically low levels. This is the result, in the first place, of the exogenous income growth assumptions that allow only rarely for the eventuality that per capita income of individual countries might in 30 years be lower than at present.

The prospect that only few countries may suffer income declines is, of course, at variance with the empirical evidence that shows quite a few countries having lower incomes today than three decades ago. The World Bank has data for 80 of the developing countries covered individually in this study (World Bank, 2001b; Table 1.4). No fewer than 28 of them have had negative growth rates in per capita GDP in the period 1965-99 (the number is larger if the transition economies are included). They include many of the countries devastated by war or civil strife at some period from 1965 to 1999. As noted, it would be foolhardy to predict or assume which countries may have similar experiences in the future. For example, in the 1988 edition of this study (Alexandratos, 1988) with projections to 2000, the collapse of food and agriculture in the formerly centrally planned economies of Europe and their virtual disappearance as large net importers of cereals had not been predicted.

Why only one scenario. In this study, only one possible outcome for the future based on a positive, rather than normative, assessment is presented. Alternative scenarios have not been explored for a number of reasons, some conceptual, some practical, and usually a mix of both. Producing an alternative scenario is essentially a remake of the projections with a different set of assumptions. On the practical side, the major constraint is the time-consuming nature of estimating alternative scenarios with the methodology of expert-based inspection, evaluation and iterative adjustments of the projections. On the conceptual side, defining an alternative set of exogenous assumptions that are internally consistent represents a challenge of no easy resolution. For example, among the major exogenous variables are the projections of population and income (GDP). As discussed, for population the medium variant demographic projections of the United Nations were used. There are also high and low variants. In estimating an alternative scenario with, say, the high variant, it would not be known how the exogenous GDP projections should be modified so as to be internally consistent with the high population variant. If the GDP growth rates were retained unchanged, projected per capita incomes would be lower, and this would mean implicitly accepting that population growth is detrimental to economic welfare. If the GDP growth rates were raised to keep projected per capita incomes unchanged, it would mean accepting that population growth made no difference. Neither of the two views can be correct for all countries. In actual life, some countries would be better off with higher population growth and some worse off (see more discussion in Box 2.3, Chapter 2). It would be impossible to define in an empirically valid manner what the relationships could be for each of the more than one hundred countries analysed individually in this study.

The one alternative scenario that it would be highly desirable to have is one that would introduce feedbacks from agriculture to the overall economy, at least for the countries in which agriculture is a substantial component of the economy. The methodology used in this study is of the partial equilibrium type, that is, interdependence is accounted for among, and balance is brought about in, the demand and supply of the individual agricultural commodities, at the country and world levels. Other aspects of interdependence and balance in the wider economy are ignored, e.g. how a more robust agricultural performance would eventually contribute to a higher GDP growth rate than originally assumed and how the latter would in turn stimulate demand for food and agriculture

itself. To introduce such general equilibrium elements in the analysis, rather sophisticated economy-wide models would have to be built and validated for the individual countries. This is a quasi impossible task, partly because of the time and resources required and partly because the data available for many of the countries for which such analyses would be most appropriate (low-income ones with high dependence on agriculture) are generally not adequate to support such an undertaking. Circumventing the problem by assuming arbitrarily the existence of linkages between agriculture and the rest of the economy (e.g. that a 1 percent increase in agricultural GDP causes the rest of the economy to grow by x percent) would not do. As noted above, in a number of countries robust agricultural growth was associated with meagre or declining growth for the rest of the economy, implying a negative link if this simplistic approach were followed. Obviously, rather sophisticated economy-wide analysis would reveal the reasons why such "perverse" relationships exist in the data, and could even lead to the conclusion that some of the data are outright wrong.

In conclusion, alternative scenarios would be certainly useful for exploring the future in the face of uncertainties about how key variables of the system may evolve. In this study, an attempt was made to contain this uncertainty by bringing to bear the expert judgement of the discipline and country specialists on the future values of the relevant variables (e.g. rates of growth of yields, land, irrigation, etc.). Running a scenario with alternative values for one or more of these variables would mean repeating this process. Much of the work for a new scenario would be devoted to the definition of plausible alternative values. It is just not a question of assuming that, for example, irrigation would expand at a higher rate than in the baseline projection in each and every country. This would be impossible for some countries because of physical water constraints. The same holds for higher yield growth rates: the potential exists in some countries and crops but not in others. If realistic alternative paths for such values cannot be defined, the results of estimating alternative scenarios with blanket assumptions about uniform changes in the values of some variables in each and every country would certainly be misleading rather than illuminating.

Statistical tables

TABLE A1   Total population

	Million				Annual growth rates (% p.a.)			
	1997/99	2015	2030	2050	1980-99	1997/99-2015	2015-30	2030-50
World	5 900.0	7 207.4	8 270.1	9 322.3	1.6	1.2	0.9	0.6
All developing countries	4 572.6	5 827.3	6 869.1	7 935.3	1.9	1.4	1.1	0.7
Sub-Saharan Africa	574.2	882.7	1 229.0	1 704.3	2.9	2.6	2.2	1.6
Angola	12.4	20.8	32.6	53.3	3.1	3.1	3.1	2.5
Benin	6.0	9.4	13.3	18.1	3.1	2.8	2.3	1.5
Botswana	1.5	1.7	1.9	2.1	2.9	0.7	0.7	0.6
Burkina Faso	11.0	18.5	29.2	46.3	2.6	3.1	3.1	2.3
Burundi	6.2	9.8	13.8	20.2	2.3	2.7	2.3	1.9
Cameroon	14.2	20.2	25.8	32.3	2.8	2.1	1.6	1.1
Central African Republic	3.6	4.9	6.4	8.2	2.4	1.8	1.8	1.2
Chad	7.4	12.4	18.7	27.7	2.9	3.1	2.8	2.0
Congo	2.8	4.7	7.2	10.7	3.0	3.0	2.8	2.0
Congo, Dem. Republic	48.4	84.0	132.6	203.5	3.4	3.3	3.1	2.2
Côte d'Ivoire	15.4	21.5	26.5	32.2	3.3	2.0	1.4	1.0
Eritrea	3.4	5.7	7.7	10.0	n.a	3.1	2.0	1.3
Ethiopia	59.9	89.8	127.0	186.5	n.a	2.4	2.3	1.9
Gabon	1.2	1.8	2.4	3.2	3.0	2.4	2.1	1.4
Gambia	1.2	1.8	2.2	2.6	3.8	2.2	1.4	0.8
Ghana	18.5	26.4	32.8	40.1	2.9	2.1	1.5	1.0
Guinea	7.9	11.3	15.6	20.7	3.0	2.2	2.2	1.4
Kenya	29.4	40.0	46.9	55.4	3.3	1.8	1.1	0.8
Lesotho	2.0	2.1	2.3	2.5	2.1	0.5	0.4	0.4
Liberia	2.5	5.6	8.9	14.4	0.8	4.8	3.2	2.4
Madagascar	15.1	24.1	34.3	47.0	2.9	2.8	2.4	1.6
Malawi	10.7	15.7	21.8	31.1	3.2	2.2	2.2	1.8
Mali	10.7	17.7	26.9	41.7	2.6	3.0	2.8	2.2
Mauritania	2.5	4.1	6.0	8.5	2.6	3.0	2.6	1.7
Mauritius	1.1	1.3	1.4	1.4	0.9	0.8	0.5	0.1
Mozambique	17.6	23.5	30.4	38.8	2.2	1.7	1.7	1.2
Namibia	1.7	2.3	3.0	3.7	3.3	1.9	1.7	1.0
Niger	10.1	18.5	30.1	51.9	3.3	3.6	3.3	2.8
Nigeria	107.9	165.3	220.4	278.8	2.9	2.5	1.9	1.2
Rwanda	6.4	10.5	14.2	18.5	0.4	2.9	2.0	1.4
Senegal	9.0	13.5	17.9	22.7	2.7	2.5	1.9	1.2
Sierra Leone	4.2	7.1	10.1	14.4	1.4	3.2	2.4	1.7
Somalia	6.8	13.0	20.8	34.6	0.4	3.9	3.2	2.6
Sudan	29.8	42.4	52.6	63.5	2.3	2.1	1.4	1.0
Swaziland	0.9	1.0	1.2	1.4	2.6	0.8	1.0	0.8
Tanzania, United Rep.	33.5	49.3	65.6	82.7	3.3	2.3	1.9	1.2
Togo	4.2	6.6	9.1	11.8	2.8	2.6	2.2	1.3
Uganda	22.0	38.7	62.7	101.5	3.2	3.4	3.3	2.4
Zambia	9.9	14.8	21.3	29.3	3.0	2.4	2.5	1.6
Zimbabwe	12.2	16.4	19.6	23.5	3.0	1.8	1.2	0.9
Sub-Saharan Africa, other	3.0	4.3	5.8	7.5	2.7	2.1	1.9	1.3
Near East/North Africa	377.5	520.0	651.3	809.3	2.6	1.9	1.5	1.1
Afghanistan	20.8	35.6	50.5	72.3	2.4	3.2	2.4	1.8
Algeria	29.2	38.0	44.9	51.2	2.4	1.6	1.1	0.7
Egypt	65.5	84.4	99.5	113.8	2.3	1.5	1.1	0.7
Iran, Islamic Rep.	68.1	87.1	104.5	121.4	3.0	1.5	1.2	0.8
Iraq	21.8	33.5	43.1	53.6	2.9	2.6	1.7	1.1
Jordan	4.7	7.2	9.3	11.7	4.3	2.6	1.8	1.1

	Million				Annual growth rates (% p.a.)			
	1997/99	2015	2030	2050	1980-99	1997/99-2015	2015-30	2030-50
Lebanon	3.4	4.2	4.7	5.0	1.5	1.3	0.8	0.3
Libyan Arab Jamahiriya	5.1	7.1	8.4	10.0	2.6	2.0	1.2	0.8
Morocco	28.8	37.7	44.1	50.4	2.2	1.6	1.1	0.7
Saudi Arabia	18.9	31.7	44.8	59.7	3.7	3.1	2.3	1.4
Syrian Arab Republic	15.4	23.2	29.3	36.3	3.2	2.4	1.6	1.1
Tunisia	9.3	11.3	12.8	14.1	2.0	1.2	0.8	0.5
Turkey	64.6	79.0	89.9	98.8	2.1	1.2	0.9	0.5
Yemen	16.9	33.1	57.5	102.4	4.2	4.0	3.7	2.9
Near East, other	5.1	6.9	7.8	8.6	2.4	1.8	0.9	0.5
Latin America and the Caribbean	498.2	624.2	716.7	798.8	1.8	1.3	0.9	0.5
Argentina	36.1	43.5	48.9	54.5	1.4	1.1	0.8	0.5
Bolivia	8.0	11.2	14.0	17.0	2.3	2.0	1.5	1.0
Brazil	166.1	201.4	226.5	247.2	1.7	1.1	0.8	0.4
Chile	14.8	17.9	20.2	22.2	1.6	1.1	0.8	0.5
Colombia	40.7	52.6	62.1	70.9	2.0	1.5	1.1	0.7
Costa Rica	3.8	5.2	6.2	7.2	2.9	1.8	1.2	0.7
Cuba	11.1	11.6	11.7	10.8	0.8	0.3	0.0	-0.4
Dominican Republic	8.1	10.1	11.2	12.0	1.9	1.3	0.7	0.3
Ecuador	12.2	15.9	18.6	21.2	2.4	1.6	1.1	0.6
El Salvador	6.0	8.0	9.4	10.9	1.6	1.7	1.1	0.7
Guatemala	10.8	16.3	21.2	26.6	2.6	2.5	1.7	1.1
Guyana	0.8	0.7	0.7	0.5	-0.1	0.0	-0.7	-1.4
Haiti	7.9	10.2	12.1	14.0	2.1	1.5	1.2	0.7
Honduras	6.1	8.7	10.7	12.8	3.0	2.1	1.4	0.9
Jamaica	2.5	3.0	3.4	3.8	0.9	0.9	0.9	0.6
Mexico	95.8	119.2	134.9	146.7	1.9	1.3	0.8	0.4
Nicaragua	4.8	7.2	9.3	11.5	2.7	2.4	1.7	1.1
Panama	2.8	3.5	3.9	4.3	2.0	1.3	0.8	0.4
Paraguay	5.2	7.8	10.1	12.6	2.9	2.4	1.8	1.1
Peru	24.8	31.9	37.2	42.1	2.0	1.5	1.0	0.6
Suriname	0.4	0.4	0.4	0.4	0.8	0.3	0.1	-0.3
Trinidad and Tobago	1.3	1.4	1.4	1.4	0.8	0.5	0.2	-0.2
Uruguay	3.3	3.7	4.0	4.2	0.7	0.6	0.5	0.3
Venezuela	23.2	30.9	36.5	42.2	2.4	1.7	1.1	0.7
Latin America, other	1.6	1.8	2.0	2.1	0.8	0.8	0.6	0.2
South Asia	1 282.9	1 672.0	1 969.5	2 258.0	2.1	1.6	1.1	0.7
Bangladesh	131.8	183.2	222.6	265.4	2.4	2.0	1.3	0.9
India	976.3	1 230.5	1 408.9	1 572.1	2.0	1.4	0.9	0.5
Maldives	0.3	0.5	0.6	0.9	3.1	3.0	2.4	1.5
Nepal	22.0	32.1	41.7	52.4	2.3	2.3	1.8	1.1
Pakistan	133.9	204.3	272.7	344.2	2.8	2.5	1.9	1.2
Sri Lanka	18.6	21.5	22.9	23.1	1.4	0.9	0.4	0.0
East Asia	1 839.8	2 128.4	2 302.8	2 365.0	1.5	0.9	0.5	0.1
Cambodia	12.4	18.6	23.8	29.9	3.6	2.4	1.7	1.1
China	1 260.4	1 418.2	1 493.6	1 471.7	1.3	0.7	0.3	-0.1
Indonesia	206.4	250.1	282.9	311.3	1.8	1.1	0.8	0.5
Korea, DPR	21.9	24.4	26.5	28.0	1.4	0.6	0.5	0.3
Korea, Republic	46.1	50.6	52.5	51.6	1.0	0.6	0.2	-0.1
Lao PDR	5.0	7.3	9.3	11.4	2.6	2.2	1.6	1.0
Malaysia	21.4	27.9	33.0	37.9	2.5	1.6	1.1	0.7
Mongolia	2.5	3.1	3.6	4.1	2.3	1.3	1.1	0.6
Myanmar	46.5	55.3	62.5	68.5	1.8	1.0	0.8	0.5

	Million				Annual growth rates (% p.a.)			
	1997/99	**2015**	**2030**	**2050**	**1980-99**	**1997/99-2015**	**2015-30**	**2030-50**
Philippines	72.7	95.9	112.6	128.4	2.3	1.6	1.1	0.7
Thailand	61.2	72.5	79.5	82.5	1.6	1.0	0.6	0.2
Viet Nam	76.1	94.4	110.1	123.8	2.1	1.3	1.0	0.6
East Asia, other	7.3	10.1	12.9	15.8	2.4	2.0	1.6	1.0
Industrial countries	892.4	950.7	978.5	985.7	0.7	0.4	0.2	0.0
Australia	18.7	21.9	24.2	26.5	1.4	0.9	0.7	0.4
Canada	30.2	34.4	37.7	40.4	1.2	0.8	0.6	0.3
European Union	375.5	376.8	367.5	339.5	0.3	0.0	-0.2	-0.4
Iceland	0.3	0.3	0.3	0.3	1.0	0.6	0.4	0.1
Israel	5.8	7.7	8.9	10.1	2.5	1.7	0.9	0.6
Japan	126.5	127.5	121.3	109.2	0.4	0.0	-0.3	-0.5
Malta	0.4	0.4	0.4	0.4	0.9	0.4	0.1	-0.2
New Zealand	3.7	4.1	4.4	4.4	1.0	0.6	0.4	0.1
Norway	4.4	4.7	4.9	4.9	0.5	0.3	0.3	0.0
South Africa	42.1	44.6	43.9	47.3	2.1	0.3	-0.1	0.4
Switzerland	7.2	7.0	6.6	5.6	0.8	-0.2	-0.4	-0.8
United States	277.5	321.2	358.5	397.1	1.0	0.9	0.7	0.5
Transition countries	413.2	397.9	381.4	349.0	0.4	-0.2	-0.3	-0.4
Albania	3.1	3.4	3.8	3.9	0.9	0.5	0.6	0.2
Armenia	3.8	3.8	3.7	3.2	n.a.	0.0	-0.3	-0.7
Azerbaijan	7.9	8.7	9.1	8.9	n.a.	0.6	0.3	-0.1
Belarus	10.3	9.7	9.1	8.3	n.a.	-0.4	-0.4	-0.5
Bosnia and Herzegovina	3.7	4.3	4.1	3.5	n.a.	0.9	-0.4	-0.8
Bulgaria	8.1	6.8	5.8	4.5	-0.5	-1.0	-1.1	-1.2
Croatia	4.7	4.6	4.5	4.2	n.a.	0.0	-0.2	-0.3
Czech Republic	10.3	10.0	9.5	8.4	n.a.	-0.2	-0.4	-0.6
Estonia	1.4	1.2	1.0	0.8	n.a.	-1.1	-1.2	-1.4
Georgia	5.3	4.8	4.2	3.2	n.a.	-0.6	-0.9	-1.3
Hungary	10.1	9.3	8.5	7.5	-0.4	-0.5	-0.5	-0.7
Kazakhstan	16.3	16.0	16.0	15.3	n.a.	-0.1	0.0	-0.2
Kyrgyzstan	4.8	5.8	6.7	7.5	n.a.	1.2	0.9	0.6
Latvia	2.5	2.2	2.0	1.7	n.a.	-0.6	-0.7	-0.7
Lithuania	3.7	3.5	3.3	3.0	n.a.	-0.3	-0.4	-0.6
Macedonia, The Former Yugoslav Rep.	2.0	2.1	2.0	1.9	n.a.	0.2	-0.1	-0.4
Moldova, Republic	4.3	4.2	4.0	3.6	n.a.	-0.2	-0.3	-0.5
Poland	38.6	38.0	36.6	33.4	0.4	-0.1	-0.3	-0.5
Romania	22.5	21.4	20.1	18.1	0.0	-0.3	-0.4	-0.5
Russian Federation	146.8	133.3	121.4	104.3	n.a.	-0.6	-0.6	-0.8
Slovakia	5.4	5.4	5.2	4.7	n.a.	0.0	-0.2	-0.6
Slovenia	2.0	1.9	1.8	1.5	n.a.	-0.2	-0.5	-0.8
Tajikistan	6.0	7.1	8.5	9.8	n.a.	1.0	1.2	0.7
Turkmenistan	4.5	6.1	7.2	8.4	n.a.	1.7	1.2	0.8
Ukraine	50.5	43.3	37.6	30.0	n.a.	-0.9	-0.9	-1.1
Uzbekistan	24.1	30.6	35.7	40.5	n.a.	1.4	1.0	0.6
Yugoslavia	10.6	10.3	9.9	9.0	n.a.	-0.2	-0.3	-0.4

TABLE A2   Per capita food supplies for direct human consumption

| | Calories/day | | | | All cereals, including milled rice (kg/p.a.) | | | |
	1969/71	1979/81	1989/91	1997/99	1969/71	1979/81	1989/91	1997/99
World	2 413	2 552	2 709	2 803	149	160	171	171
All developing countries	2 113	2 312	2 525	2 681	146	162	174	173
Sub-Saharan Africa	2 108	2 089	2 109	2 195	116	115	119	123
Angola	2 105	2 109	1 726	1 879	78	88	56	76
Benin	1 996	2 044	2 338	2 498	82	96	116	115
Botswana	2 144	2 124	2 386	2 278	165	148	152	143
Burkina Faso	1 763	1 683	2 096	2 294	173	156	202	220
Burundi	2 105	2 027	1 850	1 660	46	53	50	43
Cameroon	2 308	2 326	2 170	2 260	115	110	106	107
Central African Republic	2 353	2 319	1 912	1 969	49	42	48	49
Chad	2 076	1 642	1 735	2 120	153	107	109	133
Congo	2 062	2 221	2 152	2 174	30	49	58	75
Congo, Dem. Republic	2 144	2 082	2 119	1 712	41	43	43	40
Côte d'Ivoire	2 512	2 830	2 437	2 566	95	122	102	120
Eritrea	n.a	n.a	n.a	1 709	n.a	n.a	n.a	147
Ethiopia	n.a	n.a	n.a	1 809	n.a	n.a	n.a	134
Gabon	2 198	2 422	2 447	2 517	47	85	93	101
Gambia	2 180	1 805	2 421	2 574	167	132	159	163
Ghana	2 289	1 710	2 032	2 547	76	64	78	85
Guinea	2 203	2 213	2 012	2 197	98	109	106	103
Kenya	2 218	2 179	1 915	1 933	155	149	112	121
Lesotho	1 937	2 220	2 267	2 308	197	206	206	213
Liberia	2 387	2 543	2 286	2 084	115	134	114	91
Madagascar	2 432	2 372	2 105	2 004	143	136	113	107
Malawi	2 358	2 273	1 936	2 116	196	174	157	153
Mali	2 018	1 765	2 319	2 239	169	146	203	193
Mauritania	1 938	2 123	2 584	2 690	107	122	166	171
Mauritius	2 333	2 671	2 844	2 951	144	153	158	161
Mozambique	1 912	1 907	1 767	1 924	74	77	72	98
Namibia	2 191	2 295	2 175	2 091	101	117	126	136
Niger	2 039	2 129	2 044	2 010	224	210	221	205
Nigeria	2 232	2 026	2 410	2 814	122	110	144	154
Rwanda	2 210	2 287	2 054	2 018	51	51	45	55
Senegal	2 282	2 276	2 276	2 284	179	180	185	158
Sierra Leone	2 236	2 109	1 987	2 079	124	123	119	118
Somalia	1 824	1 831	1 761	1 554	78	89	94	56
Sudan	2 055	2 192	2 159	2 366	133	128	150	158
Swaziland	2 312	2 458	2 603	2 551	171	169	168	157
Tanzania, United Rep.	1 733	2 245	2 168	1 926	74	136	126	115
Togo	2 328	2 266	2 324	2 512	114	114	139	154
Uganda	2 388	2 108	2 299	2 185	102	79	72	73
Zambia	2 261	2 271	2 031	1 937	193	199	166	161
Zimbabwe	2 270	2 272	2 115	2 084	202	192	170	163
Sub-Saharan Africa, other	1 852	2 042	2 284	2 268	104	124	136	132
Near East/North Africa	2 371	2 839	3 024	3 006	177	199	212	209
Afghanistan	2 203	2 132	1 890	1 800	211	196	156	144
Algeria	1 841	2 639	2 901	2 934	151	196	214	228
Egypt	2 348	2 915	3 158	3 317	176	212	239	251
Iran, Islamic Rep.	2 094	2 740	2 856	2 928	146	182	194	191
Iraq	2 274	2 849	3 051	2 417	157	198	227	166
Jordan	2 246	2 616	2 815	2 812	151	158	169	174
Lebanon	2 352	2 731	3 182	3 231	135	134	136	136

	Calories/day				All cereals, including milled rice (kg/p.a.)			
	1969/71	1979/81	1989/91	1997/99	1969/71	1979/81	1989/91	1997/99
Libyan Arab Jamahiriya	2 459	3 452	3 252	3 291	148	197	193	197
Morocco	2 474	2 749	3 050	3 031	225	232	253	250
Saudi Arabia	1 898	2 842	2 912	2 957	132	155	166	173
Syrian Arab Republic	2 346	2 974	3 191	3 327	161	175	223	221
Tunisia	2 361	2 833	3 156	3 342	173	204	223	222
Turkey	3 038	3 302	3 547	3 487	210	225	229	222
Yemen	1 769	1 949	2 051	2 040	151	151	166	165
Near East, other	2 901	3 112	2 906	3 201	134	128	119	124
Latin America and the Caribbean	2 475	2 702	2 699	2 824	119	130	131	132
Argentina	3 271	3 207	2 962	3 166	138	132	144	144
Bolivia	1 999	2 133	2 162	2 223	98	104	113	109
Brazil	2 427	2 677	2 783	2 972	100	117	115	114
Chile	2 659	2 665	2 541	2 856	159	156	145	141
Colombia	1 948	2 294	2 410	2 578	76	84	92	102
Costa Rica	2 373	2 587	2 717	2 767	107	112	113	119
Cuba	2 661	2 884	3 027	2 453	123	131	124	100
Dominican Republic	2 022	2 273	2 273	2 322	59	86	83	86
Ecuador	2 160	2 359	2 492	2 702	82	91	108	112
El Salvador	1 848	2 296	2 446	2 492	117	141	153	152
Guatemala	2 082	2 291	2 418	2 230	142	152	156	142
Guyana	2 283	2 504	2 374	2 558	123	146	150	144
Haiti	1 947	2 042	1 778	1 926	91	91	87	100
Honduras	2 145	2 120	2 307	2 367	135	129	133	124
Jamaica	2 524	2 647	2 594	2 740	115	113	110	108
Mexico	2 697	3 136	3 101	3 148	173	183	187	188
Nicaragua	2 334	2 272	2 226	2 235	127	119	129	128
Panama	2 333	2 273	2 305	2 464	128	102	117	124
Paraguay	2 571	2 545	2 452	2 575	96	85	86	95
Peru	2 250	2 131	2 026	2 552	90	106	102	127
Suriname	2 240	2 397	2 487	2 611	141	141	157	142
Trinidad and Tobago	2 507	2 957	2 712	2 696	136	145	139	130
Uruguay	2 964	2 866	2 576	2 844	137	144	136	134
Venezuela	2 342	2 760	2 391	2 280	131	150	134	132
Latin America, other	2 448	2 683	2 735	2 766	100	110	109	109
South Asia	2 067	2 084	2 334	2 403	151	151	165	163
Bangladesh	2 122	1 975	2 065	2 123	173	167	177	181
India	2 041	2 083	2 367	2 434	147	150	164	160
Maldives	1 624	2 165	2 366	2 365	92	152	143	125
Nepal	1 848	1 892	2 443	2 292	165	163	208	190
Pakistan	2 240	2 194	2 352	2 477	161	147	153	159
Sri Lanka	2 289	2 348	2 217	2 351	141	139	141	142
East Asia	2 016	2 321	2 628	2 921	152	181	199	199
Cambodia	2 089	1 714	1 836	1 975	182	157	165	170
China	1 995	2 327	2 684	3 040	153	189	211	208
Indonesia	1 860	2 215	2 642	2 903	123	153	185	202
Korea, DPR	2 200	2 442	2 518	2 083	171	186	181	164
Korea, Republic	2 761	2 960	3 008	3 051	228	229	218	214
Lao PDR	2 079	2 074	2 106	2 149	209	200	198	203
Malaysia	2 530	2 723	2 721	2 925	160	151	125	142
Mongolia	2 225	2 379	2 204	2 004	140	154	136	125
Myanmar	2 040	2 327	2 620	2 788	163	191	217	216
Philippines	1 808	2 229	2 329	2 332	115	133	147	138

| | Calories/day | | | | All cereals, including milled rice (kg/p.a.) | | | |
	1969/71	1979/81	1989/91	1997/99	1969/71	1979/81	1989/91	1997/99
Thailand	2 146	2 268	2 176	2 414	156	155	122	121
Viet Nam	2 130	2 128	2 218	2 503	175	158	166	186
East Asia, other	2 132	2 337	2 368	2 375	63	78	84	98
Industrial countries	3 043	3 135	3 293	3 380	132	139	154	159
Australia	3 239	3 059	3 212	3 141	129	116	114	109
Canada	2 893	2 899	3 004	3 145	123	115	118	143
European Union	3 168	3 275	3 396	3 431	133	133	132	132
Iceland	2 952	3 286	3 115	3 260	93	91	102	107
Israel	3 141	3 147	3 377	3 561	153	158	156	152
Japan	2 697	2 709	2 821	2 779	157	142	142	139
Malta	3 158	3 276	3 264	3 475	171	165	153	170
New Zealand	2 974	3 100	3 191	3 130	101	106	106	105
Norway	3 028	3 315	3 172	3 386	101	117	130	137
South Africa	2 760	2 822	2 864	2 837	185	192	193	191
Switzerland	3 440	3 457	3 308	3 279	130	125	132	131
United States	3 034	3 193	3 475	3 711	115	145	194	207
Transition countries	3 323	3 389	3 285	2 906	201	189	179	173
Albania	2 400	2 695	2 565	2 683	203	224	199	169
Armenia	n.a	n.a	n.a	2 158	n.a	n.a	n.a	159
Azerbaijan	n.a	n.a	n.a	2 133	n.a	n.a	n.a	182
Belarus	n.a	n.a	n.a	3 206	n.a	n.a	n.a	166
Bosnia and Herzegovina	n.a	n.a	n.a	2 934	n.a	n.a	n.a	301
Bulgaria	3 501	3 616	3 462	2 795	267	253	229	173
Croatia	n.a	n.a	n.a	2 539	n.a	n.a	n.a	121
Czech Republic	n.a	n.a	n.a	3 242	n.a	n.a	n.a	164
Estonia	n.a	n.a	n.a	3 075	n.a	n.a	n.a	200
Georgia	n.a	n.a	n.a	2 399	n.a	n.a	n.a	182
Hungary	3 318	3 448	3 670	3 414	185	172	178	143
Kazakhstan	n.a	n.a	n.a	2 610	n.a	n.a	n.a	194
Kyrgyzstan	n.a	n.a	n.a	2 734	n.a	n.a	n.a	217
Latvia	n.a	n.a	n.a	2 930	n.a	n.a	n.a	170
Lithuania	n.a	n.a	n.a	3 013	n.a	n.a	n.a	196
Macedonia, The Former Yugoslav Rep.	n.a	n.a	n.a	2 858	n.a	n.a	n.a	175
Moldova, Republic	n.a	n.a	n.a	2 719	n.a	n.a	n.a	252
Poland	3 426	3 528	3 379	3 343	201	190	166	170
Romania	2 973	3 205	3 016	3 258	203	188	185	220
Russian Federation	n.a	n.a	n.a	2 862	n.a	n.a	n.a	158
Slovakia	n.a	n.a	n.a	3 079	n.a	n.a	n.a	158
Slovenia	n.a	n.a	n.a	3 019	n.a	n.a	n.a	153
Tajikistan	n.a	n.a	n.a	1 975	n.a	n.a	n.a	169
Turkmenistan	n.a	n.a	n.a	2 660	n.a	n.a	n.a	212
Ukraine	n.a	n.a	n.a	2 829	n.a	n.a	n.a	165
Uzbekistan	n.a	n.a	n.a	2 912	n.a	n.a	n.a	232
Yugoslavia	n.a	n.a	n.a	2 910	n.a	n.a	n.a	129

TABLE A3 Cereal sector data (including rice in milled form)

	Production ('000 tonnes)			Net trade ('000 tonnes)		
	1979/81	1989/91	1997/99	1979/81	1989/91	1997/99
World	1 441 855	1 732 115	1 888 558	3 017	3 878	9 107
All developing countries	649 212	868 214	1 025 962	-66 458	-89 120	-102 543
Sub-Saharan Africa	40 698	58 255	70 855	-7 974	-8 410	-13 670
Angola	367	297	536	-363	-318	-461
Benin	362	563	876	-68	-205	-164
Botswana	35	60	21	-122	-125	-194
Burkina Faso	1 152	1 961	2 442	-98	-177	-236
Burundi	216	283	258	-28	-29	-27
Cameroon	850	885	1 261	-199	-380	-405
Central African Republic	99	100	147	-18	-39	-45
Chad	495	645	1 124	-24	-48	-51
Congo	14	11	3	-74	-103	-225
Congo, Dem. Republic	821	1 343	1 540	-446	-371	-491
Côte d'Ivoire	716	1 021	1 480	-531	-584	-848
Eritrea	n.a	n.a	288	n.a	n.a	-237
Ethiopia	n.a	n.a	8 359	n.a	n.a	-487
Gabon	10	23	32	-49	-80	-112
Gambia	57	92	113	-47	-88	-150
Ghana	697	1 121	1 648	-209	-352	-314
Guinea	532	490	714	-138	-291	-283
Kenya	2 268	2 878	2 808	-137	-175	-1 028
Lesotho	198	170	184	-168	-212	-283
Liberia	169	128	129	-108	-134	-179
Madagascar	1 494	1 748	1 879	-209	-95	-151
Malawi	1 328	1 543	1 944	-39	-190	-277
Mali	1 026	1 995	2 316	-98	-76	-126
Mauritania	44	114	147	-150	-249	-543
Mauritius	1	2	0	-169	-208	-244
Mozambique	625	602	1 618	-360	-521	-455
Namibia	73	103	100	-55	-98	-152
Niger	1 692	2 095	2 473	-68	-117	-187
Nigeria	7 085	17 098	21 009	-2 073	-616	-2 064
Rwanda	270	286	196	-20	-30	-191
Senegal	818	939	762	-500	-632	-801
Sierra Leone	374	396	264	-106	-168	-308
Somalia	299	492	241	-258	-216	-158
Sudan	2 928	2 771	4 306	3	-676	-578
Swaziland	91	126	117	-74	-108	-68
Tanzania, United Rep.	2 927	3 907	3 728	-211	-69	-437
Togo	296	495	682	-60	-89	-201
Uganda	1 166	1 580	1 758	-36	-7	-145
Zambia	990	1 463	994	-342	-107	-245
Zimbabwe	2 273	2 391	2 180	139	458	190
Sub-Saharan Africa, other	104	152	183	-166	-244	-311
Near East/North Africa	57 921	76 871	83 494	-23 922	-39 088	-48 518
Afghanistan	3 922	2 645	3 511	-91	-239	-228
Algeria	1 957	2 481	1 812	-2 979	-5 854	-5 865
Egypt	7 343	11 641	16 723	-5 949	-7 916	-9 761
Iran, Islamic Rep.	8 391	12 285	15 502	-2 702	-6 272	-7 191
Iraq	1 749	2 469	2 025	-2 687	-3 324	-3 169

Self-sufficiency ratio (%)			Domestic use ('000 tonnes)			Food	Feed
						% of domestic use	
1979/81	1989/91	1997/99	1979/81	1989/91	1997/99	1997/99	1997/99
100	100	101	1 437 144	1 728 298	1 864 356	54	35
91	91	91	711 956	951 675	1 128 746	70	20
85	87	82	47 838	67 297	86 256	82	5
52	49	51	707	608	1 044	90	2
81	74	87	445	757	1 013	68	3
23	28	9	153	211	240	90	2
94	96	91	1 221	2 048	2 680	90	0
89	91	89	244	312	292	91	0
80	66	77	1 066	1 332	1 645	92	0
87	65	77	113	155	192	91	0
87	86	96	569	749	1 174	84	2
17	8	1	84	132	216	98	0
68	80	75	1 213	1 690	2 041	95	0
58	65	66	1 245	1 577	2 245	82	3
n.a	n.a	53	n.a	n.a	541	93	0
n.a	n.a	94	n.a	n.a	8 872	91	1
15	22	21	68	105	154	77	18
59	50	45	97	184	253	79	3
81	77	84	864	1 457	1 962	80	3
82	63	72	653	780	997	81	0
83	99	73	2 720	2 922	3 863	92	3
59	43	38	337	395	490	86	5
61	49	41	277	262	313	73	15
88	95	92	1 704	1 843	2 037	79	5
100	90	92	1 330	1 720	2 121	78	10
90	98	94	1 144	2 032	2 459	84	0
22	31	21	203	371	689	62	6
1	1	0	162	201	239	77	22
63	54	80	985	1 106	2 020	85	9
57	53	36	128	195	281	81	9
107	93	91	1 576	2 248	2 724	76	4
77	97	91	9 154	17 716	23 073	72	8
93	87	49	289	328	397	90	0
61	60	48	1 336	1 561	1 606	88	1
74	71	43	505	559	610	81	0
52	70	56	571	708	432	88	0
104	69	83	2 816	4 008	5 195	90	3
57	59	58	160	212	202	69	15
96	101	85	3 050	3 883	4 405	88	3
85	83	79	350	597	860	76	2
94	99	88	1 245	1 597	1 994	80	9
77	99	58	1 286	1 483	1 723	93	2
120	116	88	1 892	2 069	2 471	80	12
39	38	37	270	401	497	81	7
72	69	63	80 342	111 934	132 743	59	26
98	92	94	4 013	2 883	3 739	80	5
40	33	21	4 865	7 638	8 655	77	14
55	60	69	13 330	19 286	24 305	68	21
74	65	69	11 314	18 939	22 425	58	33
42	40	40	4 206	6 238	5 061	71	18

	Production ('000 tonnes)			Net trade ('000 tonnes)			
	1979/81	1989/91	1997/99	1979/81	1989/91	1997/99	
Jordan	88	105	72	-494	-1 114	-1 684	
Lebanon	41	80	92	-580	-607	-759	
Libyan Arab Jamahiriya	225	284	232	-713	-2 127	-2 029	
Morocco	3 575	7 453	4 853	-2 044	-1 637	-3 692	
Saudi Arabia	303	4 214	2 331	-3 219	-3 919	-6 417	
Syrian Arab Republic	3 069	2 601	4 292	-675	-1 700	-390	
Tunisia	1 146	1 611	1 514	-890	-1 336	-1 851	
Turkey	25 127	28 198	29 726	648	-515	-517	
Yemen	897	693	726	-656	-1 405	-2 370	
Near East, other	88	111	83	-891	-1 124	-2 595	
Latin America and the Caribbean	86 648	96 664	125 499	-8 456	-11 631	-14 254	
Argentina	24 323	19 775	35 597	14 371	9 758	22 297	
Bolivia	632	804	1 083	-322	-241	-202	
Brazil	27 964	34 601	41 279	-6 295	-4 424	-9 292	
Chile	1 700	2 949	2 753	-1 144	-212	-1 432	
Colombia	2 729	3 429	2 486	-662	-794	-3 365	
Costa Rica	262	199	208	-123	-378	-688	
Cuba	399	388	379	-2 146	-2 256	-1 456	
Dominican Republic	320	379	396	-377	-703	-1 206	
Ecuador	560	1 138	1 360	-333	-458	-697	
El Salvador	700	764	796	-130	-281	-347	
Guatemala	1 109	1 398	1 092	-216	-330	-725	
Guyana	179	146	381	31	-1	219	
Haiti	380	363	407	-193	-291	-492	
Honduras	481	648	627	-111	-199	-397	
Jamaica	6	3	3	-394	-389	-512	
Mexico	20 208	23 412	28 390	-5 893	-6 735	-10 996	
Nicaragua	349	415	515	-129	-154	-215	
Panama	194	266	232	-95	-158	-364	
Paraguay	456	789	1 222	-74	19	206	
Peru	1 234	1 664	2 383	-1 468	-1 557	-2 706	
Suriname	172	153	130	56	-1	29	
Trinidad and Tobago	10	12	10	-251	-259	-209	
Uruguay	916	1 077	1 733	181	523	973	
Venezuela	1 338	1 862	1 988	-2 536	-1 888	-2 432	
Latin America, other	28	32	50	-207	-224	-248	
South Asia	147 495	203 035	239 380	-1 764	-3 221	-3 412	
Bangladesh	14 281	19 047	22 332	-1 375	-1 801	-2 857	
India	113 354	158 418	185 762	383	416	1 642	
Maldives	0	0	0	-26	-36	-34	
Nepal	2 853	4 557	5 187	-5	-11	-7	
Pakistan	15 574	19 419	24 332	109	-808	-913	
Sri Lanka	1 433	1 594	1 767	-850	-983	-1 243	
East Asia	316 452	433 393	506 738	-24 344	-26 776	-22 696	
Cambodia	918	1 751	2 487	-187	-15	-69	
China	238 016	328 034	386 157	-16 783	-15 471	-2 657	
Indonesia	23 758	36 319	42 608	-2 810	-2 130	-6 585	
Korea, DPR	5 033	7 002	3 073	-280	-830	-1 374	
Korea, Republic	6 194	5 847	5 132	-5 742	-9 904	-11 482	
Lao PDR	715	984	1 304	-80	-15	-34	

Self-sufficiency ratio (%)			Domestic use ('000 tonnes)			Food	Feed
						% of domestic use	
1979/81	1989/91	1997/99	1979/81	1989/91	1997/99	1997/99	1997/99
16	9	4	548	1 155	1 788	45	49
7	12	10	634	671	943	49	39
23	13	10	961	2 170	2 375	42	40
60	90	55	5 958	8 293	8 879	81	5
11	68	23	2 793	6 173	10 044	33	64
93	54	77	3 312	4 845	5 608	61	25
58	56	45	1 978	2 885	3 366	61	30
104	104	99	24 100	27 218	30 177	48	22
64	33	24	1 412	2 130	3 035	92	5
10	8	4	921	1 412	2 346	27	47
93	86	88	93 645	112 365	141 977	46	42
223	176	243	10 892	11 220	14 624	36	44
70	76	83	908	1 057	1 312	66	22
84	86	81	33 487	40 197	51 059	37	53
61	90	66	2 812	3 272	4 149	50	44
83	80	42	3 299	4 270	5 914	70	28
66	36	23	398	556	910	50	45
16	15	21	2 545	2 643	1 845	60	34
47	34	26	685	1 107	1 497	47	49
65	72	66	868	1 588	2 057	66	22
84	75	69	834	1 016	1 155	79	16
82	85	57	1 357	1 647	1 902	80	16
126	98	248	142	150	154	71	23
68	55	45	560	662	906	88	6
82	79	61	587	826	1 037	73	18
2	1	1	413	411	470	58	38
81	75	72	24 839	31 064	39 416	46	40
80	70	71	436	593	730	84	5
71	67	39	272	398	595	57	37
93	98	109	490	803	1 126	44	36
45	51	47	2 721	3 255	5 028	62	32
144	102	115	120	149	112	52	25
4	4	5	264	277	210	79	30
130	156	204	703	692	849	52	34
35	44	43	3 786	4 259	4 625	66	23
12	12	17	229	257	300	58	29
98	100	102	150 728	203 503	233 800	89	1
91	90	87	15 670	21 115	25 661	93	0
98	102	106	116 147	156 091	175 526	89	1
0	0	0	25	34	36	96	0
102	100	101	2 798	4 557	5 159	81	5
112	102	99	13 931	19 131	24 501	87	5
66	62	61	2 158	2 577	2 917	90	3
93	95	95	339 409	456 583	534 020	68	22
78	95	100	1 174	1 848	2 482	85	2
93	97	98	255 143	339 323	394 727	66	24
92	95	88	25 785	38 449	48 577	86	4
92	92	69	5 459	7 603	4 479	80	4
53	38	30	11 693	15 402	17 136	57	44
97	103	108	740	954	1 204	85	6

	Production ('000 tonnes)			Net trade ('000 tonnes)			
	1979/81	1989/91	1997/99	1979/81	1989/91	1997/99	
Malaysia	1 378	1 270	1 408	-1 529	-2 833	-4 325	
Mongolia	320	719	202	-133	-48	-138	
Myanmar	8 778	9 562	11 846	650	200	111	
Philippines	8 362	11 129	11 274	-926	-2 191	-4 171	
Thailand	14 666	17 165	19 974	5 193	5 870	5 564	
Viet Nam	8 286	13 588	21 251	-1 328	1 162	3 240	
East Asia, other	31	27	25	-390	-571	-777	
Industrial countries	550 609	581 479	652 495	110 787	129 505	110 746	
Australia	20 921	21 110	32 479	14 612	14 973	22 676	
Canada	42 727	52 915	5 444	20 063	22 883	20 730	
European Union	160 790	191 289	207 627	-8 391	29 643	23 741	
Iceland	0	0	0	-50	-62	-79	
Israel	239	331	166	-1 686	-2 167	-3 101	
Japan	9 883	9 721	8 605	-24 465	-28 105	-28 149	
Malta	8	8	11	-120	-147	-194	
New Zealand	789	783	908	5	-150	-210	
Norway	1 130	1 410	1 333	-780	-491	-591	
South Africa	14 188	12 733	11 117	3 308	1 386	-110	
Switzerland	843	1 331	1 181	-1 407	-814	-653	
United States	299 095	289 850	337 625	109 702	92 559	76 688	
Transition countries	242 070	282 483	210 207	-41 309	-36 501	904	
Albania	912	790	583	-23	-172	-328	
Armenia	n.a.	n.a.	291	n.a.	n.a.	-400	
Azerbaijan	n.a.	n.a.	1 018	n.a.	n.a.	-663	
Belarus	n.a.	n.a.	4 609	n.a.	n.a.	-1 202	
Bosnia and Herzegovina	n.a.	n.a.	1 281	n.a.	n.a.	-579	
Bulgaria	8 105	8 863	5 540	-375	-491	446	
Croatia	n.a.	n.a.	3 091	n.a.	n.a.	-38	
Czech Republic	n.a.	n.a.	6 869	n.a.	n.a.	212	
Estonia	n.a.	n.a.	543	n.a.	n.a.	-179	
Georgia	n.a.	n.a.	750	n.a.	n.a.	-380	
Hungary	12 989	14 594	12 858	769	1 296	3 344	
Kazakhstan	n.a.	n.a.	10 919	n.a.	n.a.	3 653	
Kyrgyzstan	n.a.	n.a.	1 609	n.a.	n.a.	-127	
Latvia	n.a.	n.a.	928	n.a.	n.a.	-85	
Lithuania	n.a.	n.a.	2 570	n.a.	n.a.	98	
Macedonia, The Former Yugoslav Rep.	n.a.	n.a.	662	n.a.	n.a.	-320	
Moldova, Republic	n.a.	n.a.	2 670	n.a.	n.a.	161	
Poland	18 466	27 594	26 103	-7 340	-1 473	-1 194	
Romania	18 093	18 268	18 196	-910	-884	599	
Russian Federation	n.a.	n.a.	62 407	n.a.	n.a.	-3 588	
Slovakia	n.a.	n.a.	3 373	n.a.	n.a.	395	
Slovenia	n.a.	n.a.	523	n.a.	n.a.	-551	
Tajikistan	n.a.	n.a.	479	n.a.	n.a.	-443	
Turkmenistan	n.a.	n.a.	1 193	n.a.	n.a.	-159	
Ukraine	n.a.	n.a.	27 991	n.a.	n.a.	3 965	
Uzbekistan	n.a.	n.a.	3 942	n.a.	n.a.	-2 039	
Yugoslavia	n.a.	n.a.	9 212	n.a.	n.a.	306	

Self-sufficiency ratio (%)			Domestic use ('000 tonnes)			Food % of domestic use	Feed % of domestic use
1979/81	1989/91	1997/99	1979/81	1989/91	1997/99	1997/99	1997/99
48	32	25	2 893	3 988	5 726	53	43
70	99	56	456	730	359	86	1
117	94	101	7 520	10 186	11 755	85	6
90	86	75	9 267	13 006	15 041	67	28
155	144	142	9 475	11 891	14 054	53	37
88	108	120	9 392	12 634	17 701	80	8
7	5	3	414	571	780	91	2
129	127	124	428 028	459 763	524 961	27	63
340	287	335	6 161	7 365	9 701	21	59
184	218	164	23 202	24 321	31 447	14	77
95	123	115	168 642	155 831	180 333	27	62
0	0	0	50	62	79	37	47
12	13	6	1 935	2 470	2 854	31	57
28	25	24	35 215	38 310	36 646	48	46
6	5	5	133	158	205	32	58
103	85	92	765	916	992	39	53
59	81	70	1 920	1 747	1 903	32	64
150	110	88	9 465	11 601	12 583	64	31
37	62	62	2 255	2 150	1 897	50	46
168	135	137	178 287	214 834	246 321	23	66
81	89	100	297 258	316 972	210 729	34	50
96	81	61	951	980	954	55	26
n.a.	n.a.	40	n.a.	n.a.	734	82	9
n.a.	n.a.	61	n.a.	n.a.	1 681	86	7
n.a.	n.a.	79	n.a.	n.a.	5 802	29	55
n.a.	n.a.	69	n.a.	n.a.	1 860	60	16
99	96	111	8 169	9 230	4 988	28	56
n.a.	n.a.	100	n.a.	n.a.	3 086	18	74
n.a.	n.a.	103	n.a.	n.a.	6 655	25	63
n.a.	n.a.	70	n.a.	n.a.	775	37	51
n.a.	n.a.	66	n.a.	n.a.	1 130	86	7
107	114	136	12 134	12 804	9 475	15	73
n.a.	n.a.	167	n.a.	n.a.	6 527	49	15
n.a.	n.a.	93	n.a.	n.a.	1 736	60	28
n.a.	n.a.	92	n.a.	n.a.	1 013	41	44
n.a.	n.a.	97	n.a.	n.a.	2 662	27	51
n.a.	n.a.	67	n.a.	n.a.	993	35	26
n.a.	n.a.	106	n.a.	n.a.	2 511	43	45
72	98	93	25 612	28 196	28 058	23	63
94	103	102	19 338	17 736	17 765	28	64
n.a.	n.a.	94	n.a.	n.a.	66 515	35	44
n.a.	n.a.	112	n.a.	n.a.	3 024	28	60
n.a.	n.a.	49	n.a.	n.a.	1 064	29	59
n.a.	n.a.	44	n.a.	n.a.	1 084	93	3
n.a.	n.a.	87	n.a.	n.a.	1 369	70	24
n.a.	n.a.	116	n.a.	n.a.	24 242	34	51
n.a.	n.a.	64	n.a.	n.a.	6 125	91	4
n.a.	n.a.	103	n.a.	n.a.	8 907	15	67

TABLE A4.1 Wheat: area, yield and production (countries with more than 10 000 ha in 1997/99)

	Harvested area ('000 ha)			Yield (kg/ha)			Production ('000 tonnes)		
	1979/81	1989/91	1997/99	1979/81	1989/91	1997/99	1979/81	1989/91	1997/99
World	234 836	227 060	226 426	1 863	2 462	2 637	437 442	558 997	596 973
All developing countries	96 032	102 379	111 027	1 633	2 284	2 524	156 803	233 798	280 235
Sub-Saharan Africa	1 030	1 278	1 557	1 304	1 591	1 619	1 343	2 033	2 521
Burundi	9	11	11	674	764	782	6	9	9
Eritrea	n.a.	n.a.	19	n.a.	n.a.	822	n.a.	n.a.	16
Ethiopia	n.a.	n.a.	903	n.a.	n.a.	1 250	n.a.	n.a.	1 129
Kenya	106	120	130	2 011	1 746	1 794	212	210	234
Lesotho	28	40	22	936	584	1 183	26	23	26
Togo	10	53	41	2 400	1 063	2 134	24	57	88
Sudan	205	295	249	998	1 515	1 888	205	447	470
Tanzania, United Rep.	57	49	70	1 605	1 738	1 526	91	85	108
Zambia	3	12	11	3 473	4 477	6 608	9	56	75
Zimbabwe	37	51	56	4 785	5 717	5 389	179	290	300
Near East/North Africa	25 353	26 786	27 217	1 346	1 703	1 830	34 122	45 628	49 795
Afghanistan	2 065	1 623	2 112	1 240	1 063	1 270	2 561	1 725	2 681
Algeria	1 943	1 463	1 592	654	859	847	1 270	1 257	1 347
Egypt	577	799	1 021	3 193	4 980	5 973	1 844	3 978	6 096
Iran, Islamic Rep.	5 858	6 364	5 739	997	1 195	1 782	5 843	7 605	10 224
Iraq	1 215	1 200	1 521	703	879	674	854	1 055	1 026
Jordan	110	54	36	606	1 224	808	67	66	29
Lebanon	26	26	28	1 260	2 117	2 102	32	56	58
Libyan Arab Jamahiriya	251	146	160	497	1 013	1 020	125	148	163
Morocco	1 673	2 663	2 757	897	1 562	1 070	1 500	4 160	2 950
Saudi Arabia	71	816	423	2 254	4 523	4 390	160	3 689	1 858
Syrian Arab Republic	1 383	1 283	1 695	1 358	1 359	1 934	1 878	1 743	3 278
Tunisia	887	830	896	944	1 337	1 349	837	1 109	1 210
Turkey	9 208	9 419	9 130	1 853	2 005	2 050	17 058	18 887	18 717
Yemen	77	93	100	1 066	1 497	1 441	82	139	145
Latin America and the Caribbean	10 080	10 115	8 885	1 495	2 076	2 529	15 073	21 001	22 469
Argentina	5 245	5 210	5 650	1 537	1 976	2 508	8 060	10 292	14 167
Bolivia	98	92	169	661	793	886	65	73	149
Brazil	2 958	2 671	1 395	883	1 443	1 720	2 613	3 855	2 399
Chile	513	530	380	1 721	3 191	3 996	882	1 691	1 519
Colombia	36	50	20	1 397	1 851	2 113	50	93	42
Ecuador	33	38	28	1 042	709	696	35	27	20
Mexico	723	1 020	727	3 813	4 040	4 571	2 755	4 122	3 321
Paraguay	59	196	172	1 368	1 633	1 240	80	320	214
Peru	98	100	123	1 012	1 113	1 194	99	112	147
Uruguay	281	190	215	1 341	2 018	2 237	377	383	480
South Asia	30 030	32 937	36 339	1 550	2 103	2 459	46 558	69 277	89 370
Bangladesh	430	584	798	1 869	1 665	2 190	803	972	1 748
India	22 364	23 926	26 660	1 545	2 217	2 582	34 550	53 031	68 825
Nepal	372	599	649	1 195	1 404	1 637	444	840	1 063
Pakistan	6 865	7 829	8 231	1 568	1 844	2 155	10 760	14 433	17 734
East Asia	29 539	31 263	37 030	2 021	3 066	3 135	59 707	95 860	116 081
China	28 930	30 515	36 569	2 046	3 113	3 162	59 196	94 998	115 632
Korea, DPR	85	87	69	1 286	1 462	2 182	109	127	151

	Harvested area ('000 ha)			Yield (kg/ha)			Production ('000 tonnes)		
	1979/81	1989/91	1997/99	1979/81	1989/91	1997/99	1979/81	1989/91	1997/99
Mongolia	414	531	296	617	1 144	673	255	607	199
Myanmar	90	129	93	923	974	986	83	126	92
Industrial countries	70 721	67 600	64 553	2 377	2 928	3 345	168 121	197 963	215 947
Australia	11 440	8 468	11 441	1 265	1 568	1 914	14 468	13 279	21 901
Canada	11 386	13 992	10 819	1 794	2 116	2 319	20 430	29 613	25 087
European Union	16 744	17 485	17 223	3 790	5 137	5 728	63 468	89 810	98 655
Israel	96	89	59	2 095	2 519	1 493	201	224	89
Japan	188	261	163	3 032	3 442	3 533	571	898	575
New Zealand	85	37	50	3 643	4 579	6 302	309	168	313
Norway	15	46	60	4 230	4 383	4 755	63	203	287
South Africa	1 781	1 614	949	1 107	1 210	2 086	1 972	1 954	1 981
Switzerland	88	99	96	4 655	6 113	5 938	409	604	572
United States	28 898	25 508	23 691	2 292	2 399	2 806	66 229	61 204	66 477
Transition countries	68 083	57 082	50 848	1 653	2 229	1 982	112 518	127 236	100 792
Albania	196	185	129	2 514	2 741	2 734	492	508	352
Armenia	n.a.	n.a.	113	n.a.	n.a.	1 869	n.a.	n.a.	211
Azerbaijan	n.a.	n.a.	488	n.a.	n.a.	1 755	n.a.	n.a.	857
Belarus	n.a.	n.a.	359	n.a.	n.a.	2 085	n.a.	n.a.	748
Bosnia and Herzegovina	n.a.	n.a.	95	n.a.	n.a.	3 124	n.a.	n.a.	295
Bulgaria	986	1 167	1 107	3 937	4 346	2 836	3 881	5 071	3 138
Croatia	n.a.	n.a.	207	n.a.	n.a.	3 894	n.a.	n.a.	804
Czech Republic	n.a.	n.a.	868	n.a.	n.a.	4 420	n.a.	n.a.	3 838
Estonia	n.a.	n.a.	61	n.a.	n.a.	1 728	n.a.	n.a.	106
Georgia	n.a.	n.a.	139	n.a.	n.a.	1 584	n.a.	n.a.	221
Hungary	1 187	1 207	1 055	4 043	5 177	4 044	4 800	6 249	4 265
Kazakhstan	n.a.	n.a.	9 508	n.a.	n.a.	875	n.a.	n.a.	8 314
Kyrgyzstan	n.a.	n.a.	494	n.a.	n.a.	2 420	n.a.	n.a.	1 196
Latvia	n.a.	n.a.	150	n.a.	n.a.	2 520	n.a.	n.a.	377
Lithuania	n.a.	n.a.	356	n.a.	n.a.	2 834	n.a.	n.a.	1 010
Macedonia, The Former Yugoslav Rep.	n.a.	n.a.	115	n.a.	n.a.	2 928	n.a.	n.a.	336
Moldova, Republic	n.a.	n.a.	370	n.a.	n.a.	2 759	n.a.	n.a.	1 020
Poland	1 525	2 305	2 590	2 746	3 870	3 447	4 189	8 919	8 927
Romania	2 154	2 242	2 022	2 494	3 063	2 802	5 371	6 868	5 665
Russian Federation	n.a.	n.a.	21 211	n.a.	n.a.	1 607	n.a.	n.a.	34 088
Slovakia	n.a.	n.a.	380	n.a.	n.a.	4 286	n.a.	n.a.	1 627
Slovenia	n.a.	n.a.	33	n.a.	n.a.	4 250	n.a.	n.a.	142
Tajikistan	n.a.	n.a.	340	n.a.	n.a.	1 155	n.a.	n.a.	393
Turkmenistan	n.a.	n.a.	490	n.a.	n.a.	2 351	n.a.	n.a.	1 153
Ukraine	n.a.	n.a.	6 027	n.a.	n.a.	2 595	n.a.	n.a.	15 642
Uzbekistan	n.a.	n.a.	1 404	n.a.	n.a.	2 442	n.a.	n.a.	3 428
Yugoslavia	n.a.	n.a.	739	n.a.	n.a.	3 573	n.a.	n.a.	2 641

TABLE A4.2 Rice: area, yield and production (countries with more than 10 000 ha in 1997/99)

	Harvested area ('000 ha)			Yield (kg/ha)			Production ('000 tonnes)		
	1979/81	1989/91	1997/99	1979/81	1989/91	1997/99	1979/81	1989/91	1997/99
World	143 662	147 403	161 398	2 742	3 513	3 640	393 930	517 894	587 408
All developing countries	138 818	143 062	157 190	2 656	3 445	3 575	368 692	492 848	561 877
Sub-Saharan Africa	4 537	5 879	7 164	1 347	1 659	1 629	6 112	9 750	11 667
Angola	12	4	22	919	833	871	11	3	19
Benin	8	7	16	1 172	1 371	2 075	10	10	34
Burkina Faso	39	21	50	1 141	2 090	1 853	44	43	92
Burundi	4	12	17	2 367	3 249	3 186	10	40	55
Cameroon	21	14	21	2 273	4 888	3 249	48	69	69
Central African Republic	14	7	13	996	1 517	1 449	14	10	19
Chad	43	41	85	895	2 349	1 425	39	97	121
Congo, Dem. Republic	293	479	481	806	805	740	236	385	356
Côte d'Ivoire	383	581	767	1 171	1 138	1 584	448	661	1 215
Gambia	22	14	16	1 676	1 501	1 523	37	21	24
Ghana	107	70	126	838	1 451	1 588	89	102	200
Guinea	486	377	494	900	1 136	1 505	438	428	743
Kenya	8	13	18	4 654	3 456	2 742	39	45	50
Liberia	203	179	151	1 252	1 069	1 277	254	191	193
Madagascar	1 182	1 150	1 202	1 738	2 070	2 119	2 055	2 381	2 547
Malawi	37	29	42	1 074	1 740	1 788	39	51	76
Mali	165	230	337	1 026	1 556	2 093	169	358	706
Mauritania	3	15	22	3 681	3 375	4 233	12	50	94
Mozambique	92	109	172	811	758	1 080	74	83	186
Niger	20	22	30	1 535	3 367	1 890	31	75	57
Nigeria	517	1 504	2 051	1 988	2 001	1 596	1 027	3 010	3 273
Senegal	74	75	72	1 300	2 312	2 486	96	173	179
Sierra Leone	403	390	271	1 250	1 305	1 212	504	508	329
Tanzania, United Rep.	262	380	469	959	1 830	1 450	251	694	679
Togo	18	22	46	800	1 437	1 826	15	31	85
Uganda	12	39	64	1 343	1 379	1 380	16	53	88
Zambia	5	12	13	497	993	893	2	12	11
Sub-Saharan Africa, other	82	72	83	817	1 868	1 797	67	134	150
Near East/North Africa	1 170	1 285	1 554	4 005	4 648	5 633	4 686	5 974	8 751
Afghanistan	190	173	167	2 179	1 907	2 540	415	329	423
Egypt	416	437	607	5 708	7 086	8 661	2 377	3 098	5 257
Iran, Islamic Rep.	434	542	588	3 211	3 809	4 231	1 394	2 064	2 490
Iraq	56	79	126	2 887	2 731	1 910	162	217	241
Turkey	67	51	58	4 722	4 976	5 314	314	253	310
Latin America and the Caribbean	7 983	6 685	5 951	1 945	2 633	3 466	15 525	17 597	20 629
Argentina	89	103	243	3 244	4 112	5 356	288	422	1 300
Bolivia	60	111	132	1 506	2 098	1 879	91	232	248
Brazil	5 932	4 441	3 320	1 438	2 097	2 796	8 533	9 313	9 284
Chile	40	35	22	3 154	4 161	3 862	125	146	84
Colombia	428	491	409	4 277	4 047	4 713	1 831	1 986	1 929
Costa Rica	73	53	67	3 058	3 591	4 068	224	192	273
Cuba	146	158	127	3 105	3 030	2 800	455	479	356
Dominican Republic	111	95	111	3 534	4 783	4 674	392	456	519

	Harvested area ('000 ha)			Yield (kg/ha)			Production ('000 tonnes)		
	1979/81	1989/91	1997/99	1979/81	1989/91	1997/99	1979/81	1989/91	1997/99
Ecuador	123	277	336	3 074	3 077	3 377	378	852	1 135
El Salvador	15	15	12	3 725	4 039	5 370	56	62	65
Guatemala	14	15	14	2 772	2 979	2 256	37	46	30
Guyana	91	65	139	2 924	3 303	4 077	266	215	567
Haiti	51	59	57	2 322	2 106	2 111	119	125	120
Honduras	20	19	11	1 736	2 613	2 778	35	48	30
Mexico	151	114	99	3 636	3 714	4 442	550	423	441
Nicaragua	37	42	71	3 545	2 737	3 359	130	115	239
Panama	96	94	79	1 834	2 257	2 588	175	211	203
Paraguay	26	33	28	1 836	2 606	3 744	47	87	105
Peru	132	186	273	4 462	5 161	6 061	587	957	1 655
Suriname	65	61	55	3 977	3 771	3 504	258	229	194
Uruguay	63	92	181	4 566	4 983	6 047	289	459	1 092
Venezuela	214	120	158	2 985	4 378	4 700	638	525	742
South Asia	54 477	57 195	59 314	1 910	2 602	2 917	104 021	148 833	173 044
Bangladesh	10 310	10 386	10 362	1 952	2 598	2 969	20 125	26 980	30 762
India	40 091	42 501	44 225	1 860	2 619	2 915	74 557	111 290	128 921
Nepal	1 275	1 433	1 511	1 852	2 352	2 438	2 361	3 372	3 683
Pakistan	1 981	2 106	2 419	2 465	2 309	2 928	4 884	4 862	7 081
Sri Lanka	819	770	797	2 557	3 028	3 258	2 093	2 330	2 596
East Asia	70 652	72 019	83 209	3 374	4 314	4 180	238 348	310 693	347 787
Cambodia	1 154	1 763	1 990	1 081	1 432	1 838	1 248	2 524	3 657
China	34 323	33 238	40 042	4 241	5 614	5 026	145 561	186 597	201 249
Indonesia	9 064	10 438	11 607	3 263	4 298	4 292	29 570	44 864	49 814
Korea, DPR	660	629	590	4 420	5 927	3 490	2 917	3 730	2 059
Korea, Republic	1 230	1 237	1 056	5 512	6 231	6 746	6 780	7 705	7 121
Lao PDR	722	606	645	1 419	2 276	2 810	1 025	1 379	1 812
Malaysia	722	676	686	2 844	2 739	2 964	2 053	1 852	2 034
Myanmar	4 684	4 689	5 556	2 698	2 913	3 043	12 637	13 661	16 909
Philippines	3 513	3 414	3 694	2 205	2 833	2 852	7 747	9 672	10 536
Thailand	8 986	9 241	9 964	1 888	2 099	2 331	16 967	19 398	23 226
Viet Nam	5 579	6 075	7 370	2 117	3 174	3 983	11 809	19 281	29 354
Industrial countries	4 141	3 649	3 732	5 469	6 238	6 517	22 650	22 760	24 321
Australia	111	97	141	6 210	8 649	8 794	688	839	1 243
European Union	300	359	410	5 570	5 919	6 514	1 671	2 124	2 671
Japan	2 384	2 073	1 847	5 587	6 119	6 352	13 320	12 688	11 733
United States	1 345	1 118	1 332	5 179	6 354	6 511	6 968	7 106	8 670
Transition countries	703	692	477	3 683	3 306	2 542	2 588	2 287	1 211
Kazakhstan	n.a.	n.a.	76	n.a.	n.a.	3 040	n.a.	n.a.	230
Russian Federation	n.a.	n.a.	146	n.a.	n.a.	2 702	n.a.	n.a.	395
Tajikistan	n.a.	n.a.	15	n.a.	n.a.	2 655	n.a.	n.a.	41
Turkmenistan	n.a.	n.a.	27	n.a.	n.a.	901	n.a.	n.a.	25
Ukraine	n.a.	n.a.	22	n.a.	n.a.	2 939	n.a.	n.a.	64
Uzbekistan	n.a.	n.a.	169	n.a.	n.a.	2 276	n.a.	n.a.	385

TABLE A4.3 Maize: area, yield and production (countries with more than 10 000 ha in 1997/99)

	Harvested area ('000 ha)			Yield (kg/ha)			Production ('000 tonnes)		
	1979/81	1989/91	1997/99	1979/81	1989/91	1997/99	1979/81	1989/91	1997/99
World	125 731	132 343	144 365	3 347	3 663	4 168	420 814	484 769	601 667
All developing countries	75 617	86 028	96 606	1 964	2 402	2 775	148 484	206 615	268 110
Sub-Saharan Africa	12 195	19 610	20 775	1 137	1 193	1 253	13 862	23 393	26 023
Angola	600	756	659	506	301	659	303	228	434
Benin	407	467	594	709	903	1 162	289	422	690
Botswana	42	43	19	278	355	306	12	15	6
Burkina Faso	123	195	254	876	1 420	1 435	108	277	365
Burundi	127	124	115	1 109	1 369	1 175	141	170	135
Cameroon	495	219	366	844	1 905	2 005	418	417	733
Central African Republic	108	69	90	373	856	984	40	59	89
Chad	32	33	122	842	982	1 025	27	32	125
Congo, Dem. Republic	745	1 225	1 463	811	814	816	604	997	1 194
Côte d'Ivoire	514	683	700	684	730	819	352	499	573
Eritrea	n.a	n.a	28	n.a	n.a	664	n.a	n.a	19
Ethiopia	n.a	n.a	1 606	n.a	n.a	1 696	n.a	n.a	2 724
Gabon	6	14	18	1 674	1 583	1 722	10	22	31
Gambia	6	13	10	1 344	1 232	1 382	9	16	14
Ghana	390	547	683	974	1 339	1 487	380	733	1 015
Guinea	87	73	86	1 000	1 010	1 028	87	74	88
Kenya	1 273	1 447	1 502	1 346	1 673	1 530	1 714	2 420	2 298
Lesotho	116	126	134	968	948	959	112	119	128
Madagascar	124	153	192	982	1 004	889	122	154	170
Malawi	1 077	1 336	1 342	1 184	1 109	1 361	1 275	1 481	1 826
Mali	52	177	231	1 173	1 280	1 684	61	226	389
Mauritania	8	4	13	573	648	782	5	2	10
Mozambique	674	1 007	1 185	569	368	960	383	370	1 137
Namibia	25	27	31	1 253	1 281	867	31	35	27
Nigeria	443	4 612	4 016	1 371	1 199	1 316	607	5 529	5 286
Rwanda	73	90	73	1 160	1 117	899	84	100	66
Senegal	75	100	61	885	1 223	931	66	122	57
Somalia	151	213	200	794	1 116	702	120	238	140
Sudan	67	80	69	585	492	632	39	39	44
Swaziland	66	90	63	1 292	1 358	1 842	85	122	116
Tanzania, United Rep.	1 350	1 820	1 786	1 305	1 448	1 323	1 762	2 635	2 362
Togo	147	273	409	1 020	982	1 056	150	268	432
Uganda	263	417	607	1 368	1 433	1 202	360	598	730
Zambia	523	808	553	1 799	1 665	1 481	941	1 345	818
Zimbabwe	1 097	1 143	1 437	1 667	1 626	1 190	1 829	1 859	1 710
Sub-Saharan Africa, other	31	49	44	721	586	559	23	28	25
Near East/North Africa	2 349	2 230	2 218	2 392	3 719	4 658	5 619	8 295	10 332
Afghanistan	447	264	187	1 649	1 713	1 554	738	453	290
Egypt	800	847	795	3 949	5 687	7 671	3 159	4 817	6 095
Iran, Islamic Rep.	35	32	158	1 500	3 905	6 345	52	126	1 004
Iraq	22	75	60	2 431	2 289	2 022	53	171	122
Morocco	396	389	327	618	1 006	725	245	391	237
Syrian Arab Republic	21	59	65	2 080	2 909	3 794	43	171	248
Turkey	583	512	582	2 168	4 087	3 885	1 263	2 093	2 260
Yemen	37	47	38	1 642	1 273	1 420	61	60	54

	Harvested area ('000 ha)			Yield (kg/ha)			Production ('000 tonnes)		
	1979/81	1989/91	1997/99	1979/81	1989/91	1997/99	1979/81	1989/91	1997/99
Latin America and the Caribbean	25 185	25 819	26 779	1 839	1 994	2 790	46 326	51 473	74 711
Argentina	2 895	1 715	3 067	3 224	3 496	5 260	9 333	5 995	16 132
Bolivia	295	270	283	1 430	1 629	2 023	422	439	572
Brazil	11 430	12 459	11 585	1 686	1 915	2 721	19 265	23 854	31 529
Chile	124	109	94	3 796	7 983	8 728	471	866	816
Colombia	620	806	528	1 401	1 460	1 709	868	1 177	902
Costa Rica	43	40	15	1 779	1 700	1 787	77	68	26
Cuba	77	74	92	1 239	909	1 529	95	67	141
Dominican Republic	32	28	27	1 276	1 617	1 195	41	45	32
Ecuador	230	452	456	1 075	1 082	1 198	247	490	546
El Salvador	281	288	288	1 840	1 961	1 974	517	565	569
Guatemala	627	635	611	1 511	1 971	1 659	947	1 251	1 013
Haiti	207	228	283	868	802	808	179	183	229
Honduras	339	383	409	1 201	1 424	1 270	407	545	520
Mexico	6 675	6 919	7 479	1 739	1 920	2 426	11 607	13 280	18 142
Nicaragua	179	215	245	1 022	1 230	1 163	183	265	285
Panama	62	75	53	953	1 327	1 485	59	99	78
Paraguay	216	206	365	1 553	1 921	2 506	335	396	916
Peru	341	384	440	1 665	1 999	2 136	569	767	940
Uruguay	124	62	60	1 016	1 657	3 360	126	102	203
Venezuela	372	451	373	1 471	2 180	2 863	547	983	1 069
Latin America, other	13	17	20	1 730	1 668	2 116	23	29	42
South Asia	7 101	7 532	8 012	1 144	1 504	1 685	8 124	11 330	13 499
India	5 887	5 893	6 301	1 102	1 509	1 718	6 486	8 893	10 822
Nepal	455	754	798	1 516	1 607	1 683	690	1 212	1 343
Pakistan	736	852	882	1 257	1 396	1 475	925	1 189	1 301
Sri Lanka	21	29	28	1 018	1 121	1 078	21	33	30
East Asia	28 787	30 836	38 822	2 590	36 363	3 698	74 553	112 125	143 547
Cambodia	87	48	44	981	1 403	1 094	85	67	48
China	19 986	21 188	29 869	3 038	4 337	4 086	60 720	91 891	122 044
Indonesia	2 761	3 004	3 549	1 461	2 129	2 644	4 035	6 394	9 382
Korea, DPR	690	674	576	3 865	5 932	2 324	2 667	4 000	1 338
Korea, Republic	34	24	21	4 441	4 341	4 013	150	105	82
Lao PDR	29	38	42	1 062	1 725	2 272	31	65	95
Malaysia	7	20	27	1 143	1 763	1 938	8	35	52
Myanmar	128	124	170	1 295	1 538	1 762	166	191	299
Philippines	3 267	3 699	2 594	972	1 264	1 637	3 174	4 677	4 247
Thailand	1 412	1 551	1 263	2 198	2 559	3 389	3 103	3 969	4 280
Viet Nam	383	463	667	1 071	1 570	2 508	410	727	1 672
Industrial countries	39 408	36 280	38 307	5 898	6 600	7 731	232 426	239 452	296 143
Australia	54	51	63	3 067	4 125	5 362	164	210	336
Canada	1 039	1 057	1 101	5 685	6 642	7 656	5 904	7 017	8 431
European Union	3 987	3 942	4 277	5 575	6 950	8 859	22 229	27 395	37 892
New Zealand	20	17	19	8 280	9 400	10 045	163	161	189
South Africa	4 539	4 124	3 717	2 525	2 447	2 312	11 462	10 092	8 592
Switzerland	17	28	21	7 230	8 602	9 334	121	239	196
United States	29 749	27 054	29 103	6 466	7 180	8 261	192 366	194 239	240 433

	Harvested area ('000 ha)			Yield (kg/ha)			Production ('000 tonnes)		
	1979/81	1989/91	1997/99	1979/81	1989/91	1997/99	1979/81	1989/91	1997/99
Transition countries	10 707	10 037	9 453	3 727	3 856	3 958	39 904	38 703	37 416
Albania	100	57	58	3 175	3 892	3 415	318	221	197
Azerbaijan	n.a	n.a	15	n.a	n.a	3 197	n.a	n.a	49
Bosnia and Herzegovina	n.a	n.a	218	n.a	n.a	3 919	n.a	n.a	855
Bulgaria	605	516	465	4 344	4 046	3 354	2 627	2 087	1 561
Croatia	n.a	n.a	378	n.a	n.a	5 560	n.a	n.a	2 100
Czech Republic	n.a	n.a	38	n.a	n.a	6 573	n.a	n.a	249
Georgia	n.a	n.a	210	n.a	n.a	2 309	n.a	n.a	486
Hungary	1 270	1 114	1 065	5 528	5 757	6 295	7 022	6 414	6 707
Kazakhstan	n.a	n.a	63	n.a	n.a	2 515	n.a	n.a	159
Kyrgyzstan	n.a	n.a	47	n.a	n.a	4 997	n.a	n.a	236
Macedonia, The Former Yugoslav Rep.	n.a	n.a	39	n.a	n.a	4 225	n.a	n.a	166
Moldova, Republic	n.a	n.a	420	n.a	n.a	3 344	n.a	n.a	1 403
Poland	26	60	89	3 886	4 849	5 674	102	291	504
Romania	3 226	2 592	3 060	3 168	3 096	3 512	10 218	8 023	10 748
Russian Federation	n.a	n.a	630	n.a	n.a	2 416	n.a	n.a	1 521
Slovakia	n.a	n.a	128	n.a	n.a	5 831	n.a	n.a	745
Slovenia	n.a	n.a	46	n.a	n.a	7 250	n.a	n.a	332
Tajikistan	n.a	n.a	10	n.a	n.a	3 361	n.a	n.a	34
Turkmenistan	n.a	n.a	14	n.a	n.a	495	n.a	n.a	7
Ukraine	n.a	n.a	1 078	n.a	n.a	2 900	n.a	n.a	3 126
Uzbekistan	n.a	n.a	48	n.a	n.a	2 726	n.a	n.a	130
Yugoslavia	n.a	n.a	1 328	n.a	n.a	4 582	n.a	n.a	6 084

TABLE A4.4 Barley: area, yield and production (countries with more than 10 000 ha in 1997/99)

	Harvested area ('000 ha)			Yield (kg/ha)			Production ('000 tonnes)		
	1979/81	1989/91	1997/99	1979/81	1989/91	1997/99	1979/81	1989/91	1997/99
World	81 221	74 534	58 554	1 893	2 288	2 390	153 759	170 552	139 920
All developing countries	16 684	19 476	16 911	1 289	1 316	1 420	21 504	25 633	24 014
Sub-Saharan Africa	918	901	1 019	1 214	1 149	1 062	1 114	1 035	1 082
Eritrea	n.a.	n.a.	38	n.a.	n.a.	1 046	n.a.	n.a.	39
Ethiopia	n.a.	n.a.	948	n.a.	n.a.	1 022	n.a.	n.a.	969
Kenya	49	24	22	1 302	1 405	2 581	64	33	58
Near East/North Africa	10 942	14 782	11 619	1 111	1 146	1 312	12 155	16 940	15 242
Afghanistan	296	207	193	1 057	1 071	1 217	313	221	235
Algeria	875	1 234	558	677	928	778	592	1 144	434
Egypt	41	56	58	2 689	2 393	2 224	111	134	129
Iran, Islamic Rep.	1 727	2 558	1 576	902	1 238	1 649	1 558	3 166	2 599
Iraq	858	1 382	1 202	846	793	593	726	1 095	712
Jordan	48	46	58	434	748	532	21	34	31
Lebanon	6	11	13	1 000	1 706	2 110	6	19	28
Libyan Arab Jamahiriya	284	270	125	342	495	486	97	134	61
Morocco	2 190	2 390	2 164	782	1 170	735	1 712	2 796	1 589
Saudi Arabia	7	56	49	1 178	6 616	5 173	8	371	252
Syrian Arab Republic	1 220	2 618	1 510	926	259	503	1 129	678	759
Tunisia	457	491	346	609	949	852	279	466	294
Turkey	2 846	3 365	3 673	1 926	1 942	2 178	5 480	6 533	8 000
Yemen	50	49	45	995	968	1 075	49	48	48
Near East, other	38	51	49	1 950	1 985	1 427	74	100	69
Latin America and the Caribbean	935	960	1 009	1 360	1 722	1 870	1 272	1 653	1 887
Argentina	178	182	238	1 289	2 323	2 620	229	422	623
Bolivia	80	83	88	653	660	605	53	55	53
Brazil	84	105	140	1 120	1 637	2 082	94	172	291
Chile	51	28	25	1 998	3 433	3 699	103	95	93
Ecuador	29	56	46	840	850	764	24	48	35
Mexico	280	270	242	1 722	1 860	1 860	483	502	449
Peru	127	103	140	903	1 018	1 128	115	105	158
Uruguay	48	81	82	1 474	1 944	2 046	71	158	168
South Asia	2 046	1 216	992	1 073	1 467	1 745	2 196	1 784	1 732
India	1 802	1 011	798	1 121	1 595	1 924	2 020	1 613	1 537
Nepal	26	30	34	875	928	1 051	23	27	36
Pakistan	199	157	151	709	842	1 021	141	132	154
East Asia	1 752	1 533	2 268	2 698	2 700	1 794	4 726	4 138	4 069
China	1 295	1 317	2 131	2 420	2 582	1 723	3 133	3 400	3 671
Korea, DPR	72	60	58	2 163	2 472	1 942	155	148	112
Korea, Republic	386	155	77	3 728	3 802	3 710	1 438	589	284
Industrial countries	26 940	24 365	21 542	2 990	3 449	3 661	80 556	84 030	78 863
Australia	2 540	2 470	3 092	1 291	1 711	1 888	3 279	4 227	5 837
Canada	4 631	4 491	4 347	2 418	2 735	3 024	11 199	12 281	13 144
European Union	16 022	13 576	11 371	3 471	4 137	4 487	55 615	56 167	51 023

	Harvested area ('000 ha)			Yield (kg/ha)			Production ('000 tonnes)		
	1979/81	1989/91	1997/99	1979/81	1989/91	1997/99	1979/81	1989/91	1997/99
Japan	120	105	55	3 261	3 180	3 300	392	333	181
New Zealand	71	86	63	3 612	4 418	5 597	255	381	352
Norway	186	178	176	3 424	3 774	3 735	636	672	657
South Africa	81	114	115	1 258	2 114	1 411	102	241	163
Switzerland	48	58	49	4 589	6 046	6 222	220	352	303
United States	3 214	3 272	2 266	2 750	2 863	3 179	8 838	9 367	7 202
Transition countries	37 598	30 693	20 101	1 375	1 984	1 843	51 699	60 889	37 042
Armenia	n.a.	n.a.	62	n.a.	n.a.	1 051	n.a.	n.a.	65
Azerbaijan	n.a.	n.a.	81	n.a.	n.a.	1 261	n.a.	n.a.	102
Belarus	n.a.	n.a.	845	n.a.	n.a.	2 037	n.a.	n.a.	1 721
Bosnia and Herzegovina	n.a.	n.a.	23	n.a.	n.a.	2 626	n.a.	n.a.	59
Bulgaria	425	368	279	3 386	4 043	2 606	1 439	1 487	726
Croatia	n.a.	n.a.	40	n.a.	n.a.	3 115	n.a.	n.a.	126
Czech Republic	n.a.	n.a.	589	n.a.	n.a.	3 800	n.a.	n.a.	2 238
Estonia	n.a.	n.a.	162	n.a.	n.a.	1 585	n.a.	n.a.	257
Georgia	n.a.	n.a.	32	n.a.	n.a.	1 205	n.a.	n.a.	39
Hungary	265	312	358	3 203	4 554	3 428	848	1 421	1 226
Kazakhstan	n.a.	n.a.	2 080	n.a.	n.a.	952	n.a.	n.a.	1 980
Kyrgyzstan	n.a.	n.a.	82	n.a.	n.a.	1 998	n.a.	n.a.	164
Latvia	n.a.	n.a.	172	n.a.	n.a.	1 774	n.a.	n.a.	305
Lithuania	n.a.	n.a.	462	n.a.	n.a.	2 191	n.a.	n.a.	1 013
Macedonia, The Former Yugoslav Rep.	n.a.	n.a.	52	n.a.	n.a.	2 494	n.a.	n.a.	129
Moldova, Republic	n.a.	n.a.	115	n.a.	n.a.	1 993	n.a.	n.a.	229
Poland	1 362	1 195	1 162	2 616	3 453	3 120	3 563	4 128	3 626
Romania	833	845	518	2 786	3 577	2 670	2 321	3 022	1 382
Russian Federation	n.a.	n.a.	8 778	n.a.	n.a.	1 564	n.a.	n.a.	13 729
Slovakia	n.a.	n.a.	247	n.a.	n.a.	3 382	n.a.	n.a.	835
Slovenia	n.a.	n.a.	11	n.a.	n.a.	3 533	n.a.	n.a.	38
Tajikistan	n.a.	n.a.	27	n.a.	n.a.	899	n.a.	n.a.	24
Turkmenistan	n.a.	n.a.	68	n.a.	n.a.	252	n.a.	n.a.	17
Ukraine	n.a.	n.a.	3 580	n.a.	n.a.	1 834	n.a.	n.a.	6 567
Uzbekistan	n.a.	n.a.	146	n.a.	n.a.	711	n.a.	n.a.	104
Yugoslavia	n.a.	n.a.	128	n.a.	n.a.	2 626	n.a.	n.a.	337

TABLE A4.5 Millet: area, yield and production (countries with more than 10 000 ha in 1997/99)

	Harvested area ('000 ha)			Yield (kg/ha)			Production ('000 tonnes)		
	1979/81	1989/91	1997/99	1979/81	1989/91	1997/99	1979/81	1989/91	1997/99
World	37 684	37 362	37 885	673	755	737	25 367	28 216	27 917
All developing countries	34 765	34 279	36 518	678	733	724	23 580	25 116	26 427
Sub-Saharan Africa	11 586	15 727	19 870	655	673	660	7 589	10 588	13 120
Angola	80	120	177	613	536	475	49	64	84
Benin	13	38	39	503	624	741	7	24	29
Burkina Faso	803	1 169	1 216	486	555	715	390	649	870
Cameroon	130	60	70	753	1 061	1 014	98	64	71
Central African Republic	16	10	12	681	933	1 000	11	9	12
Chad	360	527	740	505	363	429	182	191	318
Congo, Dem. Republic	36	52	73	668	641	633	24	34	46
Côte d'Ivoire	64	77	94	580	613	657	37	47	62
Eritrea	n.a.	n.a.	69	n.a.	n.a.	482	n.a.	n.a.	33
Ethiopia	n.a.	n.a.	343	n.a.	n.a.	910	n.a.	n.a.	312
Gambia	17	53	74	1 031	974	934	18	52	69
Ghana	182	192	175	641	636	904	117	122	158
Guinea	35	19	12	1 410	1 102	826	49	21	10
Kenya	80	110	90	1 048	590	622	84	65	56
Malawi	11	18	38	588	554	497	7	10	19
Mali	643	1 124	993	716	732	786	461	823	780
Mauritania	12	16	20	255	407	260	3	7	5
Mozambique	20	20	96	250	250	550	5	5	53
Namibia	140	152	267	245	359	237	34	55	63
Niger	3 011	4 187	5 082	435	393	393	1 311	1 644	1 999
Nigeria	2 366	4 468	5 562	1 043	1 046	1 068	2 467	4 672	5 939
Senegal	932	899	825	595	644	550	555	579	453
Sierra Leone	9	26	17	1 393	888	843	13	23	14
Sudan	1 098	1 113	2 657	350	166	227	384	185	604
Tanzania, United Rep.	450	245	251	800	954	992	360	233	249
Togo	121	134	91	364	510	471	44	68	43
Uganda	297	379	391	1 592	1 534	1 493	473	582	583
Zambia	34	51	90	637	557	711	22	28	64
Zimbabwe	353	271	252	432	501	289	153	136	73
Sub-Saharan Africa, other	18	20	35	698	1 020	914	12	21	32
Near East/North Africa	190	186	173	989	600	802	188	112	139
Afghanistan	39	30	27	862	865	815	33	26	22
Iran, Islamic Rep.	8	7	13	1 123	1 846	1 103	9	12	14
Yemen	81	122	103	1 140	365	640	92	45	66
Latin America and the Caribbean	203	55	37	1 211	1 408	1 215	245	78	45
Argentina	203	55	37	1 211	1 400	1 214	245	77	45
South Asia	18 561	15 832	13 190	518	647	785	9 619	10 237	10 360
Bangladesh	61	91	80	648	709	706	40	64	57
India	17 845	15 096	12 429	515	646	790	9 189	9 758	9 817
Nepal	122	197	262	989	1 161	1 101	121	228	289
Pakistan	509	438	412	500	410	469	255	180	193

	Harvested area ('000 ha)			Yield (kg/ha)			Production ('000 tonnes)		
	1979/81	1989/91	1997/99	1979/81	1989/91	1997/99	1979/81	1989/91	1997/99
East Asia	4 225	2 479	3 248	1 406	1 655	851	5 939	4 102	2 763
China	3 981	2 253	2 995	1 455	1 739	862	5 790	3 918	2 582
Korea, DPR	62	50	15	1 065	1 200	909	66	60	13
Myanmar	179	174	236	447	695	701	80	121	166
Industrial countries	134	172	210	1 170	1 288	1 469	156	221	309
Australia	26	31	30	998	877	1 034	26	27	31
South Africa	22	22	21	682	682	581	15	15	12
United States	81	118	159	1 314	1 501	1 665	107	177	265
Transition countries	2 786	2 912	1 156	585	989	1 022	1 631	2 879	1 182
Kazakstan	n.a.	n.a.	86	n.a.	n.a.	495	n.a.	n.a.	42
Russian Federation	n.a.	n.a.	826	n.a.	n.a.	1 048	n.a.	n.a.	866
Ukraine	n.a.	n.a.	231	n.a.	n.a.	1 091	n.a.	n.a.	252

TABLE A4.6 Sorghum: area, yield and production (countries with more than 10 000 ha in 1997/99)

	Harvested area ('000 ha)			Yield (kg/ha)			Production ('000 tonnes)		
	1979/81	1989/91	1997/99	1979/81	1989/91	1997/99	1979/81	1989/91	1997/99
World	44 742	43 097	43 859	1 462	1 326	1 394	65 403	57 157	61 135
All developing countries	38 400	38 110	39 590	1 141	1 050	1 107	43 819	40 001	43 831
Sub-Saharan Africa	13 037	17 859	22 267	858	729	817	11 187	13 013	18 198
Benin	90	140	163	649	762	819	59	107	134
Botswana	98	148	73	209	280	177	21	41	13
Burkina Faso	1 051	1 337	1 398	590	743	810	620	993	1 132
Burundi	53	58	53	1 000	1 117	1 230	53	65	65
Cameroon	374	465	381	806	769	1 078	301	357	411
Central African Republic	57	25	34	674	1 027	1 008	39	25	34
Chad	414	458	717	507	584	659	210	268	472
Congo, Dem. Republic	36	76	79	900	633	658	32	48	52
Côte d'Ivoire	40	45	55	606	580	348	24	26	19
Eritrea	n.a.	n.a.	185	n.a.	n.a.	908	n.a.	n.a.	168
Ethiopia	n.a.	n.a.	1 155	n.a.	n.a.	1 288	n.a.	n.a.	1 488
Gambia	6	12	15	793	881	923	5	10	14
Ghana	223	255	319	628	775	1 069	140	197	341
Kenya	168	127	137	951	908	915	160	116	125
Lesotho	58	32	32	1 020	815	888	59	26	28
Malawi	30	31	81	667	590	583	20	18	47
Mali	434	763	657	827	888	988	359	677	649
Mauritania	102	121	156	279	590	440	28	71	69
Mozambique	288	422	464	630	397	651	182	167	302
Namibia	29	34	21	214	262	248	6	9	5
Niger	822	2 011	2 074	422	193	205	347	389	424
Nigeria	2 683	4 892	6 634	1 224	980	1 122	3 284	4 794	7 444
Rwanda	159	142	118	1 123	1 162	990	178	165	117
Senegal	130	135	196	1 014	897	657	131	121	128
Sierra Leone	7	34	25	1 571	634	766	11	22	19
Somalia	478	450	310	350	540	316	167	243	98
Sudan	3 067	3 887	5 425	748	540	587	2 293	2 099	3 184
Tanzania, United Rep.	713	489	590	762	992	933	543	485	551
Togo	122	190	205	714	716	699	87	136	143
Uganda	175	239	277	1 784	1 495	1 356	312	357	376
Zambia	31	44	39	535	562	698	16	25	27
Zimbabwe	140	136	163	611	588	535	85	80	88
Sub-Saharan Africa, other	28	13	27	639	915	925	18	11	25
Near East/North Africa	1 142	776	793	1 239	1 543	1 908	1 414	1 197	1 514
Egypt	172	133	159	3 739	4 745	5 492	644	631	870
Morocco	45	27	28	436	552	628	20	15	18
Saudi Arabia	281	128	173	437	1 092	1 182	123	139	204
Yemen	622	470	423	987	854	978	614	402	414
Latin America and the Caribbean	4 379	3 466	3 740	2 806	2 673	3 006	12 285	9 262	11 242
Argentina	1 834	682	732	3 055	2 836	4 318	5 602	1 934	3 161
Bolivia	5	15	49	4 089	3 954	2 372	21	58	115
Brazil	81	158	323	2 128	1 544	1 763	172	243	569
Colombia	220	256	76	2 223	2 877	3 150	488	737	240

	Harvested area ('000 ha)			Yield (kg/ha)			Production ('000 tonnes)		
	1979/81	1989/91	1997/99	1979/81	1989/91	1997/99	1979/81	1989/91	1997/99
El Salvador	126	124	113	1 153	1 271	1 625	145	158	184
Guatemala	39	59	41	2 033	1 406	1 171	80	83	48
Haiti	158	120	130	762	809	760	121	97	98
Honduras	61	73	80	809	976	1 080	49	71	86
Mexico	1 464	1 607	1 915	3 362	3 172	3 174	4 921	5 096	6 077
Nicaragua	51	48	45	1 560	1 545	1 568	80	74	71
Paraguay	7	11	16	1 274	1 277	1 470	9	14	23
Uruguay	56	31	32	2 016	2 468	3 420	112	77	109
Venezuela	227	248	174	1 605	2 136	2 433	365	529	424
South Asia	16 766	14 266	10 838	693	781	830	11 616	11 136	9 000
India	16 361	13 852	10 461	696	786	839	11 380	10 893	8 773
Pakistan	403	413	377	582	586	601	235	242	227
East Asia	3 078	1 742	1 952	2 377	3 095	1 986	7 316	5 393	3 877
China	2 828	1 544	1 838	2 488	3 326	2 013	7 034	5 135	3 700
Thailand	220	186	105	1 074	1 285	1 573	237	239	165
Industrial countries	6 197	4 827	4 210	3 451	3 511	4 091	21 383	16 945	17 225
Australia	549	461	546	1 976	2 126	2 684	1 084	980	1 466
European Union	129	114	104	4 576	4 949	5 900	589	563	614
South Africa	244	197	130	2 239	1 956	2 598	547	385	339
United States	5 273	4 055	3 430	3 633	3 703	4 317	19 157	15 017	14 806
Transition countries	145	161	59	1 383	1 316	1 347	201	212	79
Albania	25	21	16	1 200	1 124	934	30	24	15
Russian Federation	n.a.	n.a.	17	n.a.	n.a.	1 033	n.a.	n.a.	17

References

ABARE. 1999. *WTO agricultural negotiations: important market access issues,* by T. Podbury & I. Roberts. ABARE Research Report No. 99.3. Canberra, Australian Bureau of Agricultural and Resource Economics.

ABARE. 2001. *The impact of agricultural trade liberalisation on developing countries,* by F. Freeman. ABARE Research Report No. 01.6. Canberra, Australian Bureau of Agricultural and Resource Economics.

Abramovitz, J. & Mattoon, A. 1999. *Paper cuts: recovering the paper landscape.* Worldwatch Paper No. 149. Washington, DC, Worldwatch Institute.

Abreu, M. 1996. Trade in manufactures: the outcome of the Uruguay Round and developing country interests. *In* W. Martin & L. Winters, eds. *The Uruguay Round and the developing economies.* Cambridge, UK, Cambridge University Press.

Adams, R. & He, J. 1995. *Sources of income inequality and poverty in rural Pakistan.* IFPRI Research Report No. 102. Washington, DC, IFPRI.

Agcaoili, M., Perez, N. & Rosegrant, M. 1995. *Impact of resource degradation on global food balances.* Paper presented at the Workshop on Land Degradation in the Developing World: Implications for Food, Agriculture and the Environment to the Year 2020, 4-6 April, Annapolis, United States.

Agosin, M., Bloom, D. & Gitli, E. 2000. *Globalization, liberalization and sustainable human development: progress and challenges in central American countries.* Geneva, UNCTAD/UNDP.

Agrodata. 2000. *Structures, performances et stratégies des groupes agroalimentaires multinationaux.* Montpellier, France, CIHEAM. 483 pp.

Ahlburg, D. 1998. Julian Simon and the population growth debate. *Population and Development Rev.,* 24(2), June.

Alberts, B. 1999. Creating a global knowledge system for food security. In *Proceedings of the UNESCO/FAO Session on Science, Agriculture and Food Security.* World Conference on Science, Budapest, Hungary, 27 June.

Alderman, H., Behrman, J., Ross, D. & Sabot, R. 1996. The returns to endogenous human capital in Pakistan's rural wage labour market. *Oxford Bull. Economics and Statistics,* 58(1): 30-55.

Alderman, H., Appleton, S, Haddad, L., Song, L. & Yohannes, Y. 2001. *Reducing child malnutrition: how far does income growth take us?* World Bank. (processed)

Alexandratos, N., ed. 1988. *World agriculture: towards 2000. An FAO study.* London, Belhaven Press and New York, New York University Press.

Alexandratos, N., ed. 1995. *World agriculture: towards 2010. An FAO study.* Chichester, UK, John Wiley and Rome, FAO.

Alexandratos, N. 1996. China's projected cereals deficits in a world context. *Agricultural Economics,* 15: 1-16.

Alexandratos, N. 1997. China's projected cereals consumption and the capacity of the rest of the world to increase exports. *Food Policy,* June.

Alexandratos, N. 1999. World food and agriculture: outlook for the medium and longer term. *Proceedings of the US National Academy of Sciences,* 96: 5908-5914. May.

Alexandratos, N. & Bruinsma, J. 1998. Europe's cereals sector and world trade requirements to 2030. In M. Redclift, J. Lekakis & G. Zanias, eds. *Agriculture and world trade liberalisation: socio-environmental perspectives on the CAP.* Wallingford, UK, CAB International.

Alexandratos, N. & Bruinsma, J. 1999. Land use and land potentials for future world food security. In *Proceedings of the UNU/IAS/IGES International Conference on Sustainable Future of the Global System*, Chapter 15, p. 331-351, 23-24 February, Tokyo.

Alexandratos, N., Bruinsma, J. & Schmidhuber, J. 2000. China's food and the world. In *Meeting the food challenge of the 21st century*. Special issue of *Agrarwirtschaft*, 49 (9/10). September.

Allen, R., Pereira, L., Raes, D. & Smith, M. 1998. *Crop evapotranspiration: guidelines for computing crop water requirements*. FAO Irrigation and Drainage Paper No. 56. Rome, FAO. 300 pp.

Amjadi, A., Reinke, U. & Yeats, A. 1996. *Did external barriers cause the marginalization of sub-Saharan Africa in world trade?* World Bank Policy Research Working Paper No. 1586. Washington, DC.

Anderson, K., Francois, J., Hertel, T., Hoekman, B. & Martin, W. 2000. *Potential gains from trade reform in the new millennium*. Paper presented at the Third Annual Conference on Global Economic Analysis. 27-30 June, Monash University, Melbourne.

Anobah, R. 2000. Development of organic markets in western Africa. *In* T. Alföldi, W. Lockeretz & U. Nigli, eds. *IFOAM 2000 — the world grows organic*. Proceedings of the 13th International IFOAM Scientific Conference, Hochschulverlag AG an der ETH, Zürich.

Arcand, J. 2001. *Undernourishment and economic growth: the efficiency cost of hunger*. FAO Economic and Social Development Paper No. 147. Rome, FAO.

Arnason, R. 2001. *The economics of ocean ranching: experiences, outlook and theory*. FAO Fisheries Technical Paper No. 413. Rome, FAO.

Arnell, N. 1999. Climate change and water resources. *Global Environment Change*, 9: 31-49.

Ash, R. & Edmunds R. 1998. China's land resources, environment and agricultural production. *The China Quarterly*. September.

Asian Development Bank. 1994. *Climate change in Asia: regional study on global environment issues*. Manila.

Asian Development Bank. 2000. *Environments in transition: Cambodia, Lao PDR, Thailand, Viet Nam*. Manila.

Asman, A. 1994. Emission and deposition of ammonia and ammonium. *Nova Acta Leopold*, 70: 263-297.

Asthana, V. 1993. *Report on impact of sea level rise on the islands and coast of India*. Ministry of Environment and Forests, Government of India, Delhi.

Avery, D. 1997. Saving nature's legacy through better farming. *Issues in Science and Technology*, 14(1).

Avnimelech, Y. 1986. Organic residues in modern agriculture. In Y. Chen & Y. Avnimelech, eds. *The role of organic matter in modern agriculture*. Dordrecht, the Netherlands, Martinus Nijhoff Publishers.

Baarse, G. 1995. *Development of an operational tool for global vulnerability assessment (GVA): update of the number of people at risk due to sea-level rise and increasing flooding probabilities*. CZM-Centre Publications No. 3, Ministry of Transport, Public Works and Water Management, the Hague, the Netherlands.

Backmann, P. 1997. Pesticide inputs now and into the twenty-first century. In B. White & L. Chuang, eds. *Crop and livestock technologies*. RCS III Symposium. Ames, Iowa, USA, Iowa University Press.

Baily, M. & Gersbach, H. 1995. *Efficiency in manufacturing and the need for global competition*. Brookings Papers on Economic Activity. Microeconomics (1995): 307-347.

Baker, R., Sansford, C., Cannon, R., Macleod, A. & Walters, K. 1999. *The role of climate mapping in predicting the potential distribution of non-indigenous pests under current and future climates*. Paper for the GCTE Focus 3 Conference on Food and Forestry: Global Change and Global Challenges, 20-23 September, University of Reading, UK.

Baldock, D., Beaufoy, G., Brouwer, F. & Godeschalk, F. 1996. *Farming at the margins: abandonment or redeployment of agricultural land in Europe*. Ministry of Housing, Spatial Planning and the Environment, the Hague, the Netherlands.

Baron, P. 2001. *Roundup of key developments and issues that will shape the future of world and Asian sugar industries*. Paper for the 7th Asia International Sugar Conference, Bangkok, 29-31 August 2001. London, International Sugar Organization.

Bartley, D. 1999. Marine ranching: a global perspective. *In* R. Bri, ed. *Stock enhancement and sea ranching*. UK, Fishing News Books.

Batjes, N. 1999. *Management options for reducing CO$_2$ concentrations in the atmosphere by increasing carbon sequestration in the soil*. Report No. 410-200-031, Dutch National Research Programme on Global Air Pollution and Climate Change, Bilthoven and Technical Paper No. 30. ISRIC, Wageningen, the Netherlands. 114 pp.

Bautista, R. 1995. Rapid agricultural growth is not enough: the Philippines, 1965-1980. *In* J. Mellor, ed. *Agriculture on the road to industrialization*. Baltimore and London, Johns Hopkins Press.

Bazzaz, F. & Sombroek, W., eds. 1996. *Global climate change and agricultural production: direct and indirect effects of changing hydrological, pedological and plant physiological processes*. Rome, FAO and Chichester, UK, John Wiley.

Begon, M., Harper, J. & Townsend, C. 1990. *Ecology: individuals, populations and communities*. Oxford, UK, Blackwell Scientific.

Behnke, R., Scoones, I. & Kerven, C., eds. 1993. *Range ecology at disequilibrium. New models of natural variability and pastoral adaptation in African savannas*. London, Overseas Development Institute.

Behrman, J. & Deolalikar, A. 1989. Is variety the spice of life? Implications for calorie intake. *Rev. Economics and Statistics*, 71(4): 666-672.

Bhagwati, J. & Srinivasan, T. 1999. *Outward orientation and development: are revisionists right?* http://www.columbia.edu/~jb38/Krueger.pdf/

Bhalla, G., Hazell, P. & Kerr, J. 1999. *Prospects for India's cereal supply and demand to 2020*. Food, Agriculture and the Environment Discussion Paper No. 29. Washington, DC, IFPRI.

Bhargava, A. 1997. Nutritional status and the allocation of time in Rwandese households. *J. Econometrics*, 77: 277-295.

Blandford, D. 2001. Are disciplines required on domestic support? *The Estey Centre J. International Law and Trade Policy*, 2(1).

Borlaug, N. 1999. Feeding a world of 10 billion people: the miracle ahead. Lecture presented at De Montfort University, UK.

Boserup, E. 1965. *The conditions of agricultural growth: the economics of agrarian change under population pressure*. London, Earthscan Publications Ltd.

Bot, A., Nachtergaele, F. & Young, A. 2000. *Land resource potential and constraints at regional and country levels*. World Soil Resources Report No. 90. Rome, FAO.

Bouis, H. 1994. The effect of income on demand for food in poor countries: are our databases giving us reliable estimates? *J. Development Economics*, 44(1): 199-226.

Bourke, I. & Leitch, J. 1998. *Trade restrictions and their impact on international trade in forest products*. Rome, FAO.

Bouwman, A. 2001. *Global estimates of gaseous emissions from agricultural land*. Rome, FAO.

Bouwman, A., Lee, D., Asman, A., Dentener, F., van der Hoek, K. & Olivier, J. 1997. Global high-resolution emission inventory for ammonia. *Global Biogeochemical Cycles*, 11(4): 561-587.

Branca, G. 2001. The effect of land degradation on crop yields. A literature survey. Background paper prepared for *Agriculture: towards 2015/30*. Rome, FAO. (mimeo)

Bremen, H., Groot, J. & van Keulen, H. 2001. Resource limitations in Sahelian agriculture. *Global Environmental Change*, 11: 59-68.

Brouwer, F. & van Berkum, S. 1996. *CAP and the environment in the European Union*. Wageningen, the Netherlands, Wageningen University Press.

Brown, L. 1995. *Who will feed China, wake-up call for a small planet*. New York, W. W. Norton.

Brown, L. 1996. *Tough choices: facing the challenge of food scarcity*. New York, W.W. Norton.

Brown, D., Deardorff, A. & Stern, R. 2001. *CGE modeling and analysis of multilateral and regional negotiating options*. University of Michigan School of Public Policy Research Seminar in International Economics, Discussion Paper No. 468.

Bull, G., Mabee, W. & Scharpenberg, R. 1998. *Global Fibre Supply Model.* Rome, FAO.

Burger, K. & Smit, H. 2000. *Natural rubber in the coming decades – policies and projections.* Paper for the International Rubber Forum, International Rubber Study Group, 9-10 November 2000. Antwerp.

Bussolo, M. & Lecomte, H. 1999. *Trade liberalisation and poverty.* ODI Briefing Note No. 6. December. UK, Overseas Development Institute.

Byerlee, D. 1996. Modern varieties, productivity and sustainability: recent experience and emerging challenges. *World Development,* 24(4): 697-718.

Cai, S., Zhao, Y. & Du, Y. 1999. The evolution of the Jianghan lakes and regional sustainable development. *J. Environmental Sciences,* 11(3).

Cannell, M., Milne, R., Hargreaves, K., Brown, T., Cruickshank, M., Bradley, R., Spencer, T., Hope, D., Billet, M., Adger, W. & Subak, S. 1999. National inventories of terrestrial carbon sources and sinks: the UK experience. *Climatic Change,* 42: 505-530.

Carson, B. 1992. *The land, farmer and future: a soil fertility management strategy for Nepal.* Kathmandu, Nepal, Department of Soil Conservation.

Caves, R. 1981. Intra-industry trade and market structure in the industrialised countries. *Oxford Economic Papers,* 33: 203-23.

CCICED. 1999. *Land use and sustainable development of the Loess Plateau.* China Environmental Press for the China Council for International Cooperation on Environment and Development, Beijing.

CEC. 1980. *Effects on the environment of the abandonment of agricultural land.* Information on Agriculture No. 62. Brussels, Commission of the European Communities.

CGIAR. 2000. *A food secure world for all: towards a new vision and strategy for the CGIAR.* TAC Paper No. SDR/TAC:IAR/00/14.1 Rev. 2. Rome, FAO.

Chen, R. & Kates, R. 1996. Towards a food secure world: prospects and trends. *In* T. Downing. *Climate change and food security.* Berlin, Springer Verlag.

Cocking, E. 2001. The rhizosphere and microbial plant interactions. *In* Proceedings of the FAO Expert Consultation on Increasing the use of Biological Nitrogen Fixation (BNF) in Agriculture, 13-15 March. Rome, FAO.

Cohen, J. 1995. *How many people can the earth support?* New York, W. Norton.

Colby, H., Zhong, X. & Giordano, M. 1998. *A review of China's meat production statistics: new estimates and implications for trade.* Paper for the Workshop on Agricultural Policies in China and OECD Countries: Review Outlook and Challenges, November 19-20. Paris, OECD.

Cole, V., Cerri, C., Minami, K., Mosier, A., Rosenberg, N. & Sauerbeck, D. 1996. Agricultural options for mitigation of greenhouse gas emissions. *In* R. Watson, M. Zinyowera & R. Moss, eds. *Climate change 1995: impacts, adaptations and mitigation of climate change: scientific-technical analyses.* Contribution of Working Group II to the Second Assessment of the IPCC. Cambridge, UK, Cambridge University Press.

Commission for Environmental Cooperation. 1999. *Assessing the environmental effects of the NAFTA, an analytic framework and issue studies.* Montreal, Canada.

Conway, G. 1987. The properties of agro-ecosystems. *Agricultural Systems,* 24: 95-117.

Conway, G. 1997. *The doubly green revolution: food for all in the 21st century.* London, Penguin Books.

Cooper, D. & Anishetty, M. 2002. International treaties to the management of plant genetic resources. Rome, FAO. (mimeo)

Crook, F. 1998. *Soybean consumption and trade in China: a trip report.* Washington, DC, USDA/Economic Research Service.

Croppenstedt, A. & Muller, C. 2000. The impact of farmers' health and nutritional status on their productivity and efficiency: evidence from Ethiopia. *Economic Development and Cultural Change,* 48 (3): 475-502.

Crosson, P. 1997. Will erosion threaten agricultural productivity? *Environment,* 39(8). October.

Daberkov, S., Isherwood, K., Poulisse, J. & Vroomen, H. 1999. *Fertilizer requirements in 2015 and 2030.* Paper presented at the IFA Agricultural Conference on Managing Plant Nutrition towards Maximum Resource Efficiency. Barcelona, Spain.

Dalal, R. & Mayer, R. 1986. Long-term trends in soil fertility of soils under continuous cultivation and cereal cropping in southern Queensland: overall changes in soil properties and trends in winter yields. *Australian J. Soil Research,* 24: 265-279.

Dånmark, G. 1998. The multifunctional role of sustainable agriculture. In *Sustainable Agriculture.* Proceedings from a Nordic Seminar, Stockholm, 10-12 September. Stockholm, Swedish Environmental Protection Agency.

Dasgupta, P. 1997. Nutritional status, the capacity for work and poverty traps. *J. Econometrics,* 77: 5-37.

Datt, G. & Ravallion, M. 1998. Farm productivity and rural poverty in India. *J. Development Studies,* 34: 62-85.

Deaton, A. 2001. Counting the world's poor: problems and possible solutions. *World Bank Research Observer,* 16(2). Washington, DC.

De Haan, C., Steinfeld, H. & Blackburn, H. 1998. *Livestock and the environment: finding a balance.* CEC/FAO/WB.

Deininger, K. & Minten, B. 1999. Poverty, policies and deforestation: the case of Mexico. *Economic Development and Cultural Change.* January.

De Janvry, A. & Sadoulet, E. 2000. *Income strategies among rural households in Mexico: the role of off-farm activities.* University of California at Berkeley and the World Bank. (processed)

Delgado, C., Hopkins, J. & Kelly, V. 1998. *Agricultural growth linkages in sub-Saharan Africa.* IFPRI Research Report No. 107. Washington, DC, IFPRI.

Delgado, C., Rosegrant, M., Steinfeld, H., Ehui, S. & Courbois, C. 1999. *Livestock to 2020: the next food revolution.* Food, Agriculture and the Environment Discussion Paper No. 28. Washington, DC, IFPRI.

Delgado, C., Rosegrant, M. & Meijer, S. 2001. *Livestock to 2020: the revolution continues.* IFPRI, Washington DC. Paper presented at the Annual Meeting of the International Agricultural Trade Research Consortium (IATRC), 18-19 January, Auckland, New Zealand.

Demery, L. & Walton, M. 1999. Are poverty and social goals for the 21st century attainable? *IDS Bulletin,* 30(2).

Deolalikar, A. 1988. Nutrition and labour productivity in agriculture: estimates for rural south India. *Rev. Economics and Statistics,* 70: 406-413.

Derpsch, R. 1998. *Historical review of no-tillage cultivation of crops.* GTZ. (mimeo)

Derpsch, R. 2000. *Frontiers in conservation tillage and advances in conservation practice.* Rome, FAO.

DETR. 1997. *Climate change and its impacts: a global perspective – some recent results from the UK research programme.* London, Meteorological Office for the Department of the Environment, Transport and the Regions.

DETR. 1999. *Design of tax/charge schemes for pesticides.* ECOTEC Research and Consulting for the Department of the Environment, Transport and the Regions, London.

DFID. 2000a. *Halving world poverty by 2015, economic growth, equity and security. Strategies for achieving the international development targets.* DFID Strategy Paper. London.

DFID. 2000b. *Eliminating world poverty: making globalisation work for the poor.* White Paper on International Development. London.

Diels, J., Coolen, E., van Lauwe, B., Merckx, R. & Sanguina, N. 1999. *Modelling soil organic carbon changes under intensified land use in tropical conditions.* Paper presented to the GCTE Focus 3 Conference on Food and Forestry: Global Change and Global Challenges.

DME. 1999. *The Bichel report.* Copenhagen, Danish Ministry of the Environment.

Dollar, D. & Kraay, A. 2000. *Growth is good for the poor.* Washington, DC, World Bank.

Dollar, D. & Kraay, A. 2001. *Trade, growth and poverty.* Paper for the Asia and Pacific Forum on Poverty, February, Manila, ADB. 52 pp.

Dorman, P. 2001. *The free trade magic act.* Washington, DC, Economic Policy Institute.

Douglas, C., Allmaras, R., Rasmussen, P., Ramig, R. & Roager, N. 1980. Wheat straw composition and placement effects on deposition in dryland agriculture of the Pacific Northwest. *Soil Science Society of America*, 44: 833-837.

Dregne, H. & Chou, N. 1992. Global desertification: dimensions and costs. In H. Dregne, ed. *Degradation and restoration of arid lands*. Lubbock, Texas Tech. University.

Duwayri, M., Tran, D. & Nguyen, V. 1999. Reflections on yield gaps in rice production. *International Rice Commission Newsletter*, 48: 13-26. Rome, FAO.

Dyson, T. 1996. *Population and food, global trends and future prospects.* London and New York, Routledge.

Easterling, D., Meehl, G., Parmesan, C., Changnon, S., Karl, T. & Mearns, L. 2000. Climate extremes: observations, modeling and impacts. Science, 289(5487): 2068-2086.

Edwards, S. 1998. *Openness, productivity and growth: what do we really know?* Washington, DC, National Bureau of Economic Research.

EEA. 2001. *Environmental signals 2001.* Copenhagen, European Environment Agency.

El-Hage Scialabba, N. & Hattam, C., eds. 2002. *Organic agriculture, environment and food security.* Environment and Natural Resources Service Series No. 4. Rome, FAO.

EROS Data Center. 1998. *The global 30 arc-second digital elevation model.* USA.

ETC/KIOF. 1998. *On-farm agro-economic comparison of organic and conventional techniques in high and medium potential areas of Kenya.* Netherlands, ETC and Kenya Institute of Organic Farming, Leusden/Nairobi, March.

European Commission. 2001. *Prospects for agricultural markets, 2001-2008.* Brussels, EU Commission, Directorate General for Agriculture.

European Commission. 2002. *Prospects for agricultural markets, 2002-2009.* Brussels, EU Commission, Directorate General for Agriculture.

Evenson, R. 2002. *Technology and prices in agriculture.* Paper for the FAO Consultation on Agricultural Commodity Prices, 25-26 March, Rome, FAO (Commodities and Trade Division).

Evenson, R. & Gollin, D. 1994. *Genetic resources, international organizations and the rice varietal improvement.* Working Paper. Economic Growth Center, Yale University, USA.

Evenson, R. & L. Westphal. 1994. Technological change and technology strategy. In J. Behrman & T. Srinivasan, eds. *Handbook of development economics*, Vol. 3A, p. 2209-2229. Amsterdam, Elsevier.

Evers, G. & Agostini, A. 2001. *No-tillage farming for sustainable land management: lessons from the 2000 Brazil study tour.* Investment Centre Division. Rome, FAO.

Fan Jiangwen. 1998. Grassland resources in southern China: the potential, problems and exploitation strategies. *In* Liu Jian & Lu Qi, eds. *Grassland management in China*. Beijing, China Environmental Science Press.

FAO. 1970. *Provisional indicative world plan for agricultural development.* Rome.

FAO. 1978. *Report on the agro-ecological zones project.* Vol. 1. Methodology and results for Africa. World Soil Resources Report No. 48. Rome, FAO.

FAO. 1981a. *Agriculture: towards 2000.* Rome.

FAO. 1981b. *Report on the agro-ecological zones project.* Vol. 3. World Soil Resources Report No. 48/3. Rome, FAO.

FAO. 1988. *Integration of environmental aspects into agricultural, forestry and fisheries policies in Europe.* Paper for the 16th FAO Regional Conference for Europe, Cracow, Poland.

FAO. 1990. *Roots, tubers, plantains and bananas in human nutrition.* Rome.

FAO. 1993a. *The use of saline waters for crop production.* FAO Irrigation and Drainage Paper No. 48. Rome.

FAO. 1993b. *The World Food Model. Model Specification.* Document ESC/M/93/1. Rome.

FAO. 1995a. *Digital Soil Map of the World and derived soil properties.* Version 3.5. CD-ROM. Rome.

FAO. 1995b. *Irrigation in Africa in figures.* FAO Water Report No. 7. Rome.

FAO. 1996a. *Food, agriculture and food security: developments since the World Food Conference and prospects.* Technical Background Document No. 1 for the World Food Summit. Rome.

FAO. 1996b. *The Sixth World Food Survey.* Rome.

FAO. 1996c. Mapping undernutrition: an ongoing process. Poster for the World Food Summit. Rome.

FAO. 1996d. *Environment, sustainability and trade linkages for basic foodstuffs.* Commodities and Trade Division. Rome.

FAO. 1996e. *World livestock production systems: current status, issues and trends.* FAO Animal Production and Health Paper No. 127. Rome.

FAO. 1997a. *Irrigation potential in Africa: a basin approach.* FAO Land and Water Bulletin No. 4. Rome.

FAO. 1997b. *Irrigation in the Near East in figures.* FAO Water Report No. 9. Rome.

FAO. 1997c. *Water resources of the Near East Region: a review.* Land and Water Development Division. Rome.

FAO. 1997d. *Numbers of fishers 1970-1997.* FAO Fisheries Circular No. 929, Rev. 2. Rome.

FAO. 1997e. *Irrigation in the countries of the former Soviet Union in figures.* Rome.

FAO. 1998a. *Rural women and food security: current situation and perspectives.* Rome.

FAO. 1998b. *Crop evapotranspiration: guidelines for computing crop water requirements,* by R. Allen, L. Pereira, D. Raes & M. Smith. FAO Irrigation and Drainage Paper No. 56. Rome.

FAO. 1998c. *Asia-Pacific forestry towards 2010: report of the Asia-Pacific forestry outlook study.* Rome.

FAO. 1998d. *The State of Food and Agriculture1998.* Rome.

FAO. 1998e. *Guide to efficient plant nutrition management.* Rome.

FAO. 1999a. *The State of Food Insecurity in the World 1999.* Rome.

FAO. 1999b. *Irrigation in Asia in figures.* FAO Water Report No. 18. Rome.

FAO. 1999c. *Bridging the rice yield gap in the Asia-Pacific region.* FAO Expert Consultation, 5-7 October, Bangkok, Thailand.

FAO. 1999d. *Valuation of animal genetic resources.* ILRI-FAO Planning Workshop, 15-17 March. Rome.

FAO. 1999e. *Review of the state of world fishery resources: inland fisheries.* FAO Fisheries Circular No. 942. Rome.

FAO. 1999f. Guidelines for the production, processing, labelling and marketing of organically produced foods. Codex Alimentarius Commission Document CAC/GL 32-1999. Rome, FAO/WHO.

FAO. 1999g. *People's Republic of China: assessment of flood damages and formulation of project proposals for rehabilitation of the agriculture, livestock and fisheries sector.* Document OSRO/CPR/801/SWE. Special Relief Operation Service. Rome.

FAO. 1999h. *Organic agriculture.* Committee on Agriculture, 15th Session, 25-29 January 1999. COAG/99/9. Rome.

FAO. 2000a. *The State of Food Insecurity in the World 2000.* Rome.

FAO. 2000b. *The State of World Fisheries and Aquaculture 2000.* Rome.

FAO. 2000c. *Carbon sequestration options under the clean development mechanism to address land degradation.* World Soil Resources Reports No. 92. Rome.

FAO. 2000d. *Food for the cities: food supply and distribution policies to reduce urban food insecurity.* Food into Cities Collection. DT/43-00E. Rome.

FAO. 2000e. *Agriculture, trade and food security: issues and options in the WTO negotiations on agriculture from the perspective of developing countries.* Vol. I and II. Rome.

FAO. 2000f. *Genetically modified organisms, consumers, food safety and the environment: some key ethical issues.* Ethics in Food and Agriculture Series No. 2. Rome.

FAO. 2001a. *The State of Food Insecurity in the World 2001.* Rome.

FAO. 2001b. *Medium-term projections to 2010.* Document CCP:GR-RI-ME-OF. Rome.

FAO. 2001c. *Global Forest Resources Assessment 2000: main report.* FAO Forestry Paper No. 140. Rome.

FAO. 2001d. *Farming systems and poverty.* Rome, FAO and Washington, DC, World Bank.

FAO. 2001e. *FAO yearbook: fishery statistics. Aquaculture production.* Vol. 88/2. Rome.

FAO. 2001f. *Climate variability and change: a challenge for sustainable agricultural production.* Document for the Committee on Agriculture, COAG/01/5. Rome.

FAO. 2001g. *Reducing agricultural vulnerability to storm-related disasters.* Document for the Committee on Agriculture, COAG/01/6. Rome.

FAO. 2001h. *Soil carbon sequestration for improved land management.* World Soil Resources Reports No. 96. Rome.

FAO. 2001i. *Contract farming,* by C. Eaton & A. Shepherd. Rome.

FAO. 2001j. *Towards improving the operational effectiveness of the Marrakesh decision on the possible negative effects of the reform programme on least developed and net-food importing developing countries.* Discussion Paper No. 2. 21 March 2001, FAO Geneva Round Table on Selected Agricultural Trade Policy Issues, Geneva.

FAO. 2001k. *Potential impacts of genetic use restriction technologies (GURTs) on agricultural biodiversity and agricultural production systems.* Commission on Genetic Resources for Agriculture, 2-4 July 2001, Rome.

FAO. 2002a. *Commodity notes.* Paper for the Consultation on Agricultural Commodity Prices, 25-26 March, Rome.

FAO. 2002b. *Long-term prospects for fish and fishery products: preliminary projections to the years 2015 and 2030 based on FAO food balance sheets.* FAO Fisheries Circular No. 972. Rome.

FAO/COAG. 1999. *Organic agriculture.* Report to the Committee on Agriculture COAG/99/9. Rome.

FAO/GIEWS. 2000a. *FAO/WFP crop and food supply assessment mission to Cambodia.* Special Report. Rome.

FAO/GIEWS. 2000b. *FAO/WFP crop and food supply assessment mission to Mozambique.* Special Report. Rome.

FAO/IFA/IFDC. 1999. *Fertilizer use by crop.* 4th ed. Rome.

FAO/UNDP/UNEP. 1994. *Land degradation in South Asia: its severity, causes and effects upon the people.* World Soil Resources Report No. 78. Rome.

FAO/WHO. 1999. *Guidelines for the production, processing, labelling and marketing of organically produced foods.* Codex Alimentarius Commission, CAC/GL 32-1999. Rome, FAO/WHO.

FAPRI. 2002. *World agricultural outlook briefing book 2002.* USA, Iowa State University. (http://www.fapri.org/brfbk02/)

Faudry, D. & Tauveron, A. 1975. Désertification ou réutilisation de l'espace rural? Paris, Institut de recherche economique et de planification.

Feenstra, R., Yang, T. & Hamilton, G. 1999. Business groups and product variety in trade: evidence from South Korea, Taiwan and Japan. *J. International Economics,* 48.

Feng, Lu. 1998. *Output data on animal products in China: how much are they overstated? An assessment of Chinese statistics for meat, eggs and aquatic products.* China Centre for Economic Research, University of Beijing, China.

Fforde, A. 2002. *Viet Nam: in-depth view.* 1/2002. Internet publication available at http://www.developments.org.uk/data/0301/id_vietnam.htm/

FiBL. 2000. *Organic farming enhances soil fertility and biodiversity. Results from a 21-year old field trial.* Dossier No. 1. August 2000, Research Institute of Organic Agriculture (FiBL), Frick, Switzerland.

Finger, J. & Yeats, A. 1976. Effective protection by transportation costs and tariffs: a comparison of magnitudes. *Quarterly J. Economics.* February.

Fischer, G., van Velthuizen, H. & Nachtergaele, F. 2000. *Global agro-ecological zones assessment: methodology and results. Interim report,* IR-00-064. Laxenburg, Austria, IIASA and Rome, FAO.

Fischer, G., Shah, M., van Velthuizen, H. & Nachtergaele, F. 2001. *Global agro-ecological assessment for agriculture in the 21st century.* Laxenburg, Austria, IIASA.

Fleischauer, E., Bayer, W. & Lossau, A. 1998. Assessing and monitoring environmental impact and sustainability. *In* Proceedings of the International Conference on Livestock and the Environment, June 1997, Wageningen. IAC for FAO and the World Bank.

Fogel, R. 1994. Economic growth, population theory and physiology: the bearing of long-term processes on the making of economic policy. *American Economic Rev.,* 84(3): 369-95.

Friedrich, T. 1996. *From soil conservation to conservation agriculture: the role of agricultural engineering in this process.* Rome, FAO.

Frink, C., Waggoner, P. & Ausubel, J. 1998. *Nitrogen fertilizer: retrospect and prospect.* Paper presented at the NAS Colloquium Plants and Population: Is there time? 5-6 December. Irvine, CA, USA.

Frohberg, K., Fritzsch, J. & Schreiber, C. 2001. *Issues and trends in agricultural development in transition countries.* IAMO comments on FAO's AT2015/30 projections. Halle, Germany, IAMO.

Frye, W. 1984. Energy requirements in no tillage. In R. Phillips & S. Phillips, eds. *No-tillage agricultural principles and practices.* New York, Van Nostrand Reinhold.

Fuller, F., Fabiosa, J. & Premakumar, V. 1997. Trade impacts of foot and mouth disease in Taiwan. Paper 97-BP 16. FAPRI, Iowa State University.

Fuller, F., Hayes, D. & Smith, D. 2000. Reconciling Chinese meat production and consumption data. *Economic Development and Cultural Change,* 49(1): 23-43.

Funke, M. & Ruhwedel, R. 2001. *Product variety and economic growth.* Empirical evidence for the OECD countries. IMF Staff Papers. Washington, DC.

Gaiha, R. 2001. *Are DAC targets of poverty reduction useful?* Faculty of Management Studies, Delhi. (mimeo)

Galloway, J. 1995. Acid deposition: perspectives in time and space. *Water, Air, Soil Pollution,* 85: 15-24.

Gallup, J., Sachs, J. & Mellinger, A. 1999. *Geography and economic development.* CID Working Paper No.1. March.

Gaspar, J. & Glaeser, E. 1998. Information technology and the future of cities. In *J. Urban Economics,* 43, 136-156.

GCSI. 1999. *Soil carbon sinks potential in key countries.* Final report for the National Climate Change Sinks Issue Table, March 1999. Ottawa, Global Change Strategies International Inc.

Gereffi, G. 1994. The organization of buyer-driven global commodity chains: how US retailers shape overseas production networks. *In* G. Gereffi & M. Korzeniewicz, eds. *Commodity chains and global capitalism,* p. 96-122. Greenwood Press, Westport, USA.

Giampietro, M., Ulgiati, S. & Pimentel, D. 1997. Feasibility of large-scale biofuel production. *Bioscience.* October.

Gibbon, P. 2000. *Global commodity chains and economic upgrading in less developed countries.* Centre for Development Research Working Papers No. 0.2. Copenhagen.

Gilbert, C. 1995. *International commodity control: retrospect and prospects.* Background Paper for 1994 Global Economic Prospects. Washington, DC, World Bank.

Gleick, P. 1994. Water and conflict in the Middle East. *Environment,* 36(3).

Goklany, I. 1999. Meeting global food needs: the environmental trade-offs between increasing land conversion and land productivity. *Technology,* 6: 107-130.

Gommes R. & Fresco, L. 1998. *Everybody complains about climate. What can agricultural science and the CGIAR do about it?* Rome, FAO.

Gommes, R., du Guerny, J., Nachtergaele, F. & Brinkman, R. 1998. *Potential impacts of sea level rise on populations and agriculture.* Rome, FAO.

Goulding, K. 2000. Nitrate leaching from arable and horticultural land. *Soil use and management,* 16: 145-151.

Government of Viet Nam. 2001. *Interim poverty reduction strategy paper.* Ministry of Finance, Hanoi.

Graff, G., Zilberman, D. & Yarkin, C. 2000. The roles of economic research in the evolution of international agricultural biotechnology. In A*gricultural biotechnology in developing countries: towards optimizing the benefits for the poor.* Bonn, ZEF.

Green, D. & Priyadarshi, S. 2002. *Proposal for a "development box" in the WTO Agreement on Agriculture.* FAO Geneva Round Table on Special and Differential Treatment in the Context of the WTO Negotiations on Agriculture, 1 February 2002.

Griliches, Z. 1957. Hybrid corn: an exploration in the economics of technological change. *Econometrica,* 25(4): 501-522.

Gross, E. 1998. Harmful algae blooms: a new international programme on global ecology and oceanography. *Science International,* 69.

GTZ. 1993. Integrierter Pflanzenschutz in Projekten der Technischen Zusammenarbeit mit Entwicklungsländern. *Ein Leitfaden.* Eschborn/Rossdorf.

Gudosnikov, S. 2001. *Russia – world's leading importer for ever?* Paper for the 10th International ISO Seminar, London, 27-28 November 2001. International Sugar Organization.

Haddad, L. & Bouis, H. 1991. The impact of nutritional status on agricultural productivity: wage evidence from the Philippines. *Oxford Bull. Economics and Statistics,* 53(1): 45-68.

Hadley Centre. 1999. *Climate change and its impacts.* (http://www.met-office.gov.UK/sec5/CR_div/COP5/)

Hammer L., Healey J. &. Naschold, F. 2000. *Will growth halve gobal poverty by 2015?* 8 July, ODI Poverty Briefing.

Hanson, G. 2001. Should countries promote foreign direct investment? G-24 Discussion Paper Series. Geneva.

Harris, G. 1997. *Fertilizer application rates: an international comparison.* Paper presented at the Fertilizer Institute's Fertilizer 1998 Outlook Conference, 6-7 November, Washington, DC.

Hausmann, R. & Rodrik, D. 2002. *Economic development as self-discovery.* February. Harvard University, John F. Kennedy School of Government.

Heffernan, W. 1999. *Consolidation in the food and agriculture system.* Report to the National Farmers Union. February. 17 pp.

Henao, J. & Baanante, C. 1999. *Nutrient depletion in the agricultural soils of Africa.* Vision 2020 Brief No. 62. Washington, DC, IFPRI.

Henderson, D., Handy, C. & Neff, S. 1996. *Globalisation of the processed foods market.* Agricultural Economic Report No. 742., Washington, DC. ERS USDA.

Hendrickson, M. & Heffernan, W. 2002. *Concentration of agricultural markets.* Missouri, USA, Department of Rural Sociology, University of Missouri.

Hendy, C., Nolan, J. &. Leng, R. 1995. *Interactions between livestock production systems and the environment: impact domain: concentrate feed demand.* FAO consultancy report for livestock and the environment study, Rome.

Henry, G., Westby A. & Collinson, C. 1998. *Global cassava end-uses and markets: current situation and recommen-dations for further study.* Report to FAO by the European group on root, tuber and plantains coordinated by Dr Guy Henry, CIRAD.

Heumeuller, D. & Josling, T. 2001. *Trade restrictions on genetically engineered foods: the application of the TBT agreement.* Paper presented at the 5th International Conference on Biotechnology, Science and Modern Agriculture: a New Industry at the Dawn of the Century, 15-18 June, Ravello, Italy.

Hohenheim. 2000. *Economic performance of organic farms in Europe.* 5th Vol. Organic farming in Europe: economics and policy. Department of Farm Economics, University of Hohenheim, Germany.

Hopper, G. 1999. Changing food production and quality of diet in India, 1947-98. *Population and Development Review,* 25(3): 443-447.

Horton, S. 1999. Opportunities for investments in nutrition in low-income Asia. *Asian Development Rev.,* 17(1,2): 246-273.

Houghton, J., Meira Filho, L., Callander, B., Harris, N., Kattenberg, A. & K. Maskell, eds. 1995. *Radiative forcing of climate change and an evaluation of the IPCC IS92 emission scenarios.* Cambridge, UK, Cambridge University Press.

Howeler, R., Oates, C. & A. Costa Allem. 2000. *An assessment of the impact of smallholder cassava production and processing on the environment and biodiversity. In* IFAD/FAO (2000).

Huizing, H. & Bronsveld, K. 1991. The use of geographical information systems and remote sensing for evaluating the sustainability of land-use systems. In *Evaluation for sustainable land management in the developing world,* p. 545-562. IBSRAM Proceedings No. 12. Vol. 2, Bangkok.

Hulme, M., Barrow, E., Arnell, N., Harrison, P., Johns, T. & Downing, T. 1999. Relative impacts of human-induced climate change and natural climate variability. *Nature,* 397: 688-691.

Hydro. 1995. *Important questions on fertilisers and the environment.* Brussels, Hydro Agri Europe.

ICSU/UNEP/FAO/UNESCO/WMO. 1998. *GTOS implementation plan.* Version 2.0. GTOS Report No. 17. Rome, FAO.

IFAD. 2001. *Rural poverty report 2001 – the challenge of ending rural poverty.* Oxford, UK, Oxford University Press.

IFAD/FAO. 2000. Proceedings of the Validation Forum on the Global Cassava Development Strategy. Vol. V. Strategic Environmental Assessment. 26-28 April, FAO, Rome.

IFMRC. 1994. *Sharing the challenge: floodplain management into the 21st century.* Report to the Administration Floodplain Management Task Force, Interagency Floodplain Management Review Committee, Washington, DC.

IFPRI. 1977. *Food needs of developing countries: projections of production and consumption to 1990.* IFPRI Research Report No. 3. Washington, DC.

IMF. 1999. *International financial statistics.* CD-ROM. Washington, DC.

Immink, M., Viteri, F., Flores, R. &. Torun, B. 1984. Microeconomic consequences of energy deficiency in rural populations in developing countries. *In* E. Pollit & P. Amante, eds. *Energy intake and activity.* New York, Alan R. Liss.

INRA. 2001. *Prospective: les protéines végétales et animales, enjeux de société et défis pour l'agriculture et la recherche.* Paris, Institut national de la recherche agronomique.

IPCC. 2001a. *Climate change 2001: synthesis report,* by R. Watson & the Core Writing Team, eds. Cambridge, UK, Cambridge University Press.

IPCC. 2001b. *Third assessment report: report of working group I.* Cambridge, UK, Cambridge University Press.

IPCC. 2001c. *Climate change 2001, impacts, adaptation and vulnerability. Third assessment report: report of working group II.* Cambridge, UK, Cambridge University Press.

IPMEurope Web site. 2002. http://www.nri.org/IPMEurope/homepage.htm/

ISAAA. 2001. *Global preview of commercialized transgenic crops: 2001.* ISAAA Briefs Nos 21-24. Ithaca, USA, Cornell University.

ITC. 1999. *Organic food and beverages: world supply and major European markets. Product and market development.* Geneva, International Trade Centre of UNCTAD/WTO.

Johnson, D. 1997. Agriculture and the wealth of nations. *American Economic Rev.,* 87(2). May.

Jolly, L. 2001. *The commercial viability of fuel ethanol from sugarcane.* Paper for the F.O. Licht 2nd World Sugar By-Products Conference, International Sugar Organization, London.

Josling, T., Roberts, D. & Orden, D. 2002. *Food regulations and trade: toward a safe and open global system.* Washington, DC, Institute for International Economics.

Juo, A. & Lowe, J. 1986. *The wetlands and rice in sub-Saharan Africa.* Ibadan, Nigeria, International Institute of Tropical Agriculture.

Just, R. & Hueth, D. 1999. Multi-market exploitation: the case of biotechnology and chemicals. *American J. Agricultural Economics,* 75(4): 936-945.

Kanbur, R. & Squire, L. 1999. *The evolution of thinking about poverty: exploring the interactions.* Department of Agricultural, Resource and Managerial Economics Paper No. 99-24. Ithaca, New York, Cornell University.

Katyal, J., Ramachandran, K., Reddy, M. & Rao, C. 1997. Indian agriculture: profile of land resources, crop performance and prospects. *In* V. Ravinchandran, ed. *Regional land cover changes and sustainable agriculture in South Asia.* Proceedings of the COSTED/EU International Workshop on Regional Land Cover Changes, Sustainable Agriculture and their Interactions with Global Change, Chiang Mai, Thailand, December 1996.

Keyzer, M., Merbis, M. & Pavel, F. 2001. *Can we feed the animals? Origins and implications of rising meat demand.* Amsterdam, Centre for World Food Studies.

Koerbitz, W. 1999. Biodiesel production in Europe and North America: an encouraging prospect. *Renewable Energy,* 16: 1078-83.

Kristensen, N. & Nielsen, T. 1997. From alternative agriculture to the food industry: the need for changes in food policy. In *IPTS report, special issue: food.* No. 20. Institute for Prospective Technological Studies, in cooperation with the European S&T Observatory Network. December.

Krugman, P. 1986. *Strategic trade policy and the new international economics.* Cambridge, UK, MIT Press.

Lal, R. & Bruce, J. 1999. The potential of world cropland soils to sequester carbon and mitigate the greenhouse effect. *Environmental Science and Policy,* 2: 177-185.

Lal, R., Kimble, J. & Follet, R. 1998. Land use and soil C pools in terrestrial ecosystems. In R. Lal, J. Kimble, R. Follet & B. Stewart, eds. *Management of carbon sequestration in soil.* Boca Raton, USA, CRC Lewis Publishers.

Lal, R., Kimble, J., Follet, R. & Cole, C. 1999. *The potential of US cropland to sequester carbon and mitigate the greenhouse effect.* Sleeping Bear Press Inc.

Lampkin, N. & Padel, S., eds. 1994. *The economics of organic farming – an international perspective.* Wallingford, UK, CAB International.

Lanjouw, P. & Shariff, A. 2001. *Rural non-farm employment in India: access, incomes and poverty.* Paper prepared for the Programme of Research on Human Development of the National Council of Applied Economic Research, sponsored by UNDP, New Delhi, India.

Lassey, K., Lowe, D. & Manning, M. 2000. The trend in atmospheric methane and implications for isotopic constraints on the global methane budget. *Global Biogeochemical Cycles,* 14: 41-49.

Lavorel, S., Lambin, E., Flannigan, M. & Scholes, M. 2001. Fires in the earth system: the need for integrated research. *Global Change Newsletter,* 48: 7-10.

Leemans, R. & Cramer, W. 1991. *The IIASA data base for mean monthly values of temperature, precipitation and cloudiness on a global terrestrial grid.* Research Report No. RR-91-18. Laxenburg, Austria, International Institute for Applied Systems Analysis.

LID. 1999. *Livestock in poverty-focussed development.* Crewkerne, Somerset, UK, Livestock in Development.

Lindert, P. 2000. *Shifting ground: the changing agricultural soils of China and Indonesia.* USA, MIT Press.

Lindert, P. & Williamson, J. 2001. *Globalisation, a long history.* Paper prepared for the Annual Bank Conference on Development Economics, 25-27 June, World Bank Europe Conference, Barcelona.

Lindwall, C. & Norse, D. 2000. Role of agriculture in reducing greenhouse gases: challenges and opportunities. *In* J. Liu & Q. Lu, eds. *Integrated resource management in the red soil area of South China.* China Environmental Science Press for the China Council for International Cooperation on Environment and Development, Beijing.

Lipper, L. 2000. *Dirt poor: poverty, farmers and soil resource investment.* Rome, FAO. (mimeo)

Lipton, M. 1999. *Reviving global poverty reduction – what role for genetically modified plants?* 1999 Sir John Crawford Memorial Lecture, CGIAR International Centers Week, 28 October 1999, Washington, DC.

Lopez, R. 1998. The tragedy of the commons in Côte d'Ivoire agriculture: empirical evidence and implications for evaluating trade policies. *The World Bank Economic Review,* 12(1): 105-31.

Luce, E. 2002. How to find favour with a billion consumers. *Financial Times,* 25 April 2002, p. 9.

MAFF. 1999. *Tackling nitrate from agriculture: strategy from science.* London, Ministry of Agriculture, Food and Fisheries.

Mareida, M. & Pingali, P. 2001. *Environmental impacts of productivity enhancing crop research: a critical review.* Doc. No. SDR/TAC:IAR/01/14 presented to the Mid-Term Meeting, 21-25 May, Durban, South Africa.

Maxwell, S. 1999. International targets for poverty reduction and food security: a (mildly) sceptical but resolutely pragmatic view, with a call for greater subsidiarity. *IDS Bulletin,* 30(2).

Mazzucato, V. & Niemeijer, D. 2001. *Overestimating land degradation, underestimating farmers in the Sahel.* IIED Issue Paper No. 101. May.

McCalla, A. 2001. *The long arm of industrialized countries: how their agricultural policies affect food security.* Presentation to IFPRI. Conference on Sustainable Food Security for all by 2020, 4-6 September, Bonn, Germany.

McKinsey&Company. 1997. *Modernising the Indian food chain.* Food and Agricultural Development Action (FAIDA), New Delhi, India.

Mellor, J. 1995, ed. *Agriculture on the road to industrialization.* Baltimore, USA, The Johns Hopkins University Press.

Metcalfe, M. 2001. *Environmental regulation and implications for competitiveness in international pork trade.* Paper presented at the IATRC Symposium on International Trade in Livestock Products, Auckland, New Zealand.

Milanovic, D. 1998. *Income, inequality and poverty during the transition from planned to market economy.* World Bank Regional and Sectoral Studies. Washington, DC, World Bank.

Mines, R., Gabbard, S. & Steirman, A. 1997. *A profile of US farmworkers.* Washington, US Department of Labor. March.

Mohanty, S., Alexandratos, N. & Bruinsma, J. 1998. *The long-term food outlook for India.* Technical Report No. 98-TR 38. Iowa State University, Ames, Food and Agricultural Policy Research Institute (FAPRI).

Morrod, R. 1995. *The role of pest management techniques in meeting future food needs: improved conventional inputs.* Paper presented at the IFPRI Workshop on Pest Management, Food Security and the Environment: the Future to 2020. May. Washington, DC.

Mortimore, M. & Adams, W. 2001. Farmer adaptation, change and 'crisis' in the Sahel. *Global Environmental Change,* 11 : 49-57.

Mosier, A. & Kroeze, C. 1998. A new approach to estimate emissions of nitrous oxide from agriculture and its implications for the global change N_2O budget. *IGBP Global Change Newsletter,* 34 (June): 8-13.

Mosier, A., Duxbury, J., Freney, J., Heinemeyer, O. & Minami, K. 1996. Nitrous oxide emissions from agricultural fields: assessment, measurement and mitigation. *Plant and Soil,* 181: 95-108.

Mosier, A., Kroeze, C., Nevison, C., Oenema, O., Seitzinger, S. & van Cleemput, O. 1998. Closing the global N_2O budget: nitrous oxide emissions through the agricultural nitrogen cycle. *Nutrient Cycling in Agro-ecosystems,* 52: 225-248.

Moyer, H. & Josling, T. 2002. *Politics and process in the EU and US in the 1990s.* Ashgate.

Moyo, A. 1998. The effect of soil erosion on soil productivity as influenced by tillage: with special reference to clay and organic matter loss. *In* H. Blume, H. Eger, E. Fleischhauer, A. Hebel, C. Reij & K. Steiner, eds. *Towards sustainable land use.* Vol. 1. *Advances in geoecology.* Reiskirchen, Catena Verlag.

Murgai, R., Ali, M. & Byerlee, D. 2001. Productivity growth and sustainability in post-green revolution agriculture: the case of the Indian and Pakistan Punjabs. *World Bank Research Observer,* 16(2). Washington, DC.

Murphy, C. 2000. *Cultivating Havana: urban agriculture and food security in the years of crisis.* Oakland, CA, USA, Institute for Food and Development Policy.

National Research Council. 1999. *Sustaining marine fisheries.* Washington DC, National Academy Press.

Nelson, M. & Mareida, M. 2001. *Environmental impacts of the CGIAR: an assessment.* Doc. No. SDR/TAC:IAR/01/11 presented to the Mid-Term Meeting, 21-25 May, Durban, South Africa.

Nelson, R. 2000. On technological capabilities and their acquisition. *In* R. Evenson & G. Ranis, eds. *Science and technology: lessons for development policy,* p. 71-80. Boulder, CO, 2000, Westview Press.

New, M., Hulme, M. & Jones, P. 1999. Representing twentieth-century space and time climate variability. Part I. Development of a 1961-90 mean monthly terrestrial climatology. *J. Climate,* 12 March.

Nicholls, R., Hoozemans, M. &. Marchand, M. 1999. Increasing flood risk and wetland losses due to global sea level rise: regional and global analyses. *Global Environment Change,* 9: 69-87.

Nicholson, M. 1976. Economy, society and environment: 1920-1970. *The Fontana Economic History of Europe,* Vol. 5, Section 14. London, Fontana.

Nicholson, S. & Tucker, M. 1998. Desertification, drought and surface vegetation: an example from the West African Sahel. *Bull. American Meteorological Society,* 79: 815.

Norse, D. 1976. Food production in the SARUM global resources model. In *Proceedings of the Fourth IIASA Symposium on Global Modelling.* Oxford, UK, Pergamon Press.

Norse, D. 1988. *Policies for sustainable agriculture: getting the balance right.* Paper for the International Consultation on Environment, Sustainable Development and the Role of Small Farmers. October 1988, Rome, IFAD. Republished by IFAD in 1992 as Staff Working Paper No. 3.

Norse, D., Li, J., Jin, L. & Zhang, Z. 2001. *Environmental costs of rice production in China: lessons from Hunan and Hubei.* Bethesda, USA, Aileen International Press.

Nweke, F., Spencer, D. & Lyman, J. 2000. *The cassava transformation: Africa's best kept secret.* Paper presented at FAO, 5 December 2000, based on the book of same title, published by Michigan State University Press, 2002.

OAU/AEC. 2000. *Report of the Secretary-General on the US Trade and Development Act of 2000.* Third Ordinary Session of the Organization of African Unity/African Economic Commission Ministers of Trade, 16-20 September 2000, Cairo, Egypt.

O'Brien, C., Fox, C., Planque, B. & Casey, J. 2000. Climate variability and North Sea cod. *Nature,* 404(6774): 142.

OECD. 1991. *The state of the environment.* Paris, OECD.

OECD. 1996. *Shaping the 21st century.* The Contribution of Development Cooperation, Development Assistance Committee (DAC). Paris, OECD.

OECD. 1998. *The agricultural outlook.* Paris, OECD.

OECD. 1999. *China's participation in the multilateral trading system: implications for cereal and oilseed markets.* Doc. No. AGR/CA/APM/CFS/MD(2000)4. Paris, OECD.

OECD. 2000a. *Environmental indicators for agriculture: methods and results – stocktaking report.* Doc. No. COM/AGR/CA/ENV/EPOC(99)125. Paris, OECD.

OECD. 2000b. *A forward-looking analysis of export subsidies in agriculture.* Doc. No. COM/AGR/TD/WP(2000)90/FINAL. Paris, OECD.

OECD. 2001a. *OECD environmental outlook to 2020.* Paris, OECD.

OECD. 2001b. *Environmental indicators for agriculture.* Paris, OECD.

OECD. 2001c. *Economic survey of Brazil.* Paris, OECD.

OECD. 2001d. *Market concentration in the agro-food sector: selected economic issues.* Doc. No. AGR/CA/APM(2001)18. Paris, OECD.

OECD. 2001e. *Agricultural policies in OECD countries – PSE tables and supporting material.* Doc. No. AGR/CA/APM(2001)2/PART 3. Paris, OECD.

OECD. 2001f. *Agricultural policies in transition and emerging economies, monitoring and evaluation.* Paris, OECD.

OECD. 2001g. *The Uruguay Round Agreement on Agriculture: an evaluation of its implementation in OECD countries.* Paris, OECD.

OECD. 2002. *Report on policy reform in the sugar sector.* Doc. No. AGR/CA/APM(2001)32/Rev. 1. Paris, OECD.

Oldeman, L. 1998. *Soil degradation: a threat to food security?* Report No. 98/01. Wageningen, the Netherlands, International Soil Reference and Information Centre.

Oldeman, L., Hakkeling, R. & Sombroek, W. 1991. *World map of the status of human-induced soil degradation,* second revised ed. Wageningen, the Netherlands, ISRIC and Nairobi, UNEP.

Oliveira Martins, J. 1992. Export behaviour with differentiated products: exports of Korea, Taiwan and Japan to the US domestic market. In M. Dagenais & P. Muet, eds. *International trade modelling.* London and New York, Chapman & Hall.

Oxfam. 2001. *Bitter coffee: how the poor are paying for the slump in coffee prices.* Oxford, UK.

Oxfam. 2002. *Rigged rules and double standards.* Oxford, UK, Trade, Globalisation and the Fight against Poverty.

Pagiola, S. 1999. *The global environmental benefits of land degradation control on agricultural land.* World Bank Environment Paper No. 16. Washington, DC, World Bank.

Pain, D. & Pienkowski, M., eds. 1996. *Farming and birds in Europe: the common agricultural policy and its implications for bird conservation.* London, Academic Press.

Pain, D., Hill, D. & McCracken, D. 1997. Impact of agricultural intensification of pastoral systems on bird distributions in Britain 1970-1990. *Agriculture Ecosystems and Environment,* 64: 19-32.

Paoletti, M. 1997. *Impacts of agricultural intensification on resource use, sustainability and food safety and measures for its solution in highly populated subtropical areas in China.* Final report of EC contract STD TS3 CT92 0065.

Parry, M., ed. 2000. *Assessment of potential effects and adaptations for climate change in Europe – the Europe acacia project.* Jackson Environment Institute, University of East Anglia, Norwich, UK.

Parry, M., Rosenzweig, C., Iglesias, A., Fischer, G. & Livermore, M. 1999. Climate change and world food security: a new assessment. *Global Environment Change,* 9: 51-67.

Patterson, L. & Josling, T. 2001. *Biotechnology regulatory policy in the United States and the European Union: source of transatlantic trade conflict or opportunity for cooperation?* European Forum Working Paper. Stanford University, Institute for International Studies.

Persley, G. & Lantin, M., eds. 2000. *Agricultural biotechnology and the poor.* Proceedings of an International Conference, 21-22 October 1999, CGIAR, Washington DC.

Phillips, P. & Stovin, D. 2000. The economics of IPRs in the agricultural biotechnology sector. In *Agricultural biotechnology in developing countries: towards optimizing the benefits for the poor.* Bonn, ZEF.

Pieri, C., Evers, G., Landers, J., O'Connell, P. & Terry, E. 2002. *No-till farming for sustainable rural development.* Washington, DC, World Bank.

Pimentel, D., Harvey, C., Resosudarmo, P., Sinclair, K., Kurz, D., McNair, M., Christ, L., Shpritz, L., Fitton, L., Saffouri, R. & Blair, R. 1995. Environmental and economic costs of soil erosion and conservation benefits. *Science,* 267: 1117-1123.

Pingali, P., Bigot, Y. & Binswanger, H. 1987. *Agricultural mechanization and the evolution of farming systems in sub-Saharan Africa.* Washington, DC, World Bank.

Pinstrup-Andersen, P., Pandya-Lorch, R. & Rosegrant, M. 1999. *World food prospects: critical issues for the early twenty-first century.* Food Policy Report. Washington, DC, IFPRI.

Pitcairn, C., Leith, I., Sheppard, L., Sutton, M., Fowler, D., Munro, R., Tang, S. & Wilson, D. 1998. The relationship between nitrogen deposition, species composition and foliar nitrogen concentrations in woodland flora in the vicinity of livestock farms. *Environmental Pollution,* 102: 41-50.

Plucknett, D., Phillips, T. & Kagbo, R. 2000. *A global development strategy for cassava: transforming a traditional tropical root crop.* http://www.globalcassavastrategy.net/

Ponte, S. 2001. *The "latte revolution"? Winners and losers in the restructuring of the global coffee marketing chain.* CDR Working Paper No. 1.3. Copenhagen, Centre for Development Research.

Poonyth, D., Westhoff, P., Womack, A. & Adams, G. 2000. Impacts of WTO restrictions on subsidized EU sugar exports. *Agricultural Economics,* 22: 233-245.

Population and Development Review. 2001. On population and resources: an exchange between D. Johnson & P. Dasgupta. *Population and Development Rev.,* December 2001, 27(4): 739-747.

Pretty, J. 1995. *Regenerating agriculture: policies and practice for sustainability and self-reliance.* London, Earthscan.

Pretty, J. 2002. Lessons from certified and non-certified organic projects in developing countries. *In* N. El-Hage Scialabba & C. Hattam, eds. 2002. *Organic agriculture, environment and food security.* Environment and Natural Resources Service Series No. 4. Rome, FAO.

Pretty, J. & Hine, R. 2000. *Feeding the world with sustainable agriculture: a summary of new evidence.* SAFE-World Research Project. December.

Pretty, J., Brett, C, Gee, D, Hine, R., Mason, C., Morison, J., Raven, H., Rayment G., van der Bijl G. & Dobbs, T. 2001. Policy challenges and priorities for internalising the externalities of modern agriculture. *J. Environmental Planning and Management,* 44(2): 263-283.

Prince, S., Brown de Colstoun, E. & Kravitz, L. 1998. Evidence from rain-use efficiencies does not indicate extensive Sahelian desertification. *Global Change Biology,* 4: 359.

Purwanto, J. 1999. *The state of aquaculture in rural development in Indonesia.* Paper presented at the FAO/NACA Expert Consultation on Sustainable Aquaculture for Rural Development, 29-31 October 1999, Chiang Mai, Thailand.

Raikes, P. & Gibbon, P. 2000. Globalisation and African export crop agriculture. *J. Peasant Studies,* 27(2): 50-93.

Raneses, A., Glaser L., Price J. & Duffield, J. 1999. Potential biodiesel markets and their economic effects on the agricultural sector of the United States. *Industrial Crops and Products,* 9: 151-162.

Rao, D. 1993. *Intercountry comparisons of agricultural output and productivity.* FAO Economic and Social Development Paper No. 112. Rome.

Reardon, T., Matlon, P. & Delgado, C. 1994. Coping with household-level food insecurity in drought-affected areas of Burkina Faso. *World Development,* 16(9): 1065-1074.

Reich, C. 1970. *The greening of America.* USA, Random House.

Reid, P., Edwards, M., Hunt, H. & Warner, A. 2000. Phytoplankton change in the North Atlantic. *Nature,* 391: 546.

Reutlinger, S., & Selowsky, M. 1976. *Malnutrition and poverty: magnitude and policy options.* World Bank Occasional Paper No. 23. Baltimore, USA, Johns Hopkins University Press for the World Bank.

Rijk, A. 1989. *Agricultural mechanization, policy and strategy, the case of Thailand.* Japan, Asian Productivity Organization.

Rodrik, D. 1997. *Globalization, social conflict and economic growth.* Revised version of the Prebisch lecture delivered at UNCTAD, Geneva, 24 October 1997.

Rodrik, D. 2001. *The global governance of trade: as if development really mattered.* Paper prepared for the United Nations Development Programme (UNDP), New York.

Rosegrant, M., Paisner M., Meijer S. & Witcover, J. 2001. *Global food projections to 2020.* Washington, DC, IFPRI.

Rosenberg, N., Izaurralde, C., Brown, R. & Thomson, A. 2001. Agriculture and environment change: how greenhouse warming and CO_2 may affect water supply and irrigation requirements in the conterminous USA. In *Promoting global innovation of agricultural science and technology for sustainable agricultural development.* Proceedings of the International Conference on Agricultural Science and Technology, 7-9 November, 2001, Beijing, China.

Ross, J. & Horton, S. 1998. *Economic consequences of iron deficiency.* Ottawa, Micronutrient Initiative.

Sachs, J., & Warner, A. 1995. *Economic reform and the process of global integration.* Brookings Papers on Economic Activity, 1995: 1, 1-118.

Sandford, S. 1995. Improving the efficiency of opportunism – new directions for pastoral development. *In* I. Scoones, ed. *Living with uncertainty – new directions in pastoral development in Africa.* London, Intermediate Technology Publications.

Satyanarayana, K., Naidu, A., Chatterjee B. & Rao, B. 1977. Body size work output. *American J. Clinical Nutrition*, 30: 322-325.

Schelling, M. 1992. Some economics of global warming. *American Economic Rev.* March.

Scherr, S. 1999. *Soil degradation: a threat to developing country food security by 2015?* Food, Agriculture and the Environment Discussion Paper No. 27. Washington, DC, IFPRI.

Scherr, S. & Yadav, S. 1996. *Land degradation in the developing world: implications for food, agriculture and the environment to 2020.* Food, Agriculture and the Environment Discussion Paper No. 14. Washington, DC, IFPRI.

Schiff, M. & Valdes, A. 1997. The plundering of agriculture in developing countries. *In* C. Eicher & J. Staatz, eds. *International agricultural development*, p. 234-240.Third ed. Baltimore and London.

Schmidhuber, J. & Britz, W. 2002. *The impacts of OECD policy reform on international agricultural commodity markets: first results of a quantitative assessment based on the @2030 model.* Schriften der Gesellschaft für Wirtschafts- und Sozialwissenschaften des Landbaues e.V., Band 36, Münster-Hiltrup, 2002.

Scott, A. 1998. *Regions and the world economy.* New York, Oxford.

Scott, G., Rosegrant, M. & Ringler, C. 2000. *Roots and tubers for the 21st century: trends, projections and policy options.* Washington, DC, IFPRI and Lima, Peru, CIP.

Shane, M., Teygen L., Gehlhar M. & Roe, T. 2000. Economic growth and world food insecurity: a parametric approach. *Food Policy*, 25: 297-315.

Shaxson, F. 1998. *New concepts and approaches to land management in the tropics with emphasis on steep lands.* Rome, FAO.

Siebert, S. & Döll, P. 2001. *A digital global map of irrigated areas, an update for Latin America and Europe.* Kassel World Water Series, No. 4. University of Kassel, Germany, Centre for Environmental Systems Research. April.

Simmie, J. & Sennet, J. 1999. Innovative clusters: global or local linkages. *National Institute Economic Rev.*, 170: 87-98.

Sinclair, Th. 1998. *Options for sustaining and increasing the limiting yield-plateaus of grain crops.* Paper presented at the NAS Colloquium Plants and Population: Is There Time? 5-6 December 1998, Irvine, CA, USA.

Sinha, S. 1997. Global change scenario: current and future, with reference to land cover changes and sustainable agriculture in the South and South East Asian context. *In* V. Ravinchandran, ed. *Regional land cover changes and sustainable agriculture in South Asia.* Proceedings of the COSTED/EU International Workshop on Regional Land Cover Changes, Sustainable Agriculture and their Interactions with Global Change, Chiang Mai, Thailand, December, 1996.

SJFI. 1998. *Organic agriculture in Denmark – Economic impacts of a widespread adoption of organic management.* Copenhagen, Danish Institute of Agricultural and Fisheries Economics,

Smil, V. 2000. *Feeding the world: a challenge for the 21st century.* Cambridge, USA, MIT Press.

Smith, A. 1976. *An inquiry into the nature and causes of the wealth of nations.* University of Chicago Press, Chicago.

Smith, K. 1999. *Effect of agricultural intensification on trace gas emissions.* Paper for the GCTE Focus 3 Conference on Food and Forestry: Global Change and Global Challenges. 20-23 September, University of Reading, UK.

Smith, L. & Haddad, L. 1999. *Explaining child malnutrition in developing countries. A cross-country analysis.* FCND Discussion Paper No. 60. Washington DC, IFPRI.

Smith, L. & Haddad, L. 2001. How important is improving food availability for reducing child malnutrition in developing countries? *Agricultural Economics*, 26: 191-204.

Smith, L., El Obeid, A. &. Jensen, H. 2000. The geography and causes of food insecurity in developing countries. *Agricultural Economics*, 22: 199-215.

Smith, S., Renwick, W., Buddemeir, R. & Crossland, C. 2001. Budgets of soil erosion and deposition for sediments and sedimentary organic carbon across the conterminous United States. *Global Biogeochemical Cycles*, 15: 697-707.

Socolow, R. 1998. *Nitrogen management and the future of food: lessons from the management of energy and carbon.* Paper presented at the NAS Colloquium Plants and Population: Is There Time? 5-6 December 1998, Irvine, CA, USA.

Spurr, G. 1990. The impact of chronic undernourishment on physical work capacity and daily energy expenditure. *In* G. Harrison & J. Waterlow, eds. *Diet and disease in traditional and developing countries.* Cambridge, UK, Cambridge University Press.

Srinivasan, T. 2000. *Growth, poverty reduction and inequality.* Department of Economics, Yale University, USA. (mimeo)

Stevens, Ch. & Kennan, J. 2001. *The impact of the EU's Everything but Arms Proposal: a report to Oxfam.* Sussex University, IDS.

Stocking, M. & Tengberg, A. 1999. Erosion-induced loss in soil productivity and its impacts on agricultural production and food security. *In* H. Nabhan, A. Mashali, & A. Mermut, eds. *Integrated soil management for sustainable agriculture and food security in southern and East Africa.* Proceedings of the Expert Consultation, Harare, Zimbabwe, 8-12 December 1997. Rome, FAO.

Strauss, J. 1986. Does better nutrition raise farm productivity? *J. Political Economy,* 94(2): 297-320.

Strauss, J. & Thomas, D. 1995. Human resources: empirical modeling of household and family decisions. *In* J. Behrman & T. Srinivasan, eds. *Handbook of development economics.* Vol. 3A. Amsterdam, Elsevier.

Strauss, J. &. Thomas, D. 1998. Health, nutrition, and economic development. *J. Economic Literature,* 36(2): 766-817.

Subramanian, S. & Deaton, A. 1996. The demand for food and calories. *J. Political Economy,* 104: 133-162.

Sutherst, R., Collyer, B. & Yonow, T. 1999. *The vulnerability of Australian horticulture to pests under climate change.* Paper for the GCTE Focus 3 Conference on Food and Forestry: Global Change and Global Challenges, 20-23 September, University of Reading, UK.

Svedberg, P. 2001. *Undernutrition overestimated.* Seminar Paper No. 693. Institute for International Economic Studies, Stockholm University, Stockholm.

Talbot, J. 1997. Where does your coffee dollar go? The Division of Income and Surplus along the Coffee Commodity Chain. *Studies in Comparative International Development,* 32(1): 56-91.

Tangermann, S. 2001. *The future of preferential trade arrangements for developing countries and the current round of WTO negotiations on agriculture.* Paper prepared for FAO/ESCP, April 2001, Rome.

Taylor, J. 2000. *Migration: new dimensions and characteristics, causes, consequences and implications for rural poverty.* Paper prepared for the FAO CUREMIS project, March.

The Economist. 2000. *A survey of Nigeria.* 15 January.

The Economist. 2002a. *Ghana as economic model,* p. 44. 27 April.

The Economist. 2002b. *Ecuador's economy: banana skins,* p. 54. 27 April.

The Economist. 2002c. *The brown revolution,* p. 81. 11 May.

Then, C. 2000. A danger to the world's food: genetic engineering and the economic interests of the life-science industry. In *Agricultural biotechnology in developing countries: towards optimizing the benefits for the poor.* Bonn, ZEF.

Thomas, D. & Strauss, J. 1997. Health and wages: evidence on men and women in urban Brazil. *J. Econometrics,* 77: 159-185.

Thomas, M. 1998. *Ecological approaches and the development of truly integrated pest management.* Paper presented to the NAS Colloquium "Plants and populations: is there time?"

Thomashow, M. 1999. Statement before the US House Science Subcommittee on Basic Research, Committee Hearings – US Congress, 5 October 1999. (http://www.house.gov/science/thomashow_100599.htm/)

Thompson, R. 1997. *Technology, policy and trade: the keys to food security and environmental protection.* Presidential Address of 23rd International Conference of Agricultural Economists.

Tiffen, M., Mortimore, M. & Gichuki, F. 1994. *More people, less erosion: environmental recovery in Kenya.* Chichester, UK, John Wiley.

Tilly, R. 1981. Konjunkturgeschichte und Wirtschaftsgeschichte. *In* W. Schröder & R. Spree, Historische Konjunkturforschung, Stuttgart, Germany.

Timmer, C. 1997. *How well do the poor connect to the growth process?* Discussion Paper No. 178. Cambridge, MA, Harvard Institute for International Development (HIID).

Turner, M. 2000. Drought, domestic budgeting and wealth distribution in Sahelian households. *Development and change,* 31(5): 1009-1035.

Umali, D. 1993. *Irrigation induced salinity: a growing problem for development and the environment.* Washington, DC, World Bank.

UN. 1991. *World population prospects 1990.* Population Studies No. 120. New York.

UN. 1993. *Earth summit, agenda 21: the United Nations programme of action from Rio.* New York.

UN. 1995. Report of the UN World Summit on Sustainable Development. Available at http://www.un.org/esa/socdev/docs/summit.pdf

UN. 1999a. *Long-range world population projections: based on the 1998 revision.* New York.

UN. 1999b. Foreign direct investment and the challenge of development. In *World Investment Report.* New York and Geneva.

UN. 2000. *World economic and social survey 2000: trends and policies in the world economy – escaping the poverty trap.* Department of Economic and Social Affairs, United Nations, New York.

UN. 2001a. *World population prospects, the 2000 revision – highlights.* Doc. No. ESA/P/WP.165. New York.

UN. 2001b. *Road map towards the implementation of the United Nations Millennium Declaration.* Document A/56/326. New York.

UN. 2001c. Promoting linkages. In *World Investment Report.* New York and Geneva.

UNCTAD. 1999a. Trade facilitation and multimodal transport. Geneva.

UNCTAD. 1999b. *Examining trade in the agricultural sector, with a view to expanding the agricultural exports of developing countries, and to assisting them in better understanding the issue at stake in the upcoming agricultural negotiations.* No. TD/B/COM. 1/EM.8/2, February 1999, Geneva.

UNCTAD. 2000. *Strategies for diversification and adding value to food exports: a value chain perspective.* Geneva.

UNDP. 2000a. *China human development report 1999: transition and the state.* Beijing, China Statistical and Economic Publishing House.

UNDP. 2000b. *Poverty report 2000 – overcoming human poverty.* New York, UNDP.

UNEP. 1997. *Global environment outlook.* UNEP and Oxford University Press.

UNEP. 1999. *Global environment outlook 2000.* London, Earthscan Publications Ltd.

UNEP/RIVM. 1999. *Global assessment of acidification and eutrophication of natural ecosystems,* by A. Bouwman & D. van Vuuren. UNEP/DEIA and EW/TR, 99-6 and RIVM 402001012.

UNICEF. 1998. *The state of the world's children 1998: focus on nutrition.* Oxford, UK, Oxford University Press.

UNSO. 1994. *Poverty alleviation and land degradation in the drylands.* United Nations Sudano-Sahelian Office, New York.

USDA. 1998a. Brazil sugar semi-annual report 1998. Foreign Agricultural Service (FAS), Washington DC.

USDA. 1998b. US Department of Agriculture, producer and consumer subsidy equivalents. Washington DC. (http://www.ers.usda.gov/publications/).

USDA. 1999a. Agriculture in Poland and Hungary: preparing for EU accession. *Agricultural Outlook,* December 1999, ERS-AO-267, Washington, DC.

USDA. 1999b. The long-term boom in China's feed manufacturing industry. *Agricultural Outlook,* ERS-AO-267, December, Washington, DC.

USDA. 1999c. *Viet Nam adjusts import duties on Chapter VIII & Chapter VIII products.* GAIN Report No. VM9009, 4/9.

USDA. 1999d. *Viet Nam import tariff amendment.* GAIN Report No. VM9031, 12/7.

USDA. 2001a. Mexican cattle exports to US, current perspectives. *Agricultural Outlook,* June-July, Washington, DC.

USDA. 2001b. Viet Nam: coffee, semi-annual. GAIN Report No. VM1022, 11/16.

USDA. 2001c. Profiles of tariffs in global agricultural markets. AER-796, Washington DC. (http://www.ers.usda.gov/publications/).

USDA. 2001d. *The road ahead: agricultural policy reform in the WTO.* Summary Report, AER-797. Washington DC.

USDA. 2001e. *Agricultural outlook summary.* ERS-AO-279, Economic Research Service, USDA, Washington, DC.

USDA. 2002. *Agricultural baseline projections to 2011.* Staff Report WAOB- 2002-1.

Van Dijk, J., van Doesburg, D., Heijbroek, A., Wazir, M. & de Wolff, G. 1998. *The world coffee market.* Utrecht, the Netherlands, Rabobank International.

Vieira, M. 2001. Brazil's organic cup not yet overflowing. In *World Organic News,* 13(11). January.

Vincens, L., Martínez, E. & Mortimore, M. 1998. The international competitiveness of the garments and apparel industry of the Dominican Republic. Santiago, Chile, ECLA.

Voroney, R., van Veen, J. & Paul, E. 1981. Organic C dynamics in grassland soils: model validation and simulation and simulation of the long-term effects of cultivation and rainfall erosion. *Canadian J. Soil Sciences,* 61: 211-224.

Waggoner, P. 1994. *How much land can ten billion people spare for nature?* Task Force Report No. 21. Ames, Iowa, Council for Agricultural Science and Technology.

Walker, D. & Young, D. 1986. The effect of technical progress on erosion damage and economic incentives for soil conservation., *Land Economics,* 2(1): 83-93.

Walker, H. & Steffen, W. 1999. *The terrestrial biosphere and global change: implications for natural and managed ecosystems.* Cambridge, UK, Cambridge University Press.

Wang, B., Neue, H. & Samonte, H. 1997. Effect of cultivar difference on methane emissions. *Agriculture, ecosystems and environment,* 62: 31-40.

Wassmann, R., Moya, T. & Lantin, R. 1998. Rice and the global environment. In *Nutrient use efficiency in rice cropping systems.* Special Issue of *Field Crops Research:* 56(1-2).

Webb, S. 1989. Agricultural commodity prices in China: estimates of PSEs and CSEs 1982-87. China Agriculture and Trade Report, RS-89-5. Economic Research Service. Washington, DC, USDA. November.

Westphal, L. 1981. Empirical justification for infant industry promotion. World Bank Staff Working Paper No. 445. Washington, DC. March.

Wiebe, K., Tegene, A. & Kuhn, B. 1995. Property rights, partial interests and evolving federal role in wetlands conversion and conservation. *J. Soil and Water Conservation,* 50(6).

Wilkie, D., Morelli, G., Rotberg, F. & Shaw, E. 1999. Wetter is not better: global warming and food security in the Congo basin. *Global Environment Change,* 9: 323-328.

Willer, H. & Yussefi, M. 2002. *Organic agriculture worldwide 2001: statistics and future prospects.* Foundation Ecology and Agriculture (SÖL) in collaboration with the International Federation for Organic Agriculture Movements (IFOAM).

Wolf, M. 2000. Kicking down growth's ladder. *Financial Times,* 12 April.

World Bank. 1990. *World development report 1990. Poverty.* Published for the World Bank by Oxford University Press, New York.

World Bank. 1994. Does dependence on primary commodities mean slower growth? Chapter 2 in *Global economic prospects and the developing countries – 1994.* Washington, DC.

World Bank. 2000a. Commodities in the 20th century. *Global commodity markets,* No. 1. January. Washington, DC.

World Bank. 2000b. *World development report 2000/1.* Attacking poverty. Published for the World Bank by Oxford University Press, New York.

World Bank. 2000c. *Global economic prospects and the developing countries 2001.* Washington, DC.

World Bank. 2001a. *Global economic prospects and the developing countries 2001.* Washington, DC.

World Bank. 2001b. *World development indicators.* Washington, DC.

World Bank. 2001c. *Global economic prospects and the developing countries 2002.* Washington, DC.

World Bank. 2001d. *The international development goals: strengthening commitments and measuring progress.* Background note for the Westminster Conference on Child Poverty, 26 February. Washington, DC. (mimeo)

World Bank. 2001e. *Globalisation, growth, and poverty.* A World Bank Policy Report, Washington, DC. December.

WTO. 1994. *Uruguay Round of multilateral trade negotiations: legal instruments embodying the results of the Uruguay Round of multilateral trade negotiations: schedules.* Geneva.

WTO. 1995. *International trade: trends and statistics.* Geneva.

WTO. 2000a. *Special agricultural safeguard.* Background paper by the Secretariat, G/AG/NG/S/9, Geneva.

WTO. 2000b. *Note on non-trade concerns.* G/AG/NG/W/36/Rev.1. Geneva.

WTO. 2000c. *Negotiating proposal by Japan on WTO agricultural negotiations.* G/AG/NG/W/91, 21 December 2000. Geneva.

WTO. 2000d. *Statements by South Africa, Fourth Special Session of the Committee on Agriculture.* G/AG/NG/W/82, 15-17 November 2000. Geneva.

WTO. 2000e. *EC comprehensive negotiating proposal.* G/AG/NG/W/90. 14 December 2000. Geneva.

WTO. 2000f. *Proposal for comprehensive long-term agricultural trade reform: submission from the United States.* G/AG/NG/W/15. 23 June 2000. Geneva.

WTO. 2001a. *WTO agriculture negotiations: the issues and where we are now.* Background Paper by the Secretariat. 18 May 2001. Geneva.

WTO. 2001b. *Negotiations on WTO Agreement on Agriculture: proposals by India in the areas of (i) food security; (ii) market access; (iii) domestic support; and (iv) export competition.* G/AG/NG/W/102. 15 January 2001. Geneva.

WTO. 2001c. *WTO Africa Group: joint proposal on the negotiations on agriculture.* G/AG/NG/W/142. 23 March 2001. Geneva.

WTO. 2001d. *Regionalism.* WTO Web site. 28 August 2001. Geneva.

WTO. 2001e. *Ministerial decision.* Ministerial Conference, Fourth Session, Doha, 9-14 November 2001. WT/MIN(01)/DEC/1.

WTO. 2001f. *European communities – transitional regime for the EC autonomous tariff rate quotas on imports of bananas.* Document WT/MIN(01)/16. Geneva.

Ye, Y. 1999. *Historical consumption and future demand for fish and fish products: exploratory calculations for the years 2015/30.* FAO Fisheries Circular No. 946. Rome.

Young, A. 1998, *Land Resources: Now and for the Future,* Cambridge University Press, Cambridge.

Young, A. 1999. Is there really spare land? A critique of estimates of available cultivable land in developing countries. *Environment, Development and Sustainability,* 1: 3-18.

Yudelman, M., Ratta, A. & Nygaard, D. 1998. *Pest management and food production: looking to the future.* Food, Agriculture and the Environment Discussion Paper No. 25. Washington, DC, IFPRI.

Yussuf, S. 2001. *Globalization and the challenge for developing countries.* World Bank, DECRG. June.

Zhang, W., Tian, Z., Zhang, N. & Li, Z. 1996. Nitrate pollution of groundwater in northern China. *Agriculture, Ecosystems and Environment,* 59: 223-231.

Acronyms

ACP	African, Caribbean and Pacific Group of States	EDI	electronic data interchange
AEZ	agro-ecological zones	EEZ	exclusive economic zone
AI	artificial insemination	EIA	environmental impact assessment
AMS	aggregate measurement of support	EPA	Environmental Protection Agency (USA)
ANF	antinutritional factor	EPZ	export processing zone
ANZFA	Australia New Zealand Food Authority	ET	embryo transfer
		EU	European Union
ANZFSC	Australia New Zealand Food Standards Council	FAPRI	Food and Agricultural Policy Research Institute
APEC	Asia-Pacific Economic Cooperation Council	FBS	food balance sheet
		FDI	foreign direct investment
BMI	body mass index	FFS	farmer field school
BMR	basal metabolic rate	FRA	Forest Resources Assessment
BSE	bovine spongiform encephalopathy	GAEZ	global agro-ecological zones
CAP	Common Agricultural Policy	GATT	General Agreement on Tariffs and Trade
CBD	Convention on Biological Diversity		
CCD	Convention to Combat Desertification	GCM	global circulation model
		GDI	gross domestic investment
CCRF	Code of Conduct for Responsible Fisheries	GDP	gross domestic product
		GDY	gross domestic income
CDM	clean development mechanism	GHG	greenhouse gas
CGE	computable general equilibrium	GIS	Geographic Information System
CGIAR	Consultative Group on International Agricultural Research	GLASOD	Global Assessment of Human-induced Soil Degradation
CITES	Convention on International Trade in Endangered Species of Wild Fauna and Flora	GM	genetically modified
		GMO	genetically modified organism
		GNP	gross national product
COSCA	collaborative study of cassava in Africa	GSP	Generalized System of Preferences
		GURTs	genetic use restriction technologies
CSI	consumption similarity index	GWP	global warming potential
CV	coefficient of variation	HACCP	Hazard Analysis Critical Control Point
DAC	Development Assistance Committee		
DAP	draught animal power	HDI	human development index
DEM	digital elevation model	HFCS	high fructose corn syrup
DES	dietary energy supply	HHFCE	household final consumption expenditure
DFID	Department for International Development (UK)	HIV/AIDS	human immunodeficiency virus/acquired immunodeficiency syndrome
DNA	deoxyribonucleic acid		
EBA	Everything But Arms Initiative		

ICA	International Coffee Agreement	MEA	multilateral environmental agreement
ICP	international commodity prices		
IDG	International Development Goal	MRA	mutual recognition agreement
IFA	International Fertilizer Industry Association	MSY	maximum sustainable yield
		NAFTA	North American Free Trade Agreement
IFAD	International Fund for Agricultural Development	NFIDC	Net Food-Importing Developing Countries
IFDC	International Fertilizer Development Centre	NGO	non-governmental organization
IFF	Intergovernmental Forum on Forests	NIC	newly industrializing countries
		NTB	non-tariff barrier
IFPRI	International Food Policy Research Institute	NT/CA	no-till/conservation agriculture
		NWFP	non-wood forest product
IIASA	International Institute for Applied Systems Analysis	OECD	Organization for Economic Co-operation and Development
IIT	intra-industry trade	OIE	International Office of Epizootics
IITA	International Institute of Tropical Agriculture	PFI	prevalence of food inadequacy
		PGRFA	Plant Genetic Resources for Food and Agriculture
ILO	International Labour Office		
IP	intellectual property	PPP	purchasing power parity
IPCC	Intergovernmental Panel on Climate Change	PSE	producer support estimate
		PTA	preferential trade agreement
IPF	Ad hoc Intergovernmental Panel on Forests	PTO	Patent and Trademark Office
		RCM	regional climate model
IPM	integrated pest management	REPA	regional economic partnership agreement
IPNS	Integrated Plant Nutrient Systems		
IPPC	International Plant Protection Convention	RFMB	Regional Fishery Management Body
		RNF	rural non-farm
IPPM	integrated production and pest management	RTA	regional trade agreements
		RWEDP	Regional Wood Energy Development Programme
IPR	intellectual property rights		
IRRI	International Rice Research Institute	RWR	renewable water resources
ISI	import-substituting industrialization	SARD	sustainable agriculture and rural development
ITC	International Trade Centre		
IUCN	International Union for the Conservation of Nature and Natural Resources (World Conservation Union)	SD	standard deviation
		SDT	special and differential treatment
		SFM	sustainable forest management
		SIT	sterile insect technique
Kcal	kilocalorie	SMW	Soil Map of the World
LDC	least developed countries	SOM	soil organic matter
LEISA	low external input sustainable agriculture	SPS	sanitary and phytosanitary (measures)
LGP	length of growing period	SSG	special agricultural safeguard
LID	Livestock in Development	SSR	self-sufficiency rate
LMO	living modified organism	SUA	supply utilization account
MAS	marker-assisted selection	TBT	technical barriers to trade
MCFY	maximum constraint-free yield	TFP	total factor productivity
MDG	Millennium Development Goal	TNCs	transnational corporations

TQM	total quality management
TRIPS	trade-related aspects of intellectual property rights
TRQ	tariff rate quota
TVE	township and village enterprise
TWT	two-way trade
UNCCD	United Nations Convention to Combat Desertification
UNCED	United Nations Conference on Environment and Development
UNCLOS	United Nations Convention on the Law of the Sea
UNDP	United Nations Development Programme
UNFCCC	United Nations Framework Convention on Climate Change
UNFF	United Nations Forum on Forests
UPI	undernourishment prevalence index
UR	Uruguay Round
URAA	Uruguay Round Agreement on Agriculture
USDA	United States Department of Agriculture
VAD	vitamin A deficiency
WCMC	World Conservation Monitoring Centre
WFS	World Food Summit
WHO	World Health Organization
WIPO	World Intellectual Property Organization
WSSD	World Summit on Sustainable Development
WTO	World Trade Organization